ISLAND

ISLAND

Fact and Theory in Nature

James Lazell

UNIVERSITY OF CALIFORNIA PRESS

Berkeley   Los Angeles   London

University of California Press
Berkeley and Los Angeles, California

University of California Press, Ltd.
London, England

© 2005 by the Regents of the University of California

Passages in the sidebars "Sharpshooter" and "First Encounter" are
from the short story "Booby" by James Lazell in: Bové, Jennifer, ed.
2005. The Back Road to Crazy: Stories from the Field. The University
of Utah Press, Salt Lake City, Utah.

Library of Congress Cataloging-in-Publication Data

Lazell, James D.
   Island : fact and theory in nature / James Lazell.
      p.   cm.
   Includes bibliographical references.
   ISBN 0-520-24352-8 (casebound : alk. paper)
   1. Natural history—British Virgin Islands—Guana Island.   I. Title.
   QH109.V55L39 2005
   508.7297'25—dc22

                                        2005019827

10   09   08   07   06   05
10   9   8   7   6   5   4   3   2   1

The paper used in this publication meets the minimum requirements
of ANSI/NISO Z39.48-1992 (R 1997) (Permanence of Paper).♾

On the cover: A pompilid wasp *(Psorthaspis gloria)*, illustrated by
Marianne D. Wallace. Insets, left to right: a grass anole *(Anolis
pulchellus)*, photo by Gad Perry; an aerial view of Guana Island, photo
courtesy the Falconwood Foundation; Swainson's thrush, photo by Gad
Perry; a "tam-tam" plant *(Leucaena leucocephala)*, photo by Divonne
Holmes à Court. On the spine: Plant known locally as "white cedar"
(Tabebuia heterophylla), photo by Divonne Holmes à Court.

At first, it seemed an island like a thousand others in the vast turquoise sweep of the Caribbean: rugged, scrubby, and arid—home to seemingly countless lizards, snakes, and sea birds. It only became a most remarkable island after we decided to try to count them.

The richness and variety—in a word, the diversity—of natural ecological communities have never been more highly valued than they are now, as they become increasingly threatened by the environmental crisis. Students of what has come to be known as 'ecological diversity' realize that their work now has practical importance (indeed, urgency) in addition to the academic interest it always had.

E.C. PIELOU (1975)

. . . nothing bears more crucially upon the future of this planet than the seemingly simple matter of human attitudes toward nature.

DAVID QUAMMEN (1988)

Community ecology in a changing world is exciting, and deadly serious.

J.H. LAWTON (2000)

CONTENTS

LIST OF CONTRIBUTORS

Dr. Vitor Becker
Instituto Uiraçu
P.O. Box 241
45653-970 Ilheus BA, Brazil
vbecker@terra.com.br

Dr. Michael Ivie
Entomology Research Laboratory
Montana State University
Bozeman, MT 59717, U.S.A.
mivie@montana.edu

Dr. Lianna Jarecki
H. L. Stoutt Community College
Paraquita Bay
Tortola, British Virgin Islands
ljarecki@hlscc.edu.vg

Dr. Fred Kraus
Department of Natural Sciences
Bishop Museum
1525 Bernice St.
Honolulu, HI 96817, U.S.A.
fkraus@hawaii.edu

Dr. D. Jean Lodge
USDA Forest Products Lab
P.O. Box 1377
Luquillo, PR 00773, U.S.A.
dlodge@fs.fed.us

Dr. Scott E. Miller
Smithsonian Institution
P.O. Box 37012
NHB Rm W619, MRC 105
Washington, DC 20013, U.S.A.
millers@si.edu

Dr. Kristiina Ovaska
4180 Clinton Place
Victoria, BC V8Z 6M1, Canada
kovaska@jdmicro.com

Dr. Gad Perry
Department of Range, Wildlife,
and Fisheries Management
Texas Tech University, Box 42125
Lubbock, TX 79409, U.S.A.
gad.perry@ttu.edu

Dr. George Proctor
3 Stanton Terrace
Kingston 6, Jamaica

Dr. Peter Roberts
Royal Botanical Gardens, Kew
Richmond, Surrey TW9 3AB,
United Kingdom
p.roberts@rbgkew.org.uk

Mr. Roy Snelling
Natural History Museum
900 Exposition Ave.
Los Angeles, CA 90007, U.S.A.
rsnellin@nhm.org

Dr. Barry Valentine
5704 Lake Breeze Court, Unit 38
Sarasota, FL 34233, U.S.A.
bv@nwcs.com

PREFACE

The scientific disciplines of population biology, theoretical ecology, and biogeography have recently emerged from yesteryear's basic natural history. They are nothing less than humanity's rational attempts to understand life on Earth. The arena for study has traditionally been insular because islands have edges. Darwin had his Galápagos; Wallace had his Malay archipelago. In recent years, the Antilles—islands of the Caribbean—have become focal: They are close, and most are usually accessible.

On 24 March 1980, I landed on a small Caribbean island called Guana. It had one permanent, year-round human resident, no goats, and no mongooses. It was named for a rock formation that resembles a giant lizard's head. This is highly appropriate, for Guana is an island of lizards, and the 'guana head is characteristic of a geologic formation responsible for some of the island's most interesting forms of life. Even if you do not much care for lizards— or peculiar rock formations—Guana is a most wonderful place for the study and contemplation of life on Earth.

I have now done a good deal of work on the small island of Guana and have persuaded a lot of my friends to work there, too. I must admit, however, that we have not yet collected all of the data necessary to formulate The Theory That

Explains Everything—even about one island. But it is time to offer a preliminary report, an introduction. The approach herein is strictly scientific, stressing pragmatic methods of collecting empirical data. The data lead to the development of theories, and theories lead to philosophy. Science is inevitably philosophical.

It is tempting to think of Guana Island as a microcosm of planet Earth, but apparently it is not. The two most robust and sturdy quantitative relationships ecologists recognize are the species : area relationship (bigger areas have more species than smaller areas) and the abundance : range relationship (common species are more widespread than rare species). Both would seem to be so obvious as to be trivial. However, time and again it is found that the best-studied groups on Guana have many times more species than the island's size should allow. Our bird list stands at 109 species and continues to lengthen; we have one more species of butterfly than Tortola—our neighbor island, which is 18 times larger. Many of Guana's most abundant species are extremely rare in the world. Guana sets the far extreme, the world's record, for supporting the densest population of any terrestrial vertebrate animal known from anywhere on Earth. It is a small ground gecko and its geographic range is just from Culebra east to Anegada on the Puerto Rico Bank of islands, and south to St. Croix. A true microcosm of this planet might be a pretty wretched place. Guana Island is just much better than virtually all the rest.

There are three contributing reasons for Guana's known rich diversity. First, most parts of the world remain relatively little known, whereas Guana has been subject to intensive study. Second, Guana has been spared much of humanity's destructive onslaught. Third, scientists come to study Guana because it is ecologically well preserved: Its owners not only provide comfortable accommodations for those working there but have also avoided destructive development.

It would be unfortunate to mistake this, however, for a book about one small, unusual island. In writing this book, I have brought to bear everything I know about the world after a lifetime of biological research. I started awfully young and have worked on and intensively studied Aves, Battowia, Biaro, Ding Peng Chau, Itbayat, Kahoolawe, Kau Yi Chau, Tiritiri Matangi, and several hundred more islands. I have worked on every continent except Antarctica. A major thesis of this book is that the rest of the world would be much more like Guana Island if humans lived more lightly on the land, and—even given our history—if we examined our surroundings more closely. So this is a book about planet Earth—as it was, or might have been, and might even be again.

The writing of this book was supported by the Falconwood Corporation and The Conservation Agency. Because this book describes not just what has been done on Guana Island over the past quarter century, but also draws heavily on my experiences all over the world going back over six decades, the number of people who have contributed to it surely exceeds hundreds and approaches some thousands—plural. Only a modest number of you are mentioned in the text and can be found in the index; more, if published, are in the literature cited. I wish I could remember you all by name. There are several categories of people who deserve special mention. First, those who come to Guana Island, whether daily as staff, regularly as owners and family members, or occasionally as ecotourists, scientists, students, and dedicated naturalists, have made the most direct contributions to this work. Second, the librarians and staffs of at least a dozen libraries have found what I wanted and needed to at least attempt to convert the ramblings of an itinerant lizard hunter into some semblance of scholarship; I hope I have not let them down. Similarly, the curators and collection management staffs of a dozen museums have provided the crucial function of caring for not only the specimens we have collected but the thousands of others needed for our comparative work. And I owe a special debt of gratitude to those who actually helped make this book by their skillful word processing, mapping, photographing, critical reviewing, contributing of chapters, editing, and producing. Perhaps the most formative single experience I remember from childhood was reading a book by a most remarkable woman, Osa Johnson: *I Married Adventure*. I knew I had to find my Osa, and I eventually did. Her name, however, is not Osa; it is Wenhua, and I dedicate this book to her.

INTRODUCTION

Why are there not more different kinds of animals?

HUTCHINSON (1959)

Practice has caught up with theory in ecology.

ODUM (1971)

Empirical studies lag behind theory.

P.R. GRANT AND PRICE (1981)

Biodiversity enhances ecosystem reliability.

NAEEM AND LI (1997)

Empirical science is fundamentally a democratic process open to all who choose to acquaint themselves with the data.

J.C. MCLOUGHLIN (1979)

Two major themes run through this book. They are effectively its muscle and its blood.

First, *diversity makes for stability, and that is good for us.* A strong, positive relationship between natural diversity (ecological richness) and ecological stability (resilience)—at least for top-level consumers—is the most appealing component of ecological theory. We may live well by sustaining the natural diversity of life. However, natural stability is dynamic, not static. Change is inevitable but not problematic as long as the rate of change does not exceed the rate of evolutionary adaptation maintaining diversity.

Second, *beware of ecological theories rendered as formulas.* Our attempts to quantify ecological generalizations and theories (such as my first theme, above) in precise mathematical terms have most often been both failures and distractions. Ecology differs from the physical sciences. Although theoretical formulations aimed at describing nature may have heuristic or pedagogical value, they are not the truth or even approximately true enough to be of value in planning for conservation or sustainability. On the contrary, they imply a uniformity or homogeneity, and concomitant predictability, of nature that will prove positively dangerous if believed.

A third theme forms the skeleton—at least in proximate part—for the first two: *Guana Island is remarkably diverse.* It enjoys greater species diversity in almost every group studied than most similar sized, or even vastly larger, comparable ecosystems yet studied.

There are two additional, lesser, themes:

Research intensity elucidates richness. At least some measure of Guana's apparent richness derives from the intensity of our research effort there; other ecosystems might prove to be more similar to Guana than we currently know, once subjected to similar study.

No two places are alike. Whether one looks at islands, habitat patches, communities, ecosystems, or continents, one will find striking distinctions in natural diversity among physically similar sites (e.g., South America vs. Africa). Each site potentially offers unique lessons and examples from which we may learn more about the whole picture of life on Earth—a picture surely greater than the sum of its parts.

My initial aim in writing this book was to tell people—interested lay persons, students, and scientific colleagues—about a remarkable place: Guana Island. It rapidly became apparent that putting the information about this island in context and conveying a sense of its comparative importance would require rendering masses of theory accessible. In comprehending most of that theory (even as simple a "theory" as mark–recapture censusing and its shortcomings), virtually all of us, lay and professional alike, are amateurs. I had considerable exposure, back in the 1960s, to the beginnings of quantitative theorizing, but by 1970 I was largely involved in the nuts and bolts of site inventories, management planning, and prioritizing land acquisitions for conservation. My math—never a strong point—had atrophied from disuse. Even as a professional biologist, I was no more knowledgeable about ecological theory than a nature-loving college freshman. I was motivated by my driving interest in the diversity of life—biodiversity—but I had (I discovered) only the dimmest and most imprecise notions of what biodiversity really is: species, individuals, and genes? Interactions?

So I have written this book for people like me: those interested in the natural world and living things. "People like me" here include my dentist, my lawyer, my cousin who runs a sports bar, a high school student I know, a lot of grad-uate students, bird-watchers, ecotourists, and almost every professional biologist I have met. Perhaps my aim is too broad, but because I am absolutely convinced that our salvation as a species depends on the conservation of biodiversity—and that far too few people understand that—I have to take my shot. I do not want to just "preach to the choir," but I certainly do not want to exclude them, and of course I want the biggest and most inclusive "choir" I can get.

There is a certain autobiographical quality to this book, a fact deprecated by some of my academic colleagues. I find it unavoidable. I first began work on the islands of the West Indies in 1957, at age 17 (figure 1; Lazell 2003). All of my academic and professional life since then has been island oriented. Many of the major players in theoretical ecology and biogeography were or are my mentors, colleagues, and friends. To the extent that my views are iconoclastic, it is only fair to know how I developed them in historical context: literally, to see where I am coming from.

In 1859 the science of evolution emerged from natural history with the publication of Charles Darwin's *On the Origin of Species by Means of Natural Selection.* There had been murmurings and false starts before Darwin, of course, but no one had set forth a rational, testable scheme accounting for the diversity of life on Earth until Darwin. A century later, in 1959, the science of ecology emerged from natural history with the publication of G. Evelyn Hutchinson's *Homage to Santa Rosalia, or Why Are There So Many Kinds of Animals?*

Darwin provided the answer to the "how" of species diversity by demonstrating a method that produces such diversity. Hutchinson asked the question: What sustains such diversity? He had inklings of answers, but a single, overall answer—such as Darwin's "natural selection"—remains elusive even today.

Diversity is the glue that sticks life together. The interactions of living things, with each other and with the physical components of their environments, make a fabric, a mesh. The greater the number of kinds of living things,

THE BRITISH VIRGIN ISLANDS

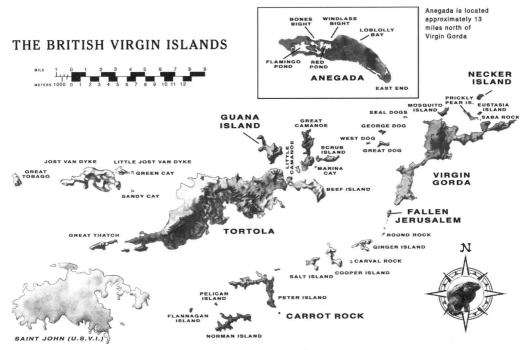

FIGURE 1. The British Virgin Islands (BVI), Puerto Rico Bank, West Indies. By John Binns.

and the more complex the environment, the greater the number of strands in the mesh. The greater the diversity, the greater the strength and stability of the whole (Naeem and Li 1997; May 2000; Raffaelli 2002; also see below).

There are several components of life's diversity: the number of kinds of living things, the number of individuals belonging to each kind, and the number of genes those individuals have. The diversity of life is greatest, as a general rule, in the tropics; the poles are relatively lifeless and void. This is the often-cited "latitudinal-diversity gradient," another quantitative ecotruism, but one that—far from being etiologically obvious—sends biologists into anguished paroxysms of uncertainty and confusion (e.g., Colwell and Lees 2000). As we travel from equator toward pole (never very close to one in my case), the number of kinds of living things tends to dwindle, and the numbers of individuals in each kind tend to fluctuate rather widely—and often wildly—over short time spans (e.g., seasons or decades): Life becomes unstable; the system becomes erratic.

Hawkins (2001) quotes the great naturalist Alexander von Humbolt, at the turn of the nineteenth century, noting ". . . the nearer we approach the tropics, the greater the increase in the variety of . . . organic life." For von Humbolt the reason was simple: "Nature undergoes a periodic stagnation in the frigid zones, for fluidity is essential to life." Well, to most life. A few species have evolved to withstand freezing, but that is enormously costly in terms of special enzymes and other biochemicals. At high latitudes, biodiversity increases seasonally as migrants return, with the heat, from lower latitudes. Surely the basic observation of seasonal change in species richness should provide the essential clue to the "why" of latitudinal gradients: temperature. Biochemical reactions that combine to create life's processes, at the most elementary level, occur most readily in warm temperatures of between about 20°C and 40°C—for most kinds of organisms at least. These are the temperatures characteristic of the tropics at sea level. Obtaining these temperatures is extremely expensive (and temporary) at

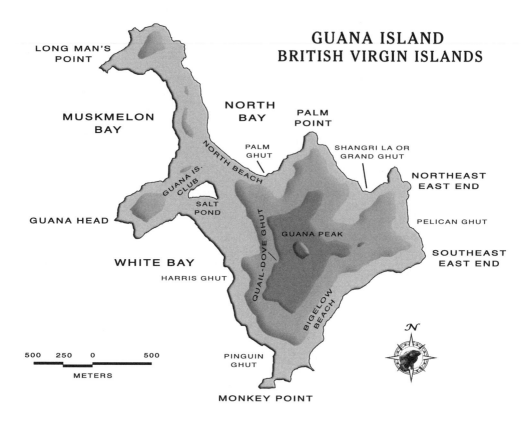

GUANA ISLAND
BRITISH VIRGIN ISLANDS

LONG MAN'S POINT

MUSKMELON BAY

NORTH BAY

PALM POINT

PALM GHUT

SHANGRI LA OR GRAND GHUT

NORTH BEACH

GUANA IS. CLUB

SALT POND

GUANA HEAD

SALT POND

GUANA PEAK

QUAIL-DOVE GHUT

NORTHEAST EAST END

PELICAN GHUT

SOUTHEAST EAST END

WHITE BAY

HARRIS GHUT

BIGELOW BEACH

500 250   0   500

METERS

PINGUIN GHUT

MONKEY POINT

N

FIGURE 2. Guana Island, BVI. By John Binns. Contours are 50 m, 100 m, and 200 m. The land is not contoured but instead steep and irregular.

the poles and no simple feat in New York in January. If you want to live at other temperatures, you will need a battery of special tricks to make things work in the cold or to prevent breakdown in the heat. Those require a lot of expensive genes.

If we want to understand the machinations of life in the broadest sense, we will have to begin with a small chunk of it—one neatly circumscribed, if possible. It would be good if it were a tropical chunk with a lot of different kinds of life-forms. It would be most instructive if it were also pretty much undisturbed; more or less the way nature made it—rather than the way engineers, agriculturists, or urban planners might have modified it. When I was much younger I thought Borneo would be just such a perfect, ideal chunk. It is more a result of education than age that I now back off. I

might have time enough left to walk across Borneo, and I might even live to tell the tale, but I would reach the other side more astounded and amazed than steeped in insight and knowledge. Borneo bewilders one with its diversity. I need a place I can walk—or at least scramble—around on in a time short enough that I can remember what I saw on the south side when I reach the north.

Rather small, then; circumscribed, as land is by water; tropical; still natural, unspoiled; and quite rich—diverse—in life. And close enough to home so that I can get there and back fairly readily (I cannot carry all the books I will need). An island, then, and in the Caribbean (figure 2).

Guana Island is a small, arid (desert, some might say) island with about two to four times as many species of plants, planthoppers, butterflies, beetles, termites, reptiles, and mammals naturally occurring on it as predicted by theoretical biogeography. These terrestrial organisms may have come to the island by any of three ways: (1) having crossed water from other islands (and ultimately from continents); (2) having been stranded here as the sea level rose after the last glacial period; or (3) having evolved here in isolation. The last alternative requires that their ancestors arrived here by one of the first two methods. It is debatable (see below), but I suspect even the stranded species are the descendants of water-crossers. There is a fourth, artificial way, too: Humans may have brought them. Of course, that necessitates crossing water as well, by boat. Guana's setting, within the Virgin Islands complex of the Puerto Rico Bank of the West Indies (sometimes referred to as the Antilles), places it in one of the most dynamic theaters of evolution on the planet today. Thus, Guana would seem to be ideal for an in-depth scientific study of natural history, population biology, and theoretical ecology.

Population biology is the cornerstone of ecology (Gotelli 2001). It is from population biology that we learn how many individuals of each species are living, what the turnover or recruitment rate is, how many will be living at some specified future time, and what the ultimate, theoretical limits to any population are. Because that turnover or recruitment rate is the mechanism by which species change, population biology is the heart of evolution. Because the changes that matter have to be heritable—capable of being passed on via genetic material—the parameters of population biology determine biochemical and biophysical properties and events. Mathematics is truly the handmaiden of population biology, but chemistry and physics have little to offer it. Biological systems are so utterly remote from the systems physical scientists use that they sometimes seem utterly magical. The physical scientist looking at the daily work of the evolutionary biologist (i.e., population biology) must utter the ultimate Yankee instruction: "You can't get there from here."

Nevertheless you can get there. The first person to do it was Charles Darwin, of course, and it is worth going over—in paraphrase—just what it was that Darwin said that made him so different. Hutchinson (1953) has said, "What we call knowledge appears to consist of a series of known relationships between unknown elements." Here, then, is the short course on populations and evolution:

1. All populations are limited.
    a. No population can increase indefinitely.
    b. Each population achieves a level of relative stability, even if that equals zero (the population is extinct). To quote Hubbell (2001, 322), "All species that manage to persist . . . for long periods . . . must exhibit net long-term population growth rates of near zero."
2. Stability (above zero) means maintenance of the same average population mean over many generations. This requires that generational deviations from the mean be small, compared to the number of offspring produced in that generation.
3. Therefore, within every population, far more offspring are produced than are needed to maintain the stable population. Most of these offspring will not survive to reproduce.
4. Evolutionary change is nothing more than the replacement of one individual with a certain genetic constitution by another individual with a different genetic constitution.

That last point is crucial: Individuals must replace individuals; genes do not move through populations by osmosis or diffusion. Let us look at an ideal situation in which there is no environmental change over time. Here, a population of organisms reaches stability at N individuals, all perfectly genetically attuned to their stable environment. Let us imagine that, over time, they have achieved such genetic perfection that no predator is able to catch one, no disease organism can strike one down, and they are even so agile and careful that no serious accident ever befalls one of them. All that happens to them is that they eventually die peacefully of old age. Would it not be possible for these organisms in this population living in this ideal environment to produce only as many offspring as were needed to replace each dying generation?

Fundamental physical science is able to supply the answer (I gave it away with that line about dying of old age): No. Not unless you can get around the second law of thermodynamics.

The second law is the law of entropy. Entropy is the inexorable process of breakdown; it is each and every system losing energy and organization to the universe outside that system. The second law covers the replication of DNA, the genetic material, just as it covers everything else. It ensures that DNA cannot replicate itself precisely, every time: There will occasionally be faults and mistakes. No matter what the environment is like, some offspring will always be less well adapted than others.

The second law guarantees mortality and wipes out our first postulate, above; no environment can possibly remain exactly the same. The second law mandates inexorable change and necessitates that, to survive, most organisms must produce far more offspring than are needed to replace them. Even to remain adaptively at zero—to simply run in place, evolutionarily speaking—the vast majority of the offspring produced must fail to reproduce. In natural populations, this means that the majority of offspring produced must die before they reach sexual maturity. Anybody who thinks nature is not blood-red in tooth and claw simply has not observed much nature.

The implications of basic evolutionary population biology for the species *Homo sapiens* are utterly staggering. Obviously, we are doing everything wrong. The senescent males who control our social systems simply do not seem to be able to figure that out. Too bad. But, as a biologist, I am no more interested in the fate of *Homo sapiens* than I am in that of starlings, Norway rats, or gonococcus bacteria. Organisms have to be fairly scarce, or at least restricted in range and habitat, to interest me (which is exactly why people who are interested in gold or diamonds are not much interested in mud and sand).

THE THREE LAWS

No doubt, some of you blessed by having missed elementary physics are wondering just what in hell that first law is. It's a good one, too. It is called the law of conservation, and, simplistically, it states that the total amount of stuff in the universe is conserved. That is, you can shuffle it around, altering its shape and order and arrangement, but there will always be just as much stuff in the universe as there ever was or ever will be. It is sort of encouraging to know that the stuff of the universe cannot really be destroyed: We can't lose any. But it's sort of discouraging to realize we can't get any, either. No gain is possible.

There is a third law, too. It is a little duller. It just says that everything in the universe has to obey the first two laws, always, forever, no matter what, no exceptions. It was Howard Platt, at Germantown Friends School in Philadelphia, half-a-century ago, who first presented me with a biologist's version of the three laws of thermodynamics: (1) You can't win. (2) You have to lose. (3) You can't get out of the game.

This is a book of recipes and examples. Guana is the grand exemplar, the chosen site. The methods of investigation have been selected from sources as diverse as the publications of the mouse men in the *Journal of Mammalogy*, those of game managers, and those of advisors to the International Whaling Commission—to name a few. Two excellent books, both available in paperback, provide the detailed mathematical insights and formulations basic to the study of ecology: *An Illustrated Guide to Theoretical Ecology* by Case (2000) and *A Primer of Ecology* by Gotelli (2001). Both are intended as working texts in theoretical ecology for undergraduate courses for biology majors. I will use and cite these often, with the caveat that, in case after case, collecting the data needed to employ a particular formula has proven extremely difficult. Therefore a real difference between theoretical and practical ecology emerges, and Guana Island has so far been a theater almost entirely of the practical. Guana provides abundant species in dense populations organized in obvious assemblages; it also provides rare species so little known that we can only glimpse them and speculate. So there will be a lot of examples of speculation, clearly labeled as such. Herein you will see just how I have applied a particular method to a sample species, or local guild of species, and, I hope, perceive how to apply such methods to the species you want to study.

As a result of following the recipes, I have generated an overview that tantalizes and torments the theoretician. This makes it possible for me to apply specific theories and assess their relative merits. I will not spare you the problems and pitfalls along the way; perhaps from seeing mine you might avoid some of your own.

Many of Guana's animals are familiar to almost any naturalist. The common snakes look rather like the grass snakes, garter snakes, and racers back home. Most Americans (at least) call the most conspicuous lizards "chameleons," revealing colloquial, if technically quite incorrect, familiarity (they are really "anoles," genus *Anolis*). In flight at night our bats on Guana look pretty much like other bats. Birds such as thrashers, flycatchers, gulls, falcons, ducks, and sandpipers resemble their counterparts on the continents. And there are butterflies, spiders, grasshoppers, and beetles.

But Guana provides more than just excellent examples of recipes and theories and speculations. It provides the bizarre. Most of the naturalists of the world have never seen or heard of an amphisbaena: It has no vernacular name accurate or descriptive enough to be worth using. It is a reptile that looks sort of like a snake. I knew amphisbaenas lived on Guana for two years before I found the first specimen. I knew because the late Oscar Chalwell, who crossed over from the large island of Tortola most days to tend the garden plot, told me, back in 1980, that there was a legless, blind creature "with rings. It bites." That is a better definition of

Amphisbaena than you could get from most taxonomists. I had had a PhD in biology for over a decade before I met my first modified flies of the family Streblidae. And you can canvass a lot of bird-watchers before you find one who has seen a bridled quail-dove, or an exquisitely elegant tropicbird (relative of the ungainly pelicans and boobies).

Of his very useful book, Gotelli (2001) states, "The purpose of this primer is to de-mystify the mathematical models used in ecology." I do not intuit math. Mathematics and I have been flung into a pit of adversity, in which I alone must struggle. Even elementary arithmetic bedevils me. For example, I balk and boggle at the fact that the product of two negative numbers is positive. That is because my mother always taught me that "negative" is a synonym for "bad" and my father always taught me that "two wrongs do not make a right." My scientific mentors taught me from early childhood never to accept anything on faith. However, to get on with life I found I had to accept many things mathematical on nothing but faith. That rankles. It makes me feel most uncomfortably akin to the mystics I deprecate. Nevertheless, I adore what math does for me. I believe computers are the most wonderful and useful inventions of humanity yet—better, even, than the thermos bottle. I have good friends who are brilliant mathematicians and computer scientists. I risk their friendship constantly by asking them stupid questions and getting them to check my arithmetic. I have worked through the formulas and calculations in this book the way a sloth crosses Brazil. I still got virtually everything wrong at least the first time. My friends patiently caught my errors and explained the truth. But there has been a major, positive result: I found out that a lot of professional biologists are not any better at math than I am. Often they pretend to be and get things wrong too. All too often I find a formula has been used contrary to its basic assumptions. All too often a colleague proceeded to do a computation incorrectly. When I catch what I believe to be examples, I proudly take them to at least three independent Mental Giants to examine my capture. Sometimes I get to gloat. Finding a couple of wrongs can be a very positive experience.

This book begins by introducing you to questions applicable to the whole world and to methods for answering them. It moves on to test the relevance of the answers found in one place to the larger picture of life on Earth. Then it swings back to introduce you to that place—the island—and its inhabitants, familiar and bizarre. Who lives on Guana? How many different species? How many individuals? Who eats whom? Which species compete? How is competition mitigated? Given predation, competition, and natural selection, how do all these species manage to coexist? How rapidly are they evolving? How did they get here? Where geographically, evolutionarily, numerically, competitively, even intellectually are we all going? I hope this book will prove useful to students, challenging and provocative to seasoned professionals, informative to interested naturalists and lay readers, and entertaining to all.

PART 1

Ways of Looking at Diversity

The ecosystem may be considered as a channel which projects information into the future. . . . The distribution of animals into species affords a preliminary measure of the width of the channel of information. One may compare an ecosystem to a message transmitted by means of a certain code . . . with a minimum value, if all individuals belong to the same species, and a maximum value, if each individual belongs to a different species.

RAMON MARGALEF (1958)

Biodiversity is the totality of hereditary variation in life forms, across all levels of biological organization, from genes and chromosomes within individual species to the array of species themselves and finally, at the highest level, the living communities of ecosystems such as forests and lakes. . . . However it is measured, biodiversity is strongly scale dependent. . . .

MICHAEL J. CRAWLEY (1997)

Diversity at one level begets diversity at other levels.

INGE ARMBRECHT, IVETTE PERFECTO, AND JOHN VANDERMEER (2004)

Guana Island is far more diverse than theory (MacArthur and Wilson 1967) predicts and more diverse than most ecosystems that were actually studied in the past. But it is small enough for us to hope to come to grips with it. If there are to be real successes in theoretical ecology, which is after all nothing less than humanity's rational effort to comprehend life on Earth, then those successes will probably emerge through the study of places like Guana Island. To the extent that we have done so at all, ecology and I have grown up together. I was born in 1939, just 12 years after Charles Elton

(1927) published *Animal Ecology*, arguably the cornerstone of this science. I was a freshman at university when Elton (1958) published his true classic *The Ecology of Invasions by Plants and Animals*. At that moment we knew we were living in a rapidly changing world. If we were ever to understand unsullied nature, free from the catastrophic disruptions of human artifice, we would have to work fast, lock up as much of nature as we could save, and only lend out the keys to our trusted colleagues. Worldwide, the chance to save the truly natural, as opposed to the artificial, was already largely past. The next,

I was a graduate student at Harvard in the mid-60s, taking formal courses from E. O. Wilson and enjoying seminars and informal discussions with him and Robert MacArthur, at a time when the latter made his frequent trips to Harvard. I did not know MacArthur well, but I knew him well enough to be certain that quantification, formulation, and physics mimicry were the stated goals of his (much-too-short) career. If he had been around longer, we might have changed his mind and benefited more from his genius. He was an affable and enjoyable companion, but I always regarded his ideas with high skepticism. For example, MacArthur and Wilson conducted a seminar on the invasion of North America by fire ants, *Solenopsis invicta* (alas, they have now invaded Guana Island: L. R. Davis et al. 2001). The centerpiece of the conversation was MacArthur's grandiose differential equation for the rate of expanding invasion. It was immediately evident to me that the available data required (and allowed) no more elegant quantification than a fifth-grade-level time–rate–distance problem. Later, he promulgated his foliage–height–density correlation to tropical bird diversity (MacArthur 1969). The idea here was simple: Observation indicates that habitats with a tree canopy, a ground cover, and—in between—shrubs or low trees, making three strata of foliage, have more birds of more species than habitats lacking one or two of these strata. For example, a forest with a solid canopy, no ground cover, and no understory will have only canopy (and trunk) dwelling birds; add some foliate ground cover, and you will get more species. That makes perfect sense. However, to quantify it, MacArthur measured leaf areas—and then formed a ratio of leaf area in the canopy to leaf area in the understory to leaf area of the ground cover. The closer to 1:1:1 this ratio came, the more bird species present, he theorized. If his formulation were true, bird species diversity would be maximal on a single tree with three withered leaves of equal size, one at the bottom, one at the top, and one halfway, standing alone on the equator: that would give you a perfect 1:1:1 ratio. (I am indebted to Robert Trivers for calling my attention to this example.)

I am proud to say that my own work (Lazell 1964a) caused MacArthur and Wilson to remove the island of Sombrero from their species-area linear graph for Antillean herpetofaunas in their famous book *The Theory of Island Biogeography* (1967). They culled those data from Phillip J. Darlington's (1957) wonderful book *Zoogeography*. Darlington was my friend and mentor. When he published his *Zoogeography*, Sombrero had only one known relevant species and served as the endpoint in his table of species-area correlation. I later discovered that Sombrero has three species, rendering it unsuitable for the kind of chop-shop theorizing that was the fashion of the times (and still goes on: Rosenzweig 1995). After the Sombrero data-culling incident, I concluded that MacArthur sometimes used his superior knowledge of mathematics to baffle and dazzle, rather than illuminate.

my sophomore, year, Hutchinson (1959) published his *Homage to Santa Rosalia*. I took my first formal course in ecology under S. Charles Kendeigh, who had published *Animal Ecology* in 1961.

The field of ecology has gone through explosive growth since those days. As noted by Lawton (2000), there has often been far more heat than light generated. In retrospect, much of the argument was semantic: Views fought over as mutually exclusive and diametric opposites are now both seen to be right, at least some of the time. I will note some of these historical battles, and the outcomes, along the way in this book. I have three major concerns, however, with the overall drift and growth of the study of ecological diversity (which certainly wholly includes what we today call biodiversity) that I must set forth now.

1. Attempts to quantify ecological generalizations, and codify them in mathematical formulas, like those of physics, are dismal failures. There are some ecological correlations that are modally true in vague, highly imprecise ways.

Larger areas (e.g., big islands) usually do have more species (greater diversities) than small ones (Munroe 1953; but see also Lazell 1976; Whittaker 1998; and below). Abundant species often do have larger geographic ranges than rare species (Anderson 1977; J. H. Brown 1995; Maurer 1999), but in the case of Guana Island, the opposite is often true. Diversity is typically greater in the tropics than at high latitudes closer to the poles (Simpson 1964; J. H. Brown 1995). But compare a rather small, high-latitude island like Nantucket with any of comparable size in Polynesia—Nantucket will win handily with respect to any terrestrial group known to me, such as vertebrates, insects, or vascular plants (Lazell 1969, 1976).

Glittering generalities—hereafter, *glitteralities*—about biodiversity look pretty at first glance but usually turn out to be soft, squishy semitruths so vague as to be trivial. You can make up dozens of your own: Big islands are higher than small islands; big rocks weigh more than small rocks; big buckets hold more than small buckets; tall men have bigger feet than short men; etc. . . . But do not even begin to mistake these sorts of glitteralities for the kinds of real truth Robert Boyle codified in his gas laws, as has often happened. And beware, accepting and believing these baubles leads to the slippery slope: Republicans are rich, and Democrats are poor; big towns have more bars than little ones; small animals raft better than large ones. . . .

2. Species diversity on small islands often is vastly greater per unit area than on large islands or continents. Very few people seem to have ever thought of this simple fact. For example, Sandy Spit, Guana Island's near neighbor, has two incumbent species of native reptiles (an anole and a gecko); it is but a half-hectare and therefore has four species per hectare (ha). Guana Island has at least 13 reptile species and one amphibian (Sandy Cay lacks amphibians), so a total herpetofauna count of 14 (or more, see below); Guana Island is 300 ha, therefore it has 0.04 species per ha. South America has a very rich herpetofauna, much of it as yet undocumented. Uetz (2000) gives 1,560 as the number

of reptiles in South America, but it is a continent of frogs, with an expanding list of salamanders. Even the reptile tally is expanding. I will bet there are at least 4,000 species of herps (reptiles and amphibians) on South America. The continent is about 24,000,000 km² (that is, 24 million square kilometers). There are 100 ha in one square kilometer, so about 2,400,000,000 ha (2 billion, 400 million ha, if I made the conversions correctly—which I often fail to do). It fairly boggles the mind! That is 4,000 species on over two billion hectares: one for every 600,000 ha or 0.0000016 per ha. Looked at per unit area, South America is practically empty. Yet, we rarely look at biodiversity that way: per unit area. We should do so more often.

3. Most of the ecological communities we study are the results of dwindling diversity, extirpations, and extinctions, not assemblages of immigrants. Postglacial sea-level rise and the blitzkrieg of human invasion have decimated—or more, obliterated—faunas, communities, and ecosystems virtually worldwide, and continue to wreak havoc. My colleagues are forever arguing over and reworking "assembly rules" for islands and communities (Diamond 1975; Weiher and Keddy 1999; Lawton 2000). Guana Island on the Puerto Rico Bank is a perfect example of the near-worldwide phenomenon of disassembly, biotic reduction, dwindling, or "relaxation." Many ecologists seem to be looking at diversity through the wrong end of the telescope. We should give more consideration to "rules" or patterns for what gets left alive after a huge island like Great Guania—Pleistocene Puerto Rico, today's Puerto Rico Bank—is inundated by a 120-m rise in sea level, loses more than one-half its total land area, and goes from one big to a couple hundred much smaller islands. And then, to reduce the fauna even more, along came the Tainos, paddling up from the Lesser Antilles three or four thousand years ago. They ate up all of the native land mammals and most of the populations of big reptiles like iguanas. Europeans can top that. In fourteen-hundred-and-ninety-two Columbus sailed the

ocean blue; by the time that gang got done, with their successors, competitors, imports, and land abuse practices, there was not much left bigger than a ground lizard (and even they were gone from many islands that now support our gift, the mongoose). Today we can talk about "assembly" on Guana Island because we have searched around and picked up some of the pieces. We have begun to reassemble at least the postglacial prehuman community: iguana, tortoise, flamingo, and white-crowned pigeon have been reestablished. But we cannot bring back the Caribbean monk seal or DeBooy's rail: They are forever extinct. (See plate 1.)

Continental communities generally fared no better and often worse. James Brown's North America, basis for much of his *Macroecology* (1995), lost most of its megafauna to the fellows who chipped Clovis points. But maybe a few communities have actually benefited. I think of John Lawton's (2000) bracken fern patches in New Mexico and at Skipwith in civilized England. I picture his neatly flagged and measured plots, his daintily marked fronds and pinnae, tiny enclosures and exclosures for the target sawflies and dipterans; and then think of him and his students returning the next day—back 15,000 years ago—after a modest herd of mammoths or aurochs trundled through. Guess you better open up a whole new computer file for dung beetles, Doc—you sure have got the habitat! By removing the megafauna, our ancestors changed the ecosystem.

Early on, when I was getting formally educated, a basic observation of field biologists and practical ecologists was that diverse, complex, and natural ecosystems are more stable than simple and/or unnatural ones. Historically, this observation has been the cornerstone of the conservation movement and underlies arguments for habitat and species preservation, no matter how pragmatically those arguments are styled (Ehrlich and Ehrlich 1970; Harcombe and Marks 1976; Myers 1976; Eisner et al. 1981). There are immediate problems, of course, in refining that observation. What is an "ecosystem"? What is "stability"? What is "natu-

ral" vs. "unnatural"? How many species make a system "diverse" vs. "simple"? A number of excellent books dealt with these questions (Odum 1971; Hutchinson 1978; Goudie 1982), but a firestorm broke out after Gardner and Ashby (1970) brought the whole stability–diversity paradigm into question. The engineers' view that the exact opposite of our field biologists' view was true was beginning to permeate: Simplicity provides stability. By the time Robert May (1973a) was done, things looked very bad for those of us who loved diversity (but see May 2000). Fortunately for conservation, the pendulum has swung. Better tests, better models, and—most important—better data now prevail. We have our stability–diversity positive correlation back again (Grimm and Wissel 1997; Naeem and Li 1997; Lawton 2000, 23; Cardinale et al. 2001).

The undoubtedly apocryphal case of the snail and the sprig of pondweed, sealed in a test tube and exposed daily to sunlight, is much too simple and much too artificial an ecosystem to maintain stability for more than a few days. (In fact, even that system is vastly more complex than you might imagine: snail, weed, and water are all necessarily hosts to a plethora of interacting, if microscopic, life-forms.) A tract of Amazonian rain forest, on the day the first field biologists arrive, is arguably an example of a diverse and natural ecosystem. New England woodlots or islands like Guana fall somewhere in between. And we can be pretty firm about deciding the island's ecosystem and community status: It has an edge.

No rational engineer or physicist would predict that a highly diversified system, complexly heterogeneous in both space and time, laced and bound by myriad interdependencies, would be more stable than a simple system: It has too many moving parts. Therein lies one of the major dilemmas in trying to reconcile the physical and biological sciences, and perhaps a major reason why biologists have so far failed to persuade a majority of the most influential decision makers that the biodiversity crisis is not only real but threatens our very continued existence. Unstable ecosystems are "bad" because they are apt to undergo increasingly turbulent oscillations in species and individual numbers, often with unpredicted corollary effects, and eventually "crash" to a greatly diminished life-supporting capacity. Humans vastly prefer stable ecosystems to unstable ones, because we rely on predictability for planning our relatively long lives, and because we are high-level consumers and do best in a life-rich environment. At simple (and very artificial) levels, we see successful investors diversify their portfolios and hear modern agriculturists constantly exhorting their colleagues in less-developed countries to diversify agriculture. We believe, with excellent reasons based on historical precedents, that diversity in investments or agriculture leads to greater stability in economics and food availability. Nevertheless, humans are the principal perpetrators of disastrous ecosystem simplification, often through agriculture. Krebs (1989) has given explicit consideration to many of the most infamous cases because they are de facto experiments in ecosystem simplification and instability.

The relationship of diversity to stability in nature has a long history of contention and theoretical discourse. Hutchinson (1959) noted of simple, natural, but highly unstable ecosystems, such as the arctic tundra, that cyclic instability "may be due in part to the communities not being significantly complex to damp out oscillations." He believed MacArthur (1955) had produced a "proof" of the damping abilities of complex systems, but MacArthur later found his own mathematical error (MacArthur 1972). P. W. Frank (1968) considered the effects of life histories on community stability. May (1973a) devoted an entire book to the problem of diversity and stability. J. H. Brown (1981) articulated the general disappointment of theoretical ecologists with their failure to demonstrate a direct relationship between diversity and stability, but pointed to brighter horizons. O. L. Smith (1980) considered the effects of environmental diversity (elevation, temperature, and humidity gradients) on animal diversity and stability; Tilman (1980) explained some aspects of stable coexistence of species competing for resources; and McNab (1980) explicitly dealt with energetics. An excellent early theoretical discussion is that of Lawler (1980), who demonstrated that Monte Carlo and other models fail because real species interact in competitive ways that are very far from random (but see Pimm 1982). May (1976) shows some of the problems inherent in mathematical models but later (1982) also argued cogently for the immediate pragmatic successes of theoretical ecology in resource management. Later yet, May (2000) complained that some of his earlier arguments had been misunderstood.

The problem of measuring diversity is thorny: It will dominate the first part of this book. Pielou (1969) was a pioneer and Magurran (1988) developed a summary synthesis. Most recent authors tend to avoid the problem and adopt no complex assessment (Case 2000; Gotelli 2001). Kelt and Brown (1999, 77–78), for example, state "...the simplest measure of community structure is species richness, a tally of the number of species present. Such a tally avoids the complications associated with indices of diversity which include measures of relative abundance...." These measures often create "... problems with statistical analysis and interpretation." Well, that is too bad. I find it unsatisfying to say one community (or ecosystem) is more diverse than another just because its species list is longer. Clearly, from a genetic viewpoint, for example, a community with 20 very abundant species has greater potential diversity than another with 20 very rare species: They are not equal. A historical perspective will indicate additional complications that, I believe, cannot be lightly dismissed. This problem is similar to many of those facing ecologists and derives from the difficulty in quantifying and codifying ecological generalizations (see Peters 1991 and Maurer 1999, 62–65).

Information Theory

THE VIEW OF THE LIVING members of an ecosystem as projecting an encoded message—information—into the future is elegantly appealing and rigorously true. The message is a statement of how things are finely tuned right now; it is encoded in the genes—the DNA—of the organisms in the ecosystem. The nature of the ecosystem of tomorrow is, in part, the direct product of the genetic message fed to it from today. Each population of living organisms in every ecosystem may be thought of as having an evolutionary role—the part it plays in the ecosystem because of its characteristics and adaptations. I have previously claimed (Lazell 1972):

> Evolutionary role is both expressed in and controlled by the characteristics of the organisms in question. That is, the characteristics of a population of organisms are to some extent a function of its evolutionary role, and its evolutionary role is to a complementary extent a function of its characteristics. This means that, while natural selection may alter the evolutionary role of a population (and thus alter its characteristics), the characteristics at any given time A directly specify the range of possibilities for natural selection to have modified by time B.

Or, as phrased by Colinvaux (1978): "Natural selection designs species. It never invents a design; it merely chooses from the range of varieties that happens to be at hand."

Margalef (1958) provides the two extreme cases, both of them imaginary and impossible in the real world. If all of the living members of the ecosystem were members of the same species, thus sharing a common gene pool, then minimum information would be projected into the future. If no two organisms were conspecific—all were genetically isolated from each other—then information content would be at a maximum, at least for the very short time before the individuals lived out their (nonreproductive) lives. All real ecosystems are projecting information organized at two levels: the gene pool of conspecific, therefore potentially gene-exchanging, individuals and the separate gene pools of each member species.

Margalef (1958) provides a way to quantify the average amount of information projected per individual, I:

$$I = \frac{1}{N} \log \frac{N!}{P_a! P_b! \ldots P_s!}$$

We will see what all of these terms mean in just a minute. But first, this formula introduces two very useful mathematical concepts worth explaining because they appear frequently in what follows. First, "log" here means *logarithm*: the exponent you must raise by a constant called the *base* to the number whose logarithm you seek. For example, the log of 100 on the base 10 is 2 because 10^2 = 100. The log of any number N is X in the equation 10^x = N. If you add two logs, you get the log of the number you would have gotten if you had multiplied the two numbers you added the logs of. For example, adding the log of 10 to the log of 100 gives you 3, the same value you get from calculating the log of 10 X 100. Similarly, subtracting one log from another will give you the log of the number you would get if you divided the first original number into the second. Plotting things on log-scaled graph paper allows you to display orders of magnitude of change on a short axis. For example, plotting area in hectares from one to a million would take a whole wall of paper if you did not use a log-scaled axis. Barnett et al. (1986, pp. 778– 791)—one of my favorite math books—provides a lucid discussion of logs.

Second, the operation indicated by that exclamation mark "!" means *factorial*. To calculate N factorial (or any number) we multiply it by every positive integer within it. Thus, if N = 10 then N! is:

$$10 \times 9 \times 8 \times 7 \times 6 \times 5 \times 4 \times 3 \times 2 \times 1 = 3{,}628{,}800$$

It is entertaining to note that zero factorial, 0!, is one— 1 (I have to accept that on faith). Fortunately for Margalef's formula, the log of one is zero, so we are not faced with a lifeless void projecting positive information (equal to one) into the future and thereby indicating diversity where none is found.

In Margalef's formula, N is the total number of individuals (of all species combined) in the ecosystem and P is the number of individuals in each species' population (thus, P_a is the number of individuals in species a, P_b the number of individuals in species b, . . . all the way to the last species, s). One can express this and other such numbers in various units of information theory, depending on whether one uses as the base of one's logarithms the binary system, 2—giving

bits; the decimal system, 10—giving *decits*, or the Naperian or natural base, *e*, 2.718—giving *nats*.

This says that the total amount of information conveyed by the ecosystem, according to Margalef, is:

$$\log \frac{N!}{P_a! P_b! \ldots P_s!}$$

(I simply did not bother to multiply by $1/N$ (or divide by N), because the average amount of information per individual is not a very useful figure, especially if no real individual approximates it, as in the case of a system with a few common and a lot of rare species.)

There are grave practical disadvantages with this formula. We must know N, the total number of individuals on the island, as well as the numbers of individuals in each species, P. To try to understand what this method is likely to tell us, however, we can invent imaginary islands with handily small numbers of N and various P values. I have invented two cays, each with three species of lizards (that at least fits reality; three species is absolutely standard for small cays: Lazell 1983a). On both cays there are a total of 12 individual lizards (that is unrealistically low, but handy).

On Cayo Uno there are four individuals of each species in the stable population, but on Cayo Dos one species is common and two are rare. Cayo Dos has eight representatives of one species, but only two of each of the others. See table 1.

Next I will attempt arithmetic to get the factorial values. (I will get at least three people to check my numbers when I am done, but I will probably never feel confident in my results. I have never been any good at arithmetic.) See table 2.

TABLE 1

	N	P_a	P_b	P_c
Cayo Uno	12	4	4	4
Cayo Dos	12	8	2	2

TABLE 2

	N!	$P_a!$	$P_b!$	$P_c!$
Cayo Uno	479,001,600	24	24	24
Cayo Dos	479,001,600	40,320	2	2

Now, if numbers are correct, the information conveyed by the Cayo Uno system is the log of:

$$\frac{479,001,600}{(24)(24)(24)} = \frac{479,001,600}{13,824} = 34,650$$

On Cayo Dos it is:

$$\frac{479,001,600}{(40,320)(2)(2)} = \frac{479,001,600}{161,280} = 2970$$

We have to look up the logs of those two end numbers, which can be done on a computer. For Cayo Uno:

$$\log 34,650 = 4.54$$

For Cayo Dos:

$$\log 2,970 = 3.47$$

Total numbers being equal, the balanced fauna of Cayo Uno earns a higher diversity index than Cayo Dos with one common and two rare species.

Let us complicate things just a little more and invent Cayo Tres. It has (like so many little cays) the standard three species, but fewer individuals; just nine. The fauna is balanced, however, and Cayo Tres has three individuals of each species. I will spare you the tortures of my arithmetic and leap to conclusions—see table 3.

TABLE 3

	N	P_a	P_b	P_c	LOG (APPROXIMATE)
Cayo Uno	12	4	4	4	4.54
Cayo Dos	12	8	2	2	3.47
Cayo Tres	9	3	3	3	3.23

We see that this method ranks Cayo Tres rather highly despite the fact that it has fewer total individuals. What happens if we consider the two extremes presented by Margalef (1958)? In an ecosystem with only one lizard species, diversity is zero—no matter what the population size; here $N = P$:

$$\log N! / P! = \log 1 = 0$$

But if there were only three lizards on the whole island, each belonging to a different species, we would get a rather high index value:

$$\log 3! / (1!)(1!)(1!) = \log 6 = 0.78$$

A thousand things that are all the same have much less diversity than three things that are all different. Of course a thousand lizards—like the little ground geckos on Carval Rock, BVI—would probably constitute a viable population and persist over time. Three individual lizards, each of a different species, would probably not survive long. If one of the three was a gravid female, it might found a population that might survive, whereas the other two lasted out only their lifetimes. Then the number of species on the island would drop down to one, and diversity would plummet to zero. Or, if two of the lizards were gravid females, and the third was not, we might end up with two species on the island, a much reduced diversity index, but with viable populations. The only way to keep the diversity index high is for all three lizards to be gravid females and to have all three establish viable populations. The probability of all three even being female is only one in eight ($1/2^3$), assuming a balanced sex ratio. Viewed in terms of genetic diversity, not simply species numbers, the descendants of a single female will be much less diverse than descendants of 20 females. Having less genetic diversity puts a population at higher risk of dying out by failing to adapt to changing conditions. Thus, the three-species island even with three gravid females as population founders is at very high risk. What does this say about the relationship between diversity and stability in ecosystems?

More Diversity Indices

Factors influential in an island's biotic richness are measures of ecological opportunity, the degree to which barriers to dispersal to an island can be overcome, and the requirement for a certain minimal area for maintenance and evolution of a population.

SHERWIN CARLQUIST (1974)

N THEIR MATHEMATICAL THEORY of communication, Shannon and Weaver (1949) codified the most widely used index of diversity:

$$H = -\Sigma P_i \log P_i,$$

where P_i is the number of individuals in the ith species divided by the total sample size; in other words, it is the proportion of all the individuals found, from all species combined, that belong to species i. For each species involved, then, we have a fraction, P_i, multiplied by its own logarithm. All logarithms of fractions are negative, so to make the index value positive we find the negative sum. In practice, using proportions rather than absolute numbers of individuals makes this approach much more appealing than calculating i, above. We can determine proportions, without counting all the individuals, by sampling a small, known area—something like a square meter, a hectare, or a square kilometer. Of course, when we extrapolate from our sample plot to the entire island, we very likely run into all the problems discussed in "Spatial Distribution," part 2, below.

Suppose we have a total sample of 100 specimens, representing 10 species, each with 10 individuals. The index value is one, because 10 times one-tenth is one, the log of one-tenth is minus one, and we found its negative sum. If, on the other hand, our sample of 100 specimens, similarly representing 10 species, contained 91 of one species and just one individual of each of the remaining nine species, the index value would be very small. Assume that 91/100 is nine-tenths, as near as it matters; nine-tenths of the log of nine-tenths is nearly zero (the log of one is zero). To that minuscule amount we must add one one-hundredth of the log of one one-hundredth nine times. The index value (I haven't done the actual arithmetic) is about two-tenths. Now we have an index that seems to make sense. It tells us what we want to hear: that the balanced ecosystem scores higher in diversity than the unbalanced one that has one very common and many very rare species.

But, alas, there are myriad ways to arrive at the same Shannon–Weaver value. The pitfalls closely approximate those demonstrated for the information index, I, discussed above—and for at least one of the same reasons: density. Because H is calculated solely from the

TABLE 4

Lizard numbers and an index of diversity (Shannon–Weaver) for three small cays in the British Virgin Islands

| Cay | Area (ha) | POPULATION ESTIMATES | | | | DENSITY PER HECTARE | | | | *H* |
		Sphaero	Anole	Ameiva	Gecko	Sphaero	Anole	Ameiva	Gecko	
Little Camanoe	16.2	40,500	2,754	32	—	2,500	170	2	—	0.106
Buck	17.0	19,312	1,314	17	—	1,136	77	1	—	0.106
Bellamy	0.6	4,167	283	—	8	2,500	170	—	13	0.108

SOURCE: Original data. Each cay supports three species.

proportion—the fraction—of the total individuals, neither absolute number of individuals nor density enters the equation. This initially seems unappealing. I strongly *want* to consider at least density (if not absolute number). I want to think of a natural ecosystem with twice as many individuals per given plot size as richer and probably more stable. As proclaimed by Balmer (2002), "abundances matter." But is it logically more diverse than the ecosystem with half the density?

Suppose we disallow density as a component of diversity but do consider absolute number of individuals. Let us now consider three real neighbor islands: Little Camanoe, Buck Island, and Bellamy Cay (table 4). The first two are nearly the same size; Buck has about one-half as many individual lizards as Little Camanoe. Bellamy Cay is tiny—literally six-tenths of a hectare—and about one-twentieth the size of the other two. But its lizards are pretty common. So Bellamy Cay has about one-twentieth the individuals of Little Camanoe and one-tenth those of Buck. All three islands have three species. Our (unformulated) novel index of diversity would rank Little Camanoe highest, then Buck, with Bellamy Cay way down at the bottom. If diversity really did promise stability, and if the equilibrium theorists were right (see "Species : Spatial Relations," part 4, below), we might really have something. Are small populations—no matter what causes their smallness—less "diverse" and less stable than big ones? What could be a more appealing notion? I have quickened a ghost that will return to haunt us.

I have real problems with this way of looking at diversity. To me, Buck Island seems certainly and clearly depauperate. Not only are its common lizards about one-half as abundant as on the other two cays, but also its third species (the ameiva or ground lizard) is much less common than is Bellamy Cay's third species (the house gecko). Buck Island is infested with cats, rats, and goats. Its vegetation and soils are all but gone. It is a wretched place and certainly seems to me to be a superb candidate for a theater of extinctions. Little Camanoe is essentially undisturbed; its lizard densities seem normal for a natural, small cay. Bellamy Cay is nearly full of buildings, including a bar and grill. It has dogs, cats, rats, and a lot of people. But it also has quite a few biggish trees and a good depth of soil and litter in several spots, and its lizards all thrive in this sort of edificarian environment.

It seems to me that tiny Bellamy Cay is the real diversity champion here, not because its geckos are slightly more numerous proportionately than the other islands' ameivas (which is why it got a slightly higher *H* value), but because it is so small compared to the much bigger islands with the same number of species (and in one case even similar densities). Bellamy Cay fairly teems with life in a seemingly balanced system. I want a diversity index that rewards Bellamy Cay for having as many species as densely abundant as does much larger Little Camanoe. Pielou (1975) notes two conditions of the *H* diversity index: First, it values evenness—equality of population sizes of included species—most highly; second, it values the presence of a greater number of species more highly if and when the populations are comparably abundant and even. However, many of us like to think of an island with more species as having the greater diversity no matter if some of those species are rare and population sizes are uneven. For example, I value the number of species present more highly than evenness of populations. I am explicitly interested in rare species. Information about rare animals seems to me much more informative than information about common ones: My prejudice is as obvious as that of the person who values diamonds more highly than pieces of glass. Often, the rarest animals in an ecosystem are among the most important, as with the very top predators at the apex of the pyramid of numbers (sensu Elton 1927). As delightfully and most informatively pointed out by Colinvaux (1978), "big fierce animals are rare." I love big fierce animals. Also, I often find that "rare" species are not really so rare after all, once one learns how and where to find them. They may appear rare to us humans because their lifestyles are utterly unsuited to producing many encounters with us in the normal course of our lives. Because they are so different from us, these animals are often the most interesting of all.

Pielou (1975) points out that when one is dealing with very large numbers of species, as in insect assemblages, other approaches may prove worthwhile. If one can reasonably expect that the population of each species in the sample—even when the sample is incomplete—fits a lognormal distribution, then one can use the Williams index, W, as discussed by Andrewartha and Birch (1982):

$$W = n_1/X,$$

derived from the series n_1, $(n_1 X)/2$, $(n_1 X^2)/3$, $(n_1 X^3)/4$, . . . , etc., where n_1 is the number of species represented in the ecosystem by one individual and X is a positive number, but less than one (because one raised to any exponent greater than one is still one, the exponential values for X in the series will never exceed one).

Andrewartha and Birch (1982) also discuss the Preston curve, fitted to the frequencies of species represented in the ecosystem by 1, 2, 3–4, 5–8, 9–16, . . . , $(2^{n-1} + 1) - 2^n$, individuals per species. In this case, ecosystem diversity is measured by the mean and variance of the curve, and the position at which it truncates or flattens out. Both of these measures have been applied to large communities of invertebrates, where "there will be a few species represented by many individuals and many species represented by few individuals" (Andrewartha and Birch 1982, 238). Both techniques provided their authors with "good," or at least appealing, fits when compared to empirical data, implying that "most communities seem to be dominated by a few species," which may be the case on Guana.

Perhaps we could marry one of these large ecosystem approaches to the novel (unformulated) index that disallows densities and includes a factor for absolute numbers. Now we can look at Mexico, with no plot boundary to bother us. But Mexico is welded to the rest of North America, which extends all the way from the Arctic to the Panama Canal, at least (and quite a lot of land animals either swim or ignore that ditch). If we have to discount density and look at whole ecosystems and absolute numbers, we will have to learn to be happy with small islands. No one could handle the computations for Borneo—let alone a small continent like Australia—even if we could census it. (See plate 2.)

Rarefaction

L ET US TURN BRIEFLY from those things that cause migraines for theoreticians and look at a highly practical approach applied by bug juggers, bird bookers, and bat snatchers. Rarefaction as a way of looking at diversity was applied to marine benthic communities by Sanders (1968) and to general ecology by Simberloff (1978). Engstrom and James (1981) and Lazell and Jarecki (1985) all seem to like it very much. I love it for islands in the South China Sea (Lazell 2002a). It is a way of looking at diversity, which, in the finest traditions of Renaissance science, enables one to clearly perceive just what it is one needs to know and where one must go next to find it out. That is exactly what I like: something definite to do about ignorance.

Rarefaction was developed by water-pollution-control experts as a way of determining the presence and relative abundance of bad things in the pipeline water. The notion is that one draws off a sample and counts the bugs in the jug. One needs to know if the sample drawn was big enough to give an adequate representation. Taking a milliliter from a water main may not tell you with any accuracy what is going through the main. How big of a sample will tell you? In order to do this one needs to know sample sizes, as in number of specimens collected. One must also have at least one place for comparison that one genuinely believes to have been adequately and fairly sampled, so that the sample sizes there reflect the truth about the absolute number of species and their relative abundance. Rarefaction, as with the Shannon–Weaver index, H, depends on the proportion of the total species diversity each species' population represents. What one does is to statistically compare the samples in hand with the best—most adequate—sampling available for a comparable habitat or island. The statistic will enable you to judge whether your sample is a simple subset of the other, or if it is somehow different.

We must make assumptions, of course. We assume that the organisms in the water (birds in the woods; bats on the island) occur in a rarefaction sequence: a most common, a most rare, and the rest spread out in between. We further assume that the sampling technique is a random one, so that the rarefaction sequence in the sample parallels that in the main pipeline. Once we run the test we will know if that assumption was true or not. This is a pragmatic method; it enables you to learn (1) if your sample is a good representation of the real, available diversity, or (2) if your sample is not representative of the diversity available in the larger system to which yours was compared, or (3) if your sample was just too small to tell one way or the other.

When comparing ecosystems, item 2 above is most important. Specifically, we will want to know if Bellamy Cay (0.6 ha) is the same basic ecosystem as Guana (300 ha) only smaller? Is Guana the same ecosystem as Tortola (5,494 ha), only smaller? In other words, do they form nested subsets? Or is the geographically smaller ecosystem fundamentally different, like Carval Rock, which does not share its single species with any other island?

As with information theory, rarefaction is best visualized as applying to a four-dimensional system projecting into time (that fourth dimension). At one time all the Virgin Islands (except St. Croix) were united with Puerto Rico and its cays in a single landmass, which I like to think of as Great Guania. We may think of Great Guania as a large pipe conveying generations of organisms, parceled into species, forward in time. As sea level rose, two things happened: The diameter of the pipe was constricted, and smaller, shunt pipelines—newly fragmented separate islands—were tapped off the main line. Today, the Virgin Islands part of the big pipe is subdivided, roughly, into five smaller ones: St. Thomas, St. John, Tortola, Virgin Gorda, and Anegada. There are more than 100 little shunt lines tapped off too—the little cays of the archipelago; our island is one of those.

We have an immediate problem: What do we know about the diversity in the original, big pipe of Great Guania? There are at least four lines of evidence:

1. The fossil record. It is far from complete, but at least there is one. We know a lot of species that once lived on Greater Puerto Rico (when it included all of the Virgin Islands except St. Croix) are gone today. Some of those occurred on the principal, larger Virgin Islands after they were separated (by the rising sea) from Puerto Rico.

2. The present fauna of Puerto Rico. This big island (a bit less than a million hectares) has many habitats and climatic diversity from rain forest to deserts, sand dunes, karst pinnacles, rivers, and caverns. What lives there today provides a clue as to what lived here yesteryear.

3. The pooled list of everything now surviving in the Virgin Island archipelago. We may hope that some of the oddities survived on some of the small islands, or at the tops of the highest remaining peaks, or off in some swamp, and that we have found and recorded them.

4. The list of everything known to occur on the best studied of the bigger remaining islands. If it is really well studied, it gives us at least a minimum diversity from which the species makeup of the other islands may have been drawn.

We have another problem: Suppose our first assumption, above—that the organisms exist in a rarefaction sequence—is untrue. This is a very fundamental problem, another quickened ghost to haunt us. Consider a group of islands of uniform relief and habitats. One island of 10,000 ha supports 10 species, each represented by 1,000 individuals. Another, only one-tenth as large—1,000 ha—supports the same 10 species, each represented by 100 individuals. There is no rarefaction sequence, so H remains identical no matter the size of the ecosystem, until size itself becomes the absolute extrinsic limiting factor. An island of 10 ha in this system probably cannot support any of the 10 species, because a population of one individual per species is not likely to be viable. This example is not so far-fetched. Some barrier islands, such as those of North Carolina's Outer Banks, or the Gulf Islands of Alabama and Mississippi, tend to be ecologically structured this way (Lazell 1979).

If there is no built-in rarefaction sequence, then the number of individuals—totaled both for all species and for each species separately—is irrelevant to species diversity, except when geographic size becomes so small as to extrinsically terminate reproduction of the species. Above a critical minimum, geographic (spatial) parameters of the ecosystem are irrelevant.

At least three types of processes tend to impose rarefaction sequences on species in most ecosystems:

1. Species interactions such as competition and niche parameters will mean that even in quite uniform environments some species will likely be numerous, others less so, and some rare.

2. Even very simple, seemingly uniform, environments tend to be patchy when looked at closely. The patches may exist in a rarefaction sequence in terms of the areas they cover. These may impose a rarefaction sequence on the organisms because each patch constitutes an ecological zone to which one species is better adapted than some other species.

3. In island systems, dispersal ability may be very important. Some species form metapopulations (Levins 1969b; Hanski 1982; Gotelli 2001; Hastings 2003) with individuals moving among islands. Species that fly (birds, bats) can colonize well; those that tolerate saltwater exposure (iguanid lizards and some geckos) can raft readily; those with permeable skins (e.g., frogs) have a hard time surviving even short trips in seawater, even if well buried in a rotten log. A rarefaction sequence may, therefore, be imposed on a given island's fauna by the average overwater dispersal abilities of available colonizers.

We (Lazell and Jarecki 1985) used rarefaction to consider the bat fauna of Guana and its neighbors in the Virgin Islands of the Puerto Rico Bank. Puerto Rico had 16 species of bats, but three are now thought to be extinct (known only as fossils or subfossil bones). Of these, six survive in the Virgin Islands today; we have no fossil evidence that there were more, but it is reasonable to assume that there were, and that the record is just poor. All six occur not only in the pooled list for all islands but also on the best-studied island, St. John. We originally found only three species on Guana, but later

(Lazell 1989a) we located a population of a fourth species, the huge fishing bat, *Noctilio leporinus*, we had suspected might be present (table 5).

Puerto Rico is about 890,000 ha and 1,338 m high; it originally had at least 16 species of bats (Koopman 1975). While Lazell and Jarecki (1985) was in press we added several more *Molossus molossus* specimens and over the years we have obtained more *Brachyphylla cavernarum*. We located a large colony of *Artibeus jamaicensis* on Anegada but did not collect more specimens. I did not modify table 5 to reflect those additions, only to include the fourth species, *N. leporinus*. See "Bats," part 5, below.

We then compared diversity as measured by rarefaction to that predicted by species–spatial approaches. Guana did not fit in terms of area: It has twice "too many" species. In terms of elevation, however, it fit better. Elevation seems to be more important than area in influencing bat species diversity in this archipelago. Still, Guana has one species more than any physical parameter (area or elevation) prediction and is right at the top of the rarefaction prediction.

Other interesting points emerged. We put in an extensive search effort on Anegada but found only the most common species. That did not make sense in view of the large number of bats seen. What is wrong with Anegada? Well, it has virtually no elevation above sea level (8.5 m is as high as it gets). That does not in itself explain the lack of diversity on Anegada, but it gives us a clue as to where to look for factors—structural and biotic—that may inform us. Then, Norman Island, at 257 ha and 131 m high, had a sample quite large enough to tell us that it just does not fit the model. It does not seem to be a shunt tapped off the old pipe at all. Norman's single species (there should be two or three) is not the commonest one in the archipelago. Perhaps the collecting method (in this case capture from a single cave roost) was too specialized and nonrandom; perhaps Norman Island is weird.

Most notable, however, was big Tortola (5,494 ha, 521 m high). The sample was properly

TABLE 5
Virgin Island Bat Faunas

	A	E	Aj	Bc	Mm	Nl	Sr	Tb	I	N	R	S_1	S_2	S_3	S_4	S_5
St. Thomas	7,660	470	16	3	5	17	1	—	42	5	4–5	5	6	3	6	5
Tortola	5,444	521	6	—	5	S	—	—	11	4	3–4	4	5	4	7	5
St. John	5,180	387	72	55	26	3	3	1	160	6	6	4	5	3	5	5
Anegada	3,872	9	1	—	—	—	—	—	1	2	1	4	5	0	0	2
Virgin Gorda	2,130	414	1	1	1	S	—	—	2	4	1–2	3	4	3	5	4
Guana	297	266	6	1	1	1	—	—	9	4	2–4	2	2	3	3	3
Norman	257	131	—	16	—	—	—	—	16	1	3–4	2	2	2	2	1
Thatch	69	146	—	1	—	—	—	—	1	1	1	1	2	2	2	1
Lovango	45	75	10	—	—	—	—	—	10	1	3–4	1	2	2	1	1
Grass	24	70	—	1	—	—	—	—	1	1	1	1	1	2	1	1

SOURCE: Lazell and Jarecki 1985. A is area in hectares. E is elevation in meters. Aj, *Artibeus jamaicensis*; Bc, *Brachyphylla cavernarum*; Mm, *Molossus molossus*; Nl, *Noctilio leporinus*; Sr, *Stenoderma rufum*; Tb, *Tadarida brasiliensis*. I is the total number of individuals in each sample. N is the total number of species per island. The remaining columns to the right give species number predictions based on various theoretical considerations and formulas: R, by rarefaction (Simberloff 1978), with range of plus or minus one standard deviation; S_1, species : area (MacArthur and Wilson 1967) using all Virgin Islands data; S_2, the same, but using only the four best-known islands: Puerto Rico, St. John, Guana, and Lovango; S_3, species : area, using only Puerto Rico, St. John, Guana, and Lovango; $S_3 = CE^z$ for all Virgin Islands; S_4, the same, using only Puerto Rico, St. John, Guana, and Lovango; S_5, the best-fit formula (Lazell 1983a) using all Virgin Islands. All numbers are rounded to the nearest whole species. In table body, S indicates a sight record not tallied in the sample.

sequenced, indicating a good representation. But it was much too small to adequately reflect diversity predicted either by area or (especially) by elevation. Tortola might easily have seven or eight species of bats revealed if we increased our sample size. This is exciting because it presages discovery of new members of the fauna—perhaps new to science and as yet undescribed or unnamed; perhaps living representatives of Puerto Rico's old bones. As I said at the outset, I love knowing not only where my pools of ignorance lie but also how to fill them in.

DeVries et al. (1999) and DeVries and Walla (2001) use rarefaction to predict species richness at any given sample size (figure 3). They looked at an ensemble (sensu Lawton 2000) of nymphalid butterflies that feed on rotting fruit—and can therefore be trapped with bait—in two Ecuadorian forests and two kinds of habitat within a forest. One forest was artificially degraded and bordered cleared cow pasture; the other was pristine old growth. Within the old growth, they chose five different areas varying from a lake edge to deep, undisturbed interior, but including an area in which a severe, very local storm had knocked down many of the largest trees some 20–25 years before their survey. They also sampled in both the forest canopy and the understory in each area. Thus they were able to consider: alpha diversity, the species numbers within small local areas; beta diversity, the change in species pool composition and numbers across a larger area still perceived as one ecosystem or community; and gamma diversity, the sum of alpha and beta diversity (sensu J. H. Brown 1995 or Lande 1996) as expressed by the total rarefaction curve (figure 3).

What DeVries and colleagues (1999) found was very interesting. The artificially disturbed forest bordering the pasture had the greatest number of species; some species that were rare in the pristine forest were common at the pasture edge. However, their sample size at the pasture edge was much larger. Rarefaction indicated that the apparent species richness in the artificially disturbed site was within the 95%

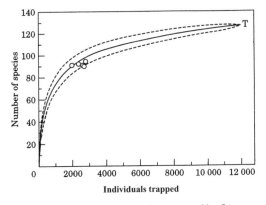

FIGURE 3. The rarefaction curve for an ensemble of Amazonian forest fruit-feeding butterflies, modified from DeVries and Walla (2001). The number of individuals trapped (with fruit bait) is on the x-axis, plotted against the number of species accumulating. T is the curve for the entire sample, approaching 130 species for 12,000 specimens collected. The dashed lines circumscribe a 95% confidence zone around the curve for this ensemble. Open circles represent five areas within the forest where traps were located. Note that each area provided about 90 species for a couple of thousand specimens and all five areas fit within the 95% confidence limits of the overall community. If one continued trapping at any one of the five areas until obtaining 4,000 specimens, then one would expect the number of species represented to increase to somewhere between 99 and 116, unless the area chosen was not representative of the larger community.

confidence limits for the pristine forest. The natural edge—lakeside, in pristine forest—had the lowest species count of all sites. I suspect that this is because a lakeside is an edge between two communities, one of which (the lake) provides no habitat or food for any species of nymphalid butterfly. The habitat that had suffered natural disturbance in the form of storm wind a couple of decades before was the richest in the pristine forest. This probably reflects a natural increase in species numbers following the "clearing" that takes place in ecological succession—at least for a short time—and is linked to the species : biomass relationship described later in this part in "Other Views" (see Grace 2001a).

Simberloff (1978) provides a computer program for calculating rarefaction. A major advantage of rarefaction is that it allows us to work with sample sizes instead of relative densities or absolute populations. Unfortunately, for this very reason, rarefaction helps us very little in tackling the theoretical and philosophical issues concerning the relationship of diversity to stability.

Recently, ecologists have begun to distinguish two types of diversity (or simple species richness): "high species richness within small local areas . . . ," alpha diversity, and "rapid turnover of species from place to place across the landscape . . . ," beta diversity (J. H. Brown 1995, 33). The possible value of this distinction may emerge when one considers ecological communities and how to delimit them, when everything we have been trying to sort out, smooth out, and straighten out gets ever so much worse.

Species Richness

WITHOUT BOGGING DOWN in the deep meaning of diversity, we do need to know how many species there are in any given assemblage (sensu Lawton 2000, 18–19) on the island. Rarefaction provides a guesstimate if we have another, well-studied, island for comparison, but we need an independent estimator. Of course, most biologists who come to Guana are inveterate collectors deeply into their specialty taxa. Most of us love to collect. Our specimens are dried or stuffed or pickled; they are catalogued and labeled; and then they are sent to our favorite scientific museums. For most of us, the first product to be derived from our collecting is a species list. For many ecologists those lists—of amphisbaenas, bats, bees, beetles, birds, . . . zorapterans, whatever—are the species richness of the island. But there are more sophisticated measures.

Here is an example of the difficulties associated with such measures. Chao (1984) proposed another index:

$$S_e = S_o + n_1^2 / 2n_2,$$

where S_e is the estimated total number of species present, S_o is the number actually observed so far in the collecting process, n_1 is the number of those collected species represented by only one specimen (singletons), and n_2 is the number of species represented by exactly two specimens (doubletons). (I have reduced Chao's letter symbols from three to two by using subscripts on S instead of different symbols for estimated and observed.) Chao adds the caveat that this formula is true only as the number of specimens collected (or observations made) approaches infinity. Well, as that number approaches infinity surely the observed number will equal the estimated number: S_o becomes S_e. But what about those n values? Why are singletons and doubletons important? And, indeed, why are singletons much more important than doubletons? What the formula says is that during the time the collection is being made, S_o—the accumulated number of species observed—is likely to be less than the real number present. Fair enough; few of us ever feel certain that we have really caught them all, even after more than 20 years. To S_o, we will add some amount to account for the ones we have not gotten yet. Chao (1984) makes no effort to justify her use of singletons and doubletons to invent this missing number, so I have had to think on it for myself. What does a singleton—a species so far represented in the collection by only one individual specimen—hint at (Novotny and Basset 2000)? Well, if you only know about its species because of one sole observation, then there might be a lot more species like it that you have not been lucky enough to observe even one of. But why square the number of singletons?

After you catch a second specimen of a given species, that species—the formula tells us—can be used to reduce the (possibly big) number of singletons-squared. In fact, with doubletons you are so much more reassured that you are getting close to the real total, S_e, that you can double their number and divide the singletons-squared by that. Do you believe that? Well, at least it is not wildly unreasonable to assume that singletons imply that there are more out there uncollected and that doubletons imply things are not so bad as the singletons seem to indicate. But singletons-squared and doubletons-doubled? Why not singletons times the shoe size of the collector and doubletons times the square root of the collector's IQ? Maybe if I worked my way through two pages of massive formulas that preceded the equation above, the answers would be revealed, you might say. I do not think so. Chao's entire justification—another two pages—is based on examples where S_e is known and her formula provides closer estimates to S_e given various S_o and n values than other formulas. Chao also uses considerable space for calculating confidence limits for S_e and even provides a second paper (Chao 1987) mostly dedicated to the question of confidence limits. This is a perfect example of a case in which I, to quote Whittaker (1998, 174), "eschew the use of inferential statistics as unwarranted." For example, among Chao's (1984, 269) seven data subset calculations where the true value of S is 420, her formula never got the right answer; her S_e values varied from 253 to 691; and her closest S_e, 421, had confidence limits of 272 to 633. All right, the formula was better than previous methods, and a lot better than nothing, but those are awfully soft numbers.

Let us consider some real Guana data. Dragonflies and damselflies make up the insect order Odonata (see "Dragonflies: Odonata," part 5, below, for list and more discussion). On Guana and elsewhere in the British Virgin Islands (BVI), they have been collected and stud-ied by Fred Sibley (1999, 2000, 2002). When Sibley began collecting in 1997, there was a single specimen of the one species of damselfly—no dragonflies—known from Guana. Therefore, S_o was one; n_1 was 1 but n_2 was zero; we cannot divide by zero so the equation cannot be used in this case. S_e is simply 1. Now, admittedly one is about as remote from infinity as we can get, so nobody would hang much on that figure. However, Sibley got three species in 1997, including one more of the singleton and several of each of the other two. This brought his S_e to three (S_o plus nothing, because the doubleton, even doubled, divided into zero added nothing). These three species are described as the "widespread and common temporary pool species" of these islands (Sibley 2000, 19), and "would appear to be the normal 'residents'" (Sibley 1999, 17). One might have guessed that this was it for Guana's odonate species richness.

Then came the most dramatic El Niño season I have ever experienced. There was heavy rain for prolonged periods in October 1997. The wind went around to the southwest, which is leeward to everyone in these islands. I never saw anything like it. Neither had old timers resident in the Antilles for eight or more decades. Usually, if the wind ever shifts at all from windward (northeast), the shift is either very slight (east or southeast), or very brief, as in a hurricane. Those strong El Niño winds brought moisture from the tropical Pacific and the high mountains of Central America. They literally blew the tops off our Caribbean towering cumulus and thus prevented hurricane formation. It was a sight to see what water blowing from "leeward" did to human habitations—all designed to greet only water blowing from east-northeast. People do not even bother to put shutters on their southwestern windows; all indoors got soaked.

And the dragonflies invaded. They blew in by the thousands. Guana's list boomed to eight species, four known only from two specimens each, but none represented by a singleton. So our estimated species richness went to eight (plus nothing). That should be plenty because obviously these odonates are metapopulational: They are dispersing on the freak winds and colonizing places where they could never survive for more than one generation by using the freak rains (and resultant extensive freshwater pools). They are not really *here* here, just having a party, taking a shot, in for the weekend.

The very next season, October 1998, things returned to normal and we had a small hurricane, Jose. With Hurricane Jose came a lot of rain, too, but of shorter duration and not from leeward. And along with it came flocks of a ninth species of dragonfly, previously known from some other, larger islands like Anegada and Tortola, but never before seen on Guana. And definitely not predicted either. After the El Niño invasion Sibley (1999, 18) said: "If I hadn't been on Guana Island when this invasion occurred is there any way to infer its existence a year later?" Well, not really. In addition to the three regulars, one of the invaders was still present (or had returned) and was represented by larvae in a temporary pool. The other four (of the previous eight) had disappeared without a trace, and a new one—the ninth, or fifth for the year—appeared. More specimens accumulated but no additional species appeared until 2001, when Sibley got a 10th species—a singleton. This singleton says there should be 10 species present. Because we have no doubleton, n_2 is again zero; the formula cannot be applied: what you have is all you can predict, and that might really be all we will ever record.

Calculating Chao's estimate of species richness is good, clean fun. It may not do much for the highly transient members of metapopulational assemblages, but it has proven especially informative—or at least very reassuring—for planthoppers and beetles. I encourage collectors to spot check from time to time as their collections accumulate and see what Chao's estimator predicts. Do not bother to calculate confidence limits, however; that is not fun, and, until your sample really does approach infinity (and you no longer need to worry because your S_o has become your S_e), I predict your data are too squishy for more than just an educated guess.

Sadly, this view of species richness does not tell us a thing about stability, or lead to any deep scientific or philosophical insights apparent to me. I still believe you could do a lot of other things with those n_1, n_2 values besides squaring or doubling them, respectively, and get predictions for S that are just as reasonable. I especially think one might use the basic natural history of the organisms to allow modifications to the mathematical treatment of singletons and doubletons. With abundant, widespread species that make up metapopulations—like our dragonflies—a good collector can almost always bag more than two specimens during a rare invasion: Singletons and doubletons will be nonexistent or meaningless. For more sedentary species with specialized ecological niches, rarity—singletons and doubletons—may be very revealing of ecologically important factors (Novotny and Basset 2000); perhaps singletons-squared *plus* doubletons should be added to S_o to get an estimate of the total species richness. (See plate 3.)

Communities

Entire communities are almost impossible to study.... Communities and their sub-components have numerous properties that cannot be deduced by studying their component species in isolation. These properties include species richness, dominance, and spatial arrangement, the length of food chains and the organization of food webs.

JOHN H. LAWTON (2000)

IT IS CUSTOMARY to think of diversity in terms of ecological communities: all of the living species of organisms occurring together at the same time in the same place. It is tempting to think of Guana Island—at least its terrestrial biota—as a community. I will get into this much more deeply later, but this is the place to introduce some vocabulary. Lawton (2000, 18) provides a diagram of categories of diversity. I have converted it to a list here with only the substantive modification of naming one of the categories "cohort" which had no name in the original developed by Fauth et al. (1996).

Here are the categories:

A. Taxa: Genetically related groups of organisms, such as orders, families, or genera, no matter where they live.

B. Communities: All species of plants and animals living together at the same time and place.

C. Guilds: Groups of species that exploit the same resources, such as insectivorous animals.

And their overlap combinations:

AB. Assemblages: Members of the same higher taxon living in the same community; for example, all squamate reptiles (lizards and snakes) living in the terrestrial habitat of Guana Island.

BC. Local guilds: Members of the same guild living in the same community; for example, all insectivorous animals living in the terrestrial habitat of Guana Island.

CA. Cohorts (my term): Members of the same taxon in the same guild; for example, all insectivorous squamate reptiles—a huge cohort of mostly lizards and some snakes of enormous ecological impact virtually throughout the tropical and warm-temperate world.

ABC. Ensembles: All members of the same taxon living in the same community that are members of the same guild; for example, the insectivorous squamate reptiles living in the terrestrial habitat of Guana Island.

Let us do a little comparison shopping for diversity. Guana and Dominica (over in the Lesser Antilles) each have 13 species of squamate reptiles; that is equal species richness for the taxon Order Squamata. Each has the gecko *Hemidactylus mabouia*, a ground lizard of the genus *Ameiva*, a skink of the genus *Mabuya*, an iguana, a boa (although Guana's is much smaller and has not been collected recently), a blind snake of the genus *Typhlops*, and each island has a pair of snakes of the family Colubridae—each with both a big and a small species. Dominica has a small ground lizard of the genus *Gymnophthalmus*; a second arboreal and house gecko of the genus *Thecadactylus* (which occurs in the Virgin Islands, but not on Guana); and two species of sphaero or ground gecko, genus *Sphaerodactylus*. On the other hand, Guana has weird *Amphisbaena*, unmatched by Dominica; a sphaero; and three species of anole lizards, genus *Anolis*; Dominica has only one anole.

Can we claim all 13 species on each island are members of one community, respectively? Certainly yes, for Guana. All of Guana's species have been found in the "herp hectare," of which I will discuss more later. All of Dominica's species are widespread; only the two different sphaeros occupy largely mutually exclusive ranges. Still, in the lower Layou River Valley one can find both of them, and I will bet all the remaining 11 species, together. So Guana and Dominica have equally diverse assemblages within their terrestrial biotic communities.

It is when we come to guilds that things get interesting. So far as we know, both boas and all four colubrid snakes eat only other vertebrate animals. They begin with lizards and frogs and shift over to birds and mammals if and when they get big enough. They belong to a terrestrial-vertebrate-eating guild that may include hawks, the large owls, some of the large frogs and toads (at least sometimes), and such invasive introductions as mongooses and feral house cats. Note that many ecologists would not include all of these species in the same guild despite their broad dietary overlap because they do not exploit food resources in a similar way (Root 1967; Simberloff and Dayan 1991). The hawks and owls, for example, usually fly to their prey; the snakes crawl or wait in ambush. Nevertheless, they all catch lizards on the ground, or in bushes, or in trees; the snakes sometimes drop on their prey. . . . To me, the methods vary and overlap too much to be significant. In any case, these snakes belong to a near-worldwide cohort that includes most snakes. The three on each island belong to their respective Guana and Dominica vertebrate-eating snake assemblages.

The two iguanas both begin life as voracious insectivores and will, throughout their lives, eat insects and even vertebrate animals if they have the chance. Nevertheless, the bulk of their diet is leaves and fruits; ecologists think of iguanas as herbivores belonging to an enormous guild worldwide but a small cohort of herbivorous squamate reptiles. Neither of our species is a member of an assemblage because neither coexists with another largely herbivorous squamate. (Actually, a lot of lizards do occasionally eat vegetable matter, especially *Anolis* and *Ameiva*; more about that in part 5 where these species are described in detail.)

And now we come to the insectivorous guild (which on Guana includes frogs and lots of birds, for example) and the squamate reptile cohort. The two remaining snake species, one member of *Typhlops* on each island, are insectivores specializing on termites. The *Ameiva* ground lizards and the *Mabuya* skinks of each island are, similarly, ecological twins. Guana's *Amphisbaena* and Dominica's *Gymnopthalmus* are quite unalike. The former is fully fossorial, a burrower rarely seen on the surface of the ground, whereas the latter is a semifossorial leaf litter and debris occupant that must compete to some extent with the other diurnal lizards (anole, skink, ameiva, and sphaeros). Can it, for example, be occupying a niche on Dominica filled by one of Guana's two "extra" anoles? No way, I say—but stomach-content analysis has not been done, and it might be surprising.

The remaining five lizard species on each island reveal interesting dichotomies. Guana has only two geckos (family Gekkonidae): one *Sphaerodactylus* of egregious abundance hunting leaf litter by day, and one "house" gecko, *H. mabouia*, that does inhabit the buildings, but also lives in trees in more natural habitats, and is nocturnal. Dominica has four geckos: two *Sphaerodactylus* that largely divide the island between them with little geographic overlap; the same *H. mabouia*; and the much larger, nocturnal *T. rapicauda*, that similarly lives in and on buildings but also lives on trees in the forest. There are, however, other species of *Sphaerodactylus* in the Virgin Islands, and also *T. rapicauda*, so perhaps Guana is just unlucky not to have retained (or acquired) a fuller complement.

But Guana has three species of *Anolis* and Dominica only one. Thus Guana has an ensemble at the generic level that is missing from Dominica and not equaled anywhere in the Lesser Antilles. No Lesser Antillean island naturally has more than two *Anolis* lizards (St. Lucia has three, but two were introduced and only live at Castries, the capital and port). Most of the Virgin Islands that are Guana's size, and even smaller, have that three species ensemble, all insectivorous and diurnal and living in the same microhabitats.

What does this tell us about diversity? Guana is 300 ha, and 246 m high. Dominica is about 600 km², which is 60,000 ha, and 1,447 m high. Thus Dominica is 200 times the area and nearly six times as high as Guana. Dominica has three distinctive montane plant associations—rain forest, elfin woodland, and caclin bush; it has lowland rain forest on the windward coast, and a zone of cactus and thornbush near-desert in the rain shadow of the high mountains, an area that is dryer than any part of Guana or the rest of the Virgin Islands. And both have that lucky 13 squamate reptiles. In many other groups, Guana will equal or exceed the species richness of vastly larger islands with a far greater spectrum of habitats. There you have two starkly disparate ways of looking at diversity. Given its relatively enormous size and its spectrum of habitats, Dominica's animal groups (e.g., squamate reptiles) seem pathetically depauperate. Little Guana obviously is the diversity champion here. However, Dominica comes close to fitting MacArthur and Wilson's (1967) classic species:area prediction—therefore, we may suspect, being the more "normal" of the two islands. In addition, Dominica's *Anolis* and *Ameiva* are endemic species, found only on Dominica and nowhere else. Both of Dominica's *Sphaerodactylus* spp., its boa, and both of its colubrid snakes are endemic subspecies. None of Guana's reptiles are endemic; all belong to widespread species. So far, we have no mathematical way of rewarding uniqueness—the possession of endemic kinds of organisms—in our assays of diversity. Could high levels of endemicity have any effect—positive or negative—on stability? (See plate 4.)

Other Views

Individuals die, populations disappear, and species become extinct. That is one view of the world. But another view . . . concentrates not so much on presence or absence as upon numbers of organisms and the degree of constancy of their numbers. These are two very different ways of viewing the behavior of systems. . . .

C. S. HOLLING (1973)

I became intrigued by the technique of path analysis and realized that it provided a potentially useful means of examining diversity patterns.

KENNETH I. MIYATA (1980)

I T SOMETIMES SEEMS THAT diverse, stable ecosystems have little resilience, and that simple, highly unstable ecosystems have a great deal of resilience. For example, the Kaibab Plateau had a very small (two or three species) ensemble (sensu Lawton 2000, 18) of large mammalian carnivores. Extirpating only one species—the wolf—in that seemingly rich, diverse system plunged it into chaos and wrecked its natural balance (Kendeigh 1961). The deer population exploded, and overgrazing by the deer wiped out many species of plants and halted reproduction of oak trees, so reducing the available mast that a second herbivore, the beautiful and endemic Kaibab squirrel, was brought to the brink of extinction and probably would not have survived without human-built bird feeders. Even today, the Kaibab Plateau requires artificial management to retain a precarious balance: It lacked resilience.

On the other hand, Arctic tundra systems are notoriously unstable but seem to persist with no species lost despite spectacular fluctuations, booms and busts. The Arctic tundra is resilient. Similarly, barrier island faunas and floras survive and bounce back from seeming catastrophes like winter-storm overwash and hurricanes (Lazell 1979). These systems are simple, rather unstable, but quite resilient.

In general, however, I agree with Naeem and Li (1997) that most stable systems also have a large measure of resilience, and most resilient systems are relatively stable. I believe there is a good general correlation between stability and resilience, just as there is between diversity and stability or diversity and resilience.

Nevertheless, it is well to bear in mind that there can be a real distinction. Holling (1973) points out the implications for management of systems in which the distinction may be real and important: "The stability view emphasizes equilibrium, the maintenance of a predictable world, and the harvesting of nature's excess production with as little fluctuation as possible. . . . A management approach based on resilience, on the other hand, would emphasize the need to keep options open, the need to view events in a regional rather than a local context, and the need to emphasize heterogeneity."

August (1983), in an illuminating paper on tropical mammal communities, clearly separates habitat complexity from habitat heterogeneity as variables affecting species diversity. A tropical forest may be incredibly complex, involving dozens of species of major, dominant trees and hundreds or thousands of additional species arrayed from canopy through understories to ground cover. Yet this forest may be uniform over broad geographic areas, at least as perceived by vertebrate animals. On the other hand, a savanna system may be extremely patchy, with clumps of tall trees along the watercourses, grasses predominating in between, and with rock outcrops scattered here and there. Such a heterogeneous, patchy system may have in total fewer species than the homogeneous, complex one—or it may have more. On our island the ravine forests are richly complex (for a xeric region), whereas the overall 300-ha area fragments into myriad patches, imposing striking heterogeneity.

Miyata (1980) attempted to correlate measurable components of environmental diversity, such as mean annual temperature, precipitation, elevation, and latitude, with diversity as demonstrated by simple species richness, that is, presence or absence. He considered only reptiles and amphibians, but those are the dominant vertebrate groups (apart from birds) in the Antilles, so his results are relevant. Miyata considered only continental areas and located specific areas with the densest counts of sympatric species.

Path analysis (Miyata 1980) involves calculating multiple linear regressions to assign quantitative values to known pathways correlated with species richness. The method requires a computer for solving sets of simultaneous equations. The value of a path, in this sense, is the portion of the standard deviation of the criterion variable (say, lizard species density at Caracas, Venezuela) that can be attributed to a measurable environmental variable (say, latitude). Miyata found that precipitation is the most important feature positively affecting amphibian diversity. He found that frog and caecilian (caecilians are peculiar, legless amphibians) species richnesses were positively correlated to mean annual temperature, but, in contrast, salamander species richness was negatively correlated. Mean annual temperature was most important in influencing reptile species richnesses, but it was a negative factor for turtles and a positive one for all other major groups—crocodilians, lizards, and snakes. The only other important path Miyata (1980) demonstrated for reptiles was elevation, which is negatively correlated to species richness for all groups.

To any naturalist familiar with reptiles and amphibians, these path analysis correlations would have been wholly predictable: we knew all that already. Well, perhaps the negative correlation of turtles to temperature is a little surprising, but I bet rather hot Australia—with only a few, rare turtles—bent the data set. But turtles are genuinely species rich in temperate North America and Asia.

Most of Guana Island's land vertebrates are lizards. Miyata (1980) found the densest concentrations of lizard species in two disparate parts of the world: the northwestern Australian desert and the tropical South American rain forest. Our island has eight or nine species of lizards (depending on whether you count *Amphisbaena* as a lizard; I do), about a fifth of the diversity Miyata found for his richest areas. Still, for a small island this richness is quite high, as I shall discuss in more detail below. Guana is indeed hot (tropical) and low in elevation; it is not really a desert, but it is certainly dry. It is close to South America, a wonderful source of colonizers. All of these may be examples of what Miyata (1980) calls "deterministic influences of broad climatic factors." Surely, too, there has been a lot of what he calls "historic noise." We shall try to comprehend, at least roughly, how much of each, as we proceed.

A very interesting and different way of looking at diversity has been developed by plant ecologists. This is the species : biomass relationship. It is not simply: more biomass, more species. Note that a square meter plot overflowingly filled by one giant redwood tree will provide metric tons of biomass but no diversity. A square meter in Guana's scrub jungle could easily produce a dozen species with a total biomass of a kilogram or so. But there is a sort of relationship. Grime (1979, and works cited therein) noted that a plot of species over biomass against time is apt to produce a distinctly hump-backed curve. If one begins with bare ground and allows succession to proceed, species and biomass grow up quickly to a species maximum. Then, although biomass continues to increase, species diversity begins to drop, eventually tailing off to a fairly stable low. You can imagine a square meter piece of cleared ground in the redwood forest: At first, many little plants sprout and grow, only one of them a redwood seedling. Come back in a century or two and that redwood will have excluded all the other plants and filled the plot.

In a series of papers, Jim Grace at the National Wetlands Research Center in Lafayette, Louisiana, and colleagues have explored the species : biomass relationship. Grace (1999) provides a historical overview including the hump-backed curve of Grime (1979), which can be seen in Guana's sampled plots (figure 4). Five major factors are elucidated (table 6). Initially, disturbance and stress are correlated with low species richness and low biomass. As these factors are mitigated with time, ecological succession, or seasonal tempering (winter to spring or dry season to rainy), colonization of suitable species and niche differentiation propel diversity to its peak. Subsequently, however, the dominance of a few species over many pushes diversity down as biomass continues to increase. Grace and Pugesek (1997) provide a structural model applied to a coastal wetland. Grace (2001a, 2001b) explicitly addresses species pools available as providers of richness for circumscribed plots that are sampled to test specific cases of the relationship.

TABLE 6

Plant data from removal plots

PLOT	1	2	3	4
Number of woody species (richness)	2	4	8	7
H (Shannon–Weaver)	0.97	0.76	2.07	2.06
Basal area, cm^2	990	1,617	1,560	2,675
Wet mass, kg	997	1,231	2,074	2,366
Herb coverage, %	21	31	10	7

SOURCE: Rodda et al. 2001b.

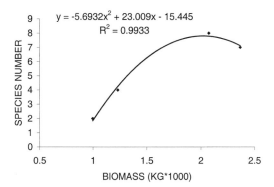

$y = -5.6932x^2 + 23.009x - 15.445$
$R^2 = 0.9933$

FIGURE 4. The typical hump-backed curve of the species : biomass relationship of plants on Guana Island (data from Rodda *et al.* 2001b).

The team led by Gordon Rodda, including botanist Renee Rondeau (Rodda et al. 2001b), collected not only the world's record-breaking lizard density data but also plant data (table 6; figure 4).

Here, the number of species includes only those with stems greater than 1 cm in diameter at breast height (dbh). Herb coverage is included because it gives a reflection of the quantity of plants whose species were not recorded. Notice that herb coverage does show a hump-backed curve if plotted against biomass, just as does species number—but the peak is shifted toward lesser biomass. Herb coverage was measured using a Daubenmire frame enclosing one-half of a square meter; the percent of ground covered by herbs is recorded for 20 Daubenmire samples in each of the 100-m², uniformly spaced, plots cleared. The figure for each plot is the average of all 20 samples. Basal area is the summed cross-sectional areas of all stems 1-cm dbh and greater. Notice that basal area gets bigger as biomass increases: The trees—seagrapes on Guana—are bigger in the later successional stages, in the more mature woods. Harking back to my discussion of information theory, note how everything increases from plot 1 to plot 2 *except H*, that Shannon–Weaver index. It tells us that plot 2, with twice as many woody species, half again as much herb coverage and basal area, and a fifth more biomass, has one quarter *less* diversity. This is not what I consider a valid measure of diversity.

On Guana's sand flats, succession proceeds to a mature woodland dominated by seagrape *(Coccoloba uvifera)*. When I first came to Guana in 1980, the North Bay Woods was just such a seagrape forest. That was one site of my pit-trap sampling of lizards, especially the *Sphaerodactylus* ground geckos, or sphaeros. Old stone walls and other edifices imply that this flat was wholly cleared for agriculture in the eighteenth century. In 1989, Hurricane Hugo devastated the North Bay Woods, toppling many of the big, old seagrape trees. It would be especially interesting to look at plant species diversity there now, more than a dozen years after the storm. There are still many big seagrapes, but my impression is that there are now more individuals of other species than there used to be. Hurricane disturbance and the diversity : biomass relationship in Nicaragua have been described by Vandermeer et al. (2000). They note of mainland rain-forest studies that the finding that ". . . more species exist in the disturbed area than in the undisturbed area appears to be a consequence of the simple fact that there are more stems in light gaps than in undisturbed forest." I presume the stems would also be smaller and with time many would be crowded out. They go on to note that ". . . the disturbed area will typically have more individual stems . . . , making the larger number of species only a statistical artifact. . . ."

Vandermeer et al. (2000) concluded that island ecosystems may be very different from mainland ones in how they respond to major storm damage. They specifically cite Puerto Rico, noting that storms on islands in the Caribbean may be so frequent that ". . . many tree species infrequently reach reproductive maturity during the intervals between storms, and thus have a reduced potential for maintaining a persistent population." On mainlands where major storms are infrequent and intervals between storms fairly long, this problem of tree maturation may not be serious, and storm disturbance may help maintain diversity in the face of long-term competitive exclusion and dominance of a few species. The main worry highlighted by Vandermeer et al. (2000) is that global warming may lead to increased storm frequency, which could turn a good thing—high diversity—into a bad thing—lowered long-term diversity, especially on oceanic islands.

At first glance it may seem that any diversity : biomass relationship that proceeds toward lower diversity with plant community succession must contradict the hopeful notion of a positive diversity : stability relationship. This, however, is just where Grace's (2001a, and works cited therein) insistence on the importance of species pools becomes important: Overall diversity must include the early successional species appropriate for the geographic area occupied by the community. It is all right if early successional species get crowded out of a given plot or subsample area within the ecosystem, but it could be catastrophic if they were lost from the whole ecosystem. Stability depends on the inevitability of individuals dying and being replaced—which may not be so great from an individual's perspective. One reason why species : area curves tend to flatten out for orders of magnitude in island size for small islands is (I have suggested) that small islands are mostly edge, dominated by the early successional species naturally abundant at edges.

McCune and Grace (2002) have gone beyond consideration of species pools to consider all of the types of measurable components considered by Miyata (1980) noted above, and many more. They explain a method for including such things as salinity, soil pH, and various minerals. This approach, called structural equation modeling (SEM), supersedes the old path analysis. McCune and Grace (2002) provide a history of the emergence of SEM. To many biologists, the expression *SEM* means "scanning electron microscope." There are probably not 10 of us in the world who might get confused; the two concepts are poles apart. But I could be one of the 10, because scanning electron microscopy has been critical to understanding the systematics of some of Guana's beetles (Lu et al. 1997). Anyway, the SEM under current consideration has become very important in the social

sciences and economics, but few ecologists are aware of it. The problem in ecology is that so many variables impinge on diversity, and many of these variables are correlated. In practice (as we have seen) path analysis—using simultaneous multiple regressions—has not proven very informative. Analysis of covariance (ANCOVA) and principal components analysis (PCA) are also very limited. McCune and Grace (2002) list nine "conditions" or situations in which SEM appears to hold more promise for understanding ecosystems and their diversity than any of those older, classic approaches. Four of these seem especially important to me (not in the original order):

> Situations where the investigator wishes to detect and study macroecological factors that are of broad conceptual interest.
>
> Field or other experiments that involve a complex set of covariates that make results conditional.
>
> An interest in studying the network of relationships among a set of correlated variables.
>
> A desire to work toward the eventual testing of a priori hypotheses about how a system works.

For SEM to be worthwhile, one needs to have developed ideas—hypotheses, judgments, and theories—about an ecosystem or community. Then one can find a balance between these ideas, or preconceptions, and statistical analysis. We have, in effect, multivariate hypotheses, and SEM will allow us to reject (one hopes) some of them. In practice, several structural models may fit the data well. We can try to collect more confirmatory data for each of these, allowing us to eventually select the best model. In a study of Gulf of Mexico coastal vegetation, Grace and Guntenspergen (1999) saw an obvious structuring of the community with distance from the sea: those few species of high-salt tolerance grew closer to the sea and species density (diversity per unit area) increased with distance as less-salt-tolerant species became more frequent. Salinity, of course, must be the critical factor. Salinity is a readily observed indicator variable that can be measured in parts per thousand in the soil.

However, observed (measured) salinity increase did not correlate strongly with the decreasing species density; there is a latent variable involved in "distance from sea" that results from storm surges that salinate soils, decreasing with distance, and that do so only occasionally. Thus the community structure is strongly influenced by rather rare events—storm surges—that only briefly affect salinity but have long-term structuring effects.

It would be fascinating to try to comprehend Guana's plant community in terms of SEM. To do this would require a preliminary examination by an experienced SEM practitioner, then sampling data collected from at least 200 plots (but the plots can be small—perhaps a square meter), and finally applying the method. For this last step Grace recommends the "how to" explanatory text of Pugesek et al. (2003).

McCune and Grace (2002) point out that this model, depicted in figure 5, is oversimplified; for example, there must be a correlation path directly from "Disturb." to "Light" because

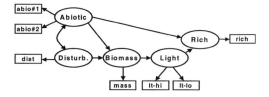

FIGURE 5. A very simplified and unquantified depiction of a structural equation model (SEM). The indicator variables are boxed and include "abio # 1" and "abio # 2," actual measurements of abiotic influencers such as soil salinity, nitrogen content, or myriad other factors; known disturbances, "dist"; weighed "mass"; and "lt-hi" and "lt-lo," measurements of sunlight intensity. The latent variables are encircled. The far right, culminating box, "rich," is counted as species diversity per unit area. Modified from McCune and Grace (2002).

disturbance creates canopy gaps that let more intense sunlight in. SEM allows one to test this correlation and potentially include it. The specific factors that regulate diversity will vary from place to place and methods such as SEM may prove useful in estimating the importance of different forces in controlling the diversity we see on Guana. (See plate 5.)

A Summary

I N LIEU OF A FORMULA—or even a solid theory or philosophy—I will sum up with a personal view solely based on observation and field experience. It seems to me that small, insular ecosystems (e.g., Guana) evolve rapidly to stability and have great resilience (unless artificially compromised). It does not seem that way at all to a lot of my colleagues. Indeed, the whole "equilibrium theory" of island biogeography depends on the turnover of species: regular, frequent extinctions balanced—or caused—by regular, frequent colonizations (Hubbell 2001). I believe demonstrable turnover is largely artificial, an aberrant deviation in the past few thousand years caused by humans and their commensal animals and plants. Islands are naturally populated by species that are physiologically very tough: able to resist and rebound from catastrophic vicissitudes such as severe droughts and hurricanes. Yet these species are generally poor competitors with continental forms, at least on continents, and they are not—were not—adapted to humans or our commensals. This is because they have not undergone rigorous selection for characteristics that mitigate the destructive effects of competition and/or predation. They are apt not to be tightly specialized and tend to have broad niches (Lack 1976). They tend to travel well and colonize other islands, but they then undergo rapid evolution into novel forms particularly adapted to each specific island. If my view is correct, it means that no two small, insular ecosystems will be much alike in close detail and that finding viable generalizations will require unusual insight (or good luck).

It seems that diversity—like relationship, or beauty (see Lazell 1972)—is not a ponderable property of physical objects but rather a value in the eye, or mind, of the observer. One can infer diversity from various quantitative measures, but one cannot measure diversity itself, any more than one can measure relationship. Determining a positive correlation between diversity and stability will depend on what sort of measure we use to infer whatever it is we call diversity. Perhaps most important, stability depends on where one's perch site—so to speak—is in the community. For humans or tigers, viewing a community from the top of the ecological pyramid, things may appear quite tranquil and stable; to plants and the myriad small organisms that live on them near the bottom of the pyramid, this same community may be as wildly inconstant as the Arctic: colonizations and extinctions, booms and busts, may occur frequently from one briar patch to the next. Stability, too, may be in the eye of the beholder.

PART 2

How to Count Snakes— and Other Things

Every specimen, whether of diatom, moth, bird or elephant, will have required a certain amount of space for its development. The number of specimens, provided we stick to a single taxonomic group, gives . . . a measure of the space needed by the successful members of that species.

G. EVELYN HUTCHINSON (1953)

"I saw a Heffalump today, Piglet."
"What was it doing?" asked Piglet.
"Just lumping along," said Christopher Robin. "I don't think it saw me."
"I saw one once," said Piglet. "At least, I think I did," he said. "Only perhaps it wasn't."
"So did I," said Pooh, wondering what a Heffalump was like.
"You don't often see them," said Christopher Robin carelessly.
"Not now," said Piglet.
"Not at this time of year," said Pooh. . . .
Where should they dig the Very Deep Pit?
Piglet said the best place would be somewhere where a Heffalump was, just before he fell into it, but about a foot farther on. . . .

A. A. MILNE (1926)

Population biology begins with the question "What is N, the total number of individuals in the population?" On Guana Island, all of the species have to produce more offspring than are needed to maintain stable populations, or they would become extinct. Because such an unexpectedly large number of species seem to manage to do that in this ecosystem, it has just got to be worthwhile and interesting to try to figure out how they do it. We can only start by learning how many of them there are. Unfortunately (for snakes), one cannot use the shepherd's simple formula: count the legs and divide by four. Initially, we will have to try to determine N at one point in time. Then we must attempt to learn if N changes with time. It might do so seasonally, or cyclically from year to year or decade to decade, or it may be in long-term flux leading to stability at some higher number—or to zero (extinction).

Spatial Distribution

THE FIRST THING to figure out when you set out to count snakes and other animals is how they are arranged in the environment. At any given time, animals may be dispersed in their environment in one of three possible ways:

1. At random. If the presence or absence of other individuals matters not at all to any given individual, then—with respect to each other—you may find them anywhere, scattered at random.

2. Regularly or uniformly dispersed. If the individuals really do not much like each other—if the influence they exert on each other is negative—they will be spread out very evenly with a highly predictable, regular amount of space in between individuals, as with swallows on a telephone line. Many organisms are apt to be this way, at least some of the time. Regular, uniform dispersion typically reflects territoriality or a defended home range.

3. In clumps or aggregations. This occurs if the animals do like each other, or are too lazy to move (resembling trees); or if they, at least, do not actively dislike each other. Mating, parental care or proximity to young, and positive social responses all tend to clump individuals.

A. John Gatz, Jr., of Ohio Wesleyan University, produced an excellent summary study on dispersion or spatial distribution (personal communication, 19 January 2005). It comes complete with a computer program (available from ajgatz@owu.edu). The basic statistical tool involved is the Poisson distribution. This statistic enables you to accept or reject the hypothesis that an observed distribution—or dispersion pattern of individuals in their environment—is random. With large animals that can be readily seen, such as ungulates in the Serengeti, you will be able to discern the dispersion pattern by simple observation. Usually you must first set up a sampling system that enables you to count individuals at a specific place and time. This could be as simple as a grid of lines or strings run through a habitat. You would then count the number of individuals in each square, or in every tenth square, or at grid intersections. Or, in a wood, you could count the number of lizards (for example) on each tree trunk at some regular distance, like every 2 m. Or, with cryptic animals that like to live under things, you could count the number under each square meter of cover, where the cover is rocks, or logs, or boards. At the most sophisticated level, you can mark individuals for unequivocal future recognition, or put radio transmitters on them (nowadays, these can be in direct communication with satellites that determine their position in real time, no matter where they are), or fit them with radioactive tags, and thus determine how they arrange and rearrange themselves in the environment.

But all you really need to know, at the most elementary level, is the mean number of individuals censused at each station. Then, use a computer to generate a Poisson distribution with the same mean as you observed for each of your species of animals. If the goodness-of-fit of your data to the Poisson distribution is sufficiently great (as determined by a small chi-square value), then you cannot reject the null hypothesis of a random pattern of dispersion. If, however, there is not a good fit between your data and the Poisson distribution generated, a large chi-square value will indicate that the null hypothesis should be rejected. In this latter case, compare the observed and expected to see what pattern of dispersion is being shown. A large number of stations with one or two individuals, combined with very few stations with zero or a large number of individuals, indicates a uniform or regular dispersion pattern. A lot of stations with a large number of individuals and/or zeros, combined with very few stations with one or two individuals, indicate a clumped pattern. Often with land vertebrates—largish animals—the sample sizes are too small for a chi-square test to show statistically significant results. In those cases, a coefficient of dispersion, CD, may help:

$CD = n/v$, where n is the mean number of individuals per station and v is the variance. If the variance is less than the mean, the CD will be greater than one, and a clumped pattern is indicated. If the variance is greater than the mean, however, the situation is not unequivocal; the pattern may be one of high dispersion, uniform or regular, or it may be random. Conspecific animals are rarely dispersed at random in natural environments. There are two weaknesses to this metric. First, it allows no statistical analysis, making it a strictly qualitative measure. Second, the choice of scale is critically important. Using 10 × 10 cm quadrants for sequoia trees would produce absurd results. So would using a 1 × 1 km grid for grasses.

It is striking that, whereas methods of marking and tracking animals have undergone technological metamorphosis, methods of counting remain just as they were decades ago (Rodda et al. 2001a). There are many other ways in which the Poisson distribution can be used to look at organism spacing in ecology and population biology. These include perceiving and defining defended territories, as in *Anolis* lizards on our island (Metzgar and Hill 1971; Davies 1978); perceiving dispersion under environmental conditions that cause crowding, such as drought (Tschinkel and van Belle 1976); understanding host–parasite relations (Mahon 1976); tracking breeding aggregations (Downhower and Brown 1977); and—most important for us—interpreting mark-recapture data (Hackney and Linkous 1978). (See plate 6.)

Direct Census

I N THEORY, the best possible way to determine N, the population size, is to count all the individuals present at one time. If one cannot count the entire population, one may be able to count the number in a known area, for instance 100 m², and extrapolate to the population's total range. This method is likely to be fraught with error because you are making the assumption that the habitat is uniformly occupied throughout the range. We can walk down to the Salt Pond, for example, and count 30 Bahama pintails (ducks, not snakes). The Salt Pond varies a good deal in size, sometimes from day to day, as a function of rainfall. At the time of our pintail count it covers 2 ha. Because Guana is 300 ha, a simplistic analysis tells us that there must be about 4,500 Bahama pintails on Guana Island. To get that four-and-a-half thousand ducks, I simply multiplied the counted number in a known area (30 in 2 ha) by the total number of hectares. Biologists do that all the time in estimating populations; what could be simpler? Of course, the fact that 2 ha of the island are covered with water, and the other 298 ha are covered with rocks, sand, cactus, trees, and other stuff that is rather inhospitable to ducks, renders my method ridiculous.

Anyway, we walk up too close to the ducks in making our count, so all of the pintails take off skittering and splashing, and fly over to Tortola. Now there are none. This is closer to the real value, but still not good enough. In truth, there are some Bahama pintails who do "live" on the island, or at least nest there, but not many—not even 30. We have just stepped into two major pitfalls: Animals are not usually spread around evenly, and animals usually do move.

In practice, there are very few species for which direct census can be done. On the island, we have few data on nests of bird species; in a few cases we have tried to determine the number of fledglings per nest, on the average. We can do it for the Bahama pintail ducks on the Salt Pond, but we must temper the result with knowledge of the species' natural history, and make many censuses at different times. Looking at nonbreeding, adult ducks on the pond gives us figures ranging from 0 to 50.

I have used this method also for the island's reptiles, including the tiny burrowing snake, *Typhlops richardi*, and various lizards. Worst first: I am unable to invent a method for effectively censusing the "blind" snakes, *Typhlops*. They are definitely not rare, for we encounter them frequently when we get to digging around; for instance, when digging pit-trap holes, or cleaning out filled pit traps, or turning boards and rocks, or just digging. But, that is the problem: We just encounter them. A pit-trap hole is about 27 cm in diameter and about 66 cm deep (it must accommodate an 8 × 24 in. piece of stovepipe easily, and then both bottom and sides are filled in). Digging 20 of them means excavating a little over 7.5 m³. Doing this we encountered two *Typhlops,* or one per 3.25 m³. What, if anything, is that datum worth?

First, it is a volumetric, not area, census; with a fossorial (burrowing) species, I like that. But where should my vertical (third) dimension stop? The 66-cm point is determined by the length of a piece of stovepipe, not by the habits of *Typhlops.* I never actually found one deeper than about 35 cm. Why include all the rest of that volume? Why not assume they go down at least a meter? No doubt they can if they want to. Or 3 m?

Second, why on earth would a *Typhlops,* which (for a nearly blind, legless creature with a microscopic brain) is exceedingly fast, agile, and resourceful, stay anywhere near where we were digging a hole? We have to scrape off leaf litter and humus, then chop—with a blunt shovel—through podzol and a dense root mat, then begin the usual stab, foot-push, rock, and heave method of humans digging holes. Any *Typhlops* who gets caught after all that action has to be pretty laid-back, even for these islands.

Third, there is the question of sample size. Two for 20 tries? Ridiculous: Dig 2,000 pit-trap holes and let's look at the data then. (The shovel is in the shed at White Bay; I'll show you where to start.) Doubtlessly, you can come up with a fourth problem that may invalidate my volumetric census.

I tried another way. I put out sheets of 3/8-in. marine plywood. D. K. Smith and Petranka (2000) got good replications of counts, and Monti et al. (2000) got higher counts under artificial cover than natural objects. I cleared the leaf litter and most of the humus, and leveled up the surface of the podzol before setting sheets, then I left them alone for 5 days. Then in early April when it was very dry, I came back, turned them over, and caught a *Typhlops:* 1 per 5 m² at a depth of about 1 cm. What is that

datum worth? Forgetting sample size and all confidence limits, we will extrapolate that to a volumetric figure for the top 0.66 m³ of ground: one *Typhlops* in about 3.33 m³. That is disconcertingly similar to the figure arrived at by the first wretched method.

I reckon most of Guana Island is unsuitable for *Typhlops* habitation, because it is too steep and rocky and has too little soil, or is open, grazed grass—nearly as hard as sandstone, hot and dry. *Typhlops* likes the shady, moist areas of deep soil on the wooded flats and in the ravines: say 100 ha maximum. That is 0.66 of a million cubic meters: 660,000 m³. We will use a rough density estimate: 1 per 3 m³. So about 200,000 of the little buggers live here: 2,000 per surface ha. . . . Do you believe that? I hope my methods will make someone—maybe you—so furious that he or she comes right down here to prove just how terribly wrong I am. When Gordon Rodda, a biologist with the U.S. Geological Survey, came to Guana in 1998 (16 years after my collection of three data points), his team found six *Typhlops*—double my sample size. This extrapolated to 150 per hectare, but they were excavating just the top 5 cm of litter and humus. So, converting their areal data to volume, they found *Typhlops* at a rate of 150 in 500 m³, which is one *Typhlops* in every 0.3 m³, or a total island population of 15,000 in the top 50,000 m³. What would have happened if Gordon and crew had gone down 66 cm? Or even 35 cm? Rodda et al. (2001b) seemed to think that their 150/ha was a low estimate, noting that "the sampling technique is considered inappropriate for this fossorial species." I have rarely seen people work as hard as they did on their removal plots (see below), so I can scarcely suggest that they should have dug deeper. Anyway, we have a lot of *Typhlops.*

Let us look at lizards: more accessible animals, making our enumeration methods more convincing. We have done a real census for four species—the three *Anolis* and the tiny ground-dwelling geckos locally called "cotton ginners," *Sphaerodactylus macrolepis.* The latter is the most abundant vertebrate animal on the island, and one of the most interesting. I will call it "sphaero." To census sphaeros we staked out plots of measured dimensions and endeavored to capture every individual in the plot. If one managed to run out and escape into the surrounding forest floor litter and jungle, we could still count it—if we saw it. I already knew, from mark–recapture results of pit trapping (see below) that sphaeros travel little and have small home ranges. We could stake out a 1-m² plot and catch sphaeros virtually every time in good habitat—usually more than one. A plot bigger than 8 m² was hard to deal with; it took at least three people to monitor its edges while clearing, so that there were not uncounted escapees. The bigger the plot, the smaller your risk of the initial stake-out and edge clearing modifying the number inside at the start of the count. So an average of 2.6/m² in the 8-m² plot might be a better overall estimate than an average of 1.6 for five separate 1-m² plots. The benefit of the small plots is that each plot is a "trial" for statistical purposes, so you can amass more trials quickly. The problem is that just preparing them for census may severely affect the count. The higher figure yields 26,000/ha. When Rodda and crew came in 1998, they sampled four 10 × 10 m plots, each 100 m². They cleared the edges and installed meter-high aluminum flashing, greased; the barriers were installed at night because the lizards of interest are diurnal. In the plot with the least leaf litter and very low vegetation just beginning to regrow on previously mowed ground, they got only six sphaeros in 100 m²—only 600/ha. In a plot dominated by the legume *Leucaena leucocephala*, a little further along in ecological succession from cleared ground, they got 262 in 100 m²—2,620/ha. In their plot in the beach strand seagrape (*Coccoloba uvifera*) woods, where I extrapolated 26,000/ha, they got 380 in 100 m²—38,000/ha, indicating that my little 8-m² plot was either poorly positioned within the habitat, or leaky, or for some other reason yielded a low count. Most impressive, when Rodda and crew did a less-disturbed seagrape wood, on the White Bay Flat but well inland from wind and storm waves that perturb

the strand wood, they got 676 in 100 m²—67,600/ha. That is the highest-known density of any nonaggregated terrestrial vertebrate (Rodda et al. 2001b).

With *Anolis* we staked out 20-m² plots in the herp hectare on the south-facing slope of the ridge surmounted by the hotel, in scrubby woods. We endeavored to catch every lizard. This requires three or four people, a lot of time and patience, and a good eye. In March 1982 we began a plot at 4 p.m. and worked it to within a half hour of dark (6:30 p.m.), giving the lizards that half hour to settle down for the night. We came back from 9 to 9:30 p.m. to get sleeping anoles. We started again at 5 a.m. (still quite dark in March) and hunted until 8 a.m.

Crested anole (*A. cristatellus wileyae*) was most numerous; followed by saddled anole (*A. stratulus*); scarcest was grass anole (*A. pulchellus*). The ratio in 10 m² was 4:3:1, extrapolating to 4,000, 3,000, and 1,000 of these lizards per hectare, respectively. A great deal of natural history data, including vegetative associations and basic habitats, would have to be considered before these figures could be extrapolated again to an area larger than the hectare in which our 20-m² plots lay. We just know, from simple observation, not to multiply these figures by 300—the number of hectares of the island—to get total populations. To use figures like these, we must compare them with figures obtained by other methods, such as mark–recapture and change-in-ratio estimators (see below).

When Rodda and crew did their four 100-m² plots in 1998, they got an average of 1,000 *A. c. wileyae*, 600 *A. stratulus*, and 50 *A. pulchellus* per hectare in their very early seral-stage legume-dominated plots. In their seagrape plots, crested anole (*A. c. wileyae*) and saddled anole (*A. stratulus*) were up some, to 1,100 and 900 per hectare, respectively; they found no grass anoles (*A. pulchellus*) at all. I believe the difference between their data (Rodda et al. 2001b) and ours, of 1982, reflects habitat: The scrub jungle of the herp hectare, where we got the higher 1982 estimates, is a richer habitat for all three anole lizards.

Salvidio (2001) counted all salamanders (no relation to lizards; none on Guana) removed from two habitats in three searches of each. Because of habitat constraints, the populations in each habitat were effectively closed: No new salamanders could readily immigrate. For analysis he used the program CAPTURE routine M_{bh} of G. White et al. (1982), which provides N, the population estimate, its standard error, and—given three removals—a measure of capture probability. Rodda et al. (2001a) did not explicitly discuss and evaluate this method. I note it could be devastating to a small, closed population; that might really be exterminated. However, Salvidio (2001) held all salamanders captured in the lab until completing the third removal, and then released them back at their respective habitats. This method could be employed, however, in fenced plots like Rodda's. It would involve far less effort than total removal; that would make it possible to increase sample size in a constrained time—an advantage that might offset the lack of precision given by plot clearance.

In a comparable study, Petranka and Murray (2000) did 21 searches and removed 2,433 salamanders from a streamside community in eastern North America—where salamanders reach staggering densities. They got no significant reduction in catch over the 21 searches. For this reason, they had to accept the fact that their estimate of 18,500/ha is very conservative. I suspect that the habitat—and thus each species' population—is far from closed and that an enormous pool of salamanders is available to invade a depleted niche. This study reflects a problem with my pitiful *Typhlops* snake count, above: When one captures the animals by checking a specific piece of cover, whether a rock or log in nature or an artificial cover board like my plywood, one really cannot say much about absolute populations. How many of the animals live in the habitat but not under the cover object turned by the human hunter? This technique is good for monitoring population trends but is not a valid direct census.

Perhaps the loveliest of all direct censusing techniques is provided by photography. The animals censused must be conspicuous to the camera, which usually means large and out in the open, for instance, ducks on a pond or wildebeest on the Serengeti. An aerial photograph is best in most cases, unless it is of birds in the air; then you can comfortably be on the ground. You need to either photograph the entire population (e.g., flock, herd) or have a size scale in the shot allowing an area calculation. If you can see them you can photograph them (maybe); then you can blow up the photograph and count at your leisure. Thus, direct censuses work well at two extremes: Little animals that do not move often or far can be counted in staked-out plots, by area or by volume (then, as noted above, the problem of extrapolation to real habitat areas remains). Big conspicuous animals can be counted (photographed) from the air or as they fly over. It is active, middle-sized creatures that cause real problems.

With passerine birds flitting through the jungle, or rodents, or big, diurnal reptiles such as snakes of the genus *Liophis* and lizards of the genus *Ameiva,* direct visual censusing is virtually hopeless. If the animals are not cryptic, they are moving—often far and fast. You can measure off a 100-m^2 plot, 10 m on each side, and place an observer at each corner—all in plain sight of each other. While you all got into position, the ameivas were zooming in and out and through it. Now they look at the four of you askance and all go off to hunt their lunch in some other 100-m^2 area. (See plate 7.)

Line Transects

IF DIRECT CENSUS IS THE most elementary way to determine N, the population, the line transect is its most elementary factor. The notion is simple: Instantaneously count all the individuals on a line of a known length (say, 100 m² long), square that number and you have the population of the area (in this case, 100 m²) whose linear dimension is the line. What could be simpler? The problems with this method can be so fundamental as to utterly invalidate it in many practical applications. First, the time factor. It is difficult to set out a line short enough to be surveyed in an instant, but long enough to have any reasonable number of individuals on it. If the organisms are tiny, so the line can be entirely seen at one time (say a meter or less), this problem is eliminated. Or, if the organisms are stationary, like trees, then time does not matter. But for most applications involving vertebrate animals we need a fairly long line, and we are deviled by the movements of the animals. We may try to travel the line as fast as possible and hope the same number of individuals, on average, move onto it as move off of it during the transect. I cannot but see myself, however, caroming along my well-laid line: I create a sort of sensory bow wave out in front of me from the point where the animals I want to count see, hear, feel, or smell me coming. Because they can

probably see, hear, feel, and/or smell far better than I, in my mind I picture them spilling off my line in all directions into the brush, leaving me to count only the halt, the lame, the dolts, and the dead.

So, second, one is sorely tempted to count those individuals visible from the line, as their tails disappear into the greensward, because one deeply believes they were on the line at the instant the transect began. Even if they were not on the line, one longs to count them if one can see them. You may, providing you devalue their contribution to the linear dimension of the area by a function of their distance from the line. In theory, both the time factor and the distance factor should be amenable to algebraic incorporation in formulas that will resolve them in your favor. A fine monograph by Burnham et al. (1980) provides just such formulas. In practice, using the formulas may not be worth your time and trouble, as you will see in the following.

Third, any transect line laid out by a human, or even a student, is by definition artificial, and as such may grossly modify the behavior and spacing of the organisms along it. The edge of a woods, for example, makes a lovely line, but the animals you see along it, and their relative numbers, may have nothing to do with the numbers of species and individuals just 10 m into the woods on one side or the field on the other. A road through a woods or a field just makes two edges. Are the animals you see and count members of species whose natural habitat is edges—neither forest nor field? Are they animals that shun the open and rarely or never actually get on the line? Are they animals that come out of the woods to bask in the sunshine? As with direct censuses, you can begin with a fourth reason and go on toward infinity thinking of things wrong with line-transect counts, but—for me—the most disquieting is the inevitable artificiality of the line. I cannot even walk through initially undisturbed brush enough times to get a fair sample size without obviously affecting the habitat by making a path. How do the animals I want to count feel about that path?

This brings up an important point: You want to run your line transect as many times as possible to increase your sample size, or number of trials, and thus tighten your statistical probabilities of having a realistic result. I have used a line-transect count in the simplest and most straightforward manner to estimate a per hectare density of the common snake on the island, *Liophis portoricensis anegadae*. Because this snake has no vernacular name except "snake," I shall call it *L. p. anegadae* (there is another member of genus *Liophis*—*L. exigua;* it is rare.) I laid out a 50-m line along a trail through fairly open scrub. I selected the site because I saw about as many snakes there, I thought, as I saw in most parts of the island. I saw denser snake numbers in the big ravines and no snakes at all out on the well-grazed flat. I made 14 counts, getting from zero to three snakes per count. I counted between 4 and 6 p.m., a time when I often see snakes about. Of course, I counted snakes only on the line; I resisted the temptation to count snakes I saw from the line but that were escaping off it. I reckoned my 14 counts of 50 m each as the same as seven 100-m counts. My average snake count was 1.7 per 100 m, or 2.9/ha: two or three, but closer to three, snakes, on the average, per hectare on Guana Island. File that figure away; unscientific as it may seem, prone to catastrophic errors involving everything from sample size and variance to my subjective notions about average habitat, it turns out not to be a bad estimate.

My favorite line-transect census was set up on Guana by Robert Chipley, then of The Nature Conservancy, while he was studying the lovely and little-known bridled quail-dove, *Geotrygon mystacea*. Chip ran a 100-m line up Quail Dove Ghut, a habitat in which we know these amiable, iridescent, waddling birds to be concentrated. He could traverse his line with little worry about the pathmaking that haunts me, because the floor of the Ghut is a boulder jumble. One steps—or hops—from rock to rock and does little damage to vegetation in the process. The Ghut remains pretty much as it was before the line was laid out. Chip used one of the most elegantly simple techniques ever developed, unmodified from Ye Olde Gamekeeper, as published by Gates (1979, 1986). Let *A* be the total area of the population's range to be sampled; *Z*, the number of individuals seen from the line; *X*, the length of the line; and *Y*, the average distance of individuals seen from the line to one side. Then:

$$N = \frac{AZ}{X \cdot 2Y}$$

What we have done is convert the line into a strip. The length of the strip we determined arbitrarily ourselves; in the case of Chip's quail-doves it was 100 m. The width of the strip is the bugbear that torments everyone trying to census by these methods. In this case we have assumed a steady rate of diminishing detectability directly proportional to distance from the line. We have assumed that 100% of the birds on the line are seen and counted; that the farthest bird seen from the line was really as far away as any could have been seen or counted; and that the birds become increasingly difficult to detect at an even and linear rate. The most distant bird Chip saw was 15 m off the line; thus the width of the strip is twice 15, or 30 m; the strip is therefore 3,000 m² (*A* in our equation above). Burnham et al. (1980) discuss at length just how unwarranted these assumptions are. First, if we are dealing with readily detected individuals—say the carcasses of dead deer—there will be a strip width greater than the line in which 100% of individuals are detected. One must decide how wide this strip is, calculate its area, and then add that to the areas of the two marginal strips along the outsides, which are calculated by the formula above. Second, there is very likely some area beyond the farthest sighting that should actually be averaged into the strip width. That is, you may really have the ability to detect at least an occasional individual farther out than the farthest one you actually saw. How much added to Y will compensate for this? Third, there are practical problems: How do you determine the distance of each sighting from the line? You want the perpendicular distance, of course. It can be easily calculated by simple trigonometry, if you know the angle at which the individual was sighted and the actual distance. In practice, one usually gets to know the area well enough so that one can recognize the sighting spot. Then one proceeds to the point on the line perpendicular to it, and simply measures.

With quail-doves on the island these difficulties seem trivial. In this rock-jumble jungle, birds very close to the line may escape detection. By repeating the transect many times, one begins to feel confident that one is detecting the occasional bird that is as far off the line as it is possible to see. And the surface features of the terrain are so varied and distinctive that measuring sighting distances perpendicular to the line is not really difficult. In Chip's case he used a 100-m, nonstretch black rope marked at 10-m intervals with labeled, bright red bands. He used a forester's tape that peels off a reel to measure the perpendicular distances of the birds, when first sighted, from the line. He walked the line 15 times, from 20 to 25 July 1984; that is like walking one transect, 1,500 m long. He saw zero to nine birds, an average of five (to the nearest whole bird), on these 15 transects; his total tally was 65 sightings. That yields an estimate of 150 birds, but we have to divide by 15—the number of transects—to get 10 birds in the 3,000-m² strip. A hectare is 10,000 m², so the estimate is 33 birds per hectare. Chip

reckoned at that time, July 1984, about 121 of Guana's 300 ha were good quail-dove habitat. That means Guana's bridled quail-dove population could be 3,993 birds; that is amazing—well nigh beyond belief! Keep those numbers in mind: 10 in the transect strip; 33/ha; well over 3,000 on the island. We are not done with quail-doves.

The best thing that can happen in a case like this is to have a batch of skeptical herpetologists come along and decide to apply their methods to quail-doves. That is just what happened to Chip. In addition to the line-transect census, we used marked birds to check the count. We mistnetted two quail-doves and put lizard tape (see "Mark," part 2, below) on them. We cut dime-sized circles of tape and stuck them on the birds' heads and central tail feathers. We had a red-marked bird and a white-marked bird. Each received a bright yellow plastic leg band. Having two known, recognizably marked, birds out there in the brush is going to add a whole new dimension to our census technique and provide a way to check our line-transect estimates. I strongly recommend marking a few individuals in each population to be censused, if at all possible.

Gates (1979) suggests three simple ways to modify Z, the number of birds counted. These tend to improve the veracity of the method, and I recommend trying all three just to get an idea of how much variation it is reasonable to assume is indicated by your data. The first is simplest: make Z equal *only* to those individuals seen within Y, the average sighting distance from the line. This reduces the sightings to 43, with an average of three (to the nearest whole bird) per transect. That yields nearly seven birds in the strip, or 23/ha, or 2,783 total. Keep those numbers in mind, too. The second and third require that you graph your sightings in the form of histograms, plotting individuals counted in each of a series of equal-width strips, or zones (say, 5 m in width) going outward from the line. Then, have Z equal all individuals counted either inside the zone of maximum sightings or inside the zone before an appreciable dropoff in sightings. The latter requires you to make a somewhat subjective assessment of your data. Gates (1979) found that both methods tended to overestimate known populations, and I have not used these for quail-doves.

A variation on the theme of line-transect censusing has been developed by ornithologists wanting to determine the density of territories inhabited by adult males that sing or display. The notion is that the number of territories is about the same as the number of nesting pairs, a statistic that is highly meaningful—of more use, really, than simple density of individuals, at least for birds. Morrison et al. (1992) describe the circular-plot census method, in which a fixed observer records direction and distance of territorial birds visible (audible?) from a point. They provide a way to assess double-count problems inherent in scanning proximate circles. The method is utterly simple and especially useful as a backup for more elaborate types of density studies. It is also the method of choice in extremely rough terrain and is excellent for locating areas of especially high density, and for delineating territories.

The best method of this sort I have seen is presented by Christman (1984). He combined circular-plot mapping technique with line-transect censusing to determine densities of defended territories. One traverses the line as usual, counting the singing or otherwise territorially displaying individuals. If in the course of x traverses one locates what seems to be the same individual in the same territory y times, then the probability of detection [$P(d)$] of that individual is:

$$P(d) = y/x$$

After many traverses one has enough data to find a good mean rate of detection for a particular individual per traverse; this is d. Then, the density of territories (D) is simply:

$$D = d/P(d)$$

Christman (1984) provides an elegant and simple way to assess variance with this method by running adjacent traverses or transects. In counting bird territories it is best to conduct two or more traverses per day during a few days at the peak of the breeding season. If you are going to use a line-transect method it is well to consider all of the theoretical ramifications discussed by Gates (1979) and Burnham et al. (1980). I recommend first using one of the simple approaches formulated above (and modified, if needed, for the demonstrable 100% detectable strip width). Then you will want to use some other method of estimation. Finally, I recommend plugging into a sophisticated computer program for line-transect estimation, such as that of Gates (1986), to get 95% confidence limits for a variety of estimators selected by you to fit your particular case.

Kristiina Ovaska and Jeannine Caldbeck used a line-transect census method for calling male frogs on Guana. They backed it up with a mark–recapture plot census. Ovaska reports the results in "The Frog" in part 5.

Mark

It was, after all, hunting that made me a zoologist.

VALERIUS GEIST (1971)

WAS INITIALLY GOING TO call this chapter "How to Catch Animals." However, I cannot explain to people, or even describe particularly well, how to catch animals. Animal catching is something either you are good at, or you are not. If you are old enough to read this, you already know which you are. If you are not good at catching animals, consider molecular biology: Then you will not even have to know what an animal is or recognize one if you see it. As I pointed out in my doctoral dissertation in 1968, and subsequently published (1972), the best general description of how to catch animals known to me is provided by Mercer Mayer (1967) in the small book *A Boy, a Dog, and a Frog.* In a mere 32 action-packed pages, within a highly informative cover, Mayer demonstrates procedure, equipment, and deployment of personnel. I cannot improve on his account, although I wish I could provide you with a photograph of Liao Wei-ping, distinguished ornithologist retired from the Academia Sinica, well up a Guana Island tree—and well out on a limb—waving a small stick with a fishing leader noose on the end at a tiny lizard, which he caught, of course.

For most species of animals, under most field conditions, your ultimate ability to come to sensible grips with elusive N, the number of individuals in the population, will be directly proportional to your ability to catch those individuals. For some of the best techniques, for instance, removal plots and change-in-ratio estimators, you may have to come to believe—and get a bunch of skeptical colleagues and bystanders to believe—that you have caught all the individuals, at least in some portion of the population's range. I do not mean to sound gratuitous about this subject, but it is obvious to those of us who have taught field biology courses and observed children outdoors that some people are apparently born with an enthusiasm and instinct for the hunt and the capture, whereas others are stumped by a simple problem like how to get an already trapped animal out of a box or a pit. If you have come to natural history and ecology later in life than, say, age six (i.e., if you can read this but do not know whether you are an animal catcher or not), the best thing I can recommend is joining an ecological expedition. There are a number of organizations that provide this opportunity, such as The Conservation Agency (6 Swinburne St., Jamestown, RI 02835, USA; http://www.theconservationagency. org/) and Earthwatch (3 Clock Tower Place, Suite

100, Maynard, MA 01754–0075, USA; http://www.earthwatch.org/). It is best to pick an expedition involving a broad spectrum ecological survey. If you pick one that deals only with hyenas on the Serengeti, you will miss out on bats and rodents, to say nothing of all of the things in the world that are not mammals—that is, the birds, insects, and reptiles. Ideally you will want to participate in a variety of approaches involving nets, nooses, box traps, pit traps, anesthetics, and everything else clever people have devised to outwit their prey, or at least to cooperate with it long enough so that they can recognize individuals repeatedly. Trapping animals, one may argue, is a distinctive subset of activities under the larger heading of catching them. Catching is often the result of the active pursuit of a specific individual, whereas trapping involves the more passive process of the would-be catcher leaving something behind in the environment in which the animal manages to catch itself. What Mayer (1967) has done for catching active, individually selected animals was long ago done for trapping by Milne (1926). I memorized that when I was six. I always had the gut feeling that I knew the best method, but it took a real scientific study to provide me with the needed citation to convince you: If you want to leave something behind in the environment to catch animals, while you go home for a cup of hot chocolate or a tall rum punch, get a shovel. Dig a hole (Williams and Braun 1983).

On any big-time, in-depth, ecological-survey-type expedition you will see an assortment of fine traps arrayed in the field. Mist nets are wonderful for flying creatures, like birds and bats. Animals that cannot—or are reluctant to—fly can be driven into nets set at ground level, as we used to catch the bridled quail-doves marked to augment Chip's study. This method relies on two characteristics of the net: relative invisibility, achieved by fineness of black threads or camouflage, and its springy, bagging quality that engulfs and entangles the animal. Getting animals out of nets is a skill you will have to learn in the field.

Williams and Braun (1983) compared snap traps, which kill the animal, and box traps, which catch it alive, to pitfalls. They found them wanting. Pitfalls caught more individuals of a greater variety of species than either other method. The trappers were explicitly interested in small mammals, including fossorial species like moles and gophers (which they caught best in pitfalls), but they point out how well the pitfalls worked for reptiles, amphibians, insects, and other invertebrates too. When initiating a field study, therefore, I believe it is most practical to start out assuming that all organisms that do not normally fly will behave exactly like Milne's heffalumps until proven otherwise. Why are pitfall traps so miraculously effective?

First, snakes, heffalumps, and most other animals do not expect pitfalls. Animals generally have learned their way around their home ranges quite well and are so familiar with the terrain that they take it for granted. The pitfall takes them by surprise. So does a rat trap or a box, but these objects in the midst of familiar ground usually require investigation on the part of the victim before they will effect its capture. Bait helps, of course, but you can easily bait a pitfall too—either by putting edibles in it or by hanging them over it. Second, even if the heffalump sees the edge of the pit before he falls, he is unlikely to be as concerned about it as he would be over a snap trap or a box. Most animals are unconcerned about going down holes, because natural holes are about as easy to get out of as into, and they often provide safety of shelter. Indeed, if you cover your pitfalls with a slightly elevated structure (rock, log, board, etc.) many wild animals will head for them with differential enthusiasm and take the plunge with seeming delight.

To work well, your pitfalls must differ from natural holes in being difficult, if not impossible, to get back out of. A standard pit liner I generally use is galvanized stovepipe. Two-foot sections in 8-in. diameter cylinders are available at most temperate-climate hardware stores. A bale of them is very heavy, but can be shipped by surface freight fairly cheaply anywhere in the world. A much cheaper structure is the ceramic flue liner. An excellent general-purpose size, readily commercially available, is ca. 33 cm square in cross section and ca. 60 cm long (roughly 13 × 13 × 24 in.). Flue liners are wonderful because they are nice and big and easy to reach (and see) into, and they last forever buried in the ground. Stovepipe on Guana Island did very well for six years, but eventually rusted away. Flue liners have the massive disadvantage of great weight (shipping 100 of them to Halmahera will not be cheap); they also break fairly easily if dinged with a shovel or off a rock. PVC pipe and even the waxed cardboard cylinders used as forms for pouring concrete pilings work well; the latter are very cheap and light, but do not last forever. We installed an experimental cardboard liner, covered with a square foot of marine plywood supported by half bricks, on Cape Cod in the summer of 1983; it was still working fine 15 months later. Big (2 L or larger) plastic soda pop bottles with the tops cut off work for small animals. An excellent modification of the pit trap, using PVC pipe and called a "pipe trap," has been described by Lohoefener and Wolfe (1984). It seems most efficacious for reptiles and amphibians.

An obvious problem with pit traps is that they can become death traps if left operating and untended. The flue liners are easily stoppered with a fitted lid (the fittings are put on the underside of a square piece of plywood). On Guana Island we simply fill the pit traps up with coconut husks, chunks of termite nests, rocks, and logs. The wonders one finds when one returns and empties the pit traps out again provide a major highlight of every expedition.

A variety of box traps are available. The most popular brands are Tomahawk, Havahart, and Sherman. Each has its advocates and detractors. The British Longworth is a superb trap. In my limited experience it has the highest per trap yield for small mammals. However, it has solid sides so you cannot see your catch (unlike with Tomahawk and Havahart) and it does not collapse (unlike some Tomahawks and Shermans); it is by far the heaviest of the lot. Box traps are most useful when you have to pull up stakes and move on, or where the terrain is unsuited to digging holes. Having selected a type of trap, the critical issue now becomes trap placement. There are two basic methods of deciding where to place the traps. The first is to lay out a regular grid with each trap precisely spaced relative to every other. This has obvious advantages as soon as recapture begins, because mark–recapture data can only tell you something about the population if you know just what area you are sampling and how much the animals move around. Grid placement is exactly what I did with my pit traps on the island on the White Bay and North Bay flats, where the ground is level and the digging moderately easy. Grid placement is the time-honored method used in virtually every study reported in those learned journals we affectionately call "J. Wild Man" (*Journal of Wildlife Management*) and the "Journal of the Mouse Men" (*Journal of Mammalogy*).

No seasoned animal trapper would dream of putting out his traps in a random grid. Even without memorizing Milne (1926), a good field person interested in actually catching animals knows to put the traps in selected spots based on where the animals are most likely to be. As examples, they should be placed in an obvious game trail or runway, in front of a burrow, under an already used piece of cover, or at a place where crumbs or scats indicate the desired individual(s) hang out. Doing this, however, will make accurate delimitation of the area sampled difficult and may complicate determining travel distance. In my studies of Tasmanian devils and rock wallabies I picked trap sites on these and other bases appropriate to the animals. I would say I let the animals pick the trap sites. I worried about delimiting the grid later. But, believe me, it was not easy when "later" arrived, either.

No matter which method you choose, you are going to have to be able to handle your animals in order to collect size, sex, age, and condition data, often to apply the marks, and—most important—to release animals exactly at the point of capture. Once the animal is in hand, most of the best methods for marking it involve mutilation. The bunny huggers are aghast: Avowed conservationists and ostensible animal lovers are out there afield mutilating little animals—and big ones too, when we can catch them. Many of the bunny huggers, however, have their ears pierced (like me) or have tattoos (like me) and recognizable scars (like me). All of those are forms of mutilation. I am so well mutilation-marked as to be instantly recognizable, even at a distance. Ferner's (1979) review gives details of most of the best marking methods. Although explicitly for reptiles and amphibians, this review generalizes readily to virtually any kind of animal because reptiles or amphibians provide just about every sort of marking problem you can think of: limblessness (no toes); frequent and entire skin shedding (ecdysis); burrowing and other cryptic activities; moist, mucoproteinaceous skins; and so on.

Most kinds of snakes are easily marked by clipping off a scale under the tail: a subcaudal. This can be done with a pair of ordinary nail clippers in most cases. In *L. p. anegadae* and most other species, the subcaudals are paired. Using a single clip, I begin on the snake's right (the left as I view the underside of its tail). Because snakes often lose the tips, and sometimes a good deal of the distal ends, of their tails, I go only to subcaudal number 30 on each side with *L. p. anegadae*. Thus, with a single clip I can mark 60 snakes. Number 61 (and all subsequent) must suffer two clips: in this case the first right (like number one) and the first left. Number 62 is first right and second left. On I go until I have marked the ninetieth snake: first right and thirtieth left. Snake number 91 is clipped second right and first left. . . . You get the picture.

Just using two clips, I can mark something like 901 snakes—more than some estimates of the total population of the island. Actually, I could double that in practice, because I can easily tell males from females. That might get me to 1,802. But I can get away without clipping about one-quarter of the *L. p. anegadae* I encounter: They are already marked. Some bear distinctive scars. Others have distinctive scale anomalies, like a single subcaudal among the first 30 pairs, or a half-ventral conveniently

close to the throat or cloacal opening (vent). Half-ventrals can be tricky though, because snakes are apt to have so many ventral plates that miscounts are easy if the anomalous scale is near midbody. Marking members of the genus *Liophis* is relatively easy, though. If you can figure out a way to mark snakes of the genus *Typhlops*, however, please let me know. I want to mark them and hope to recapture one some day. You will have to see a *Typhlops* to even try coming to grips with the problem.

Animals that have legs and feet usually also have toes. Toes are wonderful for marking purposes: You just clip them off. Toe-clipping is my favorite method for marking lizards and small mammals, and repeated studies show that it does them little or no harm. I use nail clippers again; one might say I just give them a little more permanent clip. Toe clipping is also good for amphibians, like frogs and salamanders (but not caecilians: no feet), but these will usually grow their toes back eventually. There are chemicals, however, that prevent toe regeneration (Ferner 1979).

As with scale clipping, you must devise a system that enables you to read your toe-clipped individuals' numbers unequivocally. I read my clips from the animal's left to right, because I am viewing the animal from above (dorsally) as I hold it; thus, from my left to right also. With a single toe clip I can mark 20 individuals. It is noteworthy that the forefeet of lizards and mice (for example) are oriented similarly, but the hindfeet are not. Thus—clipping or reading from left to right—lizard number 11 has no first left hind toe, corresponding to our big toe; mouse number 11 has no fifth left hind toe, corresponding to our little toe. You certainly do not have to do it my way. Using two clips, animal number 21 is missing its fifth (outer) left front toe (corresponding to our little finger) and its first (inner) right front toe (corresponding to our thumb). Animal number 35 is missing its fourth (second from left) left front toe and its first right front. . . . This will get me to 94 unequivocal clips. Once again, I can tell males from females in all the studies where I have used this system, so I can probably go to something like 188. Using this method on the island's sphaeros in the pit-trap grids, however, I can go much higher. Each of these tiny geckos has a distinctive pattern of spots and marblings. Just using the grossest features of these markings I can accommodate many recognizable individuals in each toe-clip number.

In population biology it would be folly to bias your calculations by toe-clipping animals whose chances of survival would be diminished by this mutilation. Mammals with opposable thumbs, like raccoons and opossums, or with broadly webbed toes, like devils and otters, should never be toe-clipped. Climbing lizards, too, need their toes—or at least the ones with big, broad digital pads that provide up-the-wall traction. Paulissen and Meyer (2000) found that removing one toe on each of four feet had no demonstrably ill effect on *Hemidactylus* geckos—highly specialized climbing lizards. Because we have four species of inveterate climbers on Guana, we have sought other ways to mark these lizards, despite the good news of Paulissen and Meyer (2000). Zwickel and Allison (1983) provide a way that does not even involve mutilation, although we initially combined it with a minor toe clip. It involves carefully sticking nylon tent-mending tape on the lizard's back or tail base. The stuff sticks beautifully. We found that we could start with basic fire-engine red and add colors with acrylic paint, just as Zwickel and Allison suggested. We began by marking anoles. We could locate a dot of tape on rump, midbody, or shoulders. With red, white, and blue we had nine individuals marked. Next we added a black ink dot to the tape dots, and got nine more. Then we went to two tape dots. . . . We never ran out of distinctive combinations.

We clipped off the first left toe of every tape-marked lizard in our study just to be sure we could recognize a tapeless, but previously captured animal. We were doing a two-roundup test, with the second capture session one week from the first. It worked perfectly. Not one anole—even the tiny hatchlings—shed off its tape mark in that week. Many retained the tape

for several weeks thereafter. We were so delighted with the method that we decided to use it for house geckos, *H. mabouia*. These fellows come out of the woodwork and from behind wall hangings, framed pictures, mirrors, and the like at night. They love to eat insects that are attracted to the lights inside the buildings.

An urbane and sophisticated, if somewhat acerbic, European guest of the Guana Island Hotel was ensconced in the lounge when we began our gecko roundup one evening. We eagerly seized a great, fat gecko and prominently branded his rump with a grand red dot.

"What are you doing that for?" queried the guest.

"We're marking him so we can recognize him when we see him again," we said. "Now we'll let him go." Which we did. The gecko promptly disappeared into the ceiling.

"That is ridiculous," said the guest. "That is much worse than your proverbial needle in the haystack. Now you have let it go, you will never see that lizard again."

"Hah—Silly man," thought we, and we set to catching more. We marked 11 house geckos in the lounge, bar, and dining room that evening. The next evening the walls were well decorated with house geckos, as ever. In fact, we counted 14. One was marked. After that we never saw another marked gecko, although there was no diminution of general gecko numbers. We had clipped a hallux (a tiny toe with no digital dilation or friction pad, corresponding to our "big" toe) on each tape-marked gecko, and we never caught a toe-clipped individual again either.

We tried the same technique, with nylon tape, on another species of house gecko at Aiea Heights, Oahu, Hawaii, a month later. These were a smaller species, *Lepidodactylus lugubris*, the mourning gecko. Despite their name, we found them delightfully cheerful sprites. The procedure worked very well, too. Some individuals lost their marks after eight days, but some kept them for more than 20 days (Jarecki and Lazell 1987). We also used this method with great success on the big geckos, *T. rapicauda*, of Necker Island (Lazell 1995). It really is a great method; we just do not know what happened with the house geckos on Guana. Perhaps freeze-branding would work well for the island's house geckos. This technique standardly involves writing a number on the animal with a silver-tipped copper rod frozen in liquid nitrogen (Ferner 1979). The technique has been used with excellent results on all kinds of animals, from mammals (e.g., Lazarus and Rowe 1975) to fishes (e.g., Raleigh et al. 1973) and even invertebrates such as slugs (Richter 1973). We used it on Guana *Anolis* lizards for several years.

A highly desirable aspect of any marking system would be identification of individuals without having to recapture them, with all the attendant effort and stress that accompany actually recapturing the individuals. Bird banding is a time-honored and wonderful system, but only on the larger birds can the bands with numbers be read from a distance. For this reason, bird behaviorists often color band individuals with plastic leg bands that can be read with binoculars or a telescope (e.g., Wolfenden 1975). All you need is a distinctive sequence; using four bands—two on each leg—of four colors gives 256 recognizable individuals, even in birds such as crows where the sexes are similar. The basic formula for color marking is simple enough. If B is the number of bands or marks (of paint, dye, tape, etc.) and C is the number of colors, then the number of unequivocally marked individuals, M, is: $M = C^B$. The situation is complicated if you decide not to use all of the bands or marking positions on a given individual. Having a blank—no mark or band—in the system gives you an additional "color," providing you can read the correct position of the blank. The obvious difficulty in bird banding is that you cannot read a blank as either the top or the bottom on a given leg; the bands are loose and rattle around. You can increase your number of recognizable birds by using the blank, of course, but not to the limit of the formula. The team led by Fred and Peggy Sibley on Guana used color bands on our bananaquits *(Coereba flaveola)* with excellent results. If you dye a mouse's feet, all well and good: you have four

positions; you can use four colors and no color, a blank, which makes the fifth color. Then you have $M = C^B - 1$ or $M = 5^4 - 1$. You have to subtract the one that is all blanks because you cannot tell him from all of the great uncaught. Therefore, you can identify $625 - 1 = 624$ mice.

In our study of the rock wallaby (Lazell et al. 1984), we have ear-tagged males in the right ear, females in the left, with bright yellow or red (in different years), boldly numbered tags. One can tell a marked male, therefore, from a marked female as far away as one can see that the animal is marked. It does take binoculars or a telescope to read the number, but we do it all the time. The advantages and information gains from this marking in the wallabies' vertical world of cliffs and tumbling rocks are almost literally lifesavers for the human observers. Heckel and Roughgarden (1979) used latex paint in a tree-marking gun for Antillean *Anolis* lizards. We also used latex paint on Guana's iguanas (Goodyear and Lazell 1994). The problem here is that the paint marks have to be distinctive so that individuals can be recognized; if they are not, you cannot determine travel distance. Without knowing travel distance you cannot determine the area of your grid. Thus, Heckel and Roughgarden's study was invalid as a density determination: They were certainly sampling an area larger than their measured plot (see Recapture, in this Part).

Dyes can be very useful markers in short-term studies. We used gentian violet, a vital stain, on devils in northwestern Tasmania (Lazell 1984). Devils are pretty well individually marked to begin with. They have white spots in various patterns on the chest and/or rump, as a general rule, and these are often quite distinctive. We further distinguished each individual by dyeing one of the spots, or a part of a blotch or chest band, violet. Older individuals usually had quite spectacular scars, probably because devils seem to loathe each other. We never had a problem recognizing individuals; we recaptured dozens and could still easily read the dye marks after two months.

In addition to Zwickel and Allison's (1983) tape-marking technique, two other novel solutions have been suggested. Robertson (1984) succeeded in marking frogs—wet, slimy fellows indeed—with reflective tape attached to the tops of their heads in distinctive patterns. He stuck the tape on with cyanoacrylate tissue cement, and it stayed for up to 10 days. If he freeze-branded the top of the frog's head first, to shut down the mucoprotein flow, the reflectors lasted up to 41 days. Robertson (1984) reported being able to easily identify his individual frogs, active at night, without having to actually catch them. Clark and Dillingham (1984) found that they could readily locate lizards at night by attaching short (ca. 2 cm, or <1 in.) lengths of micropipette containing Cyalume V (made by the American Cyanamid Co.) to them. This fluid glows in the dark and could be spotted up to 30 m away. They attached the micropipettes to the lizards with ordinary household cement, but Robertson's (1984) tissue cement might work even better. By putting narrow bands of electrician's tape around the micropipettes in distinctive patterns, one can readily identify individuals, if only for a limited time. The problem of where animals go at night is a thorny one because this aspect of their ecology and home range is so poorly known in most cases. The opportunity to study bats, marked with these micropipettes, during their night activity periods would predictably be most informative. Even butterflies can be individually marked with numbers, various colors, or other symbols, using regular office correction fluid that is applied to a small area of the forewing. This technique might also work for dragonflies and other flying insects.

Recapture

IF YOU TAKE a vacation trip to Acapulco, or Iponema, or Cozumel, you will find animals living there. If you walk out in your backyard, or visit a city park, you will find animals there, too. You may think that the animals found in oft-visited, seemingly ordinary, places are dull and insipid and that scientists know all there is to know about them, but you will almost invariably be wrong. We do not even know how many squirrels there are in New York's Central Park. Compared to what we could know, scientists in fact know almost nothing about life on Earth. I will guarantee you that any study of the population biology of any animal that you can manage to accomplish will be new and contain valuable information, even if you just repeat this year a study done by someone else last year, or even do one simultaneously with another researcher using a different method (or even ostensibly using the same method). If you only succeed in recapturing one marked animal, you will at least have found out that it lived X time and traveled Y distance (our house gecko lived 24 h and did not travel at all). You have made a tiny flicker of light in the darkness. What you found out, no one ever knew before.

Once you have marked an animal, you have to let it go. Once you have let it go, what hope is there that you will ever see it again? Will your marked animals be no more than tiny needles in the great haystack of life, like house geckos on Guana? Try it and see. What can you find out in a simple mark–recapture study? Well, no one can guarantee you will find out anything at all, but here are some of the most elementary things you might find out:

1. Patterns of spatial distribution

2. Home ranges

3. Territoriality or its lack

4. Longevity

5. Growth rates

6. Sex ratio

7. Births, deaths, and reproductive seasonality.

And, of course, maybe even the answer to the Big Question: How many of them are there?

How can marking and recapturing animals possibly do all of these wonderful things? First, of course, you must have meticulous and maximal data on the animals marked. You need a map of the habitat. You must know exactly where and when each marked animal was caught and released (and they must be released in exactly the same place that they were caught). You need to know as much as you can figure out about each marked animal: its sex, size, reproductive condition, etc. Then start out to answer the Big Question first, and let the data that may answer the other seven (or more—on to infinity) pile up (you will have to take an active role, of course, in the piling). So, how many are there?

The simplest first approximation involves the "corks-in-a-barrel" approach. The habitat is like a barrel; it is opaque. We cannot simply see how many lizards (or corks) there are and count them. But we can reach in and pull some out. We can do it with a lizard noose, a net, or a trap of some sort, or with a leaping grab. But because we cannot realistically expect to reach in and pull them all out (except in a very small area), what good does that do us?

Well, if we mark each cork (or lizard, or snake) that we take out, and then put it back in, and then reach in again and grab some more corks (or lizards, or snakes) will that help? You bet. I do not know just who Lincoln of the Lincoln index was; he was definitely not the founder of the Republican Party. Instead, he was the fellow who pointed out that:

$$N = Mn/X,$$

where N is the population size (e.g., number of corks in the barrel), M is the number you managed to mark, X is the number of marked ones you grabbed the second time (your recaptures), and n is the total number grabbed that second time (i.e., the number of new ones plus the number recaptured). An early paper, Schumacher and Eschmeyer (1943), deals with a situation as close to corks in a barrel as you can get: fish in a pond. Hayne (1949) applied the technique to mammals. Bailey (1952) and DeLury (1951) made improvements, and Overton (1971) sums it up with simple formulas and tables, enabling one to calculate 95% confidence limits on the values obtained for N. The basic Lincoln index of Overton (1971) is what I used on the island in 1982 because I could assume, in my very short-term study, that births, deaths, emigration, and immigration were essentially nil. Those assumptions become invalid as soon as the time of study gets long enough to include a breeding season, or a severe drought, or the sort

of predictable environmental change that can lead to animal movements or migrations. What is realistic for corks in a barrel becomes silly when applied to animals over time. In a longer-term study we need to employ more sophisticated techniques. Jolly (1965) included factors for death and immigration. Otis et al. (1978) produced a monograph with seven models taking into account many variables. Dixon and Chapman (1980) and Mares et al. (1980) dealt with the necessity of determining home range and individual movements of the animals to develop accurate population inferences. Lin and Lu (1982) provide an excellent study of the population of a Taiwanese lizard. How about our house geckos on Guana? We marked 11, saw 14 the next night, one of them marked. So, 11 times 14 divided by one is 154 house geckos in the open part of Dominica House; we cannot do a 95% confidence interval for a single datum. That datum is therefore as worthless as my one-snake *Typhlops* census discussed earlier—unless, of course, someone comes along and does another, better, study. Then that datum becomes a point of anecdotal interest.

If you only have a short time, select a known area of convenient size and do a double roundup. Catch and mark all the animals you can on the first roundup. The marks need not be individualized if the area, or habitat, is circumscribed. That is, if you know the total range of the population is one small island, or a woodlot, or a pond, and you know the animals do not travel to or away from it regularly, then all you need to recognize is marked versus unmarked individuals. Back between 4 March and 2 April of the 1982 dry season, we caught and marked 35 *L. p. anegadae* snakes as we traveled all over Guana Island. Then we scrambled all over the island again on 6 April and caught all the snakes we could: 7, only 1 of them marked. That yields 35 marked times 7 caught, which is 245, divided by 1 marked snake in that second roundup, for a point estimator 245 snakes on the island. If these snakes were evenly distributed over the 300 ha, that would be 0.8—less than one snake per hectare. I got 2.9—almost 3—on the line transect. As before, that single datum allows no confidence interval, but we can use the history of our month of snake catching to generate a modified Schnabel index sensu Krebs (1989). We caught, marked, and (rarely) recaptured snakes on 16 days during that month. To calculate results, we can program a computer as instructed by Krebs (1989) and punch in for each capture session three data: the number of snakes seen; the number caught and marked; and the number (usually zero, in this case) of recaptures. I have no idea how Schnabel (1938) did it without a computer, and I have not had the courage to look up his paper; it is in the bibliography for those of you who want to know. Anyway, our computer spit out a point estimator of 470 and—low and behold—a 95% confidence interval of 184–1,041. That would give us about 1.6 snakes per hectare. Now we have three point estimators (see table 7). All of those estimates are comfortably within our confidence limits, but the next one throws the proverbial wrench in the works.

In their 100-m² direct-census plots, Rodda et al. (2001b) extrapolated 50 snakes per hectare in the low legume plots and none at all in the seagrape plots. That surprises me: I would have expected the opposite. Anyway, their average came to 25/ha: an order of magnitude higher than any of the other three methods. Where is the truth? My gut feeling is that there are about a thousand *L. p. anegadae* on the island, and they occupy about 200 of the 300 available hectares, an average density of five per hectare in occupied habitat. That would mean even the line-transect count was low, perhaps because I ran it in suboptimal habitat (but I did

not think so), or perhaps because this method tends to underestimate densities. The mark–recapture and Schnabel estimates are probably way too low because marked animals with short memories, such as snakes, are prone to recapture by humans rewalking the same trails. If 2,000 snakes is more likely, then the density in 200 ha moves up to 10: that is almost half the removal-plot direct-census estimate. Robert Henderson, of the Milwaukee Public Museum, plans to really study our *L. p. anegadae*; maybe he knows how to count snakes.

Of course, marking individuals so that each can be unequivocally recognized again will provide more data of different sorts than just *N* from the Lincoln index. If your site is not circumscribed in some way that confines the animals to it, then individuals must be unequivocally recognizable. Otherwise, you will never be able to guess where the real edge of your sampling area is. It is all well and good to measure off 100 m² of bush and mark just the lizards within it, but if the lizards are free to come and go, and lizards that were initially outside the plot can come and go too, you will not be sampling just 100 m². There will be a zone around the edge of the 100 m² from which you are drawing individuals. How big is this zone? Is its width different for males and females? Rodda et al. (2001b) solved this problem by putting up an escape-proof fence around their plot, but that is a lot of work.

Historically, the standard method for calculating plot size was to add to your measured plot a band around the edge, the width of which depended on the travel distance of marked animals. This is why you have to be able to recognize

TABLE 7
Different Estimators of Snake Numbers

	AVERAGE PER HECTARE	ISLAND TOTAL
Line transect	2.9	870
Mark–recapture	0.8	245
Modified Schnabel	1.6	470

every individual: You have to be able to tell how far it has traveled between captures. Typically, one may use the average travel distance, or one-half the maximum recorded travel distance (Otis et al. 1978). It is especially comforting when these two numbers are similar (Lazell 1984). It is disconcerting when there is a great disparity in these numbers, or when males travel on average much farther than females, or young animals travel much farther than old ones. One must critically peruse the data to spot discrepancies of these types, because they may—if uncorrected—invalidate the method.

Mares et al. (1980) studied home range in chipmunks by simultaneously running a mark–recapture study and a radiotracking program. Radiotracking really tells you where the animal is and where it goes. Mares et al. found that they had to recapture the same chipmunk two dozen times before the travel distance data began to accurately reflect real home range as demonstrated by radiotracking: The edge zone you add on to your measured plot must equal the radius of the average home range to give you a correct approximation of the real plot size.

In studying rice rats, Spitzer (1983) discovered by radiotracking that one species, *Oryzomys argentatus,* has huge home ranges. It is found on small, subtropical islands. Its continental relative, *O. palustris,* had much smaller home ranges; in fact, Spitzer (who is now Numi Mitchell) found a paper on a population of *O. palustris* confined to a small island that seemed to have tiny home ranges. A critical perusal of the methods and data, however, revealed the fatal flaw in the reported results. The plot size was small compared to the possible real home ranges of the animals. Suppose you lay out a trap grid of one-quarter hectare, 50 m on a side, and station 25 traps in it at regular 10-m intervals. Suppose the animals you are trapping have real home ranges of three or four hectares each. You see the problem? The absolute maximum recapture distance you can get in your grid is a miserable 70 m—the diagonal across the plot. You could recapture the same individual a thousand times (if it lived so long) and get a maximum

travel distance of 70 m and an average of 35 m. You might think, therefore, that you could add a 35-m-wide zone around your plot and be taking in the radius of home range. That is just what the fellows studying *O. palustris* did, and they could not have been more wrong.

If ever your maximum actual travel distance is about the same as the maximum possible travel distance your system will allow, worry. Real home range may be large compared to your sampling plot. What gave Spitzer (1983) her first clue was that the *O. palustris* researchers were operating two grids at widely separated points on the island. They caught one marked rice rat once in the remote grid. They attributed this seemingly anomalous recapture to dispersal, but Spitzer noted it was an adult male. Adult male rice rats do not normally disperse; they normally hold home ranges quite tenaciously. The more you know about the natural history of the animals, the better you will be at evaluating data and choosing between alternative explanations. Things are not always as difficult as these small rodents make them seem. For example, sizable, diurnal lizards such as the anoles are well suited to simple mark–recapture studies. On the island we discovered that a given marked anole regularly patrols most of its home range, and sticks quite tightly to it. A dozen recaptures of as many individuals give a quite accurate notion of home range radius as determined by long-term observation of individuals over days or weeks. Rock wallaby recaptures that we used to calculate the effective grid area on the cliffs of Oahu were suspiciously close to the maximum distance between our traps (Lazell et al. 1984). So we radio-collared and tracked adult wallabies—night after night, all night long. We found out a lot of fascinating things, but our notion of basic home range, and therefore grid area, was not upset.

If a double roundup, or "two catch" experiment, can yield worthwhile data on just population size, imagine what one could do with a "triple catch" system (Begon 1979). As above, if your habitat is biologically circumscribed, you need mark animals only by group: those caught and marked the first time; those caught and marked the second time. Animals caught twice are marked the second time too and will be counted twice. To get more and better information, however, it is best to mark each individual for unequivocal recognition. A triple catch experiment can begin to yield life-history data such as survival rate over time. You can start an actuarial study of your animals. Of course, you will have to either stay at the site for a while, or return to it. For best results with most sorts of animals, do a straight Lincoln index, with your two roundups quite close together. With lizards and small mammals, I like a 48-h spacing, because I believe that it gives the animals plenty of time to settle down yet is too short for births, deaths, emigration, immigration, and such factors to affect the population significantly. Use all the animals marked in your Lincoln index experiment as F, those marked the first time. For your second catch, pick a time suitably distant.

If F is the number marked in the first catch, then let S be the number marked in the second catch. In the third catch, T, we must separate those marked the first time, T_F, from those marked the second, T_S. Then, the rate of survival from first catch to second catch, R, is

$$R = \frac{T_F(S)}{T_S(F)}.$$

On Guana, I picked one month because I wanted to know how many sphaeros survived the dry season. My first catch was in early March, my second early April, when it had begun (in 1982) to rain again. I picked a well-censused section of the White Bay Flat where I believed there were pretty consistently one or two sphaeros per square meter. There were five in 3 m². With the sphaeros, $F = 5$, $S = 5$, $T_F = 1$, and $T_S = 3$, giving $1(5)/3(5) = 0.333$. This tells us that sphaeros are rather short-lived, at least through the dry season. Only about one-third survived the 40 days. However, in the third catch we also got five in 3 m², indicating a high reproductive rate and rapid replacement. One might suspect, then, that drought is not what is killing these animals off: Their population is not dwindling. They do not

have a high metabolic rate, as do mice, shrews, and small birds, so they are probably not burning out after a few short weeks. What can the problem be?

Ah, look at that magnificent ameiva, or ground lizard—two-thirds of a meter long. Cloudy blue mottling on chestnut, he basks in the full heat of the midday tropical sun. Now he flicks his snakelike tongue; he paws the ground; he probes the litter with his crowlike face. I wonder what he has for lunch, and how much of it? And once—maybe as recently as the 1940s—there was a terrestrial, flightless rail that would surely have eaten little lizards. People exterminated the rail, and even (via their imported mongooses) the ameivas of some islands. Could people ever bring something similar back? See "Restoration," part 6, below.

Back to our sphaeros. Those of you who know elementary statistics are already infuriated because, following the old saw about science, I have extracted such a wealth of speculation from such a tiny sample of fact. And you are right. The numbers I have given are far too small for worthwhile statistical inference. Experiments of this sort must be done repeatedly, and large sample sizes generated, before one can draw any probable conclusions. Seber (1973) provides the methods of statistical analysis, inferences, and determination of confidence limits for triple-catch data. But, once again, do not get too discouraged over small amounts of data. Remember, some data are better than none: *mejor que nada*, we say. The techniques we used have largely been developed by game managers, who have samples equaling hundreds or thousands, because they work on common animals in the Great Grocery Store of life. If you work on little, inconspicuous creatures, or rare ones, or on any sort of animal in a place the game managers have not got to yet (which is about 90% of the world, including Acapulco, Iponema, and Cozumel), then any bauble of data you get is worthwhile. Begon (1979) goes on to provide an examples of how you can expand triple-catch data to ongoing, multiple catch data. Brownie et al. (1985) provide 14 multiple catch models plus methods for evaluating their applicability and statistical significance in exemplary cases. Let us take a simple case from the North Bay Woods, using sphaeros again. We will look at weekly survival rate (R) for five weekly catch times through that dry season, from early March to mid-April 1982. We stake out an initial plot 5 m on a side, or 25 m². Each time we sample it, we catch 40 sphaeros. *M* is the number marked each week; the weeks are given Roman numerals (see table 8).

To estimate *R* from the first recapture (week II) to the second recapture (week III), sum all the survivors of week I marking, beginning in week III, multiply by the number marked, *M* (in our case it is a constant, 40), and divide this product by the sum of week II's survivors (also multiplied by *M*):

$$\frac{(6 + 4 + 1)\ 40}{(15 + 9 + 3)\ 40} = \frac{11}{27} = 0.41.$$

The estimate of *R* from second recapture (week III) to third recapture (week IV) is:

$$\frac{(9 + 3)\ 40}{(13 + 10)\ 40} = \frac{12}{23} = 0.52.$$

Once again, in order to make any reasonable inferences from these kinds of data, you need a lot of them, and you need sophisticated tests based on assumptions whose validity only you, as the investigator, can assess in a specific field situation (Brownie et al. 1985).

Ideally, one wants to set up a long-term study. Then the data begin to roll in. Births, deaths, and other gains and losses (e.g., emigra-tion and immigration) occur and matter, and we need to know about them. As a first approximation, one can calculate the Jolly index, as outlined by Davis and Winstead (1980). One will want to carefully consider the models developed by Otis et al. (1978) for analysis of recapture data. Perhaps one of them will seem valuable for practical application. However, the available models are now highly complex and require far more information than I have usually been able to generate in the field.

For example, it could materially affect our estimate of devil density in the wilderness of northwestern Tasmania if males are more trap prone and range more widely than females but are less numerous (Lazell 1984). But how can I find that out? If there are just a few males, but they range widely, and they are bold and readily enter traps, I should not be lumping their re-capture data with those of females. Perhaps the females are numerous, stay-at-homes, and shy about traps. It costs tens of thousands of dollars to mount an expedition into this wilderness habitat and catch any reasonable number of these largish carnivores. Yet, here are many, many things I would love to know about devils that I have not been able to find out. And devils are dirt common compared to Tasmanian tigers (thylacines).

Let us look again at Chip's bridled quail-doves. His transect estimators of more than two or three thousand of these ostensibly rare birds on the island seem incredible. In 1984 the her-petologists, ornithologist Liao, and several other assistants helped Chip catch quail-doves in mist

TABLE 8
Weekly Survival Rate

MARKING WEEK	NUMBER RECAPTURED					
	M	**II**	**III**	**IV**	**V**	**R**
I	40	12	6	4	1	—
II	40	—	15	9	3	0.41
III	40	—	—	13	10	0.52
IV	—	—	—	—	22	0.45

nets; we marked two. Each bird got a bright yellow leg band—one right, the other left—and each got two 1.5-cm diameter nylon-tape dots—one on top of the head and one in the middle of the central tail feather. One bird was dotted red, the other white. On 24 July one of the marked birds (red) was among six sighted on a 100-m transect. Using our Lincoln index: two times six divided by one is 12. But 12 in what area? Certainly in 1984 we could do nothing with this datum. In 1985 and 1986 Chip caught a lot more quail-doves; he put radio transmitters on four in each July capture session and got good data on seven individuals. These had home ranges of 2.0–19.2 ha; he got an average of 18 fixes per bird on these seven. The bird with the biggest home range—19.2 ha—had only six fixes, the lowest number. The average home range was 6.5 ha. Can we make the leap that there are about 12 birds in the average home range? If so, we get a mere 223 total birds in the 121 ha of Guana habitat. That seems a lot more reasonable than the order-of-magnitude-higher transect estimates, but still high.

In 1985, Chip marked three more birds (not counting four carrying radio transmitters); he and Liao canvassed the entire habitat in 40 h between 18 and 22 July. They sighted 50 quail-doves, two of them marked; that is, 3 times 50 divided by 2, or 75 birds total. In 1986 they marked nine birds (not counting four with transmitters). Once again, they searched the whole habitat—121 ha—but all in 8 h on 11 July. They saw only eight birds, two marked; that is 9 times 8 divided by 2, which I hope is 36. So there you have the gist of the problem of trying to count quail-doves: estimates of 3,993, 2,783, 223, 75, and 36. I can weasel-word around, saying the high counts by line transect represent an invalid extrapolation from the very best optimal habitat in Quail Dove Ghut to the rest of the suboptimal habitat. Better, I can claim the mark–recapture data are so trivial as to be worthless: quite true, with no backup and such severe contradiction. In the end, though, I am stumped. I would guess there are about a hundred quail-doves on Guana Island, but I have no convincing data at all. This remains an exemplary mystery.

On our island there are animals such as the little snake, *Liophis exigua,* and Sir Hans Sloane's slippery-back skink, *Mabuya sloanii,* that are so rarely encountered that we have never recaptured one yet. They are cryptic and semifossorial; I am quite convinced none of them is in danger of becoming extinct, so I can keep trying. Meanwhile, my data and inferences on the other, commoner animals get better and better, or at least wilder and weirder.

Change-in-Ratio Estimators

THESE METHODS ARE frankly my fa-
vorites because I can generate so many
types of estimates from such simple data. All
one needs is a set of ratios formed from two
categories. You can use males : females; juveniles : adults; species A : species B (within an
ensemble, for example); or even microhabitat 1 :
microhabitat 2 (within the range of a species).
You can use any two relevant categories you
want or can imagine.

Then, you need three catches: a first census,
an actual removal, and a third census. You need
to know for each of the two categories the initial
percentage present, the relative percentage of
that category removed, and the remaining percentage in the final census. You also need to
know the actual numbers of each category removed. Now, let:

P = initial percentage in Category A.
K = percentage of individuals removed
   belonging to Category A.
R = remaining percentage of Category A.
X = number of individuals removed
   belonging to Category A.
Y = number of individuals removed
   belonging to Category B.

Then D, the proportion of the total (both categories) removed, is

$$D = \frac{P - R}{P - K}.$$

And C, the specific removal rate for Category
A, is

$$C = D(K/P).$$

Finally, N, the actual initial population in
Category A, is

$$N = X/C.$$

You can readily calculate the values of C and
N for Category B using the value of D found for
the total (both categories), the K and P values for
Category B, and Y.

Like all approaches to population biology,
this one is rife with problems. Begon (1979) provides a detailed exposition with walk-throughs of
all kinds of examples, including bad ones where
the method seems clearly to give wrong estimates; Begon discusses why in each specific
case, which is a huge help if you set out to try
these techniques. Paulik and Robson (1969) discuss confidence limits and statistical significance. The two biggest, baddest sources of bias
and error are (1) differential observability of
members of the two categories (are members of
Category A bigger, brighter, or bolder than members of Category B and, therefore, more easily

counted?); and (2) the arithmetic of close ratios. The slop in the system is least when the ratio is most disparate (Begon 1979). Therefore a sex ratio in many common organisms, which approximates 1:1, is a poorer pair of categories to use than a species A : species B ratio within an ensemble, where one species is common and the other rare.

On Guana Island we laid out two 50-m² plots in the scrub of herp hectare. We considered the three species of anoles that form a tight-knit ensemble: *Anolis cristatellus wileyae* (crested anole), *A. stratulus* (saddled anole), and *A. pulchellus* (grass anole). The sexes in each case are readily recognized. We estimated populations by three methods: direct census, Lincoln index, and change-in-ratio. The two plots were selected to be as similar as possible in slope, aspect, elevation, and vegetation. From one plot we removed every last anole. You may not believe we could do that, and I alone could not have. I had excellent help, and we all believe we really did it. Subsequent observations contributed to our confidence. Just imagine the series of pairs of ratios this exercise generated!

I shall discuss some of them in dealing with ensembles, guilds, and communities, below, but one example here will elucidate the method. I compared the populations of crested anole and saddled anole, the two common species, to the grass anole, which is rare. Here are the data:

P = percent of grass anoles initially: 8.6
K = percent of grass anoles removed: 12.5
R = percent of grass anoles finally: 9.1
X = number of grass anoles removed: 2

$$D = \frac{8.6 - 9.1}{9.1 - 12.5} = 0.147$$

$$C = D \, (K/P) = 0.147 \, (12.5/8.6) = 0.209$$
$$N = 2/0.209 = 9.57,$$

which is between 9 and 10 grass anoles in the initial population. We determined our plot size by two methods: travel distance of recaptured individuals, using the average of all anoles of both sexes, and direct observation of travel of individually tape-marked specimens over one full week. Our plot size was 50 m². The notion that there are 1,800 grass anoles per hectare on this particular slope accords well with other census data, direct observation, and the gut feelings of seasoned observers. It is a very interesting figure because it seems very specific to this hectare, on this island, in this ensemble, at the particular season (dry: early March) when we calculated it. Grass anole densities are wondrously different in other places and at different times. But for all that, it seems remarkably close to truth.

PLATE 1. North Bay, Guana Island, from the northern peninsula ridge, ca. 1952. The forest remains largely intact today. Photo: Bradley Fisk.

PLATE 2. In deep forest: *Myricanthes fragrans*, locally called "red Indian," trunk and buttress roots. Photo: Tricia G. Mansfield.

PLATE 3.
A seaside edge species, *Sesuvium portulacastrum* (Aizoaceae), locally called "camphor," is often the first colonizer of land. Photo: Divonne Holmes à Court.

PLATE 4.
Ipomoea pes-caprae (Convolvulaceae), a seaside beach colonizer. This tough vine stabilizes the berm where hawksbill turtles and stout iguanas nest. Photo: Divonne Holmes à Court.

PLATE 5.
Succession: *Bursera simaruba*, locally called "gumbo-limbo," "turpentine," or "tourist tree" (the latter because it is typically red and peeling), takes over the ruins of the eighteenth-century sugar mill. Photo: Jan Soderquist.

PLATE 6.
Flamingos on the Salt Pond; white mangroves in the background. Although those on Guana Island have yet to reproduce, the flock we restored to nearby Anegada has gone from 18 to over 100 in a decade. Photo: Thomas Jarecki.

PLATE 7.
Ground gecko or sphaero, *Sphaerodactylus macrolepis macrolepis*, the densest known terrestrial vertebrate on Earth. Guana Island holds the record with up to 67,000 per hectare. Adult male. Photo: Kristiina Ovaska.

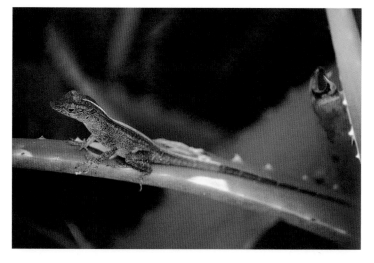

PLATE 8.
Crested anole, *Anolis cristatellus wileyae*. This is the most abundant and widespread anole in the BVI and on the entire Puerto Rico Bank. Adult female. Photo: Jan Soderquist.

PLATE 9.
Carrot Rock skink, *Mabuya macleani*. This drab, almost patternless endemic occurs within 450 m of its ancestral species, *Mabuya sloanii*, and the latter shows striking character divergence from it. Photo: Lianna Jarecki.

PLATE 10.
Tabebuia heterophylla (Bignoniaceae), locally called "white cedar." Its flowers feed iguanas, tortoises, hummingbirds, and tumbling flower beetles, among others. Photo: Divonne Holmes à Court.

PLATE 11.
Argusia gnaphalodes (Boraginaceae). Variously known locally as "bay," or "sea lavender," or "crab bush," this is a sand dune species. Photo: Divonne Holmes à Court.

PLATE 12. *Melocactus intortus* (Cactaceae), locally called "turks head." The brilliant magenta fruits are a favorite of iguanas, tortoises, birds, and even people. Photo: Divonne Holmes à Court.

PLATE 13.
The tree caper, *Capparis cynophallophora* (Capparaceae). The species name refers to the seedpods that resemble the phalli of dogs. Photo: Divonne Holmes à Court.

PLATE 14.
Elaeodendron xylocarpum (Celastraceae). Locally called "wild nutmeg," the sap of this species is used for glue. Iguanas eat the fruit. Photo: Divonne Holmes à Court.

PLATE 15.
Scaevola plumieri (Goodeniaceae). This shrub is uncommon elsewhere, but Guana has dense populations. It is found in beach strand woods. Photo: Divonne Holmes à Court.

PLATE 16.
Pictetia aculeata (Leguminosae: Faboideae). Locally called "pum tack," which we assume derives from thumb tack, this tree is indeed thorny. Photo: Divonne Holmes à Court.

PLATE 17.
Acacia macracantha (Leguminosae: Mimosoideae), locally called "crown of thorns." Although this shrub restores nitrogen to soil in early seral stages, it is the bane of field biologists. Photo: Divonne Holmes à Court.

PLATE 18.
Leucaena leucocephala (Leguminosae: Mimosoideae). Locally called "tam-tam," this is an important plant colonizing bare ground and building nitrogen in the soil. Photo: Divonne Holmes à Court.

PLATE 19.
Males of the whistling frog, *Eleutherodactylus antillensis*, often call from concealed perch sites in low vegetation. Photo: Kristiina Ovaska.

PLATE 20.
On Guana the best habitat for frogs consists of shrub forest with abundant bromeliads and agaves—plants that conserve moisture and provide retreats—and a deep layer of leaf litter used for oviposition. Photo: Kristiina Ovaska.

PLATE 21.
Male and female whistling frogs, *Eleutherodactylus antillensis*, shortly after the completion of oviposition within leaf litter. The frogs and the egg mass were exposed for photography. In this species there is no parental care of eggs or young. Photo: Kristiina Ovaska.

PLATE 22.
An adult stout iguana, *Cyclura pinguis*. This massive lizard, the largest native to the Neotropics, was once found throughout the islands of the Puerto Rico Bank. It was extirpated from all but Anegada. Restoration to Guana Island was in exchange for bringing back flamingos to Anegada, and both are among The Conservation Agency's most successful and celebrated achievements. Photo: Joseph J. Oliver.

PLATE 23.
The saddled anole, *Anolis stratulus*. But for the dewlap of the male, the sexes are quite similar in this species. Usually characterized ecologically as a "trunk-crown" tree dweller, these anoles actively forage on the forest floor on Guana Island in early morning. Photo: Gad Perry.

PLATE 24.
The grass anole, *Anolis pulchellus*. As its name implies, this is a species of low, early seral stage vegetation of edges and disturbed areas. It is our least common anole. The male has a bright red dewlap, but otherwise the sexes are quite similar. Photo: Gad Perry.

PLATE 25.
The crested anole, *Anolis cristatellus wileyae*, adult male. This species has been the subject of intensive physiological and behavioral studies as described in the text. Photo: Jan Soderquist.

PLATE 26.
Sphaero or ground gecko, adult female showing the double-bullseye "target" characteristic of her sex. Males lack this marking and have blue gray, or rarely red, heads. Photo: Bill Holzmark.

PLATE 27.
The skink, locally called "slipperyback," *Mabuya sloanii*, is fairly common on Guana but rare elsewhere. It has not been rediscovered at its type-locality, St. Thomas, in over a century. The females are live-bearers and produce two to four big babies, exemplary of a K-selected species. Photo: Gad Perry.

PLATE 28.
The strange "worm lizard," *Amphisbaena fenestrata*. "It has rings and it bites," Oscar Chalwell told me and I knew it was on Guana before I ever saw one. This species is endemic to the Virgin Islands and very different from its relatives on Puerto Rico and the other Greater Antilles. Photo: Bill Holzmark.

PLATE 29.
Snake, *Liophis* (or "*Alsophis*") *portoricensis anegadae*. This is the most abundant and wide-spread snake in the BVI and on the Puerto Rico Bank. It belongs to a recent radiation of Antillean xenodontine colubrid snakes that show some remarkable ecomor-phological diversity, but this one is generalized. Photo: George Marler.

PLATE 30.
Fruit bat, *Artibeus jamaicensis*. This bat is common on Guana Island and has remarkably specialized insect ectoparasites worthy of far more in-depth study. Photo: Jan Soderquist.

PLATE 31.
Bridled quail-dove. These pretty little pigeons are abundant on Guana Island but otherwise rare in their limited range from Puerto Rico to the northern Lesser Antilles. Finding them on Guana in numbers was seminal to the initiation of our studies back in 1980. Photo: Joseph J. Oliver.

PLATE 32.
A bird in the hand, Swainson's thrush; a major new range extension for a neotropical migrant. We believe this is an island hopper not an open ocean flier. Photo: Gad Perry.

PLATE 33.
Variegate shore crab, *Geograpsus lividus*. This beautiful species is rarely encountered on Guana Island. Photo: Gregory Mayer.

PLATE 34.
The tarantula, *Cyrtopholis bartholomei*, Guana Island's largest spider, is friendly and not known to bite. Photo: Gregory Mayer.

PLATE 35.
An unidentified jumping spider (Salticidae). Guana's spiders are as yet very little known. Photo: Tricia G. Mansfield.

PLATE 36.
A true bug (Hemiptera: Lygaeidae), *Oncopeltus aulicus*. This is one of our most colorful insect species in a very little studied group. Photo: Divonne Holmes à Court.

PLATE 38. The pompilid wasp *Psorthaspis gloria*, one of the most attractive species found on Guana Island, named for Gloria Jarecki (see "Human History"). This is a classic hunting wasp, believed to provision its larder with the large trap-door spiders of genus *Ummidia*. Illustration by Marianne D. Wallace.

PLATE 39. Caterpillars of the moth *Pseudosphinx tetrio*, called "Princeton worms," for their colors. They regularly defoliate frangipani, *Plumeria alba* (Apocynaceae).

PLATE 40. Guana Island in aerial view from the south. Monkey Point, nearly disjunct and connected by a sand tombolo, is at the bottom. White Bay beach and the largely cleared flat are conspicuous, upper left (facing southwest). Photo courtesy of the Falconwood Foundation.

PART 3

Putting Things Together

Increasing dynamical stability is not a general mathematical consequence of increasing complexity; rather, the contrary is true.

ROBERT M. MAY (1973A)

The balance of evidence would seem to suggest that, in the real world, increased complexity is usually associated with greater stability. There is no paradox here. . . . The real world is no general system.

ROBERT M. MAY (2000)

In this section I shall try to elucidate ways of building from empirical data to a search for complex patterns. The patterns may be seen in trophic levels or food webs or predator–prey relations; they may emerge from competition and guild structure; they may develop from evolution by means of natural selection. The weighty volume edited by Strong et al. (1984) addresses, redresses, discusses, stresses, and distresses the search for pattern in nature. Losos (1996b) added phylogenetic perspectives. Jablonski and Sepkowski (1996) added a paleontological perspective, going back two million years to look at Pleistocene communities. The volume edited by Weiher and Keddy (1999) reveals that most of the same difficulties are still with us. Are the animals of our island just an odd lot of stranded relicts and lucky colonizers, or are they a community surviving by means of finely tuned interactions?

Absolutely essential to any view of communities is the concept of *niche*. As developed by the early naturalists, the notion of niche was rather vaguely thought of as a species' place in the system relative to all of the other species and the physical environment. Long before the niche notion got itself quantified and codified formally, naturalists had perceived the law of competitive exclusion. This simply states that no two species can occupy the same niche at the same time. It is not very different from the physical law that no two objects can occupy the same place at the same time, except that physics deals only with three or four dimensions, whereas ecology—and niches—are n-dimensional; that is, one can easily quantify a dozen or more dimensions of a lizard's ecological niche. It has only been since the late 1950s that niche has become a measurable quantity, even though measuring perch heights, prey sizes, activity hours, temperatures, and so on might seem to you and me (in retrospect) to be simple and elementary. One of the nice things about at least the physical aspects of niches is that, although they could be infinitely complex and unfathomable, they usually sort out very quickly along rather simple, easily perceived, and measured dimensions, at least for common animals.

As I have said before, Guana Island is a wonderful potential theater for the study of community ecology because it is small enough and homogeneous (physically) enough to be comprehensible, yet it has great species diversity for its size. It is fascinating to view the history of the study of communities as it develops through the works of Mayhew (1968), Schoener (1969), and Pianka (1973, 1977). Lawlor and Smith (1976) were among the first to deal with the concept of ecosystem stability in the face of competition. Schoener and Schoener (1971a, 1971b) dealt with Antillean lizards that seem to sort out in terms of body size. Simberloff and Boecklen (1981) reviewed 28 cases where body size was claimed as evidence for niche segregation and found few that could hold up under a null hypothesis rejection test. For a time, it seemed that apparently large minimum and constant size ratios between members of an ensemble (sensu Lawton 2000, see above) might not be demonstrably indicative of niche differences, but rather just a matter of chance. Then Losos et al. (1989) took a second look at statistical method and largely rescued the idea. However, it does not seem to apply to Guana's pretty-much-same-size anole lizards. Pacala and Roughgarden (1982) looked at two species ensembles of anole lizards and found disparate sizes—a big species and a small one—showed much less evidence of competition than did two similar-sized species—just as we might expect.

In a well-patterned world, species diversity and selection against competition might combine to determine just how much of each resource a species could use; body size, physiology, predation, and competition might combine to set the parameters of each species' niche. The breadth of a species' niche may be determined for each dimension of that niche. Thus, a lizard may survive and function between a measurable low and a measurable high temperature. Below the low, it becomes torpid and too sluggish to catch food or escape predators. Above the high, it cannot thermoregulate well enough to carry on; its very biochemistry begins to break down. Within the temperature range, there will probably be an optimum, at which the species functions best. Thus, the thermal dimension of a niche, like most other dimensions, is stochastic—deviating, at any one time of measurement, within some reasonable distance from the optimal point. Determining the optima can be very difficult, but every species will have a critical thermal maximum that can be measured and compared (Lutterschmidt and Hutchison 1997).

A strictly diurnal and heliophilic (sun-loving) lizard, such as the ground lizard *Ameiva exsul*, and a strictly nocturnal species, such as the house gecko *Hemidactylus mabouia*, could easily live in the same place, eat the same food items (as long as they renewed themselves rapidly), and overlap each other in a multitude of their niche dimensions, but scarcely impinge upon or compete with each other at all, so different is their thermal niche dimension: literally, day and night.

The standard measure of niche breadth for a given species is the Shannon–Weaver index, which we have already seen above. As applied here,

$$H = -\Sigma P_{ih} \log P_{ih},$$

where P is the proportion of the niche dimension used by the species in question, i symbolizes the ith species in the system, and h is the instantaneous measure of the resource or physical feature. So for any species S_i, P_{ih} is that proportion of the whole surviving at temperature h, or eating food of size h, and so on.

In a completely censused community, diversity is determined as:

$$H = \frac{1}{N} \log \frac{N!}{\pi N_i!},$$

where N is (as before) the total number of individuals, $N!$ (N factorial) is $N(N-1)(N-2) \cdots$, and πN_i is the product of all the values of factorial $N_l \ldots N_i \ldots N_s$, where s is the total number of species present. Finally, I is a characteristic of all lines and curves, which can be plotted relative to x and y axes, such that

$$I = X (dy/dx),$$

which means that I is X times the derivative—or rate of change—of y with respect to x. One will need a computer just to deal with the numbers generated by several million reptiles belonging to the dozen or so species known from Guana.

Here is a comforting note: "In terms of niche specificities the observed local faunas . . . could be accommodated by a hyperspace with four or five dimensions, the coordinates divided into 10 lengths, each representing the possible minimum and maximum values for tolerances or requirements of a single species. This is obviously far too formal a way of trying to put nature together, but indicates the kinds of magnitudes that might be involved" (Hutchinson 1978).

Reproductive Strategies

> Every single organic being may be said to be striving to the utmost to increase in number, . . . each lives by a struggle at some period of its life, . . . heavy destruction inevitably falls . . . during each generation or at recurrent intervals. Lighten any check, mitigate the destruction ever so little and the number of the species will almost simultaneously increase to any amount.
>
> CHARLES DARWIN (1859)

THE CORNERSTONE OF population biology is reproduction. For a species to survive, or a population to remain stable, not only must a lot of offspring be produced but also some have to survive to reproduce again. There are two extremes of reproductive strategy, each garnered from the logistic equation of Verhulst, as presented by MacArthur and Wilson (1967):

$$dN/dt = rN(1 - N/K),$$

where N is the number of individuals in the population, r the Malthusian parameter or intrinsic rate of increase (a function of life history), K is the carrying capacity of the habitat or asymptotic limit of the population (a function of the environment), and t is time. Philosophically, r and K are the ends of a conceptual continuum. Evolutionarily, these extremes are called r-selection and K-selection. Species that are r-selected produce huge numbers of offspring and put as little parental investment into them as possible. An insect such as the roach is classically r-selected. It lays a zillion eggs, utterly ignores them, and does not even get around to hoping for the best or wishing them well. Species that are K-selected produce fewer offspring but invest much time and energy, relatively, in the development of those offspring. Humans are a fine example of K-selection, as are the island's skinks called slipperybacks, *Mabuya sloanii*. One can clearly see that there are advantages to each extreme. Most organisms, including those studied on Guana, fall somewhere in between these two extremes. The simple fact that the European house roach, *Blattella germanica,* and the human, *Homo sapiens,* are perched at the opposite extremes, but both are essentially worldwide in distribution, egregiously abundant, and still increasing exponentially, is proof enough that one strategy is not, per se, better than the other. In addition to the first chapter of Hutchinson (1978), excellent short articles on r- and K-selection are provided by Pianka (1970b) and Green (1980). Fitch (1970) documents reptilian reproductive patterns empirically, and Tinkle et al. (1970) consider theory. A good overview is provided by Gotelli (2001).

TABLE 9
Comparison of Reproductive Strategies

	MAINLAND	SMALL ISLAND

Vole	Sparse population
	No reproductive
	    regulation
	r-selected
		Dense population
		Reproduction intrinsically
		    regulated
		K-selected

Mouse	Dense population
	Reproduction intrinsically
	    regulated
	K-selected
		Sparse population
		No reproductive
		    regulation
		r-selected

Determination of reproductive strategy for each species is absolutely crucial not just to population biology, but also to understanding its ecology. It must be done empirically, by actual observation. Believe it or not, a fascinating study of Massachusetts rodents seems especially applicable in expanded form to Puerto Rico Bank *Anolis* lizards. Adler and Tamarin (1984) looked at two species, a mouse *(Peromyscus leucopus)* and a vole *(Microtus breweri)* on Muskeget Island off Nantucket. They compared reproductive strategies of the island populations to those of their very close relatives—and presumed direct ancestors—on the mainland (*P. leucopus* and *Mi. pennsylvanicus,* respectively). In presenting their results in tabular form, I have been less diffident than they were, and no doubt have simplified the case. They did not stress, for example, a point of extreme importance in comparing their rodents to our lizards: in both cases the rodents and the *Anolis* (three of the latter on Guana) are the only species that exist on the small island,

but they live in an ecosystem rich with close relatives on the mainland (defined as Puerto Rico for Guana's anoles). My distillate of Adler and Tamarin's (1984) results appears in table 9.

Some will question whether intrinsic reproductive regulation and the closely related trend toward *K*-selection are even possible in reptiles like anoles. Indeed, they have yet to be demonstrated in these iguanids but are characteristic of at least some skinks of the genus *Mabuya*. Comparisons of Massachusetts rodents to Guana Island anoles are appropriate here. *Anolis c. wileyae* and *A. stratulus* are both denser on Guana than on Puerto Rico. *Anolis pulchellus*, on the contrary, is downright rare on Guana compared to Puerto Rico. Are Guana's commoner anoles like Adler and Tamarin's vole? Is *A. pulchellus* like their mouse? How will we find out? The vole and the crested anole, *A. cristatellus,* and its rock-knockoff, *A. ernestwilliamsi* (Lazell 1999a), have undergone the most dramatic speciation in each case. Is there a similar reason?

Life Histories

The dictum of Galilei, "to measure what can be measured, to make measurable what can not be measured," has been applied to the product of Lachesis' loom, with interesting results.

E. S. DEEVEY, JR. (1947)

THE CULMINATION OF ANY comprehensive population biology study is the writing of a life history. One hopes, or wishes, to be able to write a life history for every member of the community, but in practice one cannot do this, even for a relatively small ecosystem like Guana Island. One can, however, rapidly accumulate good life-history statistics for the commoner and more conspicuous members, and one can hope that these members are important enough in the system to permit the building of a foundation, at least.

A good life history consists of basic demographic data: density, home range, defense of territory (if any), longevity, recruitment, reproductive strategy, and so on. It also involves ecological features such as diet, predation, and the avoidance of competition. Ultimately, one can perceive evolutionary rates. Complete life histories are available for a number of insect species, especially those of economic importance like fruit flies and mosquitoes. Often, to even begin to study, a single species will require the work of several generations of biologists.

Tinkle (1976) provided a demographic life table for a spiny lizard, an iguanid, which might be quite comparable to that of our commonest iguanid, the crested anole *(Anolis c. wileyae)*. Table 10 shows what a good life table should look like. Here, X is age in years, 1 is typically the fraction surviving (with X subscript, the age at a given year), m is the number of female offspring per female produced (similarly with X subscript), and the last column (the product of survivorship times maternity) is measured reproductive success. In our example, some lizards lived about five years. Most failed to even reach sexual maturity and were dead by age two (90%, in fact, failed to make it). Maternity was higher at age three than at two and stayed as high for the rest of the lizards' lives. This says that all of the lizards in the population died while still reproductively active. Because they probably did not die of concussions falling off their trees, you can bet somebody ate them. It is a tough world.

Data collection for the life histories of Guana Island's animals is far from complete, but some data are available. I divided the species for which I had relevant data into three groups:

1. *All-aged.* Four species, crested anole *(A. c. wileyae)*, saddled anole *(A. stratulus)*, grass

TABLE 10
Demographic Life Table

X	l_X	m_X	$l_X m_X$
0	1.0	0	0
1	0.2	0	0
2	0.1	5	0.50
3	0.04	7	0.28
4	0.02	7	0.14
5	0.01	7	0.07

SOURCE: Tinkle 1976.

anole *(A. pulchellus),* and sphaero *(Sphaero-dactylus macrolepis)* were encountered at all ages from eggs to old adults. The notion that these are year-round breeders, with each female laying one egg at a time, at an interval of something like two weeks, accords well with the published evidence (e.g., Lazell 1972). I noted, however, that there is often pronounced seasonal variation in life-history parameters even in year-round breeders (Chondropoulos and Lyakis 1983; Lazell 2002a). A close relative of our anoles from Jamaica, which occupies a niche that seems to me quite like our *A. c. wileyae,* was studied over a full year by Vogel (1984). Hatchling recruitment into the population varied by a factor of 10 according to the season and rainfall; growth rate of young varied by a factor of two. Although the seasonality of rainfall during Vogel's year of study, 1979–80, was normal, there were unusual irregularities and the year was, in total, wetter than usual. Rainfall generated arthropods, the food of anoles. The lizards responded by immediately increasing growth rates. Enhanced reproduction began about two months after the rainfall. To get Vogel's kinds of data you have to be on-site all year around.

2. *Even-aged.* Two species, the ground lizard *(Ameiva exsul)* and the common snake *(Liophis p. anegadae)* were pretty much of a size during the dry seasons of 1980, 1982,

and 1984 (March, April). This implies seasonal breeding; because both species were encountered as adults, it implies that we were not on the ground during that season in these years. We returned late June through July 1984, and every July through 1988; then we came in October each year, missing only 1989 (hurricane Hugo).

3. *Bimodal.* Two species, the house gecko *(Hemidactylus mabouia)* and the fruit bat *(Artibeus jamaicensis)* were encountered at two discrete sizes or ages: full adults or infants (includes eggs, embryos, hatchlings, and nursing young). One cannot conclude that these are seasonal breeders and we were present at that season: our encounters with these species were far too few. All one can conclude is that these two species produce young at least during the driest season of the year.

The common snake, *L. p. anegadae,* lays eggs that can sometimes be palped in the body. Of 22 females for which I accumulated March–April (dry season) data only four contained palpable eggs (see table 11).

I set up some very weak hypotheses about common snake reproduction and tried to collect data that would corroborate or refute them. Most female common snakes (ca. 80%) contained no palpable eggs. Therefore, it seemed likely that a few of these snakes would begin to lay about April. If the eggs take about a month to hatch, we should expect to see the first hatchling common snakes in May. The peak of laying will be notably later—let us guess late May or June. The peak for hatchling observation should then be about the first of July through that month, or into early August.

I came back on 27 June 1984, and remained through July. I had a battery of able assistants, and we tabulated a total of 17 *L. p. anegadae,* but not a single hatchling. We did get two of the smallest individuals I had then seen, at 44- and 47-cm snout-to-vent (SVL) length, respectively, but those may have been a year old. It happens that 1984 was exceedingly dry. There was no

TABLE 11
Egg-bearing Common Snakes

	NUMBER OF EGGS	CONDITION	SNOUT-TO-VENT LENGTH (CM)
9 March	6	Shelled (firm)	50
10 March	3	Shelled (firm)	59
20 March	5	Unshelled (soft)	54
27 March	7	Unshelled (soft)	56

SOURCE: Original data.

appreciable rain even by 31 July when I left. Did this affect snake hatching? Are hatchlings so behaviorally different from adults that we did not encounter them? My hypothesis was neither confirmed nor refuted. It is possible that the small snakes we did find had hatched in May, as predicted, and grew rapidly at first, but drought had wiped out the bulk of subsequent clutches. We failed to find in July 1984 the bloom of baby snakes I predicted. It is encouraging that 12 of the 17 snakes we did handle in late June and July were females, and not one of them contained palpable eggs. But you do not need a statistical test to see that the difference between the ratios 0:12 and 4:22 is not significant. The sample sizes are too small.

I burden you with this example because it illustrates several important facts. As biggish (to over a meter long) predatory animals go, *L. p. anegadae* is plain common. It is surely a very important member of its ecosystem, eating a lot of lizards, and eaten by kestrels and red-tailed hawks. It is diurnal and conspicuous and not very hard to catch (I catch about 50%, pretty consistently, of the ones I see). Guana Island is readily accessible and work there is well funded. Imagine the problems of trying to document genuinely rare, cryptic, or nocturnal animals in some truly remote part of the world, such as thylacines in southwestern Tasmania.

The animals in group three provided much of the late June through July data in 1984. The house geckos were obviously actively reproducing. These pallid, rather translucent lizards lay snow-white hard-shelled eggs measuring

9.5–10 mm in length (measured on 25 and 28 March: four eggs total). A female with two shelled eggs clearly visible inside her on 11 March was held in captivity and laid those eggs on 25 March. Eggs found in a drawer of a desk (they lay them in the damnedest places) hatched on 27 and 28 March. The eggs turned dark prior to hatching. The hatchlings measured 22-mm SVL and 46 mm total (with their perfect, unregenerated tails). In July we also encountered females with plainly visible shelled eggs. A juvenile caught on 4 July measured 25-mm SVL. A hatchling of 23-mm SVL appeared in the spiral binding of the bar book (where guests log their imbibitions) on 31 July at 6:20 a.m. (now what alert soul could have caught him?). I suspect *Hemidactylus mabouia* should move to group one—year-round breeders—but there may be a real and pronounced winter dry season lull, for example from December to March.

Our data on fruit bats, *Artibeus jamaicensis*, are shown in table 12. It seems unlikely that there are two breeding seasons for *Artibeus*. It seems more likely that offspring production is continuous from March through July. Some individuals must be pregnant, at least, in February. Indeed, Wilson (1979) found this species pregnant from February to July, but reproductively inactive in August. He says: "Enough information is now available to make speculation tantalizing, but not enough to make generalization rewarding." We can now add another species to this group: the mastiff bat, *Molossus molossus*. Liao Wei-ping netted a pregnant female, forearm 40 mm, embryo 26-mm crown-rump, on 8 July. He got two

TABLE 12
Data on Fruit Bat Reproduction

	OFFSPRING	FEMALE FOREARM (IN MM)	OFFSPRING, CROWN-RUMP (IN MM)
8 March	Furred, nursing	62	71
11 March	Embryo	61.5	34
18 March	Furred, nursing	62	48
19 July	Embryo	61	—
19 July	Furred, nursing	62	47
19 July	Furred, nursing	62	54
20 July	Embryo	61	40

SOURCE: Original data.

more females that we released on 14 July; both were heavily pregnant. Our first specimen was a volant juvenile collected on 6 November (Lazell and Jarecki 1985). If offspring production is seasonal in this mastiff bat, it must at least span midsummer and fall: the rainy season. I will present more comprehensive bat data in "The Cast," part 5, below.

Another factor influencing diversity of communities can be life-history changes within a species. Ballinger (1983) considers this topic in lizards. But, the amphibians, with a name that tells us they lead two lives, provide the best examples. Bruce (1980) studied a salamander comparable in size to a slippery-back Skink, *Mabuya sloanii* (and wholly carnivorous, too). He found the larvae of the species had a much broader diet than the adults and had quite variable life histories among individuals. For example, some indulged in long larval lives whereas others had short ones. Christian (1982) studied small frogs and found, similarly, that the young (not larvae in this case) had broader diets. Size was a critical factor in prey selection at all ages, with individuals specializing on larger—but rarer—prey as they grew older. In this study, stomach content analyses were quantitatively compared to relative abundance—an excellent innovation. Guana has only one known amphibian, the piping frog *(Eleutherodactylus antillensis)*. However, evidence from some reptiles,

such as common snakes and crested anoles, suggests that the young may occupy very different niches from the adults. On Guana's neighbor island of Necker we found the large, "turnip-tail" geckos *Thecadactylus rapicauda* each tended to have its own very distinctive niche: Individuals could be as different as whole species elsewhere (Lazell 1995). Dingle and Hegmann (1982) edited a symposium *Evolution and Genetics of Life Histories,* which I shall discuss later. This volume was reviewed by Stearns (1982) who effectively frames the large question: "Life-history traits are components of fitness, for any definition of fitness will involve some expression of survival and reproduction over the short or long term. Such traits usually vary continuously and are inherited in patterns best described by quantitative genetics rather than by the single-locus models of theoretical population genetics. However, the initial attempts to construct a predictive theory of life-history evolution used optimality models that ignored genetic constraints and could predict only local equilibria, not dynamical behavior. Thus it makes good sense to ask, what are the inheritabilities of life-history traits, what are their genetic correlations, can we use such information to infer past selection pressures and to predict the future course of evolution, and what are the genetic constraints on optimization?" Mostly we just do not know yet.

Trophic Levels

The complexity introduced into animal nature by the elaboration of the food web is probably the most obvious cause of biological diversity.

G. EVELYN HUTCHINSON (1978)

A niche space of dimension one suffices, unexpectedly often and perhaps always, to describe the trophic niche overlaps implied by real food webs in single habitats. Consequently, real food webs fall in a small subset of the set of mathematically possible food webs. That real food webs are compatible with one-dimensional niche spaces more often than can be explained by chance alone has not been noticed previously.

JOEL COHEN (1978)

Food webs remain the ecologically flexible scaffolding around which communities are assembled and structured.

ROBERT T. PAINE (1996)

IF BY DIVERSITY we mean something more than just a list of the species present (and we do), the first area to consider will be the elementary species interactions involved in who eats whom. There are four ways to find out what consumers and predators eat:

1. Direct observation of feeding.

2. Analysis of stomach (or bird crop) contents.

3. Analysis of scats (remnants of what has passed through the digestive tract).

4. Counts of scraps (bits of food items that got left behind, as at a nest or regular feeding site); owl and hawk pellets fit in here.

No doubt direct observation is best, if you can quantify what you see. You may watch a hawk, for example, bring food to its young. If you can identify each prey item as it comes in, and really keep an accurate tabulation of the items by species, you may develop a very precise idea of what young hawks (in that nest) really have for a diet. Scat and scrap analysis are worst and can be very inaccurate; for instance with scraps you are really counting what the animal did not eat. A skunk might eat several clutches of turtle eggs, two scarab beetles, and an apple core during the course of an evening. Scat analysis will reveal no trace of the eggs, but rather a nice batch of beetle shells and apple seeds. You end up believing the skunk's menu was half beetle and half apple, when really it was 80% turtle eggs. However, usually owl pellets are pretty good indicators, because owls tend to eat whole prey items. A skull count from their pellets could be pretty accurate. But, on Guana Island, you could be grossly misled. Small tropical owls, we have discovered, eat a lot of plant material such as fruits (or eat bats that have eaten fruit first?), and a lot of soft-bodied insects.

There has been some controversy over stomach contents, with suggestions that hindgut analysis is better, or at least should be included. The notion was that maybe large, hard prey items remain disproportionately long in the stomach and the analysis of stomach contents alone would be biased in their direction. Floyd and Jenssen (1984) have convinced me to my satisfaction that this is not the case: Stomach contents will do fine. Herpetologists are lucky because they can often find out the stomach contents of some rare or remote species just by visiting the museum and securing the curator's permission to look. Mammalogists and ornithologists usually find that what they want to look for has long since been thrown away. Stomach contents are handy because you can take your time puzzling over the identity of a (now pickled) prey item; you can quantify them by count, by weight, or by volume. You can save them in a jar of alcohol so that the next biologist can confirm or refute your counts and identifications. Stomach-contents analysis is a repeatable process.

Living things are divided into three major trophic types (trophic refers to how they get food energy; in the case of animals, they eat). Producers convert energy (especially sunlight and heat) into molecular bonds. They make their own food and, in so doing, make food for everything else, too. Plants (and some bacteria and other protists) are producers. Consumers eat producers and their products. These are the herbivores of the world, like many insects and even our iguanas. Predators eat consumers or other predators. There may be many trophic levels between the first predator (e.g., a sphaero) who eats a consumer (e.g., a beetle grub) and a red-tailed hawk (e.g., sphaero is eaten by ground lizard, ground lizard is eaten by snake, snake is eaten by hawk).

Paine (1966) goes on to make five generalizations about food webs:

1. Diversity is higher if production is uniform throughout the year, rather than seasonally peaked, because competitive displacement of consumers can be mitigated by consumer species specializing on portions of the production (see Lawton 2000, 176).

2. Therefore, stability of production is directly proportional to diversity, other things being equal.

3. The upper limit of diversity is set by some combination of stability of production and rate of production.

4. Diversity should also be directly proportional to the number of predator species present, because these will tend to specialize within the consumers and thus prevent any one prey (consumer) species from monopolizing a resource.

5. Increased stability may lead to increased capacity to sustain predators. (This is the same as my point about stability being relative to those at the top of the pyramid.)

Two generalizations seem to have repeatedly impressed ecologists who attempted to correlate number of species present and trophic levels or food-web relations. First, islands have fewer species than do some comparable-sized pieces of mainland. Second, life becomes disproportionately more abundant as one descends latitudes toward the tropics. To quote Paine (1966): "Though . . . latitudinal gradients in species diversity tend to be well described in a zoogeographic sense, they are poorly understood phenomena of major ecological interest. Their importance lies in the derived implication that biological processes may be fundamentally different in the tropics, typically the pinnacle of

most gradients, than in temperate … regions.…
Understanding of the phenomenon suffers from
both a specific lack of synecological data applied
to particular, local situations and from the difficulty of inferring the underlying mechanism(s)
solely from descriptions and comparisons of faunas on a zoogeographic scale."

The reptile community of Guana Island is
one of remarkable diversity of predators. Because the island is tropical there is good production year-round, but because Guana is dry
there is a wet-season peak. Production on the
island might never equal that of tropical wet forest because of aridity but might approach it seasonally following the rains. A basic, comparative
study of production and consumption on Guana
would be fascinating. I note, however, that the
reptile community on Sage Mountain, Tortola—a fine, moist forest community with very
little seasonality—is not known to be as diverse
as is the island's: many of Guana's reptiles are
specifically adapted to arid lowlands (Nellis
et al. 1983).

Reagan and Waide (1996) have edited a
thick volume on the food web of the tropical
rain forest near the El Verde Field Station in
Puerto Rico. Puerto Rico, the "mainland" of the
Puerto Rico Bank, is the mother isle of the approximately 150 islands on that bank, including
Guana. Thus, El Verde makes a fine comparison to Guana. Reagan, Camilo, and Waide, in
the summary chapter of Reagan and Waide's
(1996) volume, note that the El Verde community contains 214 species of autotrophic plants
(Guana has 339) and more than 2,600 enumerated animal consumers. For food-web purposes, these are reduced to 20 structural kinds
of plants (fruit, nectar, leaves, etc.) and 156
kinds of consumers. There are 40 species of
birds in the El Verde community, nine of them
neotropical migrants; this seems quite similar
to Guana's roster. There are four species of
anole lizards at El Verde; assuming that *Anolis
roosevelti* is either extinct or does not live on
Guana Island (but see Lazell 2005 in press),
that is one more than our list of three. There are
nine other species of reptiles, so El Verde's tally

is quantitatively similar to Guana's. In the case
of amphibians, however, we are far outdone: El
Verde has 11 species of frogs to our one (Stewart
and Woolbright 1996). Each place has four
species of bats and the introduced Eurasian rat
(*Rattus rattus*), but fortunately Guana lacks the
introduced Indian mongoose (*Herpestes aurop-
unctatus;* Willig and Gannon 1996). McMahon
(1996) records only four species of termites at
El Verde, compared to Guana's eight. A few
other arthropod groups, reported by Garrison
and Willig (1996) are exemplary: Odonata
(dragonflies and damselflies), El Verde 10,
Guana 10; Cerambycidae (longhorn beetles), El
Verde 19, Guana 16; Mordellidae (tumbling
flower beetles), El Verde 1, Guana 7 (Lu and Ivie
1999); and Lepidoptera, in part (butterflies), El
Verde 26, Guana 31 (Becker and Miller 1992).
The purpose of this long-winded aside is that
the 300-ha arid Guana Island is quite comparable to the 40-ha rain-forest site on Puerto Rico,
and we must expect a comparable explication of
Guana's food web to require a similarly thick
volume.

Both El Verde and Guana, being island systems, have many fewer species than mainland
systems (a point made by almost every author
in Reagan and Waide 1996), but biomasses are
comparable or even well in excess of those on
mainlands, especially among consumers and
predators. Williamson (1981) coined the term
"density compensation" for this phenomenon:
fewer species but similar numbers of individuals and, often, biomass. However, we need an
extended explanation. At El Verde, Reagan,
Camilo, and Waide (1996) point out "…den-
sity compensation only partially accounts for
these high densities" of consumers and predators. On Guana, the situation is even more extreme, at least in groups like squamate reptiles,
where densities reach levels unequaled in any
other ecosystem yet studied (Rodda et al.
2001b).

Pimm (1982) sorts out three distinct modeling approaches to food webs. Two are applicable
to situations where populations vary continuously over time, as in most cases involving ver-

tebrate animals like Guana's squamate insectivorous reptile ensemble. Both require differential equations. The first, in which predators affect prey populations (the vast majority of cases, of course) are modeled with classic Lotka–Volterra equations. The second, called "donor-controlled" situations, do not fit Lotka–Volterra equations. These are less common—or less commonly considered—cases, but food webs of scavengers are examples: turkey vultures have no effect at all on their "prey," who are already dead before being consumed. Detritus feeders like most termites, and some benign parasites, are also examples. The third kind of model fits populations with dramatic seasonality and short generations, thus discontinuous population changes. These cases require difference equations.

Case (2000) and Gotelli (2001) provide Lotka–Volterra equations and lucid explanations of them, as well as some history of the two men. Briefly, Alfred Lotka (1880–1949) and Vito Volterra (1860–1940) did not work together. Volterra, a mathematician, got interested in his son-in-law's fishery statistics and developed predator–prey equations that had already been published by Lotka. However, Lotka had made serious errors that Volterra did not. Using Volterra's correct equations, Lotka was able to beat Volterra into print with a major summary work by two years (1924 vs. 1926). History credits both men equally for what are the most fundamental equations in ecology. Here is the Lotka–Volterra equation for a simple one-predator, one-prey species situation:

$$dV/dt = rV - aVP$$

Here, V represents the prey population and P is the predator population; r is the intrinsic rate of increase in the prey; a is capture efficiency of the predators and is measured as the number of prey consumed divided by the product of the number of prey, times the time interval considered, times the number of predators. What the equation says is that the change in prey numbers with respect to time is equal to the number of prey individuals (V) times their intrinsic rate of increase (r) minus the product of capture efficiency of the predators times the number victims times the number of predators. Would you figure that out for your son-in-law?

Of course, the equation is a classic glitterality and depends on all sorts of things that cannot be true, such as that the predators are the only thing limiting the number of prey (i.e., prey would increase exponentially at rate r if the predators were removed). Case (2000) and Gotelli (2001) add bits of reality and multispecies complexity to this basic formula, as did Lotka and Volterra. I hasten to point out that in a balanced, stable system the equation must equal zero. If there is a positive change, an increase in prey numbers (V) with respect to time, their numbers will eventually overwhelm the system. If there is a negative change, the prey will eventually disappear and the predators will starve. Over time, dV/dt has got to fluctuate around a mean of zero.

Optimal foraging theory made a grand splash in the previously calmer seas of ecological theory just about the time Pimm (1982) wrote; he mentioned it but did not fully anticipate its impact. The basic idea seems just another truism: Organisms will do the best they can to be as effective as they can be. The two obvious ways to optimize foraging—that is, to get the most food for the least effort—are to maximize the quantity and quality of the resource (food) intake and minimize the amount of time and energy spent getting it. If organisms really do that, then simple predictions emerge that should be amenable to verification (or refutation) by observing animals in the field, or even in the lab. For example, Guana's insectivorous lizards use two extremes of foraging behavior. The anoles are ambush hunters most of the time. Characteristically, they sit and wait for prey to wander close enough for them to grab. The ameivas, or ground lizards, on the other hand, are rarely still, seemingly always on the move, and cover large areas in an active search for things to eat. Which is better? We should be able to quantify, for example, time spent moving, distances covered, food-items consumed, ambient temperature, and time spent in the sun well enough to get an idea. When organisms overlap widely in food-items consumed, as the ameiva and crested anole appear to on Guana, a lot of these types of data are needed to understand niche differences; see below.

Perry and Pianka (1997) reviewed the theory and its results. They cite literature that analyzed over 100 "conclusive tests" of optimal-foraging theory. Fewer than 7% showed close quantitative agreement with predictions; nearly two-thirds were "inconsistent" with the model or only "partially or qualitatively consistent" with the model. Perry and Pianka (1997) note two predictions of optimal-foraging theory: First, in lean times—droughts or dry seasons—diets should broaden, animals should have to eat more different kinds of things, than in "rich" times. Second, the "compression hypothesis" predicts that having larger numbers of competing species should result in "contractions" of habitat use—spatial niche specialization—but "little or no change in diets." This would seem linked to Williamson's (1981) "density compensation" notion. We might test this hypothesis in a comparison of Guana and Dominica (small and large islands with the same species numbers) or Monkey Point and the Guana "mainland" (small, partially isolated community vs. a much larger area with much higher diversity). Perhaps each strategy is optimal and some difference other than food determines which lizard forages which way. In fact, Perry (1999) showed just that. The foraging behavior of a lizard species is much easier to predict from knowledge of its phylogeny than from any ecological datum.

Rubenstein (1982) presented a review of a symposium on feeding strategies, including ecological, ethological (behavioral), and even psychological approaches: Kamil and Sargent (1981). As with life histories, it has been assumed that feeding strategies are prone to optimization by means of natural selection. However, elementary optimal-foraging theory set up models often violated by real animals in the field. For example, we might rank prey species

according to the energetic or nutritive profit they provide a predator, or by relative abundance (ease of capture), or both. It does not take long, however, observing members of the lizard ensemble on Guana, to realize that real predators often rank their favored prey on other grounds that are less obvious to the human observer. In one of the most important papers presented in this symposium from our point of view, considering Guana Island, Arnold (1981) documented actual natural selection in action. A high-level predator, a snake not unlike our own *Liophis p. anegadae* in many ethological and ecological respects, varies genetically in prey preferences. Times for actual evolutionary divergence can be estimated. Can such an approach help explain the remarkable differences in population densities and apparent success we see in the several populations of *L. portoricensis* within the Virgin Islands archipelago? On some islands these snakes are common and on others quite scarce; Beef Island and Tortola are, respectively, good examples. Does environmental stability lead to more specialized foraging behavior?

To start to learn about trophic levels on Guana Island (or anywhere else), you will need a lot of natural-history information: who eats whom. I recommend using Joel Cohen's (1978) *Food Webs and Niche Space* (with updates: Cohen 1993). I am told no one ever uses this approach, at least all the way through interval graphing (below), but I am very fond of it. There are two reasons: It is the way I learned about food webs, so I understand it. Most important, it tells me exactly what I want and need to know: Where are the problems? Which species, if any, overlap completely in diet so that some other aspect of their niches must be discovered to explain their coexistence? If I were going to hire someone for a job with the title "ecologist" I would set up an interview with each applicant one week away. I would tell them to bring to their interview one example of a real habitat on which they had done a Cohen-style analysis. One will need powers of concentration and a tidy mind, because it *is* a complicated process.

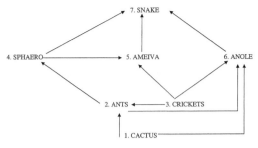

FIGURE 6.

For the sake of simplicity, I will skip most of the intermediate steps in the process and only present stages that are conceptually important.

For my habitat I have picked Monkey Point— that lovely little bit of craggy rock and cactus attached to Guana by a *tombolo*, a wave-washed sandbar. I picked it because it is simpler than the whole island, but we do have a good list of its occupants. I shall use only known eating relationships based on firsthand knowledge from direct observation or stomach contents, or on the testimony of a demonstrable expert. More details of the species are given in "The Cast," part 5, below.

First we draw a food web (see figure 6). In our food web the arrows point in the direction of energy and nutrient flow: from those eaten to those who eat. Thus cactus is eaten by ants, crickets, and even the crested anole (which eats the fruits), but not by the sphaero, ameiva lizard, or snake. The top predator in this system, the snake *L. p. anegadae*, eats all of the lizards but not insects or plant material. Crickets and ants are consumers, drawing most of their food from the cactus, but the ants also eat crickets (never vice versa) and each other (I cannot show that in this diagram). Of course, this food web is a gross simplification. Sphaeros eat many things in addition to ants, such as spiders, but I do not know what the spiders eat (it must be insects, but maybe not ants), so I left them out. Cohen (1978) does deal with hypothesized food relations, and Cohen and Newman (1988) have constructed a model explicitly incorporating gaps in the data, but all that complicates the illustrative and exemplary nature of the scheme above.

Next we would need to present the food-web relations in tabular form. Cactus, which eats no one, is omitted. After a visit to Monkey Point, you may wish to leave it in the food web because it is apparent that the cactus is at least trying to kill you. Also, nothing in this system is known to eat the snake *L. p. anegadae*. That is surely artifact, for we know both kestrels and red-tailed hawks eat these snakes on the main part of the island. I suspect ameivas eat at least snake eggs and probably hatchling snakes when they can get them. Anyway, on Monkey Point we do not know of anyone who eats snakes, so snakes are never considered as prey. I will not present this tabulation stage, nor the next, which involves converting the data in our food-web table into a food-niche overlap matrix.

We next translate our just-mentioned matrix into a food-niche overlap graph (see figure 7). As noted before, cactus eats no one, so it cannot share its food niche. Snakes, however, do share their food niche partially with ameivas, as they both eat sphaeros. Anoles share part of their food niche with both ants and crickets, as all three eat cactus. Crickets (we claim) eat nothing but cactus, so they share their food niche with ants and anoles, but no one else. Ants, however, eat crickets as well as cactus, so—as cricket eaters—they share their food niche with ameivas. Because ants eat cactus *and* each other, they triply share their food niche with anoles: both eat cactus, ants, and crickets.

The final stage we will look at involves identifying sets of species that share dietary elements.

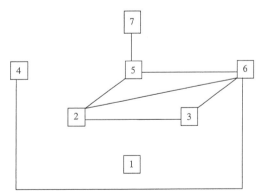

FIGURE 7.

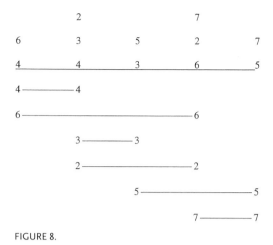

FIGURE 8.

Cohen (1978) calls each of these a "dominant clique." I call them trophic levels. If the biology is right and the calculations were performed correctly, then the trophic levels in the system can be represented by an interval graph (see figure 8). At the top are all the eating species in the whole system. Below that I have dissected out each organism's portion of the line. The overlaps on the interval graph correspond to the food-niche overlaps of the organisms. Where one organism's food niche is entirely included within another's, as the sphaero's (4) is included within the anole's (6), we know we must search more diligently to explain their occurrence together in the same habitat. Perhaps (as in this case) we know too little about the whole diet of the sphaero. But perhaps the niche overlap in food is real and total. That tells us to look for another niche dimension [thermal, spatial, daily activity (diel)], which allows them to obey the law of competitive exclusion. That is easy for the sphaero and the crested anole because the former is a tiny beastie of shady leaf litter and the latter a largish lizard that perches on trees and rocks, often in the sun. It is harder on the Guana "mainland" with its three species of anoles, but we think we know the differences (see "The Lizard Genus *Anolis*" by Perry, part 5, below).

If we find a situation in which the "consecutive ones property" described by Cohen (1978) is not fulfilled, then we cannot construct an interval graph. This means all the food niches in the community in question cannot be represented in a single (linear) dimension. Such a case may imply that we have compounded or confounded two different communities. Cohen (1978) could not find a case of a single community or habitat that could not be represented in one dimension with respect to food niche, but at least theoretically they may occur.

Pimm (1982) discusses two important aspects of food webs: connectance and complexity. When organisms in the web interact, as when one eats another, they are connected; if they have zero interaction, they are not connected. To quantify web connectance, then, we simply divide the number of actual interactions by the number of possible interactions. A simple, straightforward, and (almost) real case would be Carval Rock, BVI, with seagrape bushes, ants, and a sphaero ground gecko. The ants eat the seagrapes, and the sphaeros eat the ants. Thus, there are two actual interactions. There is only one other possible interaction: The sphaeros eat the seagrapes. Because that never happens, interaction is a zero. Thus, of a total of three possible interactions, two occur. The web connectance value is then two divided by three, or two-thirds (0.67). Back on Monkey Point, I count 21 possible interactions of which 12 actually occur, rendering a connectance of 0.57. Actually, I do not have to count. The number of possible interactions is the number of species or "kinds" of organisms minus one, $N - 1$, added to $N - 2$, added to $N - 3$, added to $N - 4, \ldots$, all the way to the last member in the set. Pimm (1982) uses connectance as the index of complexity, which is counterintuitive. How can a simple system like that at Carval Rock be more complex than the seven species set on Monkey Point? Well, because lack of interaction among members of a group of species that live in the same place simplifies life there, it detracts from complexity, at least in Pimm's (1984) sense of the term. The higher the percent of zero interactions (noninteractions) the lower the connectance and complexity.

To go from the crystalline simplicity of the little community on Monkey Point to even a rough approximation of the complexity bestowed by diversity on the whole island may be bewildering (or, as I suggested above, require hundreds of pages).

Cohen (1978) and Auerbach (1984) distinguish three sorts of food webs. Ours depicted above for Monkey Point is a community food web because I selected the organisms to include on the basis of their occurrence in a specific habitat; the selection was intended to be broad and was made with no a priori knowledge of eating relations. A sink food web is a subset of a community web that begins with a predator, or group of predators, and traces the energy–nutrient flow upstream (all the way to producers if possible). A source food web is the opposite of a sink. It is a subset of a community in which we pick the producer or prey organisms and trace eating relations downstream along the energy–nutrient flow. When one examines the anole ensemble on Guana Island one is dealing with a sink.

That diversity begets stability in natural ecosystems is not a notion based on theory or mathematical models. It is an empirical observation grounded in the cumulative centuries of outdoor work by field biologists, naturalists, and ecologists. Mathematical models are wonderful things, and often genuinely useful too, but they are the carts of science, not the horses.

Ptolemy did a brilliant job of modeling the universe with Earth at the center. Numerous competent scientists of their day "proved" Earth was flat; engineers and physicists could readily show that a bumblebee's wing surface was too small to support its weight in flight. Let us not throw out from our introductory texts true and repeatable descriptions of observations for which there are, as yet, no convincing mathematical models. As Robert May points out: "It is lunacy to imagine that the dynamics of behavior of real communities bears anything but the vaguest metaphorical relation" to "conventional" models (May 1982); he is a convincing proponent of pragmatism in ecology: the need to get on with the job of shepherding our resources while we strive for fundamental understanding, demonstrable theories, and a complete overview.

May (2000) notes "...real ecosystems 'develop' by adding, and losing, species over time, not by randomly sampling ecological possibilities."

Auerbach (1984) discusses the problems in trying to assess the stability of food webs to get at the diversity–stability relation. From the works of Robert May and others, Auerbach presents a predictive index in which a value of less than one would indicate stability and a value greater than one would indicate instability:

$$i\,(CS)^{1/2}.$$

Here, S is the total number of organisms enumerated in the food web—usually species; C is the connectance matrix value: the percentage of nonzero elements in the food-niche overlap matrix. If there are hypothesized relations in the web (we excluded them for simplicity, above), then these will be nonzero values, too. At Monkey Point there were seven species in the niche overlap matrix and 30 possible connections. Of that 30, 16 are nonzeros (ones), which is 53% (here, note that I have left the cactus out, but not in my original connectance calculation; that alters the percentage a little in this case). This means we have to know if the following value is less than or greater than one:

$$i\,\sqrt{(53.0)(7)} = 19.3i.$$

For that value to *equal* one, i would have to be about 0.052. What is this thing called i anyway? How do we find its value? Well, i is a lovely figure indeed. It is the average strength of interaction between the enumerated (seven) members of the food web (see Neutel et al. 2002; discussed below).

Oh come on! How are we going to figure that out? Determine the stability and solve for i? Question the lizards? Look in the bible? Ask our father-in-law? As Auerbach (1984) points out: because S and C are real, fixed values, "i becomes the critical parameter if current models are to be viewed as valid representations of the real world." He also notes "...there is a stronger mathematical than biological basis for some current predictions about food webs." That criticism is broadly expanded by Paine (1988), who notes the almost insuperable barriers to accurately quantifying what we most need to know in real-world situations.

The most compelling theoretical support for a strong, positive diversity–stability relationship has recently emerged from analyses of trophic levels and food webs. Neutel et al. (2002) review the history of these analyses and extend the reach of the correlation by perceiving "food loops." Raffaelli (2002) elucidates the thesis, which is complex and initially turgid. At the simplest level, any two species that interact trophically (one or both eats the other) form a loop. In my Monkey Point food web, above, I have obscured this by putting in the energy-flow arrows unidirectionally, from eaten to eater. Consider the sphaero and the ants: Sphaeros eat a lot of ants, and there are a lot of sphaeros, so clearly they exert a strong influence on ant populations. By being a food supply, conversely, the ants also influence the sphaeros. But, because sphaeros can eat things other than just ants, the influence of the ants on the sphaeros is weaker than that of the sphaeros on the ants. We have a nice little two-member food loop, but one of the links—ants to sphaeros— is relatively weak. Let us consider another two-member loop, one much celebrated in classical ecology (Kendeigh 1961): lynx and hare on the arctic tundra. The predator, lynx, is so dependent on the prey, hare, in this ecosystem that hares fairly control the lynx population: hares increase, lynx thrive and increase too; hares decrease and the lynx do not just starve, they withhold reproduction—their population plummets. Here, the loop consists of two strong links (no pun intended).

Neutel et al. (2002) quantified links in loops and found mean values between pairs; the means are called loop weights. The longer the loop—that is, the more species or trophic levels involved—the lighter the loop weights. In other words, the weaker the links, the less destabilizing effect a perturbation would have. Going back to Monkey Point, only two of the animals, sphaero and cricket, have a single food source (and even that is not true; just an oversimplifica-

tion I have made for ease of communication). However, viewing their loops, their food sources' influence on them is weak. Therefore, mean loop weight is still fairly light. All the other animals in the web have two or three potential food sources, so each link is relatively weak. Even the snake at the pinnacle of the pyramid eats all three kinds of lizards, so its link to any one of them is only about one-third as strong as the one connecting the lynx to the hare in the Arctic. What this means, in practical terms, is that fluctuations—even major crashes—in one lizard's numbers at Monkey Point will not much perturb the snakes. Suppose a hawk moves in and systematically picks off all the large ameivas: the potential food biomass of ameivas for snakes to eat drops. However, this takes predation pressure off the sphaero and relieves the anoles of competition: They will predictably prosper; the snake will hardly know the difference—there are still plenty of lizards to eat. Kondoh (2003) provides a mathematical model indicating that a predator or forager with a wide range of prey (food) options will add stability to a system. Indeed, the greater the diversity of items consumed, the lighter, on average, any loop weight will be. Note that this also meshes well with my notion that diversity generates stability, especially for the species at the top—like the snake on Monkey Point, who eats all kinds of lizards. Lacking such diversity, the lynx are put in a nosedive by a reduction in hares; this can and will be terminal for many individual lynx.

Raffaelli (2002) points out that all of this extrapolates well to the pyramid of biomass, which decreases with ascending trophic levels. The more levels in the system, and the greater the biomass at each level, the more gradual the slope of the pyramid. The biomass of producer vegetation on the Arctic tundra is not great (compared by unit area to Monkey Point); the lynx consume all the hares they can catch. We have a small-based, steep pyramid. You would not want to have to live at the top of it. Raffaelli (2002) notes: "A factor of 10 decrease in biomass with increasing trophic level would provide patterns of interaction strengths necessary for stability," which I view as a hypothesis, not revealed truth. He adds, "these pyramids are not uncommon in nature." Well, maybe not common enough.

We have biomass data collected on Guana's White Bay Flat by Rodda et al. (2001b). They weighed all the vegetation and counted and weighed all reptiles on four 100-m² plots. This ecosystem has four trophic levels: producers (plants), consumers (herbivores), first-level predators (herbivore eaters), and second-level predators (first-level predator eaters). The team weighed a total of 6,668 kg of plant biomass on the 400 m². This yields 166,700 kg of plant matter on a hectare (1 ha = 10,000 m²). They did not quantify insects and these are the principal herbivores in this ecosystem, so we are missing the second level in our pyramid. For the sake of simplicity, I will count all five species of lizards (sphaero, ameiva, and crested, saddled, and grass anoles) and the blind snake *(Typhlops)* as insectivores (first-level predators; some of these lizards do eat some plant matter, but I discount that). Rodda et al. (2001b) got about 25 kg of insectivores per hectare. The only second-level predator here, as on Monkey Point, is the snake *L. portoricensis*: a mere 3 kg/ha. A pyramid like that shown by Raffaelli (2002, Figure B) would go like this:

$$3$$
$$25$$
$$.......$$
$$166,700$$

To calculate the slope of this pyramid, one would have to plot one-half of each level's biomass on the *x*-axis and decide on quantification for the *y*-axis (for easy picturing, just plot the trophic levels in descending order —4, 3, 2, 1— on the *y*-axis from the bottom up). If you try to envision such a graph, with a slope ascending from 1.5 to 83,350 in four steps, you can imagine that you will need a wall to draw it on and—with any reasonable units on the *y*-axis—it is going to look like a virtually flat, horizontal line (the slope is almost zero). Because Raffaelli (2002) argues that the more gradual the pyramid's slope the more stable the ecosystem, this

is comforting. However, the top two levels are less so: Raffaelli (2002) does not think we are safely stable unless the biomass diminishes by a factor of 10 with each ascending level. We have a half kilogram of snakes too many per hectare. Because each snake only weighs an average of 0.06 kg, that is more than eight snakes per hectare "too many." Some theoretical ecologist could hypothetically use data like these to justify a snake-reduction program, which is exactly why I believe espousing econumerology is dangerous (see "A Unified Theory," part 4, below).

Winemiller and Polis (1996) note that food-web studies, as with all of systems ecology, have tended to consider more and more the physical and chemical aspects of systems and to downplay the roles of natural selection and stochastic environmental influences. They note five generalizations that have enjoyed considerable currency but that engender skepticism. First, that network linkage density and connectance are independent of the numbers of species involved. Second, food chains are usually limited to three or four levels. Third, ratios of predator to prey species tend to be between 0.8 and 1.0. (These figures surprise me; I would expect many prey species for each predator species and, therefore, much smaller predator to prey ratios—perhaps averaging 0.2, *or* one to five.) Fourth, the fraction of species at the bottoms, middles, and tops of food chains does not vary with the numbers of species. And fifth, the frequency of omnivory in nature is less than would be expected by chance. All five of these generalizations seem to be turning out to be classic glitteralities; that is, simplistic and infrequently true (Williams and Martinez 2000; Brose et al. 2003; Gibbs 2003). However, they pose a question particularly appropriate for small tropical islands like Guana: "Can we recognize a web structured by a hurricane or fire, and how would it differ from the web of a mature community structured by density-dependant biotic interactions?"

Holt (1996) considers food-web theory in the context of island biogeography theory, with all the attendant problems that come with the latter.

Do area and food-chain length tend to increase together? As suggested above, Guana to Dominica and Monkey Point to Guana "mainland" comparisons would be appropriate here. Is there a meaningful equilibrium between colonizations and extinctions? Are there assembly rules? Or are (were) most islands structured like Guana by the "relaxation" or dwindling of a recently significantly larger community? Island food webs (and communities) are especially prone to destabilizing predators—typically introduced exotics—such as the mongoose and feral cat, now so prevalent in the West Indies. Holt (1996) seems to think islands are fundamentally different from continents (except for patch habitats within continents) because continents are sources and not biotic assemblies determined by equilibrium turnover. Well, so is Guana Island (at least with respect to vertebrates). Furthermore, continents have suffered and continue to suffer both catastrophic extinctions and devastating invasions: The rest of the world, including oceanic islands, provides a species pool for continents (e.g., Meshaka 2001).

Cousins (1996) asserts that: "A central, or, general theory of ecology would require that the key components of the science could be derived solely from the study of food chains or food webs." As pointed out by Holt (1996), there is real sequential dependency in nature: The snake of Monkey Point could not exist there without the lizards it eats. I will discuss this further in "A Unified Theory," part 4, below.

Lawton (2000), whose work won him the Ecology Institute prize, took on all the critiques and criticisms of food-web theory, which as Peters (1988) pointed out are basically the same for all theoretical ecology. His communities live in bracken patches, and I have teased him, above, for not explicitly noting that almost everything about these patches today depends on the absence of the Pleistocene megafauna we humans exterminated. He would no doubt embrace my taunt, as he states: "Experiments in community ecology may not generalize to the longer term." Lawton (2000, 89) notes: "Changes in the larger world will almost always modulate and some-times overwhelm the local rules of engagement, altering average population abundances and the frequency and amplitude of population fluctuations." Here comes a whole flock of blackpoll warblers down from Cape Cod, all absolutely famished. On Guana Island, they eat just what the insectivorous reptile ensemble eats, but they are much faster, more agile, smarter, and much, much needier than any lizard. Perhaps no real island is ever "an island" after all.

I have always assumed that there is absolutely nothing we cannot find out about the universe given time. That assumption makes me a scientist. If you have followed through the webs of who eats whom this far, you are ready for a bit of fieldwork. Energetics is a topic I have scarcely noted that is surely a major factor in species diversity, especially for animals like reptiles who must maintain their body temperatures by behavior (they do not generate much metabolic heat physiologically, like we do). Times of activity can tell us a great deal about lizard energetics, because insolation is the animals' source of heat. I have graphed some of these in the following chapter, "Communities, Local Guilds, and Ensembles."

Huey (1974), Hertz (1979), and Henderson (1982) have produced comparative studies dealing with lizards and snakes with very close relatives on Guana. Snyder (1975) provides a model for the study of heating and water loss in a single species. Lawton (1981) provides a broad overview of the questions of endotherms versus ectotherms and r-selected versus K-selected species. He points out that ectotherms (e.g., reptiles) are an order of magnitude more efficient energetically than endotherms (e.g., us) because they are not using food consumption as a source of heat energy (hence the misnomer "cold-blooded"). Eventually a food-web theory developed that seemed to drift farther and farther away from reality (Hastings 1988; Paine 1988; Strong 1988). Pimm and Kitching (1988) advocate a solution to this drift by getting on with experimental designs; this would be relatively easy at Monkey Point on Guana Island and could make a superb study or doctoral dissertation.

The various critiques and complaints were as water on a duck's back to Henry Wilbur, of Mountain Lake Biological Station in Virginia; he went to work on the amphibians that live in (sometimes) and around (the rest of the time) temporary ponds on the U.S. southeastern coastal plain. Wilbur (1997), whose work won him the MacArthur Award, was able to duplicate and manipulate temporary pond communities with artificial tanks—and thus get the replicates necessary for clear observational inferences of causes and effects. Alas, none of his species have relevant analogues on Guana. It is hard to imagine how we could do anything like what he did on Guana Island because we lack discrete, segregated, comparable habitats such as temporary ponds. It might be feasible with Salt Pond communities, once again using artificial tanks. This may be an excellent task for Lianna Jarecki of the H. L. Stoutt Community College on Tortola (Jarecki 2003).

Communities, Local Guilds, and Ensembles

The . . . guilds that I have described demonstrate that intuitively reasonable and frequently used descriptions of nature may be dynamically insufficient for community ecology. . . . The insufficiency seems to be related to the implicit assumptions that are the basis for the forms of descriptions. . . . It has been assumed that taxonomic affinity implies ecological impact, and furthermore that the impact is interspecific competition. . . . It is further assumed that the ways in which the species differ are the result of natural selection operating through the impact of the competitive relationships. I believe I have shown that these assumptions are neither necessary nor sufficient to explain the observations.

NELSON G. HAIRSTON, SR. (1981)

Lacking tangible progress, people turn upon one another. If there is agony in community ecology, . . . much of it appears to be self inflicted.

EVAN WEIHER AND PAUL KEDDY (1999)

THE BASIC DILEMMAS IN community ecology revolve around what communities, as opposed to chance groupings of organisms indulging in ecological activities, really are. Does competition truly modify community members? What kinds of tests are needed to differentiate between chance groupings and potentially competitively interacting members of communities? Do other sorts of ecological relationships, such as predator–prey, exert stronger influence? Chave et al. (2002) explore all of this and find, as one might expect, that different patterns and processes occur in different apparent communities. To the brew, I add my own question: Why would anyone expect competition to be a major factor within a stable, coevolved community of species?

The controversy that has produced symposium volumes—Strong et al. (1984), Salt (1983, 1984), Weiher and Keddy (1999)—is directly relevant to Guana Island. For example, the insectivores of Guana Island make one of the finest cases imaginable of an ecological subcommunity called a local guild (Lawton 2000, 18). Root (1967) developed the concept of a guild while studying gnatcatchers—small, insectivorous birds that fit into a group of such species exploiting a particular habitat. A guild is a group of comparable species, though not necessarily closely related, that compete for a particular resource. In the original conception, Root (1967) added the requirement that all members must exploit the resource "in a similar way." Simberloff and Dayan (1991) reiterate this, but Vitt (1983) points out myriad small ways in which undisputed members of the same guild differ in practices and techniques. Indeed, the two sexes of a single species may exploit the same resource in different ways; this will become important in my description of Guana's anole lizards, below. At one place, for instance Guana Island, one has a local guild in Lawton's sense. Considering just related species in the same local guild, one has an ensemble (Lawton 2000, 18). The proper taxon in our case is Squamata, the squamate reptiles, because all snakes are really lizards, but we will leave out the blind snake *Typhlops* for now. There are seven species of insectivorous lizards on Guana. Because of their differences in activity patterns, prey size, foraging zones, and so on, they epitomize a guild system. Hutchinson's great question can here be honed to a keen edge: Why are there so many species of insectivorous lizards on Guana Island?

Pianka (1970a) studied local guilds of widely separated desert lizards (western United States, southern Africa, and Australia). He found fine examples of niche segregation within each guild but was surprised that the local guilds in the different places were not composed of ecologically comparable species. Each local guild seemed to have evolved independently and dissimilarly to the others. Pianka et al. (1979) expanded this work to include more area and species but continued to find virtually incomparable species in seemingly closely comparable habitats. Pianka (1977) provides a formulation for diversity within a local guild (it does not fit a whole community because it does not consider, e.g., primary production). Here, D_s is total species diversity:

$$D_s = D_r/D_u (1 + C\alpha),$$

where D_r is the diversity of resources, D_u the diversity of resource utilization by the average species (thus, average niche breadth per niche

In the 1950s I was an animal catcher and an undergraduate. I was so enamored of the first role that I kept my satchel packed and headed out, to islands, mountains, and caverns, as often and as rapidly as I could. I was so bored with the latter role that I spent, by actual count of hours, more time underground in caves than I did in the classroom. I learned far more in the three months of summer about everything except biology than I learned in the nine months of the academic year. That biology could win my attention in the nine months was hardly surprising, because I took so many courses in it; the length of the win, however, was very short. My priorities were simple: Find new kinds of animals, unknown and undescribed in the annals of science; find alive species other scientists believed to be extinct; and rediscover populations of animals others believed had been extirpated. I was good at what I did and loved doing it. My notion was to use education to improve on the performance of previous animal hunters. I took an educated eye to islands, up mountains, and down caverns—places where previous eyes had failed, through want of relevant knowledge, to see what was important. Quite often, I went to places no biologist had ever been before.

Among my colleagues later in graduate school were students such as George Gorman, Daniel Simberloff, and Thomas Schoener. One of my professors was a young man, struggling for tenure—Edward O. Wilson. They were all nifty fellows, full of bright ideas, but not very interested in rare species. They were searching for patterns and generalized processes and broad explanations. That was fine, I supposed, even though it was obvious to me that their views were somewhat simplistic. There was not one kind of biochemical difference that signaled species versus subspecies level between populations; species : area relationships were vague at best; but, on the other hand, one hardly needed calculus to solve time, rate, and distance problems. It seemed to me that mathematical modeling, biochemistry, and biophysics were related to the subject of biology about the same way yachting

continued

continued

was related to the subject of transportation. In fact, I did not really think those three disciplines were part of biology at all; I viewed them as parts of math, chemistry, and physics instead. (The analogy held: I did not regard yachting as part of transportation, either.)

I provide this digression into my personal history because it is directly relevant to ecological theory as that pertains to communities. It would be decades before I was drawn back toward the theoreticians by the inexorable needs of wildlife conservation—the arena into which chasing rare animals had led me. What I found was what George Salt (with characteristic academic understatement) called "turbulence" (Salt 1984). It is good, healthy turbulence, too, from my point of view. I need good theory about ecological communities so that I can get on with my job of preserving rare species, and understanding both the processes that produce new species and species' preserving behavior traits. I need to know the truth about diversity and stability at community and habitat levels so that I can effect long-term conservation. I agree with Salt (1983) that the major divisions within ecology and all their disparate factions are all right—at least some of the time.

dimension), C is a measure of the number of neighbors in a given niche space, and α is the average niche overlap.

Hairston (1981, 1984) attempted to study an ensemble of small carnivores (Appalachian salamanders) quite comparable in some ways (size, diet) to Guana's lizards. He removed the commonest species or two from selected plots and looked at what happened to those remaining. He decided that competition was insufficient to explain the results. It seemed that predation on some members of the ensemble (from outside of it) better explained some sorts of niche differences than competition. Hairston concluded (1981) that "experimental tests are needed even in widely accepted examples of guild organization" and (1984) that "It is my contention that community ecology will never escape from its

agony until we begin the self-conscious application of a rigorous scientific method. . . . The task will be particularly difficult because we have only unfounded ideas about what to look for. The greatest need is for legitimate means of identification of interacting groups of species."

We have begun study on Guana of the interactions in a very close-knit ensemble of three species of primarily insectivorous lizards: the iguanids called *anoles*. Two of the species are common virtually all over the island: *Anolis cristatellus wileyae,* the crested anole, and *A. stratulus,* the saddled anole. The third is severely restricted and localized, nowhere as abundant as the other two: *A. pulchellus,* the grass anole. The same ensemble occurs on about 20% of the islands on the Greater Puerto Rico Bank; Guana is close to the smallest (Norman Island has all three and is about 14% smaller; Lazell 1983a). As far as I know, these are the smallest islands in the world with three native species of that heavily studied group, the anoles. (See plate 8.)

The first question one might ask about the island's three-species anole ensemble is whether it exists by pure chance. Grant and Schluter (1984) address this question. If S is the total number of available species (in our case members of the genus *Anolis*) and T is the number occurring together as an ensemble, then C, the number of possible combinations involving T species, is:

$$C = \binom{S}{T} = \frac{S!}{T!\,(S-T)!}.$$

There are five members of the genus *Anolis* in the BVI: the three in our group; the very rare *A. roosevelti,* a giant known (at least in the past) from Culebra, Vieques, Tortola, and St. John; and *A. ernestwilliamsi* (Lazell 1983a). The latter is not quite a giant, but it is much bigger than the three in our ensemble. It occurs abundantly only on Carrot Rock, all by itself. It was recorded on Peter Island, with our three species, in 1960 (MacLean 1982), but has not been found there since (I looked, but only for a few hours, so more needs to be done).

We plug into the equation above 5—the total number of anole species in the BVI—for S, and

3—the number in our ensemble—for T. Then the total possible number of three-species ensembles is:

$$C = \frac{5!}{3!\,(2)!} = 10.$$

There are 48 islands in the BVI inhabited by anoles; most of them (30) harbor only a single species, either *A. cristatellus* or its very closely related replacement *A. ernestwilliamsi*. Only five islands harbor just two species and those are always *A. cristatellus* and *A. stratulus*; no other two-species combination occurs. Thirteen islands have our three: *A. cristatellus*, *A. stratulus*, and *A. pulchellus*. But two of these (Tortola and Peter Island) have (or had) a fourth species; for simplicity I will leave them out. I hardly need them. Hence 11 islands have our three-species ensemble and no island has any other of the nine possible combinations. The probability that a second island has the same three-species ensemble as Guana is $1/C$, so 1 in 10. The probability that a third island has this same three-species ensemble is $(1/C)^2$, or 1 in 100. With each island added to the list with our same ensemble the power to which $1/C$ is raised goes up by one. If there are a total of n such islands, then that probability is:

$$(1/C)^{n-1}.$$

Because there are 11 such islands, $n - 1$ is 10. Thus it is wildly improbable that all 11 islands have our ensemble simply by chance. But it gets even worse: because there are 10 possible combinations that Guana might have had to begin with, we have to multiply that improbability by 10 (which is C): $10 \times (1/10)^{10}$. When I try to deal with numbers like this, my pitiful lack of quantitative intuition causes tachycardia, blurred vision, and order-of-magnitude errors. Fortunately Daniel Simberloff and Martin Michener are made of sterner stuff; they walked and talked me through all of this. It seems that:

$$C\,(1/C)^{n-1} = C^{2-n}.$$

In our case, this is 10^{-9}. Negative powers (like dark powers and supernatural powers) are things I do not voluntarily deal with, but I am reliably informed that the actual number is our $1/10$ preceded by $2 - n$ zeros, therefore nine zeros, only one of which lies left of the decimal point:

0.000000001

All right, there is essentially no possibility that our three species BVI ensemble occurs by chance. What can be the glue that sticks these three little lizards together? Is it no more than simple history? Our three occurred together virtually throughout Great Guania when it was all one big island, before the Holocene sea-level rise fragmented it into dozens of islands of which the BVI comprise 48. Is the three-species ensemble just what is left after chance extinctions knocked out one or two of the species on most of the islands? That is an unexciting answer that begs more questions. Simberloff and Connor (1981) present methodology for dealing with missing species in ensembles, as in those five cases where *A. cristatellus* and *A. stratulus* occur together without *A. pulchellus*. And over in the U.S. Virgin Islands (USVI), we have the island of Little Saint James that has *A. cristatellus* and *A. pulchellus* but not—as far as we can see—any *A. stratulus* at all. In cases more complex than ours, one can repair to Kincaid and Bryant (1983) or Colwell and Winkler (1984) for a method of evaluating the null hypothesis of random sympatry: Are the species occurring together by chance?

Vitt (1983) frames the next set of questions I want answers to, and I paraphrase him:

1. Do the members of the ensemble divide up ("partition," ecologists say) the available food items, and, if so, do they do it by prey size, by kind (species), or by some other method (e.g., one sees and eats active prey, another smells and eats stationary prey)?

2. If partitioning does occur, is it reflected in the sizes of the ensemble members (e.g., big members eat big prey) or in microhabitats occupied (e.g., specialists on one sort of prey live in a different part of the environment than specialists on another sort)?

3. Is there seasonal, diel, or other temporal differentiation in activity?

4. Is reproduction seasonal and are there differences in this seasonality among the members?

5. Is sexual dimorphism evident; if so, is it associated with the reproductive potential of the females or with competition between the sexes?

See, for example, the comparison of food by taxon from *Anolis* stomach contents presented in table 13. These data are not compelling evidence for separable food niches, but there is a strong hint that better knowledge of the diet of *A. pulchellus* could be quite revealing. The percentage and count of ants is low. In my notes I recorded that *A. pulchellus* ate bigger ants, on average, than the other two, and that the difference vis-à-vis *A. cristatellus* (but not *A. stratulus*) was significant. It is that mass of unidentified Arthropoda that intrigues me. If I knew who they were I might have evidence for a very broad food niche in *A. pulchellus*. Apart from the ants, I saw nothing obviously different about food-item size.

The questions involved in assessing size differences among reptilian species are highly complex. In working with anoles, Schoener (1969, 1970) used only the largest one-third, often only males, in each sample off the museum shelf. This enabled him to get the clear sorts of categorical size differences he found in birds (Schoener 1984), but buried a world of competition. The problem is that birds grow to a given terminal size and then potentially compete. Reptiles potentially compete throughout life, while growing (Simberloff 1982a). For an arcane discussion of the mathematical complexities introduced by this kind of thing, see Thompson (1994). Reptiles not only keep on growing essentially all through their lives, but their sizes at sexual maturity, and even in old age, show great dispersion even within small populations (Lazell 1983a).

In table 14, we compare the length and weight of adult anoles on Guana Island. The differences are not impressive. Big male *A. cristatellus* stand out, but the rest overlap extensively. The situation is further complicated by the fact that *A. pulchellus,* the smallest species, has an elongated head, which means that it easily duplicates the size ranges of prey items eaten even by big

TABLE 13

Food by Taxon (Species or Higher Category) from Anolis Stomach Contents

	A. cristatellus	A. stratulus	A. pulchellus
Ants (Hymenoptera)	82, 60	90, 80	60, 57
Beetles (Coleoptera)	16, 14	8, 5	—
Crickets (Orthoptera)	1, 6	3, 3	2, 4
Bugs (Hemiptera)	—	—	2, 3
Leafhoppers (Hemiptera)	10, 3	—	—
Flies (Diptera)	—	—	5, 10
Thrips (Thysanoptera)	—	—	2, 2
Spiders (Araneae)	—	4, 2	2, 2
Whip Scorpions (Amblypygi)	1, 2	—	—
Unidentified arthropods	—	—	24, 22
Cactus fruit	5, 6	2, 9	—
Other plant matter	5, 6	1, 1	—
Pebbles	6, 3	—	—

SOURCE: Original data. The first number is item count; the second percent by volume in 10 specimens of each species, except only six of *A. pulchellus.* The counts have been corrected to be comparable.

TABLE 14
Sizes of Large Adult Anoles on Guana Island

	SNOUT-TO-VENT LENGTH (MM)		WEIGHT (G)	
	MALE	FEMALE	MALE	FEMALE
A. cristatellus	70	50	4.9	2.9
A. stratulus	55	50	2.1	1.4
A. pulchellus	50	45	2.0	1.3

SOURCE: Original data.

male *A. cristatellus*. These three species are far more similar in size than any of them are to the fourth or fifth species in the Virgin Islands. There is no overlap in size of sexually mature females of *A. cristatellus* and *A. ernestwilliamsi*, and little in males; *A. ernestwilliamsi* males average ca. 17% bigger than the largest male *A. cristatellus*. We know little about the very rare *A. roosevelti*. It is a true giant; one male I measured was 156-mm SVL. Why should the common, seemingly severely competitive species differ little in size, and the rarer ones that exist largely beyond the influence of possible competition be so different from them? Is it just evolutionary time? That could not be true for a very young species like *A. ernestwilliamsi* (Mayer and Lazell 2000). Will size differences be selected for among the members of our guild? Pacala and Roughgarden (1982) found that food-resource partitioning and competition were more intense between two anole species that were close in size than between the smaller of them and a giant (on another island). Does gigantism evolve as a result of selection to avoid competition? If so, why is *A. ernestwilliamsi*, all by itself on Carrot Rock, so big (see "Evolution," part 4, below)?

Microhabitats deserve a category of their own. Schoener and Schoener (1971b) showed *A. cristatellus* to use low tree trunks, rocks, and ground habitats. *A. stratulus* uses upper trunks and crowns, at least of small trees. *A. pulchellus* is found on grass, small bushes, and the ground. Perch heights and foraging areas are certainly

modally different for the three species. This not only helps to minimize potential competition, but it leads to striking differences in relative abundance in various parts of even a very small island like Guana. *A. cristatellus* is the common species in the cactus-and-rock open habitats; *A. stratulus* is more abundant in the ravine forests; and *A. pulchellus* is virtually confined to areas that are open, because the trees have been cut down, and from which sheep have been excluded, so grass grows tall. (See figure 9.)

I tried to quantify another aspect of microhabitat that seems important to me: amount of insolation (direct sunlight) on areas of activity during activity periods; I admit the procedure is prone to errors. Using the apparent activity periods (see below), I drew with a felt-tip pen the shaded area on a 3 × 5 in. sheet of paper at the places where the anoles were active. I then cut up the papers into sunlit and shadowed portions, and weighed each pile of pieces. At the very least, my method needs to be greatly expanded, but patterns not unlike what I expected emerged. Insolation in grass anole habitat was hardest to measure, because it is hard to quickly draw the shadows of a mass of little stems. You will have to try this in order to truly appreciate the difficulties; then maybe you can think of a better way. (See figure 10.)

Temporal differences in activity were recorded over a period of several days. The problem here is deciding when an anole is "active," sitting there alertly scanning his domain for a glimpse of food, a predator, a potential mate, or

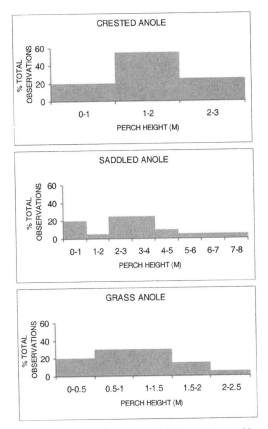

FIGURE 9. Perch heights in Guana Island's anole ensemble.

the saddled anoles moved up off the ground—typically to tree-trunk perches. For most of the day's overall activity span—8 h, from 10 a.m. to 6 p.m. (1800 h)—most saddled anoles were just perched motionless on tree trunks, often fairly high up.

Grass anoles were the least active of the three species. A few were active between 5 and 6 a.m. At least 1 in 10 were active after that until about 11 a.m., when activity tapered off again. I could find no active grass anoles after 6 p.m. (1800 h), so their total activity span was just 13 h each day.

The question of seasonal reproductive differences in this ensemble has yet to be investigated. At the height of the dry season, all three species are producing eggs and hatchlings are evident. I am sure the kinds of correlations to rainfall and arthropod food abundance discussed above occur on the island, probably in all three

a territorial encroacher, and when he has just dozed off, with heart rate down and mind a total blank. (See figure 11.)

I could find one or more crested anoles that I judged to be active during 15 h of the several days in my survey. A few, usually females or juveniles, became active between 5 and 6 a.m. From 8 a.m. to noon about 10% of crested anoles were active. Activity slowed in the afternoon, but there was a noticeable increase between 4 and 6 p.m. (1600–1800 h).

Saddled anoles were strikingly different, but some were also active 15 hours each day. They became active earlier than crested anoles. In ravine forest, some appeared foraging on the ground between 4 and 5 a.m. This ground foraging included both sexes and all ages, and peaked with up to 2 of every 10 individuals observed active between 6 and 8 a.m. Activity tapered off as crested anole activity increased, and

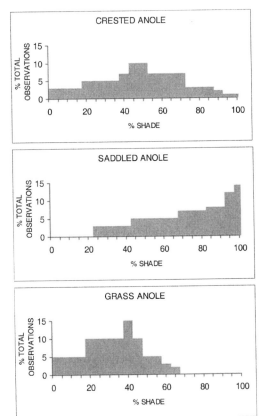

FIGURE 10. Direct sunlight on activity areas during times of anole activity on Guana Island.

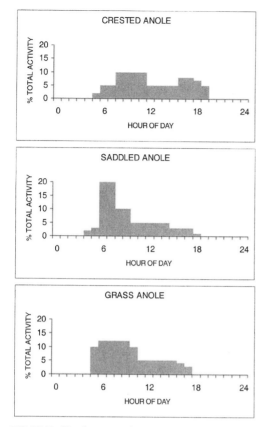

FIGURE 11. Hourly activity of anoles during a day on Guana Island.

members of this ensemble. Sexual dimorphism is a complex issue. It is striking in both size and overall color pattern in *A. cristatellus*, but both males and females have colorful throat fans and use them in display. The other two reverse this: the sexes overlap broadly in size and are not patterned distinctively, but males have big and gaudy throat fans and females almost none at all. Should we conclude that *A. cristatellus* has responded to competition between the sexes by developing size dimorphism, leading to food-resource partitioning at the intraspecific level, but the others have not? Several recent studies (e.g., Perry 1996 for *A. polylepis*) strongly suggest that sexual dimorphism is unlikely to evolve for this reason. Still, that does not say it could not be the cause in this particular case. Are there evolutionary implications of sexual dimorphism for geographic variation such as may emerge for

life-history patterns as suggested by the mice and voles mentioned above? Mice and voles show virtually no sexual dimorphism of the kind seen in anoles. Am I being overly myopic in looking at the island's three anoles, but excluding (for the moment) the other four species of insectivorous lizards?

Perhaps interactions between grass anoles and slippery-back skinks or sphaeros are more ecologically important. Certainly the skink is much rarer and the sphaero much more abundant than the grass anole. Indeed, all the anoles are common. Is differential abundance a clue to real ensemble interactions and potential competition?

On Guana Island, we have bulges of species diversity called local guilds and ensembles. I have paid most attention to the lizard ensemble (especially the three anoles), but there are clearly a bat ensemble, a shorebird ensemble, a predatory bird local guild (which includes, e.g., kestrel, red-tailed hawk, mangrove cuckoo, pearly eyed thrasher, owl, and cattle egret), and many more. Are these guilds relatively stable over time? Or are colonizations and extinctions taking place on Guana that will someday bring its fauna to a MacArthurian "equilibrium" at a diversity far lower than today's, and more in line with previous biogeographic predictions?

It seems apparent to me that looking for evidence of competition within local guilds is often futile. If evolution works—and it obviously does—then I would not expect local guild members to compete within the niche dimension used to define the local guild. J. B. Wilson (1999) quotes Dice back in 1952 saying: "All the species which are members of a given association . . . are adjusted more or less perfectly to one another." Exactly so: at least that is what I expect. I am quite surprised by the evidence of competition found by Brown and Bowers (1984), Gilpin and Diamond (1984), and especially by Pacala and Roughgarden (1982). The latter case surprises me most because I know the lizards well in the field; I named and described one of them (Lazell 1972). If a suite of species can be shown to form a local trophic guild or ensemble, such as our three anoles on the island (or, by extension, the whole insectivorous lizard guild of at least seven species), then they are behaving with respect to trophism just like a sort of giant, polymorphic species, rather like the geckos of Necker (Lazell 1995). If they are not a chance assemblage, then they stick together. If they stick together in terms of what they eat, then food is not the limiting factor in their lives. Either they specialize in portions of their general food niche (taxonomically, spatially, or temporally; see Huey and Pianka 1983) or the resource is so abundant that no competition arises, or some combination of those two situations occurs. This is very close to what Wiens (1984) is saying about "permitted" and "forbidden" species in a community. It is not far from what Connell (1980) and Roughgarden et al. (1983) call the "coevolutionary theory" or Andrewartha and Birch (1984) call the "functional environment model."

If a community is composed of many local guilds, it will be highly diverse (at least in species richness) compared to a simpler community with single or few species occupying the positions filled by those guilds. If we define our local guilds by the most obvious ecological trait characterizing their members (e.g., what they eat) then a highly diverse, strongly stable community would have most of its species organized into local guilds within which competition had evolved away to a minimum. This suggests a procedure for distinguishing between true communities and happenstance groupings. First, attempt to array the species present in local guilds; test for the reality of the guilds. Begin with the most obvious ecological niche dimension, such as food. Next, try rearranging the species into other local guilds based on less obvious, but potentially vital niche dimensions (e.g., cave dwellers, cavity nesters, pond breeders). The most stable system would be the one in which all species belonged to a rich local guild with respect to every niche dimension, but in which no two species belonged to all of the same guilds. Then, stability would be highest with greatest species diversity up to the point

where the total biomass of organisms began to reach K, the carrying capacity, and individual members of a given species were so dispersed that they nearly failed to find each other for mating. Is this a picture of the teeming, steaming tropical jungle, or what?

Instability would be caused by external, environmental changes. Competition would show up at times of stress, as indicated by Dunham (1983). Climatic changes in the tropics at sea level might be slow enough so that species turnover would only take place on a geologic time scale. In temperate and boreal regions climatic changes are probably frequent enough to make community evolution a process of constant catch-up, and chance assemblages more the rule. If the temperate-climate organisms can colonize rapidly and the climatic changes are minor, however, real community characteristics may evolve, as indicated by Jarvinen and Haila (1984), Wilbur (1997), and Lawton (2000). The most secure species in my kind of community would be the one that belonged to the richest guild with respect to every niche dimension. Indeed, Schoener (1983) cites examples where competition was least when niche-dimension overlap was greatest. Eliminating such a species would be very difficult without effecting a catastrophe upon the whole system. Invading generalists would fare poorly until divergence in behavioral or form characteristics had honed them to fit comfortably into several local guilds; they would probably not live long enough for this to happen. Invading specialists, reliant on some niche dimension available but not overexploited in the community, would predictably undergo wild population oscillations—boom and bust—to be eventually damped out only when a new balance involving adjusted local guild membership evolved, or the invader became extinct.

Br'er Mongoose *(Herpestes javanicus)* seems to fit this role in the Antilles. A voracious predator introduced into local guilds of highly evolved, noncompeting predators, it fit nowhere. It ate what everyone else ate, and then ate them too. It may have actually exterminated species and has undoubtedly extirpated populations. It is an outlaw species, outside the local guilds, comprising the stable communities. Unfortunately, it is doing rather well and has undergone significant adaptive evolution on many islands (Simberloff et al. 2000).

Ancient member specialists might be expected to fare poorly, too. As species richness in local guilds increases, there will often be a tendency for selection to favor individuals who specialize in a particular subset of the food niche (e.g., food items that are not specifically sought by other species). This is a prelude to characteristic divergence. If a species becomes, eventually, an extreme specialist in each of its guilds (i.e., along each niche dimension), its life may become precarious and it may be genuinely rare, even though in most obvious ways nothing seems to limit it. The slippery-back skink (*Mabuya sloanii*) may be just such a species: in most respects (size, diet, microhabitat, diel activity) it seems a generalist member of the insectivorous lizard ensemble, but it is relatively rare. Interestingly, it is a *K*-selected live-bearer producing few offspring. But so is its congener *M. macleani* on Carrot Rock, and it is densely abundant.

As a conservationist, I do not worry about gradual declines and eventual extinctions that seem natural and take place on a geologic timescale. These are mitigated by new speciation. By long-term stability I mean situations in which species richness and overall biomass remain pretty constant over thousands of years. I note that the natural, albeit unstable, ecosystems of temperate and boreal climes are at least resilient: their species seem adapted to survive the boom and bust. Groupings that are not evolved into communities are predictably unstable but can be expected to evolve toward stability, or at least resilience, as their species coevolve into local guilds, if they are given sufficient time. Catastrophic events that exterminate community members will predictably cause such severe perturbations that stability will be long in redeveloping (Bond 2001). Indeed, because stability is directly proportional to species richness in the community's guilds, catastrophes will decrease stability and increase time to recovery in direct proportion to the number of species they eliminate. About the worst catastrophe some pristine habitat could suffer would be the arrival of *Homo sapiens* with axes, saws, dogs, cats, rats, goats, and mongooses. It will take one hell of a strong, diverse, stable, and resilient ecosystem to withstand that. I wonder to what extent Guana has done it and to what extent I am being deluded by a picture of only the bits and pieces, the remnants of catastrophe. Steadman et al. (1984) and Wing (2001) elaborate on the solid foundation provided by Martin (1984) of evidence of massive human-caused extinctions in the recent past. These may be so pervasive as to obliterate a clear view of community processes, even on small islands of the Antilles. But I rather think not in the case of Guana: it may have escaped much human destruction because it was so small and arid that it had little to offer human colonizers and they did not stay long or destroy everything. Regarding human-caused extinction, direct research comparison to a severely degraded island like Norman will be most informative.

Consider situations in which a local guild or ensemble of several species is replaced by a single species—monospecific guild replacement. For example, there are three species of anole lizards on Guana but only one on Dominica. This would surely seem to convey potential instability. The loss of one of the three species of anole lizards on Guana would probably not be as destabilizing as the loss of the only anole on Dominica. Most likely, the single species is no less prone to extinction than is a member species of a local guild. But its loss will tend to be far more disruptive to its community, because its loss is the analogue of the loss of a whole local guild or ensemble of species in the more diverse system. Monospecific guild replacement is a regular occurrence in BVI, where the anole lizard ensemble of three species on most of the larger islands is reduced to a single species (*A. cristatellus* or its allopatric knock-off *A. ernestwilliamsi*) on the smaller islands (Lazell

1983a). Similarly, the insectivorous reptile guild of three species on most very small islands like Sombrero or Carrot Rock (Lazell 1964a; Mayer and Lazell 2000) is reduced to a single species on some bleak rocks like Watson Rock, BVI, with its sole species—a *Sphaerodactylus* gecko (Lazell 1994a). To test my notions, one might compare geographically proximate islands (or reasonably disjunct habitats) with similar geologies and geologic histories like Guana and Norman islands. We want to eliminate as many irrelevant variables as possible. Then we would set out to determine if they harbored similar species sets. If they did, we might hypothesize that these similar species sets were a true community adapted to the given geography and geology, and resulting from the same history. We could even compare species sets from islands of disparate sizes, for example, by using rarefaction. That will tell us if the smaller is a true subset of the larger.

We would agree that an assemblage constitutes a true community, rather than a chance amalgamation, if it recurs too frequently for just chance (a factor we used to identify a guild). Then we could predict that most or all species would be members of local guilds with respect to each niche dimension. *Anolis pulchellus*, for example, belongs to the anole ensemble guild primarily, the insectivorous-lizard local guild secondarily, then the heliophilic local guild energetically, the ground-foraging guild, the leaf-litter-nesting guild, the water-droplet-drinking guild, and so on. The factor *K*, carrying capacity, for this species (and for any other in the most stable communities) is not determined by competition within any one local guild. It is determined by the composite of all niche dimensions: the composite position of the species in all of its several local guilds.

For any species to approach *K* in any specific niche dimension would be disruptive to stability. That would be an indication that the "community" was really just a chance amalgamation and that the "guild" had not really evolved together locally. It would likely imbalance that guild, and in so doing threaten the community

because community stability depends on the smooth, well-evolved system of local guilds. Assemblages whose member species are not organized into coevolved guilds with respect to all niche dimensions are more prone to instability than those whose are. Communities in which a single species operates as a guild analogue are similarly prone to instability because one species is more prone to influential vicissitudes than is a local guild of several.

To determine the degree of group correspondence indicating a true community (or its lack) we need a comparative test. The Canberra-metric index of community similarity seems a good one (Nichols and Watkins 1984). Here the index, *C*, is calculated as:

$$C = 1 - \left(1/n\Sigma \frac{(X_{i_1} - X_{i_2})}{(X_{i_1} + X_{i_2})} \right),$$

where *n* is the number of species in the pool of geographically proximate communities (the metacommunity), X_{i_1} is the density of the *i*th species in the first local community, and X_{i_2} is the density of the *i*th species in the second community within the same ecosystem.

If my view of communities is true, then predator–prey relations, energetics, environmental diversity and stability, and history are all far more important than competition. Indeed, competition, in the simple, direct sense measurable along niche dimensions, would characterize intrapopulational individual interactions, on the one hand, and interspecies relations of new, chance amalgamations on the other. Anderson and Koopman (1981) looked at continental patterns and gleaned evidence that this may well be the case. Bergerud (1983), in a fascinating description of historical events on Newfoundland, shows how an ecosystem can reevolve toward new community stability after severe, human-caused disruption. Morgan Ernest and Brown (2001) and Bond (2001) discuss the roles, losses, and replacements of keystone species in communities. Jehl (1984) describes a small, arid, tropical island's disruption and readjustment. His evidence is that species turnover was nil, competition no problem, and stability quite solid until *Homo sapiens* colonized and changed everything via new species introductions and even such basic environmental components as available freshwater. Karr and Freemark (1983) note the difficulties in postdicting original communities or species sets, but their results seem to confirm my views. If my view of a community is correct we should not expect removal of one guild member to have much effect on the community, or even the local guild. There would be no population surge in the other member species, because there was no limiting competition from which they were released. One would look for effects elsewhere. In predator–prey relations, for example, loss of a species might indeed have a fairly pronounced effect on a member of some other local guild that ate, or was eaten by, or nested in the same place as the now-missing species. Lundberg et al. (2000) have generated a depressing view of what happens when communities lose species, in which restorations may be impossible (but see "Restoration," part 6, below). Similarly, loss of pollinators or seed dispersers could have catastrophic impacts on an ecosystem.

In an attempt at broad overview, J. H. Brown (1984, 1995) makes three assumptions: (1) abundance and distribution are controlled by physical and biotic factors that define the species' niche; (2) though stochastic, geographic proximity generally provides environmental similarity; and (3) closely related, ecologically similar species differ in just a few niche dimensions. All seem quite reasonable. Brown goes on to make two predictions: First, population density (abundance) should be greatest at the geographic center of a species' range, and second, within a guild, the commonest species should be the most wide ranging.

It will be fascinating to see empirical evidence brought to bear on these predictions. There is strong evidence contradicting the first: species survival is often or even usually peripheral to the center of original geographic range (Lomolino and Channell 1995; Tynan et al. 2001). Many species of sand barrier and coastal,

continental islands are a special class of exceptions to the first prediction (Lazell 1979). Some kinds of animals seem especially adapted to islands edging continents (Lazell 1979). In eastern North America these include such disparate forms as rice rats *(Oryzomys palustris)*, green snakes *(Opheodrys aestivus* and *Op. vernalis)*, mud turtles *(Kinosternon subrubrum)*, and glass lizards *(Ophisaurus ventralis)*. J. H. Brown (1984) looks for exceptions where a single environmental variable undergoes an abrupt shift or a needed habitat is patchy. These do not seem to explain the staggering abundance of some animals, like those noted above, on sand barrier and coastal islands. They are widespread forms on adjacent mainlands, with generalized and copious habitats. Sand-barrier islands are certainly very special and different sorts of ecological realms. They are both edges and ecotones in hypertrophied form.

The second prediction can be tested right in the Virgin Islands and on Guana. The three-species anole guild is widely established in lowland Puerto Rico as well as on at least 13 islands to the east of it. In these islands of Great Guania, *A. cristatellus* is by far the most widespread lizard. It is not the most abundant member of the guild on Puerto Rico—which must be considered central to the guild's range—*A. pulchellus* is. Even *A. stratulus* is more widespread (and more abundant) in the Virgin Islands (and on Guana) than *A. pulchellus*.

In my opinion, however, the relative abundance pattern we see today is highly artificial. The most abundant anole in relatively undisturbed habitats, like the island's ravine forests, is *A. stratulus*. I believe it would have dominated on the larger islands of Great Guania prior to the extensive deforestation effected by Europeans in the past five centuries. *Anolis pulchellus* would have originally been rare, confined to edges and naturally open areas on the larger islands. Its absence from small islands (see Lazell 1983a) is puzzling, because these naturally support sedge-grass communities, but is probably explained by very poor water-loss resistance. *Anolis cristatellus* was, I believe, always the most widespread form.

It not only occurs in the forest, but also it is especially abundant around island edges in open cactus-and-rock habitats. Little cays amount to nothing more than patches of "edge" in this sense, and always harbor *A. cristatellus*. *Homo sapiens* has arranged to make *A. cristatellus* vastly more common than it originally was on the bigger, forested islands by cutting down that forest. This has also aided *A. pulchellus* in many habitats where grass and graminoid plants like sugar cane now grow; sheep and goats, however, seem to limit *A. pulchellus* on small, dry islands by eating up its habitat. Nevertheless, my postdiction is that *A. stratulus* was the most abundant member of the anole guild, and it was never the most widespread.

J. H. Brown (1984) considered several evolutionary implications of his assumptions and predictions. They are pertinent to the phenomenon popularly known now as "punctuated equilibrium," which was originally described by Charles Darwin (1873). J. H. Brown (1984) hypothesizes that stabilizing selection would tend to produce long periods of little adaptive change in common species at the centers of their ranges—precisely where, because of abundance, they might be predicted to fossilize most readily. He notes that rapid evolution might be characteristic of isolated peripheral populations because they may be subject to very different types of selection pressures than are central populations. Speciation rates of peripheral isolates would likely be greatest among the most abundant and widespread species, simply because these are the ones most likely to have the greatest number of most distant peripheral populations. A peripheral isolate might evolve into a truly remarkable form. The dodo was, after all, just a remotely isolated pigeon. Because of the improbabilities of fossilization, a well-differentiated species first evolved as a peripheral isolate may not appear in the fossil record until long after its origin. Its sudden, seemingly de novo appearance in the fossil record may belie its true evolutionary history and give credence to more dramatic "punctuation" than actually occurred. This scenario will occur, for example, if *A. ernestwilliamsi* colonizes other, larger islands in the Virgin Islands, undergoes further character divergence in sympatry with its closest relative, *A. c. wileyae*, and only then—subsequent to its spurts of evolution—finds its way into the fossil record. In general, J. H. Brown's (1984, 1995) evolutionary implications of his extension of community and guild patterns seem to fit fairly well with the observed situation in the Virgin Islands.

No rational person doubts for an instant the role of competition in shaping communities. But it does just that: it shapes them. Once community coevolution has proceeded to stability, which means retention of the same species richness and biomass over long periods of time, competition ceases to function as a primary controlling mechanism. The most stable communities are the most diverse because they are characterized by having all member species also members of species-rich local guilds with respect to all niche dimensions. We are likely to find such communities mostly in the tropics at sea level, for biochemical reasons. Increasing altitude and latitude will decrease community diversity because fewer species are biochemically adapted to colder regions and environmental (climatic) vicissitudes necessitate frequent adaptive shifts, therefore causing instability. Aridity works in the same way, as does extreme heat. Artificial disruptions of tropical, near-sea-level communities may have created instability and competition in previously more diverse, more stable communities. But that may now be difficult to perceive. I finish with what I believe is Grant's (1983) most important conclusion, "it is a really difficult task in ecology to design and carry out tests that will distinguish unambiguously between alternative explanations."

Parasitism

Parasites are more of a phenomenon than I had thought! They are exceedingly numerous in species and numbers of individuals per species; some taxa having undergone the most spectacular adaptive radiations. New adaptive zones have been frequently created throughout evolutionary time and repeatedly colonized by new parasites. . . . Parasites affect the life and death of practically every other living organism.

PETER W. PRICE (1980)

PARASITES ARE LOATHSOME. They are pathological, and pathogenic. The very picture of nematodes and botfly larvae, intestinal amoebas, and elephantiasis causes one's gorge to rise. What is a chapter on parasites and parasitism doing in a nice book like this? Parasites are devastatingly important. By Price's (1980) tabulation, more than 70% of the insects in the British Isles are parasitic to one degree or another. That is just insects—no worms or germs. And that is in cool, civilized Britain—not the steaming tropics.

Parasites are fascinating. If we are going to set out to understand life on Earth, we will first have to grasp the fundaments of geology and geography and elementary chemistry; that will be difficult enough for folks like me. Next, at least an order of magnitude (more likely an exponential power) more difficult, we will have to comprehend the distribution and abundance and interactions of such mundane creatures as rabbits and mice and lizards and robins. This book is intended only to introduce one to the problems encountered at that level. But if a large percentage—probably most—of living things are parasitic, then we will have to ultimately comprehend their distributions, abundances, and interactions. For the streblid fly *Megistopoda,* the colony of fruit bats that it parasitizes on Guana Island is an archipelago on a small island in an archipelago. . . . The exponent of difficulty has greatly increased. For those who prefer to walk before they run, a sojourn into parasitism may seem quite ridiculous at this stage. I am myself still an unsteady walker in this world of population biology and theoretical ecology. However, I think I will just stagger as far as the bleachers and take a look at the track.

Price (1980) has gleaned two sets of concepts or principles about the ecology and evolution of parasites that seem to make good guidelines for avenues of study. We may think of them as hypotheses, based on the majority of observations, and seek evidence that supports or refutes them, or we may be intrigued by what seem to be peculiar exceptions. I here render Price's concepts in tabular form:

Ecological	Evolutionary
Adapted to small, discontinuous environments	Rapid evolutionary (speciation) rates
Extremely specialized resource exploitation	Extensive adaptive radiations
Existence in nonequilibrium conditions	Bizarre, not necessarily allopatric speciation

All of these concepts or principles are predictable if one considers parasites as an order of magnitude more complex ecologically and evolutionarily than their hosts. Their ecological and evolutionary processes are superimposed on the ecological and evolutionary processes governing their hosts.

An exemplary little ensemble of parasites occurs on the colony of fruit bats *(Artibeus jamaicensis)* on Guana: at least two species of the family Streblidae, of the order Diptera, class Insecta: *Megistopoda aranea* and *Trichobius intermedius.* I am indebted to Rupert Wenzel of The Field Museum for virtually everything I know about them, except my own field data, which were collected largely with the help of Lianna Jarecki. There are two families of dipteran insects called batflies because their species are largely or exclusively parasitic on bats. Nycteribiidae is a small family of 13 genera, all but two of which are confined to the Old World. One of these two genera, *Hershkovitzia,* is endemic and found only on the highly specialized bats of the genus *Thyroptera* in South America. The second genus, *Basilia,* occurs mostly on the abundant and widespread bats of the genus *Myotis;* like those bats, it occurs in both Old World and New World. The nycteribiids seem, therefore, to show a pattern consistent with fairly recent invasion of the New World and just the beginnings of adaptive radiation here (Wenzel and Tipton 1966). The Streblidae are far more complex. There are 23 New World genera; endemism is high: not even subfamilies are shared between New World and Old World. The picture is of an older, long-established set of radiations (Wenzel et al. 1966). Marshall (1981) notes streblid parasites of parrots, doves, and an opossum. Most streblids are true ectoparasites—blood suckers. One Old World genus, *Ascodipteron,* however, has become so modified as to be essentially endoparasitic. The females lose their wings and invade the host's tissues. All streblids are said to give birth to fully developed larvae, one at a time; that is about as *K*-selected as an insect can get (Marshall 1981). In a magnificent monograph, Wenzel et al. (1966) document the streblids of Panama. They note a major problem with the group: some species, like our *Trichobius intermedius,* fly well and "not only leave the host at the slightest disturbance but will often land on another nearby bat, frequently of another species." They go on to note: "Such 'disturbance transfers' are not uncommon among Streblidae of bats caught in nets. These flies often return to the original host, if time and opportunity permits. . . . While the need for the most rigorous precautions in collecting ectoparasites is well known, in the past they have rarely been observed in the case of batflies." (See figure 12.)

Webb and Loomis (1977) list 19 species of streblids known from *Artibeus jamaicensis,* our fruit bat. Most are Central American (Overal 1980). Only *Megistopoda aranea* and some species of *Trichobius* are recorded for the West Indies (both from Puerto Rico); both streblids also occur in Central America (R. Wenzel, pers. comm.). These authors list three species of nycteribiid batflies in *Basilia* on Central American *Artibeus jamaicensis,* so it is probable that nycteribiids may be found someday in the Antilles or even on Guana. Marshall (1981) describes the differences between nycteribiids and streblids, just in case we do ever need to know. Nycteribiids have their heads set back, apparently arising behind the front coxae, and capable of being folded back on the dorsum; the thorax is flat and broadly expanded; the palpi are longer than broad; and there are never wings at any stage in life. Streblidae have their heads in the normal fly position; they do not fold back; the thorax is never more than slightly flattened; the palpi are broader than long and usually project leaflike in front of the head; and they all have wings, though females of *Ascodipteron* do lose theirs.

Our second species of phyllostomid (leaf-nosed) bat is *Brachyphylla cavernarum*, the cave bat. *Brachyphylla* occur only in the West Indies. Webb and Loomis (1977) list just two species of batflies, both streblids, from them: *Nycterophilia coxata* (British West Indies) and *Trichobius truncatus* (Puerto Rico). We have found neither on Guana Island. Our third bat species, *Noctilio leporinus,* the fishing bat, is known from a single specimen; she had no apparent ectoparasites. The colony, however, should be studied.

Our fourth known bat species is *Molossus molossus* (family Molossidae), which is found on both the islands and the mainland. Wenzel et al. (1966) list few parasites of any sort for this genus. The only batfly noted, a streblid, is *Trichobius dunni* from the South American mastiff bat *Molossus bondae.* If this lack of batflies on widespread, abundant *Molossus molossus* is real, it eliminates a discernable pattern in batfly biogeography. I suspect it reflects instead a great need for additional study. Are mastiff bats, genus *Molossus,* too small to harbor large ectoparasites such as batflies? Put another way, are the habitat patches—individual mastiff bats—so small that they can support too few individual streblids to comprise viable populations? This is the sort of species : area problem that prevents big predators (e.g., bears or lions) from living on little islands.

Both *Artibeus jamaicensis* and *Brachyphylla cavernarum* occur in the same cave on Guana (Lazell and Jarecki 1985). This presents us with an opportunity to test the notion of Marshall (1981) that streblids are "roost specific," rather than host specific, parasites. Do streblids of the widespread bat *A. jamaicensis,* which supports a rich diversity of species, colonize *B. cavernarum* in the same roost? We have now examined a half-dozen living cave bats from Guana, but we have yet to catch a batfly on one. Ann Payne came to Guana Island in July 1984 and illustrated both *Megistopoda aranea* and *Trichobius intermedius* from fresh specimens subsequently

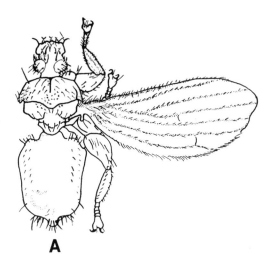

A

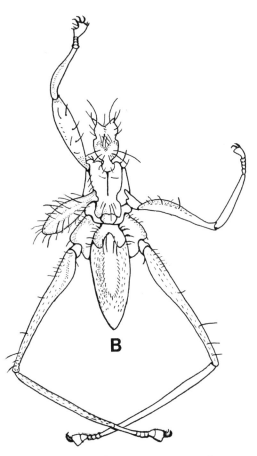

B

FIGURE 12. Streblid batflies parasitic on Guana's fruit and cave bats. (A) *Trichobius intermedius,* with all left appendages omitted. (B) *Megistopoda aranea,* with right front leg, right vestigial wing, and left middle leg omitted. By Ann Payne.

TABLE 15
Batflies on Fruit Bats

	Megistopoda aranea	Trichobius intermedius
Female with nursing young, crown-to-rump = 54 mm	—	11
Female with nursing young, crown-to-rump = 47 mm	—	8
Pregnant female (Guana)	—	—
Pregnant female (Tortola)	1	—
Female, nonreproductive	1	1
Female, nonreproductive	2	1
Female, nonreproductive	1	2
Male, old and scarred	1	—
Male, adult	1	—
Male, adult	—	—
Male, adult	—	—

SOURCE: Original data.

sent to Wenzel. On 19 July 1984, we collected some ecological data on the streblid batflies of Guana, and we added one *Megistopoda aranea* from a Tortola *Artibeus* on 20 July. Almost two decades later, on 14 October 2003, Fred Sibley, of bird and dragonfly fame, netted three more fruit bats: two females with three batflies each and a male with none. A very interesting picture emerges on *Artibeus jamaicensis* (see table 15).

Females are more heavily parasitized by *Trichobius intermedius* in our small sample, and females carrying nursing young are by far the most heavily parasitized. Male fruit bats are more likely to have only *Megistopoda aranea*, and that species is no more abundant (absent, in fact, in our sample) on females carrying young than on nonreproductive or pregnant females. If *Trichobius intermedius* fits the reproductive strategy outlined for all streblids by Marshall (1981), the females bear live larvae, which pupate on the cave walls. These would naturally infest baby bats left hanging on the walls when their mothers went out foraging (Lazell and Jarecki 1985 show that female *Artibeus* can and do sometimes carry their young in flight; we do not insist that they always or must do so). A baby bat, with its returning mother, might be a superbly rich habitat for

Trichobius and generate a small population boom. But what of *Megistopoda?* Is it simply outcompeted on female-plus-young fruit bats by the burgeoning *Trichobius?* It is far more specialized for parasitic life and cannot fly. It usually remains still and attached to the bat with its hooked feet; it may not be as capable a disperser as *Trichobius.*

I have come this far without stating just what defines a parasite; that is now important. Price (1980) states: "a parasite is an organism living in or on another living organism, obtaining from it part or all of its organic nutriment, commonly exhibiting some degree of adaptive structural modification, and causing some degree of real damage to its host." That is a pretty tight and restrictive definition. It rules out vampire bats and blood-sucking flies such as tabanids and mosquitoes and lets us humans off the hook: We do not physically live on the species we drain and enslave. The last clause interests me the most, because real damage seems to me to be an analogue of the issue of competition we have discussed above. If host and parasite coevolved long enough, I would expect natural selection to eliminate real damage. This is exactly what has happened in the case of commensalism. The ultimate adaptive nirvana is mutualism, where both species benefit.

A PARASITE

In a way, *Homo sapiens* is also parasitic. In whole cultures, many individuals have shifted the entire burden of infant food supply to a host species, *Bos taurus*, whom we milk as relentlessly as the vampire laps blood. Apples and oranges quail beneath the lash of our brutal selection. If teleologists can claim there is purpose and striving in evolutionary life, then the whole goal of these plants must be directed toward indulging our whims. The most pathetic case of human parasitism I have personally witnessed involves the once-lovely Guangdong thatch palm. Individuals of this species are permitted to live in tropical China in regulated rows. Each is allowed a couple of leaves, but as soon as a new leaf unfurls, an older one is chopped for thatch. When an individual, through incredible effort and perseverance, finally attains a height that might relieve it from the knifely attentions of its masters, its heart is ripped out for dinner and its trunk cut up for fuel. One in a thousand is allowed to reach sexual maturity and thus to fruit and provide the next generation of slaves.

With the fruit bats and their streblids on the island, I do not, with a casual inspection, see real damage. Nevertheless, I believe it must be there. The streblids live on the bats' blood, and flies must be a detriment to the bats. But the bats certainly look healthy enough, despite their streblids. There is a further problem with the streblid's ecosystem, stated by Marshall (1981): "The larva after being discharged from the vagina . . . forms a puparium thus the only food taken by the species throughout its life history is blood of the vertebrate host. It is interesting that the food factors which cannot be supplied by vertebrate blood are compensated by the activity of the micro-organisms like mycetomes present in the puparia." Ah yes, and "big fleas have little fleas. . . ." One wonders if the streblids' mycetomes are parasites, commensals, or mutualists?

A TICK TALE

In July 1984, we began moving some *Cyclura pinguis* from Anegada over to the security and protection on Guana. I picked off the ticks from the iguanas and noted that they were the boldly patterned, seemingly eyeless, sorts of reptile ticks I had seen on other iguanas before. I suspected I had a new species because I did not know of anyone who had paid attention to ticks on the rare, very restricted BVI endemic species *Cyclura pinguis* before. We sent them off to the U.S. National Museum and received the following in a letter: "Scott Miller recently gave me some ticks collected by you from *Iguana pinguis*. . . . He said that you thought that this tick collection was the first you had ever made from this species of *Iguana*. However, you were the collector of the material from which the species was described. I am including a reprint of that publication for you." The letter was signed by James Keirans of the U.S. National Tick Collection and included a reprint of Kohls' (1969) description of *Amblyomma antillarum*, a tick of the family Ixodidae (similar to *Ixodes* that vectors Lyme disease). Actually, they do have eyes that are "flat" and "inconspicuous" (Kohls 1969). It turns out that, nearly 20 years before, Michael Carey had been working on *Cyclura pinguis* of Anegada (Carey 1975) while I had been working on *Iguana delicatissima* in the Lesser Antilles (Lazell 1973). We had both collected *Amblyomma antillarum* on each species. Mine were sent to Glen Kohls, then at the Public Health Service's Rocky Mountain Laboratory, in Hamilton, Montana, by Kerv Hyland, a parasitologist then at the University of Rhode Island. I never heard what became of them, because I was hardly ever around the University of Rhode Island after completing my doctoral dissertation in 1968 until my return to Rhode Island as a resident in 1980. Kierans (1985) went on to describe the nymphal stages of *Amblyomma antillarum* and extended its range from *Cyclura pinguis* and *Iguana delicatissima* to *Cyclura carinata* of the Caicos Islands, biologically and geologically part of the Bahamas. *Amblyomma antillarum* is closely related to *Amblyomma albopictum* on Hispaniola.

TABLE 16
Helminth Parasites of Anolis Lizards

		A	B	G	N	R	T	V
1	*Parapharyngodon cubensis* (N, I)	67	50	67	83	58	55	80
2	*Porrocaecum* sp. (N, P)	33	25	67	83	8	27	90
3	*Centrorhynchus* sp (A, P)	17	13	67	25	75	36	10
4	*Spauligodon anolis* (N, I)	—	38	—	—	—	—	—
5	*Trichospirura teixeirai* (N, G)	—	25	—	8	—	27	10
6	Oligacanthorhynchidae (A, P)	—	38	—	8	—	18	60
7	*Physaloptera* sp. (N, S)	—	—	—	8	—	—	—
8	*Oochoristica maccoyi* (C, i)	—	—	—	—	—	9	—
9	*Rhabdias* sp. (T, L)	—	—	—	—	—	18	—
10	*Mesocoelium monas* (T, i)	—	—	—	—	—	—	60

SOURCE: Goldberg et al. 1998. There are two letters in parentheses after each helminth's name. The first gives its taxonomic order: N, nematode (round worm); A, acanthocephalan (spiny-headed worm); C, cestode (tapeworm); and T, trematode (flat worm). The second letter in parentheses gives the location of the parasite in the lizard: I, large intestine; i, small intestine; P, peritoneum (the internal body wall); G, gallbladder; S, stomach; and L, lungs. The islands from which the lizards came are initialed across the top of the table: A, Anegada; B, Beef Island; G, Guana Island; N, Necker Island; R, Norman Island; T, Tortola; and V, Virgin Gorda. The number in each column is the percent of lizards on the respective island carrying parasites.

"Hard ticks are parasitiform mites, obligate hematophagous ectoparasites of terrestrial vertebrates" (Beati and Keirans 2001). The genus *Amblyomma*, which is found on our iguana, was determined by Beati and Keirans (2001) in a study combining mitochondrial DNA and morphological data to be close to the base of the Metastriata, one of the two major divisions of the family Ixodidae. The genus is widespread in Africa, where these ticks parasitize large mammals such as bovines and rhinos; all four species used by Beati and Keirans (2001) are African, leaving us to wonder about the origin of American species that mostly parasitize reptiles and big amphibians such as marine toads (de la Cruz 2001). Hispaniola, de la Cruz (2001) notes, ". . . is geographically between the Bahamas and the Virgin Islands. How did the species [*Amblyomma*] *antillarum* move from one group of islands to the other without interacting with the Hispaniolan population of [*Amblyomma*] *albopictum*? It is a problem that needs to be answered in the future."

Next, there are questions of host specificity and relationships. Although de la Cruz (2001) does not explicitly say so, *Amblyomma albopictum* must occur on *Iguana iguana*, the species present in the Swan Islands (Lazell 1973). de la Cruz lists "*Iguana* sp.," as well as *Cyclura cornuta, C. nubila,* two species of smaller iguanid lizards in genus *Leiocephalus*, and two snakes (a colubrid in *Alsophis* (or *Liophis*) and the Cuban boa, *Epicrates angulifer*). It immediately occurs to me that there is a second species of iguana, *Cyclura ricordi*, on Hispaniola (Ottenwalder 2000) and three species of *Epicrates* boas (Tolson and Henderson 1993). Although none are known to support an *Amblyomma* tick (de la Cruz 2001), these should be checked before assuming that *Amblyomma antillarum* has a discontinuous range. It is notable that *I. iguana* hosts *Am. albopictum* but *I. delicatissima* hosts *Am. antillarum;* each tick is found on some *Cyclura* species. The possibility that the tick, showing host specificity, indicates a close relationship between, for example, *I. delicatissima* and *C. pinguis* is discussed under "The Iguana," Part 5, below.

Stephen Goldberg, of California's Whittier College, and colleagues (1998) checked the helminth parasites of our crested anole lizard, *A. cristatellus wileyae*, from seven BVI, including Guana. Helminths are "worms" to most of us. Goldberg et al. (1998) found them in the intestinal tract, gall bladder, liver, body cavity, and lungs. Two-thirds of Guana's crested anoles harbored up to 3 of the 10 species recorded. Up to 90% of Virgin Gorda crested anoles were parasitized (see table 16). All of the helminths found had been previously recorded in *Anolis* lizards of other species, but *Oochoristica maccoyi* had not been found in *A. cristatellus* before.

For most of these anole parasites the lizards are definitive hosts: the parasite can complete its life cycle in the lizard. For three, however, anole lizards are paratenic hosts. To quote Goldberg et al. (1998): The "... helminths occur only as immature stages and have no chance of completing their life cycles ..." in the lizard. These three are the nematode *Porrocaecum* sp., the acanthocephalans *Centrorhynchus* sp., and the oligacanthorhynchids. All of these were found encysted in the peritoneum. What does this mean? Trouble for somebody else. Those three are just waiting there, curled up in their little "... connective tissue cysts ... constructed of several layers of fibrocytes ..." (Goldberg et al. 1998) waiting for somebody to eat the lizard. Then they will burst forth and begin making their fortunes off that poor somebody. I recall from my course in parasitology, taken nearly half-a-century ago, that snakes are often loaded with acanthacephalan spiny-headed worms. An obvious next step in Guana parasitology would be to check our snakes for grown-up *Centrorhynchus,* and possibly oligacanthorhynchids so far missed in the anoles (here at least).

Parasites on *A. cristatellus* from Puerto Rico are also summarized by Goldberg et al. (1998): Nine species are known—four nematodes, three trematodes, and two acanthocephalans. Five of these occur in the BVI (from Table 16: 1, 3, 4, 9, and 10), two of them on Guana: *Parapharyngodon cubensis* and *Centrorhynchus* sp. We probably will not know the species of the latter until we locate its definitive host and thus the adult life stage. Goldberg et al. (1998) finishes up: "Because sample sizes ... have been small, more individuals will need to be examined before biogeographic patterns ... can be evaluated."

Combes (2001) has produced a highly recommended text on parasitism. In a review of another recent major text on the evolution of parasites, Hudson (2001) states that "the exciting challenge must be to produce a major synthesis of parasite epidemiology and evolutionary ecology that incorporates an understanding that stretches from host–parasite molecular interactions, through the host–parasite population dynamics and reaches to the consequences for biodiversity." Current parasite research is failing to meet even dire human medical needs (Tao and Lewis 2001; Ash 2004).

PART 4

Great Guania and the Isles of Yesteryear

The enlarged Puerto Rico, Caja de Muertos and all of the Virgin Islands, except St. Croix, were united into a single land-mass which was approximately twice the area of present-day Puerto Rico.

HAROLD HEATWOLE AND FAUSTINO MACKENZIE (1967)

Heatwole and MacKenzie (1967) were describing a great island, occupying the entire area of what we call the Puerto Rico Bank today. This great island existed a mere 25,000 years ago, at the last low stadium of the Wurm (or Wisconsinan) glacial maximum, which was the most recent of a series of glaciations—ice ages—going back a couple of million years, spanning Pleistocene time. Prior to the Pleistocene, characterized by its glacial–interglacial cycles, was the Pliocene: the last epoch of what is called the Tertiary. Geologists used to speak of "primary" time—the Paleozoic era, beginning some 600 million years ago (mya) and characterized by fossil fishes—or "secondary" time—the Mesozoic era dominated by amphibians and those huge reptiles called dinosaurs. Modern geologists no longer do so, but it is still customary to divide the Cenozoic era, the "age of mammals," into the Tertiary, beginning about 65 million years ago and lasting right up to the rather blurry divide between Pliocene and Pleistocene, and the Quaternary, which is therefore just the last couple of million years, including only the

Pleistocene and the Holocene. The Holocene begins at the end of the Pleistocene, which is the end of the Wurm glacial maximum, but even that lacks a precise date because the ice began melting between 25,000 and 20,000 years ago and is still melting today. Most people think of the Holocene as the past 20,000 to 10,000 years.

People were all over the Old World 25,000 years ago, tromping around ice-bound Europe, spearing wooly rhinos and wild cattle called aurochs, and painting their pictures on cave walls. People had crossed the then-narrow water gap to warm Australia and had spread from their ancestral home continent of Africa to the far east of Asia. At that time people would not even have to cross water to reach the Americas: the great Bering land bridge, or Beringia, solidly spanning today's watery gap between Siberia and Alaska, was dry land—a land bridge indeed. A lot of animals and plants spread across it (Lazell and Lu 2003, and references therein), but did people as long ago as 25,000 years before present (ybp)? I strongly suspect they did, but leave the arguments to the anthropologists; they certainly had by 12,000 ybp; for sure, however, people had not reached the islands of the West Indies that soon (see "Human History," part 6, below).

No people ever saw the great island documented by Heatwole and MacKenzie (1967). The concept of a rich port—Puerto Rico—goes back only 400 years, of course, and the name of today's biggest island and its bank or shelf, now in large part under water, is inappropriate for the whole glacial maximum island. I therefore refer to this island, half gone and half fragmented into over a hundred smaller islands, as Great Guania; it had an area of about 18,000 km², or 7,000 mi.²—far more than double the size of Puerto Rico, its largest piece today. Great Guania's history goes back a hundred million years into the Cretaceous, the last period of the Mesozoic era, when the Caribbean plate pushed northeastward between the North and South American plates and produced a classic volcanic subduction arc (Meyerhoff 1933; Iturralde-Vinent and MacPhee 1999; Graham 2003).

Geology

The geological history and paleogeography of the West Indies is exceedingly complex and different authors have suggested different scenarios based on the same evidence.

S. BLAIR HEDGES (2001)

I F I HAD TO DEPEND on published literature or on my own single course in geology to report on this aspect of Guana Island's natural history I would be out of luck. Fortunately Ed Olsen, then curator of geology at The Field Museum of Natural History (now professor of geophysical sciences, University of Chicago) was on Guana during part of my 1982 stay. He collected samples of rock from the top of Sugarloaf, the Pyramid, the bat caves, and North Bay outcrops. I am indebted to Dr. Olsen for his identification of these samples and his analysis of geologic processes on Guana. He sent me a typescript report on 20 June 1982. According to Olsen, the island is largely made of Upper Cretaceous igneous extrusives—rocks resulting from volcanic activity so long ago that the classic morphologies of the volcanoes are long gone. Thus, we do not see craters or calderas; there are no fumaroles or sulfur vents; and there are no basalt spires such as one sees in the Lesser Antilles proper (where many volcanoes are still active, e.g., Montserrat). Most of Guana's rocks are more than 70 million years old. Olsen identified three major kinds of rocks:

Andesite flows. This rock is related to basalt on the one hand and rhyolite on the other. It is produced directly from eruption on the surface of magma from the Benioff or seismic zone beneath the crustal plates of Earth's surface. The normal pattern is for the sedimentary shell of continental strata to be reworked by upwelling magma (Berkner and Marshall 1972; Holmes and Holmes 1978). The greater the depth of origin of the andesite flow, the greater the amounts of silicon dioxide (SiO_2) and potash (K_2O) in the andesite (Holmes and Holmes 1978, 664–65). Andesites vary from 55% to 60% SiO_2 and from <1% to 4% K_2O. Basalts are dark, rich in ferromagnesian minerals, and poor in SiO_2. Rhyolites are pale, lacking ferromagnesian minerals, and rich in SiO_2. Andesite is intermediate (Stearns 1972; Holmes and Holmes 1978). The Guana andesite contains the mineral augite, which is basically calcium, magnesium, iron, and silicon dioxide (Ca, Mg, Fe, SiO_2) with aluminum added. Augite andesites are the abundant flow rocks of the Virgin Islands and are not rich enough in aluminum or other valuable metals to be worthy of mining effort. The question of the origin of these rocks vis-à-vis plate tectonics and biogeography will be addressed below. Andesites are porous and trap groundwater (Stearns 1972).

Tuffs. These very abundant rocks are made of volcanic ash, often with bits of larger debris. They may look like conglomerate or concrete with pebbles in it. Olsen recognized two sorts: welded tuffs, in which the ash was so hot at the time of formation that it is cemented into durable rock, and agglomeratic tuffs, which are similar in origin but contain chips of andesite and other minerals. The tuffs are younger than the andesite (Rankin 2002). Helsley (1971) reported our beds of tuffaceous wacke, the several varieties of tuffs, to be about 6,000 ft. (nearly 2,000 m) thick. Stearns (1972, and references therein) reports tuffs, especially the agglomeratic ones, to be very differentially porous to water. Thus one would expect very different rates of erosion, expansion, and contraction within areas or seams of a given bed. This feature is presumably responsible for such striking geologic features of Guana Island as the undercut ledges (e.g., Guana Head) and the numerous dome caves (e.g., the bat caves). The climate of the Antillean region—indeed of the tropics in general—was far wetter at times during a period of great glaciation, such as the Wurm glaciation of about 70,000–20,000 years ago. Since then we have experienced "climatic deterioration" (Frakes 1980, and references therein): The tropics have gotten drier and, no doubt, hotter, too. Because the last great glaciation was only one of a vast succession of similar cyclic climatic phenomena, spanning Pleistocene time—roughly the past two million years— there has been ample opportunity for the caves and ledges to form. Olsen and I reject sea-level rise and wave erosion as the architect of the caves because they occur at various elevations. Some, like the Guana bat caves at about 100 m, are much too high for sea level to have risen to them. Since the last glacial maximum, sea level has never stood higher than it is today (Woodroffe et al. 1983; Lambeck and Chappell 2001). During the last interglacial, the Sanga-mon, sea level was briefly higher than it is now. Estimates vary and land does shift, but the Sangamon level was about 3 to 10 m above the level at present (Morris et al. 1977; Lambeck and Chappell 2001). It cut the edge visible as Guana's Hen and Chickens platform and other edges and ledges several meters above current sea level. During the Sangamon, coralline algae, corals, calciferous bacteria, and sand overlay areas that are now dry land. In these areas, conglomerates could form.

Conglomerates. These are true sedimentary rocks and scarce on Guana. They are made of wave-worn, rounded rocks and pebbles cemented in a sandy or limestone matrix. They have to be younger than the andesites or tuffs. They can form very rapidly (Higgins 1968; Ginsburg and James 1974). The slow way is by simple compaction of material beneath water and an overburden of less consolidated sediments. The fast way is by cementation using organic materials in the algal and bacterial cells themselves that act, essentially, as glue (Snead 1982, 44–45). In either case, the true conglomerates visible on Guana today are all relatively close to current sea level and best developed at the edges of the flat at North Bay. They were largely both created and preserved in shallow areas of high sedimentation and low turbulence. Their current position implies that North Bay was once more enclosed: a sheltered lagoon.

The great teaser posed by the geologic literature, from the viewpoint of the evolutionary biologist or biogeographer, is: When was the rock actually land? The popular pedantic term is *subaerial*—literally, under air—as opposed to submarine. Geologists can provide an incredible wealth of detail about the composition and placement of the Virgin Islands' rocks and minerals (e.g., Renken et al. 2002) but leave us high and dry and yet totally awash when it comes to that (seemingly simple) question: When was this stuff actually land—something a lizard, or a giant ground sloth, or a tree could live on?

The most up-to-date and detailed description of the geology of a relevant land area is Rankin's (2002) geology of St. John, USVI. Most of

There are several peculiar features of Guana's geology that one might wish to ponder. For example, there is a large scree of coral on the southeast side of Long Mans Point at about 100-m elevation. Because the Pleistocene sea level never stood this high, and because any traces of the last interglacial corals would surely have washed away (rainwater is always slightly acidic and coral dissolves very quickly if exposed to it), *somebody* must have put all that coral up there rather recently. I wonder who and why. My best guess is that the Parkes, back in the mid-eighteenth century, started out to slake lime from the coral, for the purpose of making cement and mortar, preparatory to erecting another building. If so, they abandoned the project in its earliest stages (see "Human History," part 6, below).

If we have explained the erosional collapses that made the bat caves (and similar features), we have not explained what emptied the caves out. Why do we today find them with relatively smooth, often earthen floors? Why are they not just filled with boulder jumbles inside? Possibly people emptied them out; people did leave artifacts and charcoal in them (see "Human History," part 6, below).

Guana, with the rocks described above, seems to fit the Louisenhoj Formation (named for a locality just north of Charlotte Amalie, St. Thomas). Reading closely, it seemed that the Louisenhoj is indeed made of strata of disparate ages, as indicated above. Rankin (2002) describes the magmatic andesite as Early to Late Cretaceous and says "most of the section is marine," which I take to mean submarine: extruded under water. The tuffs are younger—Rankin (2002) says Eocene of Tertiary time (about 50 mya)—and, characteristically, "island-arc," which I take to mean above water. The conglomerates, limestone, and sandstone are much younger. Some may be as old as Miocene of the Tertiary (about 15 mya), but some may be of Sangamon interglacial age (about 120,000

years ago) or even more recent. All these sedimentary strata were formed in water or wetlands that were, at least frequently, inundated.

My interpretations of what was above water and when, based on Rankin (2002), were pretty much guess work. I was heartened when Graham (2003) published a wonderfully detailed history of the Antilles and stated: "The oldest rocks on the Virgin Islands belong to the marine Water Island Formation of Cretaceous (Albion) age, and the principal period of emergence was in the late Eocene." There has been land available for colonization for not much more than 40 million years. However, Graham also notes that, just to the west, parts of Puerto Rico were part of "a submerged volcanic island arc and remained mostly so until the middle Eocene," adding another perhaps 10 million years to the period of potential colonization. This arc, Graham (2003) reports, "was never a continuous or near-continuous landmass; and it never connected or nearly connected the North and South American continents." If any land was available anywhere on Great Guania for colonization of terrestrial life prior to the end of the Cretaceous and Mesozoic time, it was presumably totally inundated by the vast, probably 100 m high or more, tsunami that followed the bolide impact that made the Cretaceous–Tertiary (K/T) boundary (Hedges 2001, and citations therein).

But back there a ways, say two paragraphs, I said "most of" Guana was Louisenhoj, which is about the same age as the better-known Water Island also described by Rankin (2002). There is one mysterious bit: Monkey Point. The rock of Monkey Point is black; to me, and the geologists I have shown a sample to, it looks like basalt, a close relative of andesite. But it is totally fragmented, riddled, and veined (geologists say *intruded*) with bright, flashy white quartz. The quartz veins run in all directions, merging, branching, and crisscrossing. I do not know what it is or how it got the way it is, and neither the literature nor my geologist friends have solved the mystery.

The Puerto Rico Bank is geologically dynamic. Jansma et al. (2000) describe the Puerto

Rico–northern Virgin Islands (PRVI) microplate as the easternmost of several microplates making up the diffuse boundary zone of the much larger Caribbean plate. By "northern" Virgin Islands they mean all those on the Puerto Rico Bank as distinct from the St. Croix Bank to the south: exactly the eastern remnants of Great Guania. The evidence from global positioning system (GPS) data is that the PRVI block is moving northeast with respect to the North American plate at about 11–17 mm per year eastward and about 6–8 mm per year northward. However, the rest of the Caribbean plate is moving faster and this results in the PRVI microplate having a west-southwest velocity (relative to the Caribbean plate) of about 2.5 mm per year. This results in buckling: the Puerto Rico trench, running east–west to the north of the PRVI block, is more than 8 km deep, making it the deepest trench in the Atlantic. Running east–west, south of the PRVI block, is the Muertos trough, more than 5 km deep. According to Jansma et al. (2000), the buckling and folding of the plate boundary region is apparently accommodated largely by these offshore trenches and seismic (earthquake) activity takes place mostly within them, not within the PRVI block itself. That is good news.

Things were not always thus. Jansma et al. (2000) describe two major Eocene faults running through Puerto Rico that have since been overlain and partially filled with more recent material. They also cite evidence from Cretaceous and Eocene rock magnetism that the whole PRVI block has rotated as much as 70° counterclockwise. This rotation would mean the block was initially oriented on a long axis from northwest to southeast, including the position required to fit Iturralde-Vinent and MacPhee's (1999) picture of GAARlandia (Greater Antilles-Aves Ridge landspan) in the Eocene or Oligocene. This provides more evidence that GAARlandia, if it existed, was just a set of islands (Lazell 2002b). Recent evidence is that virtually all rotation of the PRVI block ceased four or five million years ago (Jansma et al. 2000), remarkably coincident with the closure of the Isthmus of Panama in the Pliocene.

Jansma et al. (2000) discuss evidence that the PRVI block is undergoing "tectonic escape," from between the Caribbean and the North American plates, going east, "squeezed like a pumpkin seed." This scenario was proposed for several plate edge microplates or terranes by Roughgarden (1995), but was unsupported by the geologic evidence when that was carefully reexamined (Perry and Lazell 1997b, and references therein). Similarly, Jansma et al. (2000) reject this notion; they conclude that the PRVI block "is attached to the Caribbean at its eastern edge"—that is, the BVI—"precluding eastward tectonic escape." But that certainly contributes to buckling, folding, and faulting. Although all of this may not cause earthquakes with epicenters among the islands, the quakes that occur offshore can cause spectacular tsunamis—seismic sea waves—like the one that rolled ashore in the Virgin Islands in 1867 (Reid and Tabor 1920).

Sea Level and Climate

Sea level change during the Quaternary is primarily a consequence of the cyclic growth and decay of ice sheets, resulting in a complex spatial and temporal pattern.

KURT LAMBECK AND JOHN CHAPPELL (2001)

DETERMINING JUST WHERE Guana Island fits climatically in the great scheme of terrains and habitats will require more data than I have. I think I know, but I cannot be sure. Whitford (2002) provides definitions for the dry habitats of the world and citations to the relevant literature. I am pretty sure Guana falls in the semiarid zone category, but to prove it I would have to know our evapotranspiration rate, and no one has ever attempted to measure that. In terms of precipitation rates and interannual rainfall variability we fit perfectly (see data below). Whitford (2002) also discusses ways to model arid and semiarid communities: regular ecosystem modeling or the autecological, pulse-reserve picture. In the latter view, organisms (plants or animals) are too few and/or too ephemeral to modify microclimates. This is clearly not the case with Guana Island, where deep-shade, forested ravines support utterly different species than the sun-baked rock faces, and the mangrove and salt-grass wetlands support another suite of species altogether. Guana is semiarid, but interspecific interactions help determine microclimates, habitat zones, and population densities. But Guana Island has undergone dramatic changes.

Predicting the weather is notoriously difficult and, in many parts of the world, scarcely credible more than about 48 hours ahead. Postdicting weather is only marginally easier as soon as one goes back beyond written records. Sea-level changes—at least those that are eustatic and dependent on actual ocean water volume (not just the local result of an earthquake, for example)—reflect climatic change, of course, but only in a broad-brush depiction. Rapid shifts in air temperature, involving as little time as a decade, can be recorded in the chemistry of air bubbles trapped in polar ice (Shackleton 2001); however, ocean temperatures can lag centuries behind air and land warming because of the floating ice resulting from that warming (Ruddiman and McIntyre 1981) and the effects of fresh meltwater actually reversing warming trends temporarily (Clark et al. 2001). Yale's late great chronicler of glaciations Richard Foster Flint (1971) described rainy periods called pluvials contemporaneous with glacial maxima, as with the most recent: the Wurm. This made sense because high-pressure cold fronts pushing southward from the icy north would collide with warm, moisture-laden (low-pressure) air over the tropics and generate torrential storms. Obvious erosional features visible in the Virgin Islands imply that there was a very rainy period in the not-too-distant past. However, the fossil and subfossil records of reptile and bird bones indicate a much drier climate around the transition from Pleistocene to Holocene than existed during Holocene time or exists today (Pregill 1981; Pregill and Olson 1981).

A wave of extinctions corresponds at least roughly with this shift in climate; it is tempting to believe that some extinctions were caused by climate change (Woods 1990) and not solely by the humans who began invading the islands about 4,000 years before present (ybp) (Steadman et al. 1984; see "Restoration," part 6, below. Among those extinctions certainly caused by climate change and sea-level rise is that of the giant rodent *Amblyrhiza inundata* from our neighbor bank, Anguilla; it became extinct at the Sangamon interglacial, probably more than 70,000 years before the first humans came along (McFarlane et al. 1998). It is easy to understand why these bear-sized rodents failed to survive the reduction of their habitat from more than 2,500 km²—the size of Greater Anguilla at glacial maximum—to about 150 km² at Sangamon interglacial—a factor of more than 20 (McFarlane et al. 1998). A less comprehensible case is the Puerto Rico ground sloth, *Acratocnus odontrigonus,* which seems to have winked out more than 30,000 years ago, during the Wurm glaciation when Great Guania was at its areal maximum (McFarlane 1999). What happened? No humans were around to persecute this ground sloth; could its demise have been precipitated by a climate shift from pluvial (rainy) times with lush vegetation to arid conditions? So far, no bones of this sloth have been recovered from the Virgin Islands, so it might always have been restricted to the much higher, and therefore wetter, western portions of Great Guania—Puerto Rico today. It is easy to see how rising interglacial seas, for example, would adversely affect bats: Vast tracts of limestone, riddled with caverns, would be inundated (Morgan 2001); I have given this consideration under "Bats," part 5, below.

The best illustrations of sea-level rise since the Wurm glacial maximum, in the Holocene, come from limestone oceanic islands; Morris et al. (1977) provide a detailed record of a continuous rise in sea level, over the past 9,000 years, from about 24 m below its current level. There is an abrupt rise to within 2 m of the current sea level over the span of 4,000–2,000 years ago, lagging the hypsithermal maximum (a hot spell described in Lazell 1976 and many other works). Lambeck and Chappell (2001) overlap the Morris et al. (1977) record at 5,000 ybp and take it back to 22,000 ybp at Barbados, when sea level was about 130 m below today's (and Great Guania was at its fullest extent). Heatwole and MacKenzie (1967), in their mapping of the demise of Great Guania, used fathoms (each one multiplied by six to get feet and that product divided by 3.28 to get meters). Their initial depiction of Great Guania at 10,000 ybp has sea level at 21 fathoms or 38.4 m below today's, rising at 8,000 ybp to 8 fathoms or 14.6 m below today's. Their agreement with the graphs of Lambeck and Chappell (2001) and Morris et al. (1977) is impressive.

The most detailed record of prehistoric rainfall comes from oxygen isotope ratios in the shells of freshwater mollusks from lake sediments—the closest to Great Guania at Miragoane in southwestern Haiti (Curtis et al. 2001). The record revealed in that location differs in significant details from other lake sediment records in Yucatan (three), Guatemala, and coastal Venezuela (one each) also detailed by Curtis et al. (2001). The Miragoane record goes back 10,000 ybp and is the most complete of the set; I will rely on it as a proxy for the precipitation history of Great Guania, which was already fragmenting in rising seas when the record begins. Grimm et al. (1993) take the record back to 50,000 ybp, but only for Florida, where continental conditions and a more northerly latitude make extrapolation to Great Guania risky. They do, at least, confirm the view that the late Pleistocene was much drier and as much as 8°C cooler than today.

Precipitation increased dramatically to a peak about 8,000 ybp, when sea level was about

20 m below its current level. There followed a dry spell, with precipitation falling off to today's levels for about 800 years, then increasing to set the highest Holocene record about 7,200 ybp (ca. 8,200 radiocarbon ybp). There was another drop to today's levels at about 6,500 ybp (ca. 7,500 radiocarbon ybp). Then rainfall increased, and continued to be high, right through the hypsithermal maximum, 6,000–4,000 years ago. The dramatic rise in sea level that brought the ocean up to very close to today's levels, ca. 4,000 ybp, was not complemented by changes in precipitation: Conditions remained much wetter than they are now. About 1,680 ybp, a 500-year spell of rainfall (similar to ours today), a relative drought, began. Following the 500-year drought, rainfall increased sharply after 1,200 ybp and peaked in a brief pluvial period centered about 1,000 ybp with conditions as wet as they were during the long Holocene span from 6,000 to 1,680 ybp—and much wetter than now. Since that time the climate has dried down to what we live with today.

Sea level is rising in response to global warming but there has long been real trouble in how to accurately measure it (Church 2001; Kerr 2001). Recent satellite technology was used by Cabanes et al. (2001) to improve on tide gauge data. For the period 1993–98, they found an average rise of 3.2 ± 0.2 mm/year—about one-thirteenth of an inch. At this rate, sea level will come up a meter in 312.5 years (a foot in close to a century). For the broader time span of 1978–2002, Vinnikov and Grady (2003) found both satellite and surface temperature data consistently indicate a warming of about 1°C in 40 years. This corresponds to a sea-level rise of about 128 mm, or 5 in., in those four decades. Very few people—and essentially none in positions of power and influence—seem concerned. However, the rate of increase is certainly increasing, as indicated by the data of Cabanes et al. (2001). The results of just a few centimeters of rise on coastal areas of tropical islands in cyclonic storm prone areas—like Guana Island in the Caribbean—can be catastrophic: We could easily lose much of White Bay and North Bay sand flats with their critically important freshwater lenses, not to mention the beloved Salt Pond and the endangered flamingoes in it. Sea-level rise seems to derive mostly from thermal expansion of warming water, but polar ice melt and the melting of glaciers also contribute (Alverson et al. 2001).

One insidious effect of sea-level rise is the increase in hot, equatorial surface water passing between Asia and Australia westerly from the Pacific Ocean into the Indian Ocean. Rising Indian Ocean temperatures correlate directly to decreasing rainfall and increasing desertification of North Africa (Kerr 2003b). This, in turn, produces the African dust clouds that can extend as far west again as the Caribbean—and Guana Island—as described under "Wind," part 4, below.

I can make excellent correlations of sea level and temperature, given that sea-level rise lags temperature rise, in Holocene time. I can see no clear relationship between either sea level or temperature to precipitation, as precipitation is revealed at Lake Miragoane, Haiti (Curtis et al. 2001). I can make good, clear correlations between rainfall and physiography in the Antilles and have done so (Lazell 1972).

Wind

AS A GENERAL RULE, the wind in the West Indies blows from the northeast. This is the trade wind, so named because it pushed the square-riggers before it across the North Atlantic and Pacific within the tropics. The trade wind derives from the minuscule inertia of the air mass surrounding a spinning Earth. If the air mass stood still, Earth—rotating about a thousand miles an hour eastward at the equator—would be almost uninhabitable: the trade wind would come at us at that thousand miles an hour. Fortunately, space above the atmosphere offers little resistance, so inertia is slight. It provides the easterly component of the trades. Brief deviations from northeast to east, or even southeast, are occasional responses to high- or low-pressure fields associated with the North American or South American continents.

The northerly component of the typical trades results from the convection cycle generated by hot air rising at the equator. There is, in fact, so much upward vertical motion to equatorial air that this even overrides the easterly air flow in a narrow band—the equatorial doldrums; these seem windless because there is so little lateral air motion. All that rising hot air cools as it gets up in the atmosphere; it rolls outward, northward in the northern hemisphere—

making a south wind if you were high enough up to feel it (winds are named for the direction they come from, not the direction they go). Now cold, that convected equatorial air descends again to Earth's surface; this happens at about 30°N in the northern hemisphere—about the latitude of Jacksonville, Florida. The effect of this descending cool air is similar to that of the rising hot, equatorial air: It cancels out lateral motion. Thus, a zone of several degrees around 30°N is called the "horse latitudes," because when sailing ships became becalmed in it they often had to jettison heavy, thirsty consumers like horses before they got moving again. The effects of the northeast trades on the vegetation of Guana and all the Virgin Islands are obvious to see, especially on windward slopes: The vegetation looks literally brushed southwestward, the result of wind and windblown salt-spray pruning.

Since 1970, the increasingly severe droughts in North Africa have meant that the easterly winds coming off that continent have borne quantities of dust that is deposited in the Caribbean (Prospero and Lamb 2003). The effect of this dust on vegetation, reefs, and ecosystems in general is currently unknown but certainly severe.

Rain

RAIN-BEARING CUMULUS and stratocumulus clouds are generated by direct evaporation from the tropical sea. These clouds, under the control of the fairly constant temperature and pressure regimes of the open ocean, tend to lie at about 2,000 ft.—a bit over 600 m—elevation; they are carried west or southwest by the wind until something (e.g., a land mass) disrupts them. I recognized three distinct land-mass configurations in the Antilles that disrupt or impede cloud passage (Lazell 1972): barrier, overspill, and snag islands. The term *barrier* was perhaps an unfortunate choice because it is also used for very low, sand, coral, or other sediment islands that flank gently sloping coastlines and result in lagoons, bays, or sounds that are relatively protected from the occasional violence of the ocean. A cloud-barrier island, on the other hand, effectively stops the windblown traverse of rain-bearing clouds; its mountains are high enough and continuous enough to fully impede cloud passage. The result is heavy precipitation in the mountains and in a zone to windward, even if the terrain is low, combined with a rain-shadow arid zone in the mountains' lee that receives little precipitation ever. The only remaining island of Great Guania that qualifies as a precipitation barrier today is Puerto Rico: its mountains har-

bor remnant rain forest; its northeastern slopes are well watered; and its southwest, the Guanica region, is strikingly arid and xeric. None of remaining fragments even qualify as overspill islands because to do so they would need to not only have peaks high enough to stop cloud passage regularly, but also spaced far enough apart to allow the resulting precipitation to fall between them, leeward of the peaks.

The larger Virgin Island remnants of Great Guania are at most snag islands: They have peaks high enough to slow clouds down but not stop them. Sage Mountain on Tortola is the highest peak on the Puerto Rico Bank, once Great Guania, east of Puerto Rico. It is 521 m or 1,709 ft. high. As part of Great Guania at glacial maximum, 20,000 years ago, it would have been at least 120 m higher relative to sea level, which was 120 m lower; at that elevation most rain-bearing clouds that passed Sage Mountain's way would have been impeded and thus would cause a real, if modest, cloud forest. Today, at a mere 521 m, few clouds get snagged by Sage Mountain. Enough do, however, to produce a distinctive "aridulate rain forest" remnant characterized by D'Arcy (1967). Clouds can potentially be considerably lower—right down to sea level in the case of ground fog, but that is more frequent at higher latitudes (as in

New England). In the days of Great Guania, the Sugarloaf—high point on today's Guana Island at 246 m—would have been at least 366 m; would it therefore have been significantly moister? Probably not, actually, because during the late Pleistocene the general climate was much drier. Tortola, with Sage Mountain, would have caught what clouds there were, and been wetter, I believe, but the elevational gain for Guana would have been insufficient to offset the drier climate.

Even today, at a mere 246 m, the Sugarloaf snags the occasional cloud. Enough gets snagged, indeed, to generate a tiny area—no more than a half-dozen hectares—around the peak where *Pepperomia* grows: a rare plant in dry country (see "Flora Guanae," part 5, below).

Rainfall over the period for which we have records averages somewhat more than that on the open ocean, which is about 840 mm, or 33 in., per year. The first four months of most any year are predictably very dry; many trees and shrubs, notably the conspicuous seagrapes, lose their leaves. With May, showers will usually pick up, but June and July can be fairly dry again. With August and hurricane season, we can get a deluge. This possibility often augments the otherwise increasing general propensity for showers and thunderstorms that peaks in September, October, and November, and dwindles again in December.

Gad Perry has collected all the rainfall data for the BVI that he can find and is working it up for publication. He has provided a summary for me to include here. Records at Road Town, Tortola, go back to 1901, but are typically only annual totals. The average for one whole century was 110.3 cm/year (about 44 in.). The driest year was 1973 with only 78.7 cm (31.6 in.) and the wettest was 1933 with 239.4 cm (95.8 in.). The best data set with monthly records was made by Rowan Roy at Hodges Creek (like Road Town, it is on

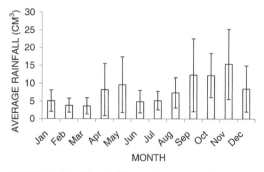

FIGURE 13. Monthly rainfall at Hodges Bay, Tortola. Vertical bars, averages; whiskers, standard deviations. Data from Rowan Roy.

the south coast of Tortola) from 1974 through 1996. For this period, the average was lower: 96.7 cm/year (38.7 in.). (See figure 13.)

We have some data for Guana, primarily 1979–84 and 1990–95—thus overlapped by Roy's Tortola data. Guana got less rain than Hodges Creek, Tortola: an average of 92.1 cm/year (36.8 in.), but the monthly pattern is similar. We installed a new rain gauge in October 2001 and have high hopes for better data now. There is also local variation within islands. Tortola is big enough and high enough to snag clouds, as noted above, so Sage Mountain should be much rainier than East End. On Guana one can sometimes actually see it raining on the Peak and much of the main mass of the island while the Hotel site and northern peninsula remain dry. There is enough within-island variation on Guana in moisture to allow some plants, like two species of piperaceous *Peperomia* and the orchid *Epidendron ciliare,* to grow at the Peak, but not naturally elsewhere (see "Flora Guanae," part 5, below). At least two species of lizards, the crested anole *(A.c. wileyae)* and the ground gecko or sphaero *(Sphaerodactylus m. macrolepis)* show geographic variation in water-loss resistance concordant with this moisture difference pattern (Perry et al. in preparation).

Land Bridges, Land Barges, and Dispersal

Discussions of West Indian biogeography have been preoccupied with the influence of long-term geological processes, such as plate tectonics and land emergence/submergence, on faunal colonization and extinction in the Caribbean islands. . . . Biogeographers have proposed several models that invoke past land bridges, cross-water dispersal, and vicariance to explain the initial arrival of insular animal taxa and their contemporary distributions on geographically isolated islands.

JASON CURTIS, MARK BRENNER, AND DAVID HODELL (2001)

THERE IS NO QUICKER way to the heart of the fray of theoretical ecology than biogeography, that is, the consideration of the geographic distributions of living things. Charles Darwin got his start as a practical biogeographer. So did Alfred Russel Wallace, Darwin's coauthor of the first evolution publication. And Hutchinson's question is a biogeographic one: for us, why are there so many different kinds of animals on Guana Island? The question has three broad facets. The first, as we have seen, Darwin handily answered. Evolution by means of natural selection can generate vast numbers of species, given diverse selection pressures, ample opportunity for spatial isolation, and time. Nothing could be better for species production, called *radiation,* than an archipelago of islands in a world where sea levels fluctuate. The West Indies in general, Great Guania in particular, and the British Virgin Islands in microcosm are great species producers. The second facet, what sustains diversity, is Hutchinson's realm. I have tried to approach it in the preceding section of this book. The third facet is the realm of biogeographers: How did they get there? Many theories of biogeography, particularly of islands, have been developed in recent years. Endler (1982a, 1982b) presents succinct discussions of the major ones that (as a biogeographer) I feel I can further refine into five main threads of thought. Threads, of course, because they will interweave.

1. *The deterministic view.* This picture of the distribution of life argues that the kinds and numbers of species in any given place (island) are determined by a set of physical parameters such as length of time of isolation, distance to nearest neighbor, latitude, elevation above sea level or depth below, and, especially, area. MacArthur and Wilson (1967) were the principal proponents of this view, which they built by culling data from Darlington (1957), who had noticed rough trends in species numbers relative to island areas. This view instantly became immensely popular because—if true—it would enable us to understand biogeography in terms of

simple, measurable, physical (as opposed to biological) things: time, distance, size. It soon became apparent that time, distance, and most spatial parameters did not matter as much as area, which primarily determines species numbers—or so it was argued. Species are highly prone to extinction on islands: just look at the historical record since the poor dodo. Similarly, species rapidly colonize islands and begin evolution toward new species. Undeniably true. So, theoretically, the longer an island (or swamp, or woodlot, or mountain) is isolated, and no matter how great its distance of isolation, sooner or later it will come to have the "right" number of species for each group of living things as determined by its area. An equilibrium between extinction and colonization will eventually develop, and the equilibrium number of species is a function of area.

2. *The biological view.* Within a few months of MacArthur and Wilson's publication (1967) many seasoned field biologists—infuriated by this simplistic view of life—rallied to show, in case after case, that the theory was not applicable. None did so better than David Lack (1976), the ornithologist who made his name studying Darwin's Galapagos finches. MacArthur and Wilson drew heavily on data for birds, reptiles, and amphibians, especially from the West Indies. Lack dealt with the land birds of Jamaica. Sauer (1969) dealt with Pacific birds. I (Lazell 1976) dealt with reptiles and amphibians on New England islands, and subsequently (Lazell 1983a) the same two groups in the British Virgin Islands. The viewpoint developed was that the presence of a particular species in a given place has to do with the special features of that species (i.e., ability to stand salinity, or cold, or dry spells), the ecological relationships of that species to others it encounters (i.e., who it eats, who eats it, who competes with it), and vast amounts of highly chancy history. This view is repugnantly unpopular because it seems to say, in effect, that theories are worthless; only empirical observations will do. No two cases are the same. Everything imaginable matters, or might. At the very least, to explain lizard distributions in the West Indies, I need an annotated timetable, with charts, for all hurricanes during the Wurm glacial maximum, ca. 70,000–20,000 years ago, when sea level was at least 120 m lower, land areas were much larger, and distances were very short. Connor and McCoy (1979) dissected the math and found that log A (for area) explained no more than half the variation in log S (for number of species): not wonderful.

3. *The classic view.* This is the view promoted by the great biogeographers of the early twentieth century, such as Simpson (1940, 1943, 1950, 1965), Darlington (1957), and Romer (1966). Darwin would have felt right at home with this school of thought. These were men with incredible knowledge of animal life (vertebrate and invertebrate), geology, and the fossil record. They acknowledged great changes on Earth: glacials and interglacials; mountain building revolutions and trench-making folds and faults; and the emergence and submergence of land areas. They believed that some land areas, such as Siberia and Alaska, would be broadly united by dry land at some times in history, but that other land areas, such as the Virgin Islands and Florida, could never have been connected by dry land. They explained the distribution of life on Earth by the origin of new, well-adapted species succeeding numerically and territorially, radiating into novel genera and families, dispersing and displacing competitors, and, in turn, giving way to newer, even better-adapted forms. Of course, because things change, those well adapted today may be hopelessly maladapted tomorrow, just by failing to change. And, all sorts of seemingly maladapted oddities might survive on islands, at the tips of long peninsulas, or up in some remote mountain range, just because their novel competitors failed to reach and displace them.

Proponents of this view would usually find it easy to distinguish between a truly oceanic island's fauna and that of an island formerly connected to the mainland by a land bridge. The former would tend to have fewer species, mostly of groups that colonize over and across water well, and would be apt to have radiations of closely related species, occupying diverse niches, paralleling, but obviously just analogous to, the mainland situation of amalgamations of distantly related species. Any area open to colonization across dry land—a land-bridge island—would tend to have a broader representation of mainland species, including some, such as primary freshwater fishes, salamanders, ungulates, carnivores, and primates, that are very poor at crossing seawater; radiations of close relatives are unlikely here.

The Galapagos Islands are, in the finest classical sense, oceanic. Their animals seem derived from a small number of single colonization events: a finch, a tortoise, an iguana, etc. These few colonizers, in a virtual competitive vacuum, underwent radiations producing an assortment of species occupying diverse ecological niches. There are ungulate-like tortoises, iguanid lizards in both the big herbivore and the small carnivore niches, and congeneric birds that are analogues of parakeets, warblers, and woodpeckers. The northeastern coastal islands off New England—Block Island, Martha's Vineyard, Nantucket, and a dozen smaller neighbors—are classic land-bridge islands. They have salamanders, deer, foxes, raccoons, etc., and previously had wolves and bears in some cases. There has been some differentiation of populations since isolation, but no great radiations.

4. *The bridge builders.* Of course, land bridges have been and are very real. The Panama land bridge has kept North America and South America united for some millions of years now. Beringia, the Bering land bridge, connects Asia and North America every time sea level drops during a glacial period. In general, all land areas separated by seas shallower than about 120 m become continuous with each other at a glacial maximum. Thus, the Virgin Islands—except St. Croix—are at times one with Puerto Rico, making Great Guania, but that large island cannot, by this method, connect to Hispaniola, St. Croix, or even Mona: The water is too deep. The dedicated bridge builders, however, are far from satisfied with such readily demonstrable land continuities. They argue for major tectonic events in the earth's crust that have produced dry land over great areas of today's deep water. They may claim that the presence of a single species on a remote island is evidence of a previous land bridge. In the West Indies the geologist Schuchert (1935) was a major architect of grand land bridges of the past. Two biologists, Thomas Barbour of Harvard (a bridge builder) and W. D. Matthew of the American Museum, New York (a classicist), developed a lengthy polemic over just what the animals of the Antilles do indicate about the geologic history of the area. A summary work (Matthew 1939) includes the major relevant papers of both. Matthew seemed clearly to have won, but he could not have known what would pop up next.

5. *The proponents of continental drift.* In the past few decades continental drift has enjoyed a vast resurgence in popularity as an explanation for the distributions of plants and animals. No one doubts that continental drift has occurred, but the timing is crucial. Hurlbut (1976) provides an early view, which based the continental unity on fossils of a reptile, *Mesosaurus,* which flourished about 250 million years ago. *Mesosaurus* represented an order of primitive aquatic reptiles attaining lengths of about a meter (Romer 1966, 116–17). Given the fact that much smaller, highly terrestrial lizards, such as some genera of skinks and geckos, as well as much larger, but less aquatic, crocodiles, occur in both the Old World and the New World tropics today, some of us do not think *Mesosaurus* a very good indicator organism. Believing congeneric species to be evidence of land continuity calls into question the existence of the Atlantic Ocean today. However, I have seen it and crossed it and genuinely believe it to be there. Pielou (1979) provides a good early account of the disparate views of the drifters and the land-barge proponents. The notions of continental drift became important in the Antillean–Caribbean region with the theories of Donn Rosen (1975) at the American Museum. He suggested that the islands of the West Indies are fragments of a Central American landmass that

broke up just a few million years ago, while drifting eastward. These chunks of the tropical portion of the North American continent, properly called terranes, functioned as loaded barges full of living organisms.

Rosen envisioned his land barges (my term: Lazell 1987b) carrying with them the mammals, reptiles, amphibians, birds, and even inshore fishes that live in the West Indies today. This view maintains that the animals did not disperse to the islands, but rather the islands—carrying animals—dispersed to where we find them today. Rosen used exactly the same evidence for his theory that the bridge builders used and that the classicists used to defeat the bridge builders. Whenever there was no evidence to support Rosen's view, which was most of the time, he merely pointed out that this lack reflected the lethargy and ineptitude of those of us who should be out in the field collecting more evidence. That worked: An astounding amount of paleontological and phylogenetic information has been garnered since Rosen (1975). Rosen was an ichthyologist and one of his strongest refutations came at the hands of his fellow fish man John Briggs (1984): There are no primary freshwater fishes native to the West Indies. Indeed, "... the Antilles should continue to be regarded as oceanic islands which have gradually accumulated their terrestrial and freshwater biota by means of overseas dispersal" (Briggs 1984). Not that that ended the argument.

I shall not be able to keep these five threads from weaving into complex patterns in what follows. All classicists believe in land bridges, but they were the very people who defeated the most ardent bridge builders. Many proponents of continental drift acknowledge the reality of dispersal along land bridges, continental corridors, and even the "sweepstakes" route—waif dispersal across water. No one doubts some sort of species : spatial correlations. Large vessels— big in surface area and/or deep in volume— hold more than small vessels, if you have a method for filling them up.

We are looking at a small island undeniably connected several times in the past to larger landmasses. Yet it owes some of its fauna to certain sweepstakes winners. The ultimate origin of even its land-bridge colonizers is hotly contested by proponents of overwater waif dispersal on the one hand and land bridges on the other. It is an island that seems to have too many species and to be far from the right hypothetical equilibrium number. Its area is small, but its elevation is rather great for its size. This elevation seems to matter, but only in a confusing way, because its diversity of species consists of lowlanders.

In my one geology course of 1952–53, I learned that the planets were originally molten globs flung from the spinning sun. Spinning themselves, they quickly took on near-spherical form (I delighted in knowing what oblate spheroid meant). Their heavier mineral constituents gravitated to their centers; the lighter stuff formed a crust. In Earth's history, this story went, something ripped off one whole half, or more, of that crust before it had solidified. It may have been the glancing blow of an asteroid. Anyway, that vast volume of crust was still liquid enough to spin into a near-sphere itself and became our Moon.

The heat loss resulting from this volumetric loss was great. The remaining one-half, or less, of the original crust fragmented into protocontinents and began to drift. I learned about Wegener's theory right then and there. Cooling resulted too from this fragmentation, so the plates of original crust that had moved rapidly at first to fill the great crustal gap began to slow down. The eventual precipitation of liquid water onto Earth's surface brought on further, rapid cooling and slowing of the land barges, which were quite empty of life. Although the magma just below Earth's extremely thin crust remains fluid today, the crust is solid enough—and cool enough—to support life (you might question this on Guana in August). My childhood notion was that land-barge movement had slowed so much since the origin of life that the continents had never drifted into complete and proper balance; the Pacific Ocean remained as the great gap from which Earth had lost its Moon.

Ah, for the simple joys of childhood. With what is known or disputed today, no one educated now will be taught this pretty picture, but the probable truth is not all that radically different (Wiechert et al. 2001). Apart from origins, however, the salient distinction between my old abiotic land barges and those envisioned by the geologists of today is not so much in configurations, continuities, fragmentations, and paths of drift (although the details of those have changed); it is mostly in the timing. Timing makes all the difference. We like to think of geology as a child of physics and chemistry—a physical science. We like to think of the physical sciences as exact and rigorously quantitative disciplines where things with firm edges are weighed and measured. The geology I have read, however, in the past few years concerning continental drift is nothing short of chaotic. Continental drift may be an accepted fact today, but beneath the sheen of that surface slick lies a turbulent and tumultuous sea of intellectual and academic discord. How can one reconcile continental accretion with continental drift as both are described in a few pages in Hurlbut (1976, 130–34)? Windley (1977) provided an early synthetic volume, but I cannot reconcile his paleomagnetic data (p. 148) with his maps (pp. 150–51). For just the Caribbean region alone, Hedges (1982) tabulates six different peer-reviewed published models and concludes virtually nothing can be surely stated about historical biogeography in view of these disparate scenarios.

Vicariance biogeography rose like the Phoenix from this intellectual pyre. As noted above, Rosen (1975) proposed the first model for the Caribbean, in which a hypothetical Central American landmass drifted eastward and fragmented to form the Antilles. Rosen used the same evidence of the land-bridge proponents and the over-water-waif dispersal proponents—a wonderful example of intellectual acrobatics. MacFadden (1980) marshals evidence to support the land-barge theory for the distribution of some Antillean mammals. Jones et al. (1983) argued for spectacular voyages of small terranes. These terranes were postulated to have drifted in many directions across the Pacific region, traversing thousands of kilometers in the past 200 million years and becoming major portions of Asia and the Pacific edge of North America. They definitely carried living things, and Jones et al. (1983) use fossils to identify them. Sykes et al. (1982) provided a detailed map and description of a terrane, or microplate—the Caribbean plate. They envisioned it wedging between North America and South America, carrying (or pushing) all of the Antilles, except Cuba, into the Caribbean at approximately 40 million ybp.

Contemporaneously, however, biochemical methods of determining the ages of living lineages were being developed, and the characteristics and causes of the dramatic paleontological shift from the Mesozoic's Cretaceous to the Cenozoic's Tertiary—known as the K/T boundary—were being elucidated. In a decade, Hedges (1996a, 1996b) could argue with strength that Briggs (1984) was right: Lineages had arrived on the islands at different times. Further, if the K/T boundary was the result of a bolide impact off Yucatan, probably very little life in the Antilles survived the resultant tsunami: all the terrane drift and vicariance biogeography in the world more than 65 million ybp meant nothing for today's Caribbean.

After its Roman Candle start, vicariance biogeography has fared poorly. Groups once touted as grand exemplars of vicariance biogeography, like chameleons, turn out to fit only an overwater dispersal scenario (Raxworthy et al. 2002; Rieppel 2002). In a tour de force, the late Arthur Meyerhoff—author in 1933 of the *Geology of Puerto Rico*—and three coauthors have demolished the dream of fitting paleontology with geologic reconstructions of plate tectonics, at least without major surgery (Meyerhoff et al. 1996). Scotese (1997) and Smith et al. (1994a) have provided a new picture, obliterating the old Laurasia versus Gondwana, Tethys Sea split of Mesozoic times, making an initial separation of Eurasia from North America, and with the latter departing Pangaea with Gondwanaland; western Eurasia—Europe—scarcely exists in this new scenario, being reduced to a few islands. All that has little to do with the Caribbean picture except that it constrains the time available for a Caribbean plate to wedge between North America and South America prior to the K/T boundary. As for what happens after that, the contest continues. Iturralde-Vinent and MacPhee (1999) have developed a deeply researched and richly documented picture of a grand "landspan"—either an old-fashioned land bridge à la Schuchert (1935) or a series of stepping-stone islands—from South America to Great Guania and beyond, all the way to Cuba, in Tertiary time, some 35–32 million ybp.

The Iturralde-Vinent and MacPhee (1999) landspan is called GAARlandia: the Greater Antilles plus Aves Ridge. The authors depicted it as continuous dry land from the Guajira region of Venezuela eastward to Tobago and northern Trinidad, with the crucial northern prong extending from just west of Tobago to today's tiny atoll of Aves, to and including St. Croix, Great Guania, at least northern Hispaniola, and eastern Cuba. A landspan is distinguished by these authors from a land bridge by two features: First, it does not connect two continents, but rather one continent to a previously wholly disjunct oceanic island or archipelago; and second, it may have been broken by narrow water gaps and not have been continuous land. Only the second serves to distinguish a landspan from Schuchert's (1935) concept of a land bridge and effectively saves Iturralde-Vinent and MacPhee from facing the same destruction of their theory as was dished out to Barbour by Matthew (1939).

The elegant GAARlandia landspan hypothesis may account for several lineages reaching the Antilles. Sloths are first known from the Oligocene of Puerto Rico (J. White and MacPhee 2001), so their geographic position and timing are perfect. Their phylogenetic position is good too, with respect to the small, arboreal, living sloths—both two-toed and three-toed—of South America today; however, the phylogenetic picture is muddied by the fact that these small, living sloths seem to have arisen from within the Antillean radiation (J. White and MacPhee

2001). Did they use GAARlandia to go the other way, from the Antilles to South America? The timing is good too for toads of the genus *Bufo;* these were widespread on Great Guania as *Bufo lemur,* and survived in the BVI on Virgin Gorda into the 1960s (Lazell 1995). Our own *Cyclura pinguis* is thought to be ancestral to the Greater Antillean iguana radiation, so it is properly positioned geographically. The timing might work for *Ameiva,* the ground lizards (Hass et al. 2001), but they seem to go back 38 to 36 million ybp—maybe a couple of million years too far. Our little blind snakes, *Typhlops richardi,* may go back about 58 million ybp and certainly need no landspan—even stepping-stone islands—to colonize oceanic islands: They are masters of waif dispersal. Our xenodontine colubird snakes, both members of *Liophis sensu lato,* belong to a very young radiation whose Antillean ancestors apparently reached the islands about 13 million ybp (Hass et al. 2001).

The biogeography of this island—small and in the Caribbean—is no rose garden. It is a jungle. Biogeography in general is a jungle of conflicting intellectual viewpoints and theories, conflicting academic disciplines, and conflicting emotions and vested interests. From where I stand, the controversies and strategies that led to the description of DNA and the resultant revolution in molecular biology seem like tame stuff. If you really want to get into the thick of it, begin with the simplest and first question most people ask about animals on islands: "How did they get there?" Once that is answered, add the Hutchinsonian twist: "And what enables them to stay?" To quote one of my greatest mentors, G. G. Simpson (1964): "The most frequent operations in historical science are not based on the observation of causal sequences—events—but on the observation of results. From those results an attempt is made to infer previous causes."

Dispersal is one of the most widely misunderstood phenomena in biology. First, as pointed out by Pielou (1979), one must not confuse dispersal, a motile process, with dispersion, the arrangement of things in the environment.

Pielou is perfectly correct, however, in maintaining that the two words are synonyms in ordinary usage, and so defined in standard lexicons. I abhor the ecologists' habit of giving special meanings to ordinary words (such as "play"—see Lazell and Spitzer 1977). Nevertheless, it seems most valuable to insist on this distinction (or, groan, invent a new term; ecology is that science in which a spade is called a geotome). The far more serious problems arise when people try to envision the motile process. Many lay people believe, for example, that when a habitat is severely disturbed or destroyed the animals move "elsewhere." That is almost never the case. Most of those rendered homeless by disturbance and destruction are killed outright or in proximate flight from the bulldozers, fires, axes, rising water, or whatever else is doing the destroying. Those that do move either enter inhospitable, uninhabitable terrain and die there, or enter habitats held as home ranges or territories by other members of their species. To survive in occupied terrain one must generally kill the incumbent. This is very difficult, because "home field advantage" is a big thing in nature (I.C.W. Hardy and Kemp 2001). As a general rule, animals do not move—at least not successfully. They die.

What happens if people move the animals? The same thing: Mostly, they die. Usually they die trying to get home. There are a number of charitable nonprofit corporations ostensibly in the wildlife conservation business that promulgate animal relocation or transport and release. A few of the more poignant cases should serve to sway people toward understanding the risk to these animals. Giger (1973) reported on displaced moles in Oregon. One of these small, nearly blind animals spent 29 days traveling back home. It had to cross a canal and other man-made obstacles. It got there, but Giger reports the others were "less successful," and we know what that means. Sunquist and Montgomery (1973) tracked a translocated silky anteater. It moved an average of 43 m a day toward home: a noble, but pathetic effort. Henshaw and Stephenson (1974) tracked five

I worked for an organization for over a decade that lent out live traps to persons who wished to "relocate" nuisance animals such as squirrels, raccoons, and woodchucks. In attempting to get them to abandon this cruel policy, I accumulated an inch stack of 3 × 5 in. card references to animal homing and the documented case histories of relocated animals. My efforts failed because I was not dealing with administrators who were either educable or interested in wildlife; they were interested in money, and catering to the whims of an ignorant public is a far quicker way of getting money than attempting to do what is right. I will not here burden you with all those references (and I have not tried to keep my file up to date since leaving that organization).

young wolves that had been relocated for their own good. All left their relocation site and struck out for home—or at least doggedly in a direction they seemed to think would take them home. None made it; they just kept trying until they died. For a similar sad case involving wolves, see Fritts et al. (1984). Schemnitz (1975) reported on white-tailed deer relocated to an island judged ideal for them by game biologists (and if there is one thing game biologists do know about it is what white-tails need and love). All left their island paradise swimming; a few reached the mainland. They would rather drown than stay. I can transport you, for example, to a lovely little suburb just outside of Anywhere—complete with flush toilets and hot showers, just like home. I bet you will not stay. I would not be surprised if you died trying to get home. For a comment on a real case involving humans, see Seymour and Girardet (1986).

Sometimes animals can be successfully relocated, of course. An island makes the best possible theater, unless (as with Schemnitz's deer) the animals can or will try to swim off of it. How, then, does dispersal work in nature? It works at a particular, usually very short, time period in each individual's life, the dispersal stage. It all comes back to what Charles Darwin said about life: many, many more offspring are produced than are needed to maintain the population. Those offspring simply cannot stay at home; their tiny minds are genetically programmed to go. Comes the day when the pull to leave outweighs the desire for the security of home, and off they go. We are all familiar with the phenomenon, and often the stay-at-homes (usually the parents) actively participate in pushing the potential dispersers out. It's time to go: take off. In many species, especially plants and marine animals, the dispersal stage is a readily recognizable one such as a seed or larval form. Dispersal may be wholly passive: the quintazillion offspring, on release or hatching, are borne by the winds and currents. This is the ultimate *r*-selection approach. In humans, as *K*-selected as species get, there are apt to be tears and recriminations, false starts and sporadic returns. The instinct for dispersal is buried in a matrix of conflicting instincts and desires to cling, protect, and be protected.

The geographic range of a species can expand by dispersal. This is the most usual, ordinary, and natural way for any species' range to expand. To quantify the rate of expansion is simple, in theory. Take the land area open to the species because of its ecological features, and determine the stable population density and, therefore, the number of individuals needed to fill up the space: *N*. Then all you need to know is the number of generations, *g*, needed to achieve that number, where *X* is the number of offspring produced per individual (remembering that only females produce offspring):

$$g = \log N / \log X$$

Of course, that is the minimum number of generations needed, or the maximum dispersal rate. If you know mitigating factors for the area to be filled, which will decrease *X*, then you can modify the formula accordingly. For example, if the organisms reproduce sexually, then only a fraction produce offspring—in our case, about one-half. Therefore, *X*, the number of offspring per individual, is only one-half of the litter or clutch size.

It is this method of dispersal, not some sort of purposeful migration, that fills the Simpsonian corridors. This is the mechanism by which camels invaded the New World—and horses invaded the Old World—across that magnificent Pleistocene corridor, the Bering land bridge. And that is how people first came to the New World too, albeit much later, at the last time when the corridor was open. Pielou (1979) provides an excellent description of the grand corridor of Beringia.

In nature nothing is likely to proceed at the maximum rate. As a general rule, the potential dispersal rate even of large, *K*-selected mammals is very fast compared to the rate at which geologic processes open up corridors for that dispersal. Most dispersing individuals, therefore, enter habitats that are either unsuitable or held by incumbents. The fate of the vast majority of these is, most of the time, death; a lucky few get to replace a recently dead predecessor, but probably only after rigorous competition with other individuals in their dispersing generation. At the edge of the species' range, dispersers are constantly moving out, attempting— but usually failing—to colonize new ground. It is environmental change that may suddenly make some of that new ground habitable: The species' range may expand. If the change reverses, the species' range will predictably contract again. However, some of the far-flung, peripheral populations might survive in isolation—relicts of the once larger continuous range. The pulsating pattern of glaciation over the past two million years has provided a perfect mechanism for expansion–contraction–peripheral isolation phenomena. Other glacial periods preceded this most recent one, the Wurm, and have probably been frequent and regular in most of life's history on Earth. It is expansion, of course, that spreads successful species, or lineages, widely over the world, and peripheral isolation (among other things) that leads to the evolution of new species. Given a bit of imagination—perhaps a time-lapse movie in the mind—you can see how great radiations within a given lineage of highly successful animals might expand to cover Earth.

To visualize the incontestable land bridges of the world, look at a map of Earth drawn with the sea level about 100 m lower. That gives you a good notion of where animals—given opportunity and taking climatic constraints into account—could have simply dispersed overland. There may certainly have been other land bridges, too, their existence obscured today by erosion, faulting, and/or folding: geologic processes that have made water more than 100 m deep today in places where it was much shallower in the past. Note that most of Earth's land is in the northern hemisphere. Assume that the desert zones of today are about where they were when sea level was lower, but expanded laterally to fit the broader exposed land. This means that the tropical rain forests of Africa, South America, and northern Australia were—as today—separated from temperate and boreal forests by arid and nearly treeless realms. Only in eastern Asia (China) was there a continuous, tropical to boreal, forest ecosystem (Lazell 1987b, 2002a).

Assume further that Beringia—the vast land under what is today the northern Pacific, which broadly united North America and Eurasia—had a climate no colder, and possibly much warmer, than today's. That warmth is made feasible by a greatly magnified Kuro Shio—the Black or Japan Current —the Pacific counterpart of the Atlantic Gulf Stream. When land expands in Southeast Asia, Australasia (mostly Australia and New Guinea), and the East Indies, this has the effect of shunting huge masses of equatorial hot water into the northern, clockwise gyre of the Japan Current. This process can make a very hospitable region of southern Beringia (Lazell and Lu 2003). Now, with conditions on Earth as I have described, let a population begin to expand. Let us assume that the lineage originated in tropical North America and that early speciation events produced more lineages adapted to both wet and dry climatic conditions, and to temperate climates too. They can spread to take over most of the world. We might expect to find only a few in sub-Saharan Africa—descendents of those that penetrated through the filter bridge of gulfs and deserts separating Africa from Eurasia. Colonization of frigid realms (e.g., Greenland and Antarctica), would be most unlikely, but the odd sweepstakes winner might make a go of it on the Palmer peninsula, for example, if its propagule arrived just at the end of a glacial maximum when the climate was rapidly warming. For members of this radiation to reach Australasia might require several sweepstakes winners: There are always several major water gaps. Colonization of the true oceanic islands, such as the Greater Antilles or Madagascar, would be chancy, but far from impossible. I would expect several successes on these sweepstakes routes, and a lot of failures. If this is the way the world was, it would produce biogeographic patterns exactly like the ones I see on Earth today, and so carefully chronicled by Darlington (1957). No other explanation for the distribution of vertebrate animals (at least) appears to be so parsimonious.

Using the sea-level curve of Morris et al. (1977), Guana Island, separated from Tortola by a channel of not more than 18.3 m (Marler and Marler 1984), probably became an island about 8,800 years ago. I believe most of the island's reptiles and amphibians dispersed across Great Guania and were stranded here by the sea-level rise. I suspect the house gecko is an exception, because this is not the same species of house gecko found on Puerto Rico; there it is *Hemidactylus brooki* (Mayer and Lazell 2000). Also, numerous people in the BVI maintain that this gecko is a recent arrival on some islands. Carey (1972) did not find it on Anegada; local people there maintain it came with prefabricated building parts about 1975. The type locality for this species is St. Vincent, whence it was described in 1818, but the species is probably of African origin (Schwartz and Henderson 1991). It may have colonized Great Guania after Puerto Rico had separated from that landmass, but before Guana separated from Tortola and St. Thomas. I suspect it is a late arrival via overwater waif dispersal. (See table 17.)

Most of my categorizations above are a matter of opinion—mine. Anyone can refute them,

TABLE 17
The 15 Species in Guana Island's Rich, Little Herpetofauna

	RAFTING POTENTIAL	PROBABLE ORIGIN	GENETIC STATUS
Piping frog (*Eleutherodactylus antillensis*)	Poor	Stranded	Isolated
Red-legged tortoise (*Geochelone carbonaria*)	Poor	Restored	Isolated
Crested anole (*Anolis cristatellus wileyae*)	Excellent	Stranded	Augmented
Saddled anole (*Anolis stratulus*)	Excellent	Stranded	Augmented
Grass anole (*Anolis pulchellus*)	Excellent	Stranded	Augmented
Ground gecko (*Sphaerodactylus macrolepis macrolepis*)	Excellent	Stranded	Augmented
House gecko (*Hemidactylus mabouia*)	Excellent	Rafted	Augmented
Slipperyback skink (*Mabuya sloanii*)	Moderate	Stranded	Isolated
Ground lizard (*Ameiva exsul exsul*)	Moderate	Stranded	Isolated
Stout iguana (*Cyclura pinguis*)	Moderate	Restored	Isolated
Amphisbaena (*Amphisbaena fenestrata*)	Poor	Stranded	Isolated
Common snake (*Liophis portoricensis anegadae*)	Moderate	Stranded	Augmented
Little snake (*Liophis exigua exigua*)	Poor	Stranded	Isolated
Blind snake (*Typhlops richardi*)	Excellent	Stranded	Augmented
Virgin Island boa (*Epicrates monensis granti*)	Moderate	Stranded	Isolated

SOURCE: Original data. Some species I believe are now genetically isolated because their close relatives do not or have not colonized overwater even from nearby Tortola since Guana's separation. Other species I believe are augmented regularly by new colonizers. Two species, the iguana and the tortoise, were extirpated and have been more than augmented; they have been restored. That seemed only fair: Modern man was responsible for their extirpation, and modern man put them back.

or at least try to. Saying that the island's fauna is simply a sample of Great Guania's fauna stranded by rising sea level begs the real question: How did the fauna of Great Guania get there? My answer is that the entire fauna of Great Guania is descended from waifs who dispersed over or across seawater. I do not believe there ever was a land bridge from the Puerto Rico Bank to any other island bank or shelf, including the proximate ones of St. Croix, Mona, and Hispaniola. I believe the notions of

land barges traveling from a primordial Central America or tropical North America, carrying living things, are spurious and stem from a misunderstanding of geologic processes: Rosen's (1975) land-barge model cannot be derived from the geologic sources he cites. For example, Malfait and Dinkelman (1972) did *not* say islands moved; they said the plate moved and islands were created as volcanics at the edge of the resulting subduction zone—right where the islands are today. One of the most notable, noticed, and remarkable land-barge theories, presented by Roughgarden (1995), is similarly flawed (Perry and Lazell 1997b).

A brief sojourn into the biological aspects of my reasoning is appropriate here. Meylan (1984) points out that amphisbaenas are poor swimmers and do not frequent the sorts of logs that would float well and thus transport them. *Amphisbaena fenestrata* demonstrably drowns easily: We have several Guana specimens found on the surface right after tropical storms swept over. But it does not follow that amphisbaenas must reach islands via land bridges or that land barges must carry them on chunks of former mainland. Meylan (1984) has not considered how excellent a raft a palm trunk (e.g., *Sabal causiarum* or *Thrinax morrisii*) can be. A palm trunk quickly decomposes to punky fiber, but retains a firm sheath and floats well. A rain-sodden *Thrinax* trunk, loaded with buoyant, fresh water and trapped air pockets, can readily carry small, fossorial, salt-intolerant animals such as frogs or amphisbaenas. The distribution of amphisbaenas is classically that of chance colonizers on oceanic islands, that is, modest radiations of close relatives confined to large islands; or, when present on small cays (i.e., Guana), clear examples of post-Pleistocene stranding. There is even evidence of two separate Antillean colonization events (Hass et al. 2001).

MacFadden (1980) makes a case for the large, endemic insectivores of the genera *Nesophontes* and *Solenodon* as land-barge travelers. In the face of their classic oceanic island radiations and the comparative lack of other mammalian families in the Greater Antilles, I believe a single event of overwater colonization is needed to explain their presence. Only one species, *Nesophontes edithae*, reached the Puerto Rico Bank (Hall 1981; Ottenwalder 2001; Whidden and Asher 2001). Overwater waif origin, followed by some insular gigantism and modest radiations resulting from eustatic sea-level fluctuations, is the most parsimonious explanation for their existence.

McDowall (2004) presents an analysis of evidence for land barges derived from the "generalized tracks" generated by vicariance biogeographers, especially Rosen (1975, 1985). These generalized tracks indicate that the Antillean terrestrial fauna was derived from either South America or tropical North America (Central America). The inshore marine fauna that is not clearly typical of the tropical Atlantic has a Pacific origin. Because the Isthmus of Panama only closed a few million years ago (Marko 2002), this Pacific component is hardly surprising. Another generalized track exists: the African connection, certainly involving demonstrably pre-Colombian crocodiles, skinks, and geckos. Generalized tracks, however, indicate nothing about how the organisms made the trip. In an effort to answer that question, one needs to consider what kinds of organisms made it and what mechanisms seem most likely to have brought them. Pregill (1981) provides the obvious answer via careful analysis of the data. The Antilles through time, as indicated by their not-inconsiderable fossil record, epitomize the pattern of radiations of close relatives, derived rarely and irregularly from one or a few colonizations, so typical of lands forever isolated in a fluctuating sea. MacFadden (1981) tries to rescue the idea of land barging but has to admit the overwhelming evidence for massive overwater waif dispersal. For me the evidence is clear: No land barges, carrying living things, came into the Antilles. Hedges (1996a, 1996b) and Hass et al. (2001) present much more evidence for various, very different, colonization times for the different ancestral stocks of what are now radiations. Hedges (2001) concludes that a land bridge is incompatible with the evidence: GAARlandia was a set of islands similar to the Lesser Antilles of today, albeit shorter. Graham (2003) concurs. I hope the war has finally been won and W. D. Matthew vindicated.

So I am a die-hard dispersalist. From February to early April—the Antarctic late summer and early autumn—of 1981, I lived on the west coast of Tasmania; I got to know the "roaring forties" the way New Yorkers know automobile traffic. I became accustomed to what the Southern Ocean bears from distant lands. I saw incredibly delicate creatures carried alive far above ordinary sea level. I have no doubt that an entire beech tree (*Nothofagus* sp.) can travel from Tierra del Fuego to Tasmania, with a load of aphids and other insects, and fetch up way beyond the berm, over the back dune, and safely in the woods. Time two storms just right—one to carry the tree, with ballast stones in its root mass, out to sea; one to fling it up—and provide the motive force of the regular wind and weather, and you do far better at seeing how life gets around this Earth than those who dream of drifting land. Which is easier to move, a tree or a continent?

Picture a sodden sloth clinging to a bobbing log, swept out to sea from South America by a mighty hurricane, finally hurled by the surf onto the slothless coast of a far Antillean isle. Make it pregnant into the bargain. That is waif dispersal. Pregnant sloths awash at sea are waifs, as are wind-blown lizards, frogs in waterspouts, and viable coconuts rolling in the surf. Most waifs never live to land on any island. Others may make it, but fail to establish a population (among sloths, only the pregnant can do that, but a single coconut will do). How likely does it seem? Well, there are quite a few of us who really believe most of the plants and animals of the West Indies reached the islands as waifs (Lazell 2002b). We really believe in waif dispersal as the original populator of the Antilles and Bahamas and even as the provider of most of the species of the Florida Keys. Preposterous? Well, go try to drown a sloth.

If you want to calculate the probability that a given type of animal will colonize across a barrier—let us say a member of the lizard genus *Anolis* across seawater—Darlington (1938) provided the formula. The probability, *P*, of an individual making a fortuitous dispersal voyage across a gap, of width *w*, can be calculated if one knows the number of individuals in the population of potential colonizers, *N*, and the width of some comparable gap the organisms have actually crossed, *m*. Then:

$$P = (N)^{m/w}.$$

This is a simplistic rule of thumb, similar to my equation for dispersal expansion rates given earlier. You can similarly modify it if you know things about the physiology, morphology, or behavior of the animals involved, such as how long they can tolerate exposure to seawater, how well they float, and how far they can purposefully swim. Only rather rarely could one individual (a gravid or pregnant female) constitute a propagule and found a viable colony. One may modify *N* accordingly, taking into account only the number of gravid or pregnant females present at any one time, or calculating the probability of two fortuitous trips, one by a female and one by a male. If, however, you use for *m* the width of a gap already crossed by successful colonizers, you have already taken all this into account. In that same classic paper, Darlington (1938) said "I think it is a mistake to apply mathematics too closely to biological phenomena of which we know as little as we do of dispersal."

To finish up general remarks on overwater dispersal, let me point out what seems an obvious correlation some enterprising biogeographer could codify into a formula and become rich and famous with: Small animals are more likely to cross water gaps than large ones. All one needs to do is plot the log of animals' masses against the log of overwater distances they have demonstrably crossed. A scatter will emerge, but the negative correlation of size to distance will be linear and spectacular. The bigger you are the worse your chances in these sweepstakes. No one, so far as I know, has ever published this simple picture. It is every bit as good a biogeographic rule as any species : spatial correlation. Beware, of course, of delusions resulting from insular gigantism: Organisms often evolve into giants in isolation, as did dodos, giant tortoises, sloths, and solenodons. Small colonizers may get much bigger before they appear in the fossil record. Do not omit notable exceptions: for example, small animals like salamanders, which are miserably poor overwater dispersers, and, big animals like boa constrictors, which disperse well. A few exceptions always make the rule look more realistic.

Three lines of evidence argue for waif dispersal to Great Guania: (1) the conspicuous absences of biotic elements easily drowned or intolerant of salt: ungulates, carnivores, primary freshwater fishes, and salamanders, to note a few; (2) the disparate times of origin of the various groups of animals that have been calibrated with molecular clocks (Hedges 1996a, 1996b; Hass et al. 2001); and (3) the evolutionary radiations of species within closely related groups to fill niches left vacant by lack of continental types. What we see, then, is a vast array of species apparently derived from only a very few ancestral types, all of which were small and likely to cross water well. The hutias, a family of rodents that are exclusively Antillean, underwent a remarkable radiation into arboreal, prehensile tailed, rock dwelling, and fossorial (burrowing) types. *Epicrates* boas, *Anolis* lizards, iguanas, spheretoed geckos, tree frogs, xenodontine snakes, and many other groups show a similar pattern: A lonely colonizer or two made the trip, and a great adaptive radiation resulted, often with giants and dwarfs, terrestrials and arboreals among the mammals, and flightless forms among the birds.

The result of such radiations is endemics: those species, genera, and even families (such as the perky little kingfisher-like birds called todies) found nowhere else on Earth. The West Indies are situated to receive the waifs of North America, South America, and Central America, and even those from Africa, such as *Mabuya* skinks, *Tarentola* geckos, crocodiles, and cattle egrets.

But another etiology may bring endemism: wholesale extinction resulting in an occasional surviving relict. David Lack (1976) presented the case against naturally occurring, rapid extinction rates for island species. High degrees of endemism such as unique radiations at the family or genus level, and bizarre adaptations, such as lumbering rock iguanas on one island and dwarf geckos on another, cannot occur if extinction is rapid. Evolution needs time to work. Insular species that evolve in situ are tenacious and long lived. They resist very well the attempts at colonization made by related—and potentially competitive—waifs.

I believe Lack's (1976) points are well taken and seriously erode a notion of equilibrium involving rapid extinction counterbalanced by frequent colonization. I agree with Whittaker (1998, 141): "We should no longer give equilibrium models primacy in island ecology." It is claimed by some that Robert MacArthur came to hold this view himself shortly before his premature death. But, Lack argued, original island ecosystems have all too often been rent asunder by the hand of man. Direct habitat destruction, such as deforestation with subsequent grazing by exotic ungulates (goats, sheep, and cattle) obliterated endemic floral elements. Pothunting wiped out spectacular birds such as parrots, macaws, and petrels. Egging pressed hard on the reproductive potentials of colonial nesting seabirds, tortoises, and marine turtles. Voracious introduced mammals such as European rats, cats, and mongooses exterminated many insular endemics that had for millennia resisted heavy competition from closely related rivals. Stephen Palumbi of Stanford University writes that human influence now constitutes the world's greatest evolutionary force, and that anthropogenic evolution is rapid indeed (Palumbi 2001). Anthropogenic destruction on Guana Island was comparatively light and some restoration has been successful; I have plans for more (see "Restoration," part 5, below).

Species : Spatial Relations

Most people who take an introductory college biology course learn of several geographic rules that account for both the diversity of species within an area and the population size of these species. . . . Most rules are made to be broken.

WILLIAM H. DRURY (1998)

. . . the present case appears to provide a very significant warning against any too simple attempts at estimating the meaning of area for species number.

ERNEST E. WILLIAMS (1983)

. . . facts that motivate the turnover theory of island biogeography are absent in the *Anolis* system of the Lesser Antilles. Specifically, there is no area effect or distance effect.

J. ROUGHGARDEN, D. HECKEL, AND E.R. FUENTES (1983)

TOLD YOU SO. The most formalistic and simplistic notion about life on Earth is the "bigger buckets" theory. It derives from the profound, but dull, truism that large vessels are potentially able to hold more than small vessels. It works for the abundance : range glitterality because, obviously, the more things you have, the more space they will take up. It is the guts of the species : area glitterality. MacArthur and Wilson (1967) extrapolated it to infer that large islands have more species than small islands—the keystone notion in what they called "*The* theory of island biogeography" (italics mine).

Theoretical ecology may not be a precise science, but neither is the world chaos. A wavewashed rock in the sea supports no land animals (e.g., the lizards, snakes, and frog of Guana). A big island supports dozens of species (e.g., Puerto Rico, where once one could hear 14 species of frogs calling at one time in the rain forest of El Yunque). Surely it is sensible to predict that Guana Island falls somewhere in between. Of course it does. MacArthur and Wilson (1967) just failed by a factor of four to figure out where. Certainly, there is a species : spatial effect; it may involve area, elevation, or volume. To deny its existence would be like denying the existence of competition in shaping batches of sympatric species into coevolved communities—or like denying that evolution can and often does proceed gradually. But regular, predictable relationships between area, for example, and diversity, or number of species, are rare. Life is just more complicated than that.

And, most important, area (or elevation) does not cause diversity; all increasing the space availability could ever possibly do is facilitate diversity.

How about this glitterality: Big islands are higher in elevation than little ones. If we plot area against elevation for the land areas of the world, we will get a scatter with a linear regression and a remarkably high correlation coefficient. The fit, or correlation, is actually better for Antillean islands than is the species : area correlation. One could codify that, publish it, and go around presumptuously referring to it as *"the* theory of topography." The implication one would have foisted off on the public is that area somehow causes or explains or, even better, determines elevation. The truth of the matter is vastly more complicated. A number of processes, especially those involved in volcanism, which generate area, are also apt to generate elevation; erosion, the most obvious process for reducing land areas, reduces elevation. Even those truths are simplistic glitteralities. We have run solidly into what Gilbert (1980) calls ". . . the innate difficulties of invoking causal explanations for correlative data."

Virtually all population biologists calculate animal (or plant) densities by planar areas. Obviously, that is much easier than getting involved in the abstruse realms of topology necessary to calculate real areas. On Guana I have done this myself. I laid out trap grids on flat plains at White and North bays; then I figured snake numbers for the whole island and densities for the planar 300 ha. However, one should not be so simplistic. From elementary geometry (take a 3, 4, 5 right triangle, for example), we see that for smooth slopes between 35° and 40° the surface area increases about 20%. At a little over 45° of slope the area has increased 40%. At a bit over 57° of slope, the actual land surface (if smooth) is double what the planar area of a map would indicate. A cliff can have a lot of surface area, and be important habitat, but we cannot compute its area from its foot length and its 90° slope; we have to know its vertical dimension.

In theory, a computer should be able to give us a reasonable figure for actual surface area given an accurate topographic map (and integral calculus). In practice, I have found it helpful to modify MacArthur and Wilson's (1967) simple formula

$$S = CA^z,$$

where S is the number of species, A is island area, and C and z are constants dependent on the kinds of animals considered—such as birds versus frogs. The relative importance of the spatial parameter in influencing species diversity is shown by the value of the exponent z in $S = CA^z$ or $S = CE^z$ (where E is island elevation). MacArthur and Wilson (1967) were pleased to find that z for the A values hovered around one-fourth. In many studies using area, z values range from 0.1 to 0.4. We have found that z values for E may approach one; the larger the z value the more important the spatial parameter. At $z = 1$ the formula becomes simply $S = CE$; that is, species number is just the island's elevation multiplied by the taxon constant. Any fractional value of z below one-half indicates that the spatial parameter is not very important; at $z = 0$ the formula becomes $S = C$, a totally uninformative declaration: My neighbor's dog knew that. I added in E for elevation (Lazell 1983a):

$$S = C_1A + C_2E + C_3AE + C_4.$$

The third item, C_3AE, implies that there is some synergistic effect of elevation and area, greater than the simple additive effect. Even so, my formulation is not any too accurate in the BVI. The most apparent synergistic effect of elevation on area that I could envision is the development of ravines on high, steep islands. Thus we would do well to consider the ravine effect on Guana. The most obvious thing the ravines on Guana have that the other slopes lack is an abundance of big trees. These produce shade, which conserves moisture. The ravines also accumulate humus and topsoil to a greater extent than the open slopes; this is both a cause and an effect of the big trees. Still, no matter how I calculated, Guana Island would not fit any general predictive formula I could devise using physical parameters (Lazell 1983a). Perhaps if I measured all the trees.

Based on Zug (1984) and Sugihara (1980), it would seem that I am carping over some trivial lack of precision: not so. I am pointing out major, systematic errors in the scheme that go from factors like two and three right on to infinity. To claim that MacArthur-Wilson's $S = CA^Z$ constitutes a meaningful quantification is worse than claiming that Ordagova's number is the number of atoms of any element in one gram-molecular weight and equals 6.02×10^z, where z is any constant between 1 and 100, where z can only be empirically determined by counting the atoms, and—furthermore—where z is apt to vary from one place to another even for the same kind of atom. Thus, a gram-molecular weight of oxygen in Oklahoma would probably have a different z value—and thus a dramatically different number of atoms—than a gram-molecular weight of oxygen in Pennsylvania. Physicists would ridicule anyone who thought such a formulation was valuable or important: We are not talking Heisenberg here, we are talking betting sports (bs).

Why is $S = CA^z$ worse than Ordagova's number? Because in the latter the analogue of C in the former really is a constant: 6.02. In the MacArthur–Wilson formula C—like z—is any number and can only be empirically determined. Consider *Anolis* lizards in the West Indies: one of the best-studied groups of animals on Earth. We have good, solid, planar values for A for all of the islands; our lists are no doubt still incomplete, but for most islands we have S values too. Putting them into the equation, we discover that Greater and Lesser Antilles have radically different optimized C and z values (Roughgarden et al. 1983; Losos 1996a). Worse, the islands of the Jamaica Bank have very different C and z values for *Anolis* species than do the islands of the Puerto Rico Bank (Losos and Schluter 2000). I have not considered Hispaniola, Cuba, and Bahama Banks, but

I expect each will be different. The equation $S = CA^z$ does not even work using the same C and z values for one genus in one archipelago. One cannot have it both ways: If ecology is to be both rendered quantitative and shown to fit deterministic principles and abiotic formulations, then its proponents are going to have to abandon glitteralities. I believe they are doomed to failure.

The species : area formula of MacArthur and Wilson (1967) usually suffers from two different sorts of systemic errors. First, the original data on which it was based were typically serious underestimates of actual species present on a given island because (a) inventories were inadequate, and/or (b) anthropogenic extinctions had already artificially reduced faunas. Both problems were true of the herpetofaunal data set MacArthur and Wilson (1967) took from Darlington (1957) and used as their fundamental exemplar. Second, log–log species : area plots are rarely or never linear but typically curve to flat or nearly so at both large and small island sizes (Crawley and Harral 2001). Connor and McCoy (1979) and Gilbert (1980) have presented cogent detractions from the theory. Gilbert (1980), especially, voices caveats against application of hypothetical species : spatial notions to conservation and applied ecology. Whittaker (1998, 129) notes: "The details as to which variables best explain variations in S in a given data set hold fascination principally in their specific context. They are difficult to generalize." That is just what Lack (1976) argued and I believe every good biogeographic study reveals. Whittaker's (1998) *Island Biogeography* is, in my opinion, the best ever written because it includes such a breadth of examples. My own approach to this issue has been to compare real species numbers to the MacArthurian ideal as I would compare (when possible) real evolutionary rates to my simple, maximal formula (see below). Then I seek to explain the discrepancies, which I have found far exceed the number of predicted fits. In my first perusal of the British Virgin Islands (Lazell 1983a), I discovered a huge systematic error: There were too many species on most of the islands. I noted a phenomenon that I had perceived decades before: Small islands (almost no matter how small) seem to have three species of reptiles including at least one *Anolis* and one gecko. I called that the "rule of three." Now why should that be? Is there a special coevolutionary bond between three species that suits them to the particular island edge vegetative zone that dominates small islands? Investigation of this almost incredibly tough, resilient, seemingly stable, and very low diversity assemblage has got to provide great insight into elementary ecological theory. Carrot Rock, with unique endemics as two of its three species (Mayer and Lazell 2000), would be a perfect site. My rule of three would be a classic "assembly rule" sensu Weiher and Keddy (1999) except that—at least in the island fragments of Great Guania—it is not the result of assembly at all: It is the result of dwindling; the product of extinctions.

If despite Roughgarden et al. (1983) there is a tendency for islands to develop a "right" number of species over time, what does this say about Guana Island with 15 species where the theory predicts three? I suggest the following:

1. Guana Island started out, 8,800 years ago, with a disproportionately large fauna because it had just been part of a much larger island, Great Guania.

2. The island's diverse species community was long coevolved; therefore, competition would not tend to push any member species to extinction.

3. If the vicissitudes of chance did eliminate a species, it might well recolonize because all of Guana's species are descendants of overwater waif dispersers: They can come back (2 of the 15 have).

Diamond (1984) considers the probability per unit time of "normal" insular extinctions. The units he uses are per year. In his simple equation, *p* is the probability per year that an individual will die and *N* is the number of individuals in the insular population. Then *e*, the probability of extinction per year, is:

$$e = p^N.$$

Diamond (1984) notes six things that mitigate extinction rates. Four are embarrassingly obvious: long generations or lifetimes; high density; large area; and high intrinsic rate of increase. The fifth, high ratio of births to deaths, is the same as the fourth. Only the sixth is interesting to me: stable population size (Diamond expresses this in reverse, i.e., unstable population size tends to increase the probability of extinction). I wonder if that is true. If a population naturally wobbles through an order of magnitude—say from 100 to 1,000—it would be less prone to extinction than a stable population of 100 individuals. It would be more prone to extinction than a stable population of 500 whenever it dropped below that mean.

If we consider again the island's snakes, *Liophis p. anegadae*, we can apply Diamond's formula. Assume the population, *N*, is 700 and stable. In each year each female produces six eggs, so with her and her mate this accounts for eight snakes. Six of those eight die during the year (or the population would be unstable and increase). Thus, the probability of any one individual dying is six in eight, or 0.75. Then, $e = 0.75^{700}$. Exponents that large boggle my mind. One deals with them by using logs: $\log_e = 700 \log 0.75$. The resulting number is vanishingly small; it has some 80 zeros to the right of the decimal point. Year to year, *Liophis p. anegadae* seems quite safe. It is important to bear in mind that we are calculating here safety from extinction solely in terms of populational stochasticity; environmental stochasticity such as catastrophes like volcanic eruption, tsunami, or bolide impact is not calculated. Nor have we calculated the effect of a biological catastrophe such as introduction of an exotic carnivore like the mongoose.

If we look at a smaller population it is more instructive. I do not know how many boas live on Guana, but, if any, they are quite scarce (maybe just hard to find?). Let's guess there are 20 and the population is stable. Assuming the sex ratio is balanced is unrealistic for boas in my experience, but I will do it in this case. Then, it is reasonable to assume each female produces a litter of 10 per year. So, of 12 (two parents plus 10 babies) alive at birthing time, some 10 must get eaten by red-tailed hawks, owls, or ground lizards (most as tiny babies). The probability of any one individual's death in a year is ten-twelfths, or 0.8333. So:

$$e = 0.833^{20} = 0.0259.$$

This says that there is a little better than a 2.5% chance of a "normal" extinction occurring in a year with a population this small. The implication is that, given 8,800 years since the island's separation, an event as improbable as a 2.5% chance of extinction in one year becomes quite likely to actually occur. Have our boas gone extinct since C. Grant (1932)? In the same vein, if we had only one pair of owls, and a pair raises only one owlet per year, they cannot possibly last long. Even over on much larger Tortola there cannot be (or cannot have been) many pairs. Are they gone? Or doomed? But what about our red-tailed hawks? There certainly is no more than one pair on Guana Island; they are no denser on Tortola, so there cannot be many. But there they are, soaring and wheeling overhead, with us year after year. Indeed, Walter (1990) describes an island endemic form of red-tailed hawk that has survived at a population size of about 20 for many, many generations. Caughley and Gunn (1996) provide many more real-life examples.

Diamond's formula cannot be applied to longer time units once they exceed the life span of the longest-lived individuals. If we consider the probability of *Liophis p. anegadae* extinction over 10 years, for example, *p* is 1: No individual lives that long (boas, on the other hand, probably can live even longer). The formula would tell you that extinction of the population was certain. Similarly, one must pick a different, much shorter, time unit when attempting to apply the formula to short-lived animals like insects. If

their life cycles are completed within a year, then *p* becomes 1 in that year: Extinction is predicted as certain. To get a realistic prediction simply select a time unit that includes births or hatchings but is much shorter than normal adult longevity.

There is a whole class of islands that seem to be radical exceptions to species : spatial constraints—the sand-barrier islands, at least of Atlantic North America (Lazell 1979). In elevation they vary little, never being more than a half-dozen meters above sea level. In area, however, they vary immensely; this seems to have almost no effect on the number of species they support. It is only since Lianna Jarecki pointed out to me the necessary relation between a lognormal or rarefaction sequence of species abundances and any species : spatial correlation (Lazell and Jarecki 1985) that I have come to suspect that I know why. I have been able to study distribution patterns of mammals, amphibians, and reptiles on three widely separated, complex sets of sand-barrier-island systems: Cape Cod and the Massachusetts coastal islands, the Outer Banks of North Carolina, and the Gulf of Mexico islands in Mississippi. Descriptions of geological and ecological features of these islands may be found in Godfrey (1976) and Leatherman (1979). Descriptions of the fauna are provided by Lazell and Musick (1973), Lazell and Conant (1973), Lazell (1976, 1979a), Spitzer (1973, 1977), and Alexander and Lazell (2000). A number of additions to the faunas are based on my own collections and observations. (See table 18.)

Two things are apparent concerning evolutionary adaptation to the barrier systems I have studied: (1) Major departures from mainland norms sometimes result in striking endemic forms. (2) Geographic isolation is not the primary cause of these departures. Sand-barrier islands are relentlessly attached to mainlands. Terrestrial vertebrates can disperse onto sand-barrier islands with ease, providing only that they can tolerate the rigors of the sand-barrier-island environment (Lazell 1979a). This fact is illustrated by the ease with which the introduced

TABLE 18
Deployment of Some Terrestrial Vertebrates on Three Complexes of Barriers

C	B	G
Procyon lotor	Procyon lotor	Procyon lotor
Mustela frenata	Mustela vison	?
Lutra canadensis	Lutra canadensis	?
Mephitis mephitis	—	—
Microtus spp.*	—	—
Peromyscus leucopus*	Peromyscus leucopus*	—
—	Oryzomys palustris	Oryzomys palustris
Sylvilagus floridanus	Sylvilagus floridanus	Sylvilagus floridanus
Plethodon cinereus*	—	—
Hyla crucifer	Hyla squirella	Hyla squirella
—	Hyla cinerea	Hyla cinerea
Rana clamitans	Rana utricularia	Rana utricularia
Bufo woodhousei*	Bufo woodhousei*	Bufo terrestris
Chrysemys picta	Chrysemys scripta	Chrysemys scripta
Chelydra serpentina	Chelydra serpentina	Chelydra serpentina
—	Kinosternon subrubrum	Kinosternon subrubrum
Malaclemys terrapin	Malaclemys terrapin	Malaclemys terrapin*
Natrix sirtalis*	—	—
Natrix saurita	Natrix saurita	—
—	—	Natrix cyclopion
Natrix sipedon*	Natrix sipedon*	Natrix fasciata*
Opheodrys vernalis	Opheodrys aestivus	?
Coluber constrictor	Coluber constrictor*	Coluber flagellum
—	Lampropeltis getulus*	Lampropeltis getulus
—	Elaphe obsoleta	Elaphe obsoleta*
—	Agkistrodon piscivorus	Agkistrodon piscivorus
—	Ophisaurus ventralis	?
—	Cnemidophorus sexlineatus	Cnemidophorus sexlineatus
—	Eumeces inexpectatus	Eumeces inexpectatus
—	—	Anolis carolinensis*
—	—	Alligator mississippiensis

SOURCES: Lazell (1976, 1979a), Spitzer (1977), Alexander and Lazell (2000) and original data. C is Cape Cod and the Massachusetts coastal islands; B is the Outer Banks of North Carolina; and G is the Gulf Islands of Mississippi. Forms with an asterisk show significant evolutionary departures from mainland populations.

cottontail (*Sylvilagus floridanus*) has colonized Massachusetts' barriers. Inlets are perpetually opening and closing, but a dispersing colonizer need never actually cross water if it expands its range along any system in the direction of longshore current and terminal accretion (Lazell and Musick 1973, Alexander and Lazell 2000).

Full speciation under these circumstances will predictably be rare; the Muskeget vole, *Microtus breweri*, is the only described case. It is remarkable because the great Smith Point–Esther Island–Tuckernuck tombolo extends to Muskeget periodically, resulting in character divergence when the Muskeget and mainland species of voles meet. If geographic isolation of the members of the various barrier systems studied has little to do with the evolution of endemics on them, what effect does it have on their species diversity? Both sand-barrier islands and sand-barrier peninsulas are exceedingly depauperate. A quick glance at species numbers within any group reveals that sand barriers differ by roughly

TABLE 19
Vascular Plant Species on Some Barriers of the Atlantic and Gulf Coasts

	SIZE (HA)	DISTANCE (KM)	SPECIES	REFERENCE
Monomoy, MA	966	1	183	Moul 1969
Muskeget, MA	113	2, 31	131	Wetherbee et al. 1972
Island Beach, NJ	930	0	267	Au 1974
Hatteras, NC	3,415	18	415	P. Hosier, pers. comm.
Ocracoke, NC	1,384	58	158	C. J. Burk, pers. comm.
Core Banks, NC	4,061	4	162	C. J. Burk, pers. comm.
Shackleford, NC	923	3	281	Au 1974
Horn, MS	1,972	10	250	A. Bradburn, pers. comm.
Petit Bois, MS	786	15	250	A. Bradburn, pers. comm.

SOURCES: As noted. Areas and distances were calculated from 1 : 250,000 USGS topographic maps revised through 1970. The two distances for Muskeget are to Tuckernuck (nearest island) and Cape Cod (nearest mainland), respectively.

an order of magnitude from adjacent mainlands: not much lives on them. Their very geologic dynamism, their perpetual motion, and their susceptibility to oceanic overwash all might be expected to contribute to spectacular waves of extinction and colonization. Sand-barrier islands, one might think, would have been one of the first subjects taken up by theoretical ecologists, but that is not the case. Perhaps size and distance are simply less relevant to sand-barrier islands than some other measurable variables that result in ecological diversity. Might not their species numbers be based on this calculable ecological diversity, resulting in nothing more than a modification of the MacArthurian view? These islands and peninsulas are all low, similarly built of sediments in formations that, albeit fascinatingly different from system to system, are not notably different in physiographic relief. There is, admittedly, quite a marked variation in obvious ecological diversity between, say, Shackleford and Core Banks, North Carolina (Alexander and Lazell 2000). This obvious difference results from the different vegetative associations present, resulting from island orientation to prevailing winds. The land animals do indeed depend on the plants, at least to some extent. It does seem apparent that those islands with greater ecological diversity, where that means more vegetative associations (i.e., more plant species), have more

land-animal species. The tautology is complete: Islands with more species have more species. (See table 19.)

It is most impressive to visit sand-barrier islands after storms. One is immediately aware of the vast numbers of storm-carried plants and animals that reach the islands as viable propagules. Lazell and Conant (1973) remarked on this occurrence regarding snakes. Anne Bradburn, Tulane University, has noted as viable propagules the many plant species in wrack lines that do not occur naturally in the floras of the islands. The irrelevance of size and distance, the abundance of propagules that fail to colonize, and the rapid evolution of endemics on geologically very young islands all argue for the basic theoretical view of island biology put forth by Lack (1976). The insular and peninsular barriers seem to possess species matrices that are highly adapted, stable, and not prone to rapid extinction–colonization equilibrium adjustments. The land vertebrates present seem to be a mix of highly specialized (even endemic) forms and mainland species with broadened niches (raccoon, water snakes, snapping turtle, toads, etc.). It will be necessary to understand the autecology of each participant species to develop a synecological view of any given island or system. The number of critically important variables affecting each member species is too great for the kind of mathematical modeling needed for scientific resource management at this time, or, probably, ever.

Connell (1978) suggests that species diversity may depend on climatic changes and other disturbances that alternately favor different species and that diversity diminishes in static systems. This may well account for the low diversities of sand-barrier-island faunas because the islands are remarkably insulated from change. Initially, that may sound ridiculous: Sand-barrier islands, it seems, are constantly subject to spectacular geomorphological change. It is just the constancy of these processes that produces stability: not physiographic stability—ecological stability. As I have pointed out (Lazell 1979a), mainland habitats may come and go, but the offshore barriers remain: They move and only species adapted to this movement live there; that is a mere handful of species, but they endure. The failure of a species : spatial correlation on sand-barrier islands results from the fact that their animal species depend on vegetative associations that are linearly arranged along the long axis of each island and set of islands. Island area depends primarily on island length; width remains pretty constant. Thus inlets breach the barriers with regularity and instantly change areas by large factors. This does nothing to disrupt the vegetative associations except right at the point of inlet breaching. The ecological communities and species numbers remain intact even when an island formerly 100 km long is fragmented into a set of islands each no more than 10 km long. Does this, by extension, explain the "rule of three" in the West Indies (Lazell 1983a)? Is the Antillean island edge ecosystem relatively unaffected by island area or even elevation? I believe so. Houston (1994) makes the point that frequent, but not too catastrophic, disturbance—intermediate disturbance—often correlates with high diversity. However, on sand-barrier islands and very small Antillean islets it correlates with resilience but not high diversity.

Finally, what of the species : elevation relationship documented for bats in South America by Koopman (1983) and in the Virgin Islands by Lazell and Jarecki (1985)? A similar pattern was seen in frogs by Ovaska et al. (2000). As far as I know, Lazell and Jarecki (1985) were the first to note that lognormal species populations (or rarefaction sequences) are a necessity for species : spatial correlations. This also seems to be the first source in which anyone has used rarefaction to look at either mammalian or insular species diversities. It seems to be the first report I know of where elevation was proved to be the most important correlative of species numbers: On the Puerto Rico Bank, one gains a species of bat with (roughly) every 100 m of elevation. But the bats all live in the lowlands. Ovaska et al. (2000) found the same elevation correlation with BVI frogs, and they all live in the lowlands too.

Weissman and Rentz (1976) found elevation the second most significant correlative factor—after area—in grasshopper diversity on the Channel Islands of California. Power (1972, 1976) found elevation second also for birds in this same archipelago. More important, he found plant species diversity the most important correlation, not island area. This brings us back to the notion that diversity begets diversity, but that did not hold up for McGuinness (1984) who studied intertidal communities and concluded "... current generalizations about the effects of habitat diversity may be unjustified." Structurally diverse habitats sometimes support fewer species.

Of course, plant species diversity is not necessarily the same as structural diversity. One suspects that they are very tightly correlated, however, from the animals' points of view. This may explain our lowland bat dilemma. Elevation causes area. This is not a loose correlation or a stochastic process. It does so very literally and precisely. On two islands with the same planar area, measured from a flat map, the higher island will probably have the greater actual surface area. That area fits topographic relief. Topographic relief is structural habitat diversity and apt to be strongly correlated to plant species diversity. Ravines, because they collect moisture and soil and are partially protected from some aspects of weather such as wind and considerable direct insolation, can support different sorts of plants than can flat plains. Sufficient elevation can provide cloud-collecting land areas with dramatically increased precipitation: Montane rain forest begins in these islands at about 600–700 m (Lazell 1972). Rainwater descends the slopes through ravines and valleys, which may hold flowing streams year round, even in the driest seasons. This brings ecological diversity to the lowlands. It may be simple factors such as fruit or roost sites that enable bat species diversity; to find out, we will have to learn the autecology of the bats involved. Until we do, that can be nothing more than a hopeful guess.

The principal problem with species : spatial correlations is their irrelevance to history. Vuilleumier and Simberloff (1980) have considered this for South American birds. It is my contention that it is in large part history that enables the remarkable species richness of Guana Island. Those species that evolved together over millennia, isolated on Great Guania, can and will continue to coexist despite the tiny size of their current home island, unless some catastrophe—probably human-caused—severely disrupts the balance there. If this is true, a secondary land-bridge island like Guana, whose stranded species are all the descendents of overwater waif dispersers, would always have more species than an island entirely colonized by overwater waifs directly. Time will tell.

Possibly the greatest ecological stability attainable on this planet results from the combination of five factors:

1. The site is in the tropics near sea level.

2. Physiography does not generate extremes of climate.

3. The fauna has long evolved together in isolation.

4. All species are members of rich local guilds with respect to all major niche dimensions.

5. There are no catastrophic disruptions such as volcanism or artificial habitat destruction.

In such an example, species richness via local guilds would correlate with and actually aid stability; modest elevation would contribute surface area and topographic relief; these would in turn contribute to species richness: a tight little circle. In considering elevation in this case, one wishes not to have such extremes as would result in harsh climates. I have previously (Lazell 1979b) pointed out a handy rule of thumb: a meter is a mile. That means that, as a rough approximation, every meter of elevation gain is climatically similar to moving a mile north (or south in the Southern Hemisphere). I did my reckoning based on summer conditions in the Northern Hemisphere. Thus Guana Island is 246 m high and about 360 miles south of the Tropic of Cancer. Its highest point is well within the climatic range of the tropics at sea level. The equator is about 1,650 miles south of the Tropic of Cancer, so a peak of 1,650 m on the equator would still be within the climatic realm of the tropics at sea level. Elevations of this sort should enhance species richness without providing the destabilizing effects of temperate and boreal climes.

I was delighted to read (of molecular biology) in B. Alberts et al. (1984): "There is a paradox in the growth of scientific knowledge. As information accumulates in ever more intimidating quantities, disconnected facts and impenetrable mysteries give way to rational explanations, and simplicity emerges from chaos. Gradually the essential principles of a subject come into focus." Perhaps we only need a few more facts now. It is disconcerting, though, to note that Lawton (2000, 51) entitles a section of his book (italics and capitals his): "*The Longer We Observe Any Environment, the More It Changes.*" Perhaps we need to hurry up collecting those facts; perhaps the counts and censuses of 20 years ago have gone the way of our screech-owls.

A Unified Theory

Understanding biodiversity and its origin, maintenance, and loss on Earth is an issue of profound significance to humanity and life. . . . In my experience, however,
too few people . . . fully grasp the enormity and urgency of this scientific and so-
cioeconomic problem.

STEPHEN P. HUBBELL (2001)

MANY SCIENTISTS APPEAR driven to
seek and explicate a grand Theory That
Explains Everything, as Albert Einstein was
driven by his dream of the unified field theory for
physics. Hubbell (2001) has developed a theory
that unifies the species : area relationship (SAR)
with relative abundances of the species. Can it
make accurate predictions or even accomodate
existing data, as discussed by Lipton (2005)? As
framed, Hubbell's theory applies to metacommunities—the regional pool of species that
might occur in a given local community. That is
perfect for our purposes in looking at Guana Island: one local community among over a hundred vegetated Virgin Islands—the metacommunity. Hubbell's theory is further restricted to one
trophic level at a time—say, all the insectivorous
animals in the metacommunity of the Virgin Islands. Hubbell's communities, therefore, are bigger than Lawton's (2000) local guilds (say, all the
insectivorous animals on Guana), but smaller
than Lawton's whole guild (all the insectivorous
animals on Earth).

Actually, Hubbell's work is based mainly on
forest trees, which are not quite all the plants
that eat CO_2 and burp oxygen. Field counts typically have a size cutoff; for example, to get
counted you may have to be bigger than 10-cm
diameter breast height (dbh). Hubbell (2001)
explicitly includes the process of new species
evolution—speciation—in his purview, but his
speciation is the instant or "point mutation" variety that actually occurs very rarely in nature as
a result of polyploidy. Hubbell (2001, 148) introduces the variable nu, a Greek letter that looks
like a v (so I will use a v for it), that equals the
rate of new species evolution per birth. Obviously, this will be a tiny number (Hubbell suggests one in a trillion: $1/10^{12}$).

The unified theory also depends on the number of individual organisms in the metacommunity, which Hubbell (2001) confusingly calls the
"size" of the metacommunity. Clearly he does not
mean size in terms of geographic area covered or
number of species—except the latter in the extreme case of every species being represented by
one individual. The total number of individuals in
the metacommunity is J. And, finally, we need to
know the mean density of individuals per unit
area covered by the metacommunity; call this p.

Right away there is a problem: in calculating p for the Virgin Islands, for example, do you include the water or just the land areas? Remember what Alfred Russel Wallace said of the East Indies: Their species diversity is greater than that of a like-sized area of continuous land; Wallace was including the water in the area of the East Indies but not the aquatic organisms living in the water (just the terrestrial ones flying, swimming, or rafting from one island to another, or actually on land). Anyway Hubbell admits that counting J is nearly impossible (except for a small forest, perhaps), so he allows us a proxy: A, the geographic area covered by the metacommunity. Once again, he does not specify with or without the water, or other uninhabitable space that falls within the linear dimensions of the place. Then Hubbell presents:

$$T = 2p A v,$$

where T (Hubbell uses the Greek letter theta) is the fundamental biodiversity number. It is equal to the mean density of individuals per unit area (p) times the whole area (A) times the speciation rate (v) times two. Hubbell found that T varied from 0.15 for a Newfoundland conifer forest to 180 for a lowland Malaysian rain forest. The fundamental biodiversity numbers calculated by Hubbell all correlate to latitudinal diversity gradients: T values are highest in the tropics at sea level and lowest near the poles or at very high elevations. Hubbell (2001) calls his theory "neutral" because he assumes all species are identical to each other in their probability of successfully colonizing any island (local community) in the archipelago (metacommunity), and all have an identical (equal) probability of going extinct—both probabilities being altered only by a given species' abundance. This neutrality underlies MacArthur and Wilson's (1967) equilibrium theory of island biogeography (ETIB) and that is the progenitor of a formal SAR. If there are real niche differences between species and those differences determine population sizes in a rarefaction sequence that remains consistent over time, then the theory will not work. For example, among

Guana's anole lizards, the grass anole is tightly adapted to only very early seral stage vegetation (e.g., grass), whereas the saddled anole is tightly adapted to high, shady sites such as trees in the forest (a building can be a good substitute). If these niche differences impose a permanent abundance rank on the species—one common, another rare—the theory will not work. So at a glance I can reject it as relevant to the groups I know best: insectivorous vertebrates (which in the Virgin Islands include many birds, most bats, most lizards, a few snakes, and all of the frogs). They are all niche restricted in coevolved sympatry in stranded local communities that were either probably not, or in only a small way, assembled by dispersal. Our species are not equal, or "neutral," ecologically and, as I have labored so long to document, no simplistic species : area formulation fits their islands or the whole Antillean archipelago.

Of course Hubbell (2001) does not claim there are no niche differences between species. What he claims, however, is that the niche differences equalize the species and the niches change over time. Similarly, MacArthur and Wilson (1967) did not claim that there were no species number differences between same-sized islands with different histories—one remote and never connected to a mainland, another close and only recently separated; they claimed those differences would disappear with time. For instance, a remote oceanic island will eventually accumulate the right number of species for its size; or on a recently isolated continental-shelf island, the number of species will dwindle by extinctions to the right number for its size. In Hubbell's neutral theory, niche determinations of species' abundances change over time with community drift (de Mazancourt 2001). Is this realistic? Given the niche structure of Guana's insectivorous lizard ensemble, it seems most unlikely that relative abundances would change in any time period short enough to matter. Erosion, sea-level changes, and tectonic events will probably eliminate or amalgamate Guana Island before the actual niches change. Might the lizards change? Well, yes. We already have the cases of *Anolis ernestwilliamsi* and *Mabuya macleani* that have undergone rapid speciation over on Carrot Rock. Both seem to now occupy niches significantly different from those of their immediate ancestors. The problem with ecological drift and equilibrium turnover on this sort of timescale is that it is useless to us when we are attempting to understand ecosystems here and now, as on Guana Island.

A neutral theory also cannot help us with the sort of fundamental adaptive differences between faunas on small tropical islands (for example) and temperate, holarctic continents (for example). McNab (2001) lists a number of fairly general adaptations of this kind, notably including reduction of body mass, evolution of flightlessness in birds, evolution of lower metabolic rates, and replacement of endotherms by ectotherms. These trends, if persistent, would predictably perpetually thwart neutrality-imposing drift. Neutral and equilibrium theories can only be applicable in a homogeneous, deterministic world where only physical parameters (e.g., land area) matter, given time. We have a pretty good paleontological record of life now, and there is nothing whatsoever about history to indicate neutral equilibria. Robert Paine (2002), University of Washington, cites Hubbell (2001) as merely adding to the "uncertainty . . . about how plants in multispecies mixtures coexist." He points out (in litt. 8 May 2002) that the unified neutral theory "is devoid of details of the mechanisms of coexistence" and amounts to little more than "modernized numerology."

If the unified neutral theory actually were true, an apparent consequence would be "that on islands common species will be commoner and rarer species rarer than in the source metacommunity" (Mangel 2002). This is the exact opposite of my observations and experience. Examples abound. Snakes of the genera *Bothrops* and *Boa* are dense on the Lesser Antillean islands of Dominica, Martinique, and St. Lucia compared to their source-mainland South America (Lazell 1964b). Skinks of the genus *Mabuya* and grass frogs, *Rana taipehensis*, are abundant on some small South China Sea islands and relatively

scarce on large islands and the mainland (Lazell 1988). The gecko *Hemiphyllodactylus chapaensis* occurs on the mainland and, rarely, on the large island of Hong Kong, but nowhere is it common except on 127-ha Shek Kwu Chau (Lazell 2002a). The weird burrowing lizard *Dibamus bogadeki* is only known from small islands and has never been seen on the mainland (Lazell 2002a). The same is true of the whipscorpion, *Mastigoproctus transoceanicus* (Lazell 2000c). On the Puerto Rico Bank, *Mabuya* skinks are vanishingly rare on Puerto Rico itself, scarce on Tortola, of frequent occurrence on much smaller Guana, and abundant on tiny Carrot Rock (Mayer and Lazell 2000). The densest nonaggregated terrestrial vertebrate known is the lizard *Sphaerodactylus macrolepis* on Guana Island (Rodda et al. 2001b); it is not rare on mainland Puerto Rico but nowhere near as abundant as the lizards *Anolis pulchellus* and *A. stratulus* (Gorman and Harwood 1977; Reagan 1992). In turn, those two anole lizards are notably less common in the Virgin Islands than is *A. cristatellus* (Lazell 1983a). I could go on and on. Indeed, I cannot believe anyone with extensive field experience on islands has not noticed the differential abundance phenomenon: Species that are rare in their source metacommunities are often abundant on small islands. There are two obvious reasons for this: first, peripheral survival (Lomolino and Channell 1995), and, second, ecological release.

Mangel (2002) states: "Neutral Theory is profound because it assumes that communities are collections of species whose biogeographic ranges overlap for historical and stochastic reasons rather than because of tight coadaptation." Of course, it would be foolish to deny the roles of history and chance in shaping communities (Drury 1998; Levine 2002), but very tight coadaptation has been of fundamental importance in the local guilds and ensembles of West Indian islands (Losos 1990a, 1990b; Beuttell and Losos 1999) and in other mainland–island metacommunites I have studied (Lazell 1999b). In my opinion neutral theory is simply profoundly wrong.

As one would expect, Hubbell's unified neutral theory immediately attracted detractors (Enquist et al. 2002; Chave et al. 2002) and staunch defenders (Bell 2001, 2002). There may be cases and places that fit the theory (Duivenvoorden et al. 2002; Condit et al. 2002): Even a stopped clock is right twice a day. The math is exquisite, but it does not apply to Guana Island.

TAUROSOD

Hubbell's unified neutral theory has inspired me to coin a new term: *taurosod*. The definition is complex. A taurosod is a verbal or symbolic construct—a theory, or postulate, or pronouncement, or formulation—that has lain about in the pasture of life for so long that wondrous growths have sprouted from it: the analogues of the magic mushrooms on the bull pats of Jost Van Dyke, BVI. The formal species : area relationship, $S = CA^z$ (SAR), and the equilibrium theory of island biology (ETIB) are palpable pats and perfect substrates ready to effloresce into taurosods.

Palpable pats and their taurosods can serve useful pedagogical purposes. Initially, they can be conceptual aids enabling the student to visualize the truth about the universe and life by perceiving something simple that can be related to real complexity, but which is clearly fiction and not truth or reality. More profoundly, pats and taurosods elucidate the all-too-human historical process of trying to see form and pattern (and ultimately even purpose) in nature, where stochasticity prevails. This process harks back at least to platonic notions of ideals, forms, and shadows—and no doubt farther back, if we had adequate records. Haeckel's *Ontogeny Recapitulates Phylogeny* and Bohr's model of the atom are fine examples. Species : spatial formulas and assembly rules need to be seen as examples, too: fictitious points of departure from which to assay reality. They will never be fine-tuned into depicting reality because they are palpably untrue to begin with. The fact that Guana Island and Dominica have the same number of reptile species, and even a roughly comparable suite of types, proves that area and elevation have nearly nothing to do with diversity. If one really wants to perceive the irrelevance of spatial parameters in any broad, general sense, begin comparing Guana and Dominica to similar-sized islands in the tropical Pacific. In fact, the extreme irregularity of species : area correlations on Pacific islands has been demonstrated and described in detail by Bauer and Sadlier (2000, 19–22).

The dreadful danger in misunderstanding a taurosod such as the unified neutral theory is the exact opposite of what its author Hubbell (2001, x) stated: "In view of the genuine possibility of a global collapse of biodiversity in the near future, it is unconscionable that we still have no serviceable general theory of biodiversity. The development of such a theory should be made a national and international research priority." Thanks to Hubbell, we have that theory now and do not need to expend additional research effort to develop it. However, if the theory is correct, then prodevelopment, antienvironmental forces can use it to argue that the world of nature is comfortingly homogeneous and saving a handful of exemplary ecosystems will suffice for long-term planetary needs. I contest the implication that a single, simple (simplistic), universal theory of biodiversity will aid efforts at nature conservation and explicitly believe the reverse. Unless policymakers and planners understand that every island, every reef, every habitat is unique in quite unpredictable ways that may, in each case, hold examples and teach lessons absent elsewhere, there will be scant impetus to conserve much of Earth's biodiversity.

Because islands such as the Virgin Islands and Sombrero (Lazell 1964a, 1983a) have many more species—by a factor of three and four—of reptiles, for example, than either MacArthur–Wilson or Hubbell theories indicate they should have, it will be easy for politicians of a certain stripe to argue that, far from restoring lost species, we can proceed with environmentally destructive policies and practices that exterminate even more species: There are currently *too many*!

In a massive inventory of 163 sites in western Amazonian forests, Tuomisto et al. (2003) found striking differences among proximate localities with different physical and chemical characteristics but strong convergence among distant localities with similar niche-determining characteristics; they concluded that the observed patterns "are not explainable by the neutral theory." McKane et al. (2002) demonstrated dramatically and compellingly a resource-based niche differentiation leading to potentially increased diversity in a plant community. In this Arctic tundra ecosystem, solid niche-assembly seems the rule; the community does not seem to resemble those that fit Hubbell's theory, but the authors do not test that or cite Hubbell. This system seems more similar to Guana's insectivorous reptiles than to Hubbell's forest or marine communities. In an analysis of herbivorous insect-feeding niches, Novotny et al. (2002) cite Hubbell's unified neutral theory as a nonequilibrium model—twice! Novotny (in litt. 17 May 2002) responded to my query: "We do not make any general claims about Hubbell's UNT [unified neutral] theory, but rather refer to his book as the latest resource . . . where such models are described and discussed." He continues, ". . . this reference contains extensive treatment of non-equilibrium situations. They are often transient stages in equilibrium models, but the stages which can—under reasonable circumstances—persist very long and as such are much more interesting and relevant than the final equilibrium, which may be rarely attained." Well, all right, but this is a theory that says that nothing matters; it will all come out the same in the end, be happy, don't worry. That contrasts strongly with Hubbell's clear concern for the biodiversity crisis our planet is undergoing and is not a helpful worldview.

Evolution

The periods during which species have undergone modification . . . have probably been short in comparison with the periods during which they retained the same form.

CHARLES DARWIN (1873)

ADAPTATION RESULTS FROM natural selection culling genetically heritable characteristics. Isolation combined with stochastic alterations of genetic material may produce visible changes in populations, called genetic drift, but it will take speciation—anagenesis caused by adaptive natural selection—to make a distinctive lineage; such a lineage may then, if it is ecologically successful, radiate further. Genetic drift is one of the most overrated phenomena in evolution, if one means selectively neutral genetic drift (McKinnon and Rundle 2002). There are three possibilities for characteristics dependent on mutations arising in a newly disjunct population: They may be beneficial, detrimental, or neutral. If neutral, they have probably already appeared in the ancestral population and these characteristics will occur, however rarely, within the range of variation of the ancestors back home. No new characteristic is likely to sweep through a newly disjunct population if it is truly neutral, and no consistent, population-level novel characteristic can remain neutral in any ecosystem for long. Genetic drift—mutation—producing either beneficial or detrimental characteristics is instantly either selected for

or selected against (respectively), and thus the characteristic's frequency or consistency in the disjunct population is the product of adaptational selection, which is not what most people mean by genetic drift. Any characteristic produced by a novel mutation in a disjunct population, no matter how neutral initially, will hit the filter of adaptive natural selection in a very short time.

I believe that, in terms of real-world evolution and speciation, we can effectively ignore genetic drift. All that said, however, genes (i.e., DNA sequences) that do not code for any characteristic truly are selectively neutral. A fundamental belief of molecular biologists who seek to determine phylogenetic relationships is that several kinds of these unexpressed genetic differences can and do accumulate with time. Examples are introns ("spacers" between expressible genes on chromosomes), portions of the mitochondrial DNA, and microsatellite DNA. To the extent that these bits really are immune to and sequestered from selection, they can serve as molecular clocks, the "moleclocks" of Lazell and Lu (2000, 2003). They are supposed to accumulate mutations at a regular rate, which can be calibrated and translated into years. However, simply because these strands are selectively neutral, they have nothing to do with evolution by means of natural selection, which is the only kind of evolution that is ecologically interesting. Sometimes they really do work as moleclocks, but for more discussion see my section "The Iguana," part 5, below.

Natural selection, the motive force of adaptive evolution both within a lineage (anagenesis) and among members of a monophyletic group of lineages (a radiation) invariably results from ecological factors. The ecological factors may be autecological (relevant to a single species), as in characteristics of the population or individuals such as sexual selection or conspecific competition, or synecological (relevant to guilds or communities), as in temperature, moisture, or predation. Roger Thorpe and Anita Malhotra, both at the University of Wales, have developed the concept of two very distinct kinds of heritable genetic characteristics: those that are phylogenetic and those that are ecogenetic (Thorpe 1987; Thorpe and Malhotra 1996; Malhotra and Thorpe 1991, 1997, 2000). Thorpe (1987) divides ". . . the causes of geographic variation into two classes . . . , current ecology and historical processes." I am amazed that anyone believes current ecology is not the direct product of historical processes. Malhotra and Thorpe (2000) state ". . . both history and variation in the strength and direction of natural selection may be responsible for patterns of geographic variation." The implication is clearly that natural selection has no history. They go on: "If there is no significant relationship between patterns of phylogenetic relationship and morphological similarity, we can conclude that the historical signal has largely been overwritten, with selection remaining as a strong contender as the main influence on morphological diversification." The implication is clear: Selection cannot, in the view of Malhotra and Thorpe (2000), produce a phylogeny.

I take the exact opposite view. I believe all phylogenies are the direct result of ecologically driven selection for heritable differences: genes.

I do not believe there are two kinds of genes, implying two fundamentally different kinds of DNA (ecogenes and phylogenes). Further, when one speaks of "phylogenies" within a panmictic species, such as the geographically varying anoles of Dominica or even Guana, one is speaking of gene or character phylogenies, never clades. No cladogenesis has yet partitioned a panmictic species or island population. If some past ecological situation did partition it (as in land-fragmenting sea-level rise), then that initiation of cladogenesis was a failure: The erstwhile isolates returned to panmixis with habitat reunion (as in sea-level retreat). The difference between gene phylogenies within species and clades of whole organisms is basic to understanding evolution (Avise 1991, 1994).

The idea that there are two fundamentally different kinds of genes is deep seated in human thought. It almost boggles the mind that the number of functional genes (codons) contained in a cell nucleus could possibly account for the diversity of life; many people tend to think that there must be something more. However, there are plenty of genes to do the job; we are entirely the products of the interactions of those genes with our environment (Szathmary et al. 2001; Claverie 2001). All of the differences among and between us (from humans to nematodes) are the products of historical factors, whether the factor be immediate selection in drought for water-loss resistance or genetic drift in isolation.

Evolution is a diachronic process; that is, it proceeds at different rates at different times. For this reason, attempting to evaluate an evolutionary relationship on the basis solely of time will generally produce false results. Evolutionary rates are of great importance to those of us interested in conservation. They tell us about how fast a given species can accommodate environmental change. Often, as demonstrated by the dodo, the auk, the quagga, and the passenger pigeon, not nearly fast enough. Because human-caused extinctions are dramatically reducing Earth's species diversity, it is important to know how rapidly new species can evolve to restore the complement. Spectacular evolution can take place within species, even when the animals are abundant and continuously distributed and no impedance to gene flow exists. The striking ecotypic subspecies of West Indian *Anolis* lizards described by Underwood and Williams (1959) and by me in a series of papers (Lazell 1962, 1964b, 1972) are examples. Full speciation, however, seems to proceed most rapidly and frequently with actual spatial, geographic isolation of stocks (Mayr 1970). Numerous ingenious schemes for achieving full species level evolution while still in geographic continuity have been proposed. Some of them might even work, especially for parasites. I am impressed by Lande's (1982) model utilizing the *A. marmoratus* forms I described from Guadeloupe (Lazell 1964b). I point out, however, that the nominal subspecies *A. m. kahouannensis,* isolated on two offshore cays, has progressed much farther in speciation than have the parapatric, still-interbreeding forms. Indeed, Breuil (2002) suggests *A. m. kahouannensis* be elevated to full species rank while documenting character approach concordant with geographic approach to the "mainland" form. This is exactly the sort of evidence I argue indicates continuity of evolutionary role in conspecific isolates (Lazell 1972).

Speciation-in-contact models are never parsimonious. That speciation in contact can occur is incontestable; theoretical models, especially ones involving linkage of genes for niche adaptation and sexual selection, abound (Bridle and Jiggins 2000). That it actually does occur is hard to prove, except in cases involving incipient speciation in isolation where stocks are reunited. Jiggins and Mallet (2000) provide good examples but the conceptual difficulties—What is a species?—are illustrated in Schilthuizen (2000). Dramatic shifts in the numbers and morphologies of chromosomes (karyotypes) have been often invoked for rapid, in-contact speciation. Karyotypic macroevolution might work, but there is little evidence for it in animals. Very closely related, freely interbreeding forms often have bizarrely distinct karyotypes (Short 1976; Paull et al. 1976). Karyotypic differences do not necessarily inhibit gene flow (Futuyma and Mayer 1980). Many things influence rates of evolution, but three major effects have shaped life on Earth in the very recent past: climate and sea-level changes; human-caused extinctions; and human introductions. Fragmentation of large land areas into archipelagoes and warming of temperate and boreal regions have provided theaters for rapid evolution. Massive extinctions have emptied a lot of the stage, however. *Homo sapiens* has spread hundreds of kinds of plants and animals around, which can now speciate in novel environments.

The pattern of evolution involving different rates at different times, described by Darwin (1873), has been called "punctuated equilibrium." Its history, reality, and underlying mechanisms have been subjects of volatile opinion (e.g., Gould and Eldredge 1983; J. H. Brown 1984; Dawkins 1986).

The inundation of Great Guania produced sets of essentially contemporaneous islands on which various disparate evolutionary rates are demonstrable. The problems inherent in distinguishing between true communities of coevolved species and chance amalgamations fundamentally involve evolutionary rates. I shall look here explicitly at questions involving island populations: How do they differentiate? How rapidly? Let us consider two isolated populations that, prior to isolation, were panmictic: freely exchanging genes and undeniably the very same sort of animal. What mechanism isolates them is not immediately important. Sea-level rise may fragment their previously continuous range into two new, smaller pieces (islands or an island off a mainland). For example, a rise of a meter or two would separate Monkey Point from the rest of Guana Island and isolate populations of crested anoles, ground lizards, common snakes, and sphaeros there. Chance or waif dispersal may take a propagule from one island to another, or to an island from a mainland (Carlquist 1981). A new population may be founded by human introduction. Four basic sorts of recognizable variation may rapidly result; the second law of thermodynamics will not allow things to remain static indefinitely.

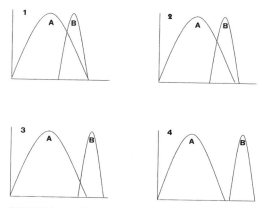

FIGURE 14. Four frequency patterns of variation in a characteristic of two populations, A and B. A measure of discernible difference in the characteristic is illustrated by the overlap zone, or the lack of it. *x* axis, the measure of the characteristic; *y* axis, number of individuals.

In figure 14, four frequency patterns of variation in a characteristic of two populations, A and B, are illustrated. In situation 1 we see classic founder effect. The smaller population, B, shows only a very small proportion of the total variation exhibited in the larger population, A. No evolution has yet taken place in this example. All of B's characteristics are included within the range of variation of the more variable population, A. Some taxonomists will call B a subspecies of A if 75% of the individuals in population A can be distinguished from the individuals in B. This is bad taxonomy, abnegating the real meaning of the "seventy-five percent rule" for subspecies (Amadon 1949; Simpson 1961; Lazell 1972). In fact, no individual from population B can be identified without recourse to locality data: All of the characteristics of B individuals are matched by some individuals from A. This is exactly the pattern one expects to see just after separation occurs. This is the moment of cladogenesis: the result of a geographic, abiological event over which the organisms involved have no control. Genetic isolation here depends on factors extrinsic to the organisms and real genetic differentiation will initially depend on entropy: the failure of DNA in each now-separate lineage to replicate exactly, every time. Genetic differences will appear; if selected against, then the lineages will remain the same, evolving in

parallel. If some of those chance changes— mutations—are selected for in one lineage but not in the other, then anagenesis begins. Anagenesis is real evolutionary change, adaptation by means of natural selection (Lazell 1996a).

In situation 2 evolution has begun. Some taxonomists would classify A and B in this example as different subspecies. Granted, some evolution has occurred, but too little for useful taxonomic differentiation. Showing a 95% probability that the means of the populations are genuinely different is nothing like saying the animals are 95% different—or 75% different, or even one-half different. To qualify as a subspecies, 75% of individuals must be identifiable without recourse to locality data. Here only a few, those outside the overlap zone between A and B, could be identified. This is a picture of incipient subspeciation: the beginning of the evolutionary process leading to taxonomically recognizable forms.

In situation 3 evolution has proceeded far. The vast majority of individuals in A can be separated from the vast majority in B (and vice versa, of course). There is still a small area of overlap: a few confusing, unidentifiable individuals occur. However, this is a good example of valid subspecies.

In situation 4 evolutionary differentiation is complete: All individuals in A are distinct from all individuals in B; there are no confusing intermediates. Taxonomic recognition is required in this case. Deciding if A and B in situation 4 are full species or just very well differentiated subspecies is not simple, but see Amadon and Short (1976). I have provided criteria that I have found handily applicable for deciding the matter (Lazell 1964b, 1972). Basically, I would call these two forms full species unless I had good evidence to the contrary. This is the classic picture of speciation.

When we ask how rapidly differentiation in isolation can occur, we are simply asking how rapidly the population turns over. For evolutionary change to occur, each individual in the original population must be replaced by a new individual with a different genotype. To even answer the question in simplest terms we will need to know what N, the total number of individuals in the stable population undergoing evolution, is. We will need to know how many offspring, X, inherit the novel genes per parent. Thus we must know how many offspring, on average, each individual produces, and how each gene is inherited. Finally, we will want to know how long a generation is in weeks, months, or years. We do not need to know this initially, if we are satisfied to just determine how many generations, g, are required for the novel genome to sweep the population. The simple relationship (Lazell 1972) is:

$$X^g = N$$

or

$$g = \log N / \log X.$$

This formula also yields the minimum number of generations required for a colonizing species to reach its optimum, stable population size, beginning with a single propagule: the number of individuals needed to found a population. The propagule could be a pair or just a gravid (or pregnant) female in sexually reproducing species. The formula is true for expanding colonies because just as with evolutionary changes they require new individuals. For evolutionary change, of course, a new individual must replace an old one. I put Lianna Jarecki, when she was in 10th grade, to the task of looking quantitatively at evolution in populations of snakes of the *Liophis portoricensis* complex, such as the one on Guana Island, *L. p. anegadae*. A number of morphologically differentiated kinds of these snakes—subspecies—occur on various islands of the Greater Puerto Rico Bank, as listed by Schwartz and Thomas (1975).

Lianna used a simple Lincoln index on our mark–recapture data for snakes on the island and assumed a similar original density on Tortola. We did a quick line-transect (road) census of mongooses on Tortola. Fortunately, there are no mongooses on Guana; the number 445 is just our guesstimate of how many there could be. Here are some figures:

	Guana	Tortola
Area (ha)	300	5,494
Number of snakes	702	13,186
Number of mongooses	445	8,247

Now, she assumed the original mongoose propagule arrived on Tortola about 1890. In four generations, the island would have been full of mongooses. If mongooses were introduced to Guana, they would predictably fill it in three generations. If just one fertilized female snake possessed the wonderful gene for mongoose avoidance and, therefore, survived, Guana could be repopulated with mongoose proof snakes in six generations. This example is illusory, of course: we have no idea how many genes are involved in successful mongoose avoidance and survival, but surely more than one. Snakes today are nowhere near as dense on Tortola as on Guana. Whether they ever were is not clear. Snakes are as scarce on many other islands without mongooses as they are on Tortola. They are more common on Guana than on most other islands.

That real evolutionary rates and actual speciation can be rapid is not surprising (McKinnon and Rundle 2002; Stockwell et al. 2003). Sea-level rise separated Muskeget Island from Nantucket and the North American mainland no more than 4,000 years ago. The Muskeget vole, *Microtus breweri*, seems undeniably to be a very distinct, full species (Lazell 1979, and references therein). Carrot Rock must have been one with Peter Island just a couple of thousand years ago, and part of Great Guania 8,000 years ago. Yet *Anolis ernestwilliamsi* (Lazell 1983a) and *Mabuya macleani* (Mayer and Lazell 2000) are strikingly distinct forms. It is possible that these forms evolved elsewhere and survive on their little islands as relicts; possible, but unlikely and not the simplest explanation. They are small populations; their islands are ecologically peculiar; selection pressures are presumably strong; and the time in generations needed for new genotypes to sweep is short. A similar case would seem to be the very distinctive

Sphaerodactylus gecko discovered on Carval Rock, BVI, by Alejandro Sanchez (specimens in U.S. National Museum) and as yet undescribed or named. (See plate 9.)

Less dramatic examples include Guana's crested anoles, which are dramatically different from other populations of *A. cristatellus wileyae* in their great ability to resist cutaneous water loss (Dmi'el et al. 1997; Perry et al. 1999, 2000), and Guana's two populations of saltgrass planthoppers, which are widely divergent in percentages of winged versus wingless morphs and have been demonstrated to be modally genetically differentiated as in situation 1, above (Denno et al. 2001). However, Guana's only unique full species known to date, the bromeliad *Pitcarnia* new species (George Proctor, pers. comm), probably is not autochthonous; that is, it probably did not evolve on Guana Island. It probably was (or is) more widespread and only survived the predation of animals like goats and donkeys where those exotics were scarce or absent or its habitat was virtually inaccessible to them. It will be very interesting to see if *Pitcarnia jareckii* turns up on other islands now that its existence is known.

There are many examples of rapid evolutionary change in populations very recently introduced to new environments by man. Ashton and Zuckerman (1950, 1951) document evolutionary divergence from African progenitors in *Cercopithecus* monkeys introduced to the Lesser Antillean island of St. Kitts. Schwarz and Schwarz (1943) long ago documented rapid evolution in the house mouse, *Mus musculus*, a commensal of *Homo sapiens*. Patton and Hafner (1983) studied divergence in European rat populations introduced to the Galapagos Islands. We have presented preliminary evidence for speciation in a population of rock wallabies (genus *Petrogale*), introduced from Australia to Hawaii just 60 generations ago (Lazell et al. 1984). Eldredge and Browning (2002) use karyotype, protein electrophoresis, and mtDNA to place Hawaiian wallabies within the many differentiated populations of *Petrogale penicillata* surviving in

Australia, but they find no match for it. They do not even mention the size, proportion, cranial, and color characteristics that separate the Hawaiian form from all other *P. penicillata*. There are no Australian wallabies known today—alive or extirpated—that look like Hawaiian rock wallabies. Grismer (2002) provides a probable case in iguanas.

When human introductions are most successful, they usually involve commensals that are simply moved with *Homo sapiens* to another *Homo sapiens* habitat. There is, therefore, scant selection pressure for rapid evolution. Hedeen (1984) provides a classic example for a European wall lizard. It has done as well on the walls of Cincinnati, Ohio, as on those of Milan, Italy; the habitats are remarkably similar. The rock wallaby in Hawaii, however, was introduced from a fairly dry, temperate climate to a very wet, tropical mountain range. It is quite remarkable that it managed to survive and less remarkable that it seems to have undergone rapid adaptation in highly visible ways. One of the most fascinating and relevant pictures of rapid evolution yet documented involves the finches studied in the Galapagos archipelago by Boag and Grant (1981). Here drought selected severely for larger-billed birds and effected measurable change in the population in just three years. McKinnon and Rundle (2002) documented rapid speciation in recently isolated populations of stickleback fishes.

Examples of inadvertent exotic introduction that have led to speciation include a remarkable case of the so-called Colorado potato beetle, *Leptinotarsa decemlineata*. The beetle originated on the Mexican plateau. Spanish cattle, sometime after the conquest of Mexico, moved burr weed, a member of the potato genus *Solanum*, northward across the Rio Grande and into the range of the bison. The burrs of the burr weed clung to the hair of both cattle and bison; bison soon moved the plant north all the way to Canada. The Mexican ancestral beetle stock, incapable of surviving on potato, moved north too, feeding on its now widespread host plant, burr weed. Potatoes, introduced from Peru to

Europe, and then back across the Atlantic to North America, were planted by expanding waves of humans of European ancestry. In the 1840s, potatoes reached Nebraska where the beetles were already in residence. In 1859, the year Darwin published his *On The Origin of Species*, the first potato-eating *Leptinotarsa* was noticed; their population exploded. Wenhua Lu, then at the University of Rhode Island, showed that a mere seven genes could account for the distinctions in appearance, physiology, and behavior between the Mexican ancestral form and the new, true potato beetle; details of the story are provided by Lu and Lazell (1996). Today the two kinds of *Leptinotarsa* are de facto species, each living on its own host plant, with a large geographic gap and little or no hybridization between them.

Studies of adaptation are ongoing on Guana Island with *Anolis* and *Sphaerodactylus* lizards, planthoppers, and other species, as described below. J. H. Brown (1984, 1995, 65–67) makes predictions about evolution in peripheral isolates of wide-ranging species that illuminate the subject of punctuated equilibrium. First of all, very common widespread species are most apt to be the progenitors of peripheral isolates simply because they are common and widespread. For the same reasons, they are apt to make it into the fossil record, especially in central areas of their range. Thus, *A. cristatellus* produced *A. c. wileyae* on eastern Great Guania (now the Virgin Islands), and *A. c. wileyae* produced *A. ernestwilliamsi* on Carrot Rock. *Anolis cristatellus* will probably appear in the fossil record of Puerto Rico over a span of millions of years. While the peripheral isolates—far flung on often tiny cays and, therefore, quite absent from the fossil record—are busily evolving into truly weird and novel forms, the stay-at-home progenitors change little over time. If and when one of the highly modified, bizarre peripheral isolates recolonizes the central part of its progenitor's range, and if and when it succeeds adaptively and begins to displace the aboriginal stock, it may suddenly appear in the fossil record. A Gould or an Eldredge, not having witnessed the novel form's classic evolution yet suddenly finding its bones in the fossil record overlying those of the ancestral type, may leap to the conclusion that evolution proceeds by miraculous jumps—saltation. Claiming that classical evolutionists are "wrong" makes wonderful ammunition for the religious fundamentalists but actually only reveals shortfalls in our data. Certainly evolution occurs in rapid spurts with long periods of relative stasis in between. Such punctuated equilibria may occur in the same lineage of organisms at different times and proceed at different rates for different characters and organ systems.

I have promulgated the notion that the magnificent, giant *A. roosevelti* evolved as a peripheral isolate very much as *A. ernestwilliamsi* did at high-stand sea level and then spread around eastern Great Guania with sea-level retreat (Lazell in press). *Anolis ernestwilliamsi* seems to have undergone rapid alteration in scale size, body size, and toe lamellae, but it has not changed with respect to the vertebral column. Another *A. cristatellus* derivative, *A. cooki*, has scarcely changed at all from the ancestral stock with respect to scalation or body size, but its vertebral column has changed and it has a disproportionately long tail (Marcellini and Jenssen 1983). There is no a priori reason why evolution has to proceed in spurts. Stabilizing selection, at work on common species such as *A. cristatellus* in the middle of its range, may allow only slight, gradual changes over long periods of time in concert with long-term changes in rainfall or average temperature. Many taxonomists have turned to molecular biology in hopes of finding some sort of accurate evolutionary clock for reading rates of change. Early on, Maxson and Wilson (1975, Maxson et al. 1975) presented a hopeful case for frogs of the genus *Hyla*. They postulated an albumin clock with a regular, gradual change of 1.7 units per million years. If this were true, it would coincide nicely with the notion that Australia, Antarctica, and South America were a continuous landmass 75 million years ago. The Australian members of *Hyla* (called *Litoria* simply because they come from

Australia) are about 120–130 albumin units different from South American forms. Only a few of the myriad forms were considered. *Hyla* are good rafters and overwater colonizers, being widespread in the West Indies (even in the Virgin islands, where they are introduced and called *Osteopilus*). Using similar techniques, Hass et al. (2001) found one line of West Indian *Hyla* that seems to have diverged from its congeners 107–109 million years ago. This would indicate that it colonized the islands about 40 million years after its lineage had differentiated elsewhere.

The very nature of evolution and its variable rates erodes the validity of the philosophy (it is philosophy, not science) of cladistics: the view that propinquity of descent—time of evolutionary separation—is the sole determinant of relationship. This means that all populations that separated at the same time are "sister groups" and must be given equal taxonomic rank. Imagine Great Guania of the Puerto Rico Bank as the rising sea level fragments it into hundreds of islands. Imagine the coast of New England at the same time in (very recent) geologic history. The populations of voles and anoles and snakes and snails separated from each other by the rising sea have hundreds of different rates and pathways for evolution—as many as there are now-isolated populations. Some will simply not change at all over many generations. Some will undergo rapid evolution to become *A. ernestwilliamsi* or *Microtus breweri*. Some, for instance, the snakes of the *Liophis portoricensis* group, will show intermediate rates and partial differentiation. Some, such as the Guana palm snail, will be clearly distinct but difficult to evaluate as species or subspecies (Lazell 1983b). To try to force this vast mosaic of evolutionary rates and processes into the same mold based on the timing of island fragmentation is antievolutionary. Brower (2000) points out that cladistics (time-based classification) has nothing to do with evolution.

A major problem is in failing to distinguish between concepts and things (Lazell 1972). Relationship is a concept, not a thing. Organisms do not have any relationship: You cannot weigh it or measure it or dissect it out. You can only infer it based on things you can measure, count, or weigh. As soon as two lineages of organisms begin to evolve separately from each other—as for example on Guana and Tortola—inference of any and all relationship between them becomes wholly subjective, relying on lines of evidence given different weights and priorities. For me, propinquity of descent is one of the least important things I can measure and add to a balanced inference of relationship. My reasons are the widely divergent evolutionary rates.

The crested anoles include *A. c. cristatellus* of Puerto Rico; *A. c. wileyae* of the islands east of Puerto Rico including Culebra (the type-locality, from which the subspecies was named), Vieques, and most of the Virgin Islands (except, of course, St. Croix) all the way to Anegada; *A. cooki* of extreme southwest Puerto Rico and some coastal cays; and *A. ernestwilliamsi* of Carrot Rock and, once at least, Peter Island, BVI. We also recognize three more unnamed geographic variants: the southwestern Puerto Rico population of *A. c. cristatellus* with a plain yellow throat fan; the West Dog, BVI, population of *A. c. wileyae* that averages very large size, has a short head, and has big scales; and the Guana Island population of *A. c. wileyae* that is detected physiologically by its unusually high resistance to cutaneous water loss. No doubt there are more. It is notable that the other two widespread common anoles in the Virgin Islands—saddled, *A. stratulus*, and grass, *A. pulchellus*—do not show a similar array of differentiated populations; they appear the same, or only very weakly modally variable, from one island to the next. I wonder why? We have a strong tendency to look for differences between populations when there are obvious ones we notice in the field: size, shape, or color. Thus, back in 1980, I was immediately struck by the large size and long head of *A. ernestwilliamsi*; this led me to start counting scales, which confirmed the distinctiveness of the form. Perhaps saddled and grass anoles have differentiated here and there in subtle ways (e.g., scale size) that have as yet gone undetected.

I know little about *A. cooki;* I have never seen one alive. Schwartz and Henderson (1991) provide an account. The species is immediately distinct from sympatric *A. c. cristatellus* because both sexes have a dark red throat fan (plain light yellow in *A. c. cristatellus* in southwestern Puerto Rico). The very long tail is also obvious if the specimen has not been damaged. My hypothesis for this species' evolution is that it originated as a "rock knock-off" sensu Lazell (1999a) of ancestral *A. c. cristatellus* on a very small, arid islet off the southwestern coast of Puerto Rico at high-stand sea level during some interglacial previous to this one we live in today. It would be fascinating to check its water-loss resistance because I believe *A. cooki* must be adapted to exceedingly dry conditions. But, because it is not distinctive in either body size or scale size, it (similar to the Guana Island *A. c. wileyae*) must be adapted to aridity in some way not visibly apparent to us humans. Throat fan color may have originally been altered from the mainland norm by genetic drift, but autecological sexual selection must have favored individuals with darker and redder fans; perhaps fan color was linked to water-loss resistance. Tail length seems inexplicable because a longer tail, one would naturally conclude, would increase evaporative water loss. So long tail would, I suppose, be what Malhotra and Thorpe (2000) would hail as a phylogenetic, as opposed to ecogenetic, difference. I believe the problem is simply that I (we) do not have the wits or field experience to perceive the ecological selection pressure for long tail in this case.

Perhaps long tail in *A. cooki* is a pleiotropic effect of the genes for some other, highly beneficial, characteristic. Pleiotropy is the condition of one set of DNA (genes) coding for two or more seemingly unrelated characteristics. An example seems to be the reduced pelvis seen in bulldogs and panda bears, which is thought to be a pleiotropic effect of genes that produce hypertrophied jaws and jaw musculature—strongly selected for in both bulldogs and pandas (D. D. Davis 1964). We probably will not know for sure until we sequence the genomes of all the lizards involved and perceive just what strip of DNA codes for which characteristics. In any case, I hypothesize that the return of a glacial cycle dropped sea levels so that *A. cooki* could invade the mainland. At times of heavy rainfall, *A. cooki* would have no advantage over *A. c. cristatellus* in southwestern Puerto Rico, a naturally arid region in the rain shadow of a high-mountain cloud barrier. At times of decreased precipitation, such as the late Pleistocene xeric period, or the Younger Dryas about 12,000 ybp, or the Little Ice Age about 275–175 ybp (Haug et al. 2001), arid-adapted *A. cooki* might outcompete *A. c. cristatellus* and expand its range. If so, a return to more pluvial conditions, as has occurred in the West Indies in the past couple of hundred years, could easily be putting *A. cooki* at a disadvantage again.

I know a lot more about *A. ernestwilliamsi,* another exemplary rock knock-off perhaps today confined to Carrot Rock, BVI. It is the best at resisting water loss among the suite of *A. cristatellus* group anoles yet tested (Dmi'el et al. 1997); marginally even better than the *A. c. wileyae* champions on Guana. However, this water-loss resistance is wholly the result of large size, which provides a high mass to surface-area ratio (the simplest possible way to reduce evaporation). This finding was unexpected because *A. ernestwilliamsi* has relatively small scales and scale size is tightly correlated to rainfall in *A. oculatus* on Dominica. Indeed, Malhotra and Thorpe (1991) regard scale size as an ecogenetic trait of no value in taxonomy because it is so evidently just the result of available water. Note, they do not argue that scale size is just the result of phenotypic adjustment or plasticity; they demonstrated that scale size is genetically determined. Based on their and my own scale-size evidence in *A. oculatus* (Lazell 1962, 1972), I expected water-loss resistance in *A. ernestwilliamsi* to exceed anything mere mass could do. Why should small scales aid in water retention? I have no idea. Neither do Malhotra and Thorpe (I have asked them). If scale size is an ecogenetic trait, none of us has a clue what its ecogenesis is.

Then why do *A. ernestwilliamsi* have small scales? For my next theory (because the first one failed), I will suggest that the genetic basis of small scales is the same as scale packing: getting more scales into the same space. *Anolis ernestwilliamsi* has many more subdigital lamellae—expanded scales that improve clinging ability—than *A. cristatellus*. It needs them: It is much bigger; that alone makes hanging on harder. Its habitat, Carrot Rock, is very windy and most of the best perches are on rock faces a lizard cannot grip as they would a stick. If this theory is right, then small body scales are just a pleiotropic effect of selecting for more subdigital lamellae, making it ecogenetic indeed: The ecology of Carrot Rock selects for both large size and high-scale (lamellae) counts.

I believe *A. ernestwilliamsi* is a product of this interglacial and began its evolutionary departure from *A. cristatellus* just a few thousand years ago. If this is so, then it must have colonized Peter Island over water; that is not remarkable because hurricanes demonstrably break down and wash away Carrot Rock's seagrape trees, and Peter Island is a very close, very large target. What is remarkable is that *A. ernestwilliamsi* managed to get from its landing stage up to the proper habitat: the boulder jumble at the top of Peter Island (now subsumed by a large house). Also, the only known Peter Island specimen, a juvenile male, had an all blue gray throat fan. *A. ernestwilliamsi* starts out with all blue gray fans and the color—red and green just as in *A. c. wileyae*—invades from the rear. In *A. c. wileyae*, at least on Peter Island and Guana, where I have paid close attention, the fan starts out colored but pale, and the colors deepen in place. Even as old adults, *A. ernestwilliamsi* never quite develops a completely colored fan; the anterior portion remains blue gray. Still, to be as big as it was and show no red or green in the fan hints at character divergence in the Peter Island *A. ernestwilliamsi* population away from *A. cristatellus*.

Losos (2000) provides an excellent review of the phenomenon of character divergence, the idea that "... natural selection will favor, in each population, those individuals whose phenotype allows them to use resources not used by members of the other species. . . . The result may be that the populations diverge in phenotype and resource use . . . , thus reducing resource competition and permitting coexistence" (Losos 2000). Darwin named this "character divergence" but W. L. Brown and Wilson (1956) renamed it "character displacement" and, I admit, did a better job of explaining and describing it. Losos (2000) discussed only what he regarded as ecological character divergence and left out sexual selection. This leaves out some of the finest examples—e.g., fan color in *Anolis cooki*—in which the resources competed for are mates. Individuals that look visibly different from the ancestral type and prefer to mate with individuals that look like they do will be selected for, if the two forms involved have speciated to the extent that hybrids between them are disadvantaged or defective. Those that indulge in mismatings, and produce defective offspring as a result, are wasting reproductive effort and energy. It is a basic tenet of my view of evolution and systematics that two very closely related forms, like *A. cristatellus* and *A. ernestwilliamsi*, simply could not live in such close proximity and retain their identities if interbreeding were successful: They can only be distinct biological species.

Losos (2000) also makes what I regard as one major mistake in his review of character divergence in his fifth postulate: rejecting clinal variation, where one or both forms become increasing dissimilar concordant with geographic approach. Character divergence actually causes such clines: the genetic influence of the divergent individuals in the geographic area where the two species come closest or actually overlap is mitigated by distance away from this area. Notice I said "come closest"—it is not necessary for the two species to actually overlap or even to be in permanent contact. Of course, it is necessary for them to be in at least occasional contact. E. E. Williams (1969) thought *A. ferreus* of Marie Galante might have evolved giant size as character divergence as a result of frequent failed invasions of the smaller anoles on surrounding islands; he called this the "nudge effect." When I initially proposed this scenario to him, several years prior to his publication of it, he dismissed it as preposterous (as have several others before and since). His example is a poor one, I believe, but the two southern Lesser Antillean giants provide a perfect case. *Anolis richardi* occupies the great Grenada Bank and gets more and more colorful and strikingly different from St. Vincent's *A. griseus* as one proceeds northward up the Grenadines, closer and closer to *A. griseus*, reaching the acme of distinction on Bequia—geologically and biologically part of greater Grenada, but within close sight of St. Vincent (Lazell 1972).

The *A. c. wileyae* on Peter Island show stronger differences in visible characters from *A. ernestwilliamsi* than do remote populations: they average small size and none show any of the four color characters that, in total combination, help make *A. ernestwilliamsi* even more distinctive than body size and scale counts do (Lazell 1983a). The skinks, *Mabuya sloanii*, similarly diverge away from Carrot Rock's *Mabuya macleani* in a clinal fashion (Mayer and Lazell 2000). A superb example occurs in the meadow vole, *Microtus pennsylvanicus*, as it approaches the beach vole, *Microtus breweri*, of Muskeget, Massachusetts. The beach vole is big and pale; the meadow vole is smaller and darker than average on Esther Point, western Nantucket, and is the smallest and darkest on Tuckernuck Island, closest land to Muskeget. A sand tombolo moves westward from Nantucket along Esther Point (sometimes Esther Island) and occasionally reaches Tuckernuck; the sand extends along the south side of Tuckernuck and on west-northwest, rarely all the way to Muskeget. Voles can travel on the sandbars, sometimes surviving for generations in the beach grass growing on them. The result is rare contact between the species; if they were not full species the exact opposite of character divergence would occur: introgression and intergradation.

The exact opposite of character divergence, in fact, happens on the islands just off the east coast, and on the east coast, of Puerto Rico: *Anolis cristatellus wileyae* intergrades with nominate *A. c. cristatellus* (Heatwole 1976). The two subspecies—they have to be subspecies in my reckoning, although some will argue they are full species with a "hybrid zone"—differ most strikingly in throat fan color: red with a bright green center in *wileyae*; orange yellow with a pea green center in nominate *cristatellus*. In fact the fan color of nominate *cristatellus* in most of Puerto Rico, as described, can be seen as intermediate between the fan colors of *wileyae* and *cristatellus* from the southwestern rain shadow region of Puerto Rico: There the fan is plain, unpatterned, pale yellow. Why not recognize the yellow-fanned anoles as a third subspecies? We looked into that in depth. In western Puerto Rico the changeover is a sharply stepped cline: a very narrow zone of intergradation, so to speak. But coming east along the south coast, around the east end, and on west along the north coast, the situation is one vast, variable, gradual cline: There is no edge; no place to draw a line, or even a band. Right out at El Morro, in fact, one can find an anole with a plain yellow fan; El Morro is for all practical purposes the type locality of *A. cristatellus*: the biggest, best-surviving population in San Juan. Thus we see striking geographic variation within the subspecies *A. c. cristatellus*—geographic variation that does not translate into taxonomy as another subspecies.

Isolated populations may actually undergo evolutionary convergence. Consider crested anoles on Guana and their conspecifics on Sage Mountain, Tortola, today obviously separated by seawater. In the recent past, with sea level 120 m lower, they were in complete genetic contact across dry land. Their ecological conditions, however, may have been very different. The 120-m sea-level drop would put the top of Sage Mountain at 641 m, high enough to constitute a "snag" sensu Lazell (1972) and have cloud–rain forest conditions much wetter than the "aridulate rain forest" (D'Arcy 1967) present on Sage today. Guana, however, though also 120 m higher, might have been even drier than it is today, if the climatic postdictions of Pregill and

Olson (1981) are correct. Even at 366 m above sea level, Guana's highest point would be well below the ca. 600-m snag line. In my scenario, the Guana land area of, say, 20,000 ybp would have been far more ecologically different from Sage Mountain than it is today. Crested anoles evolving in response to the subsequently converging environmental conditions would very likely have undergone some adaptive evolutionary convergence as well.

Throughout most of the Wurm time, from some 70,000 or 80,000 ybp up to 25,000–20,000 ybp, ice, sea-level lowering, and the land area of Great Guania were not at their maxima. It seems probable that for long periods sea level was about 50 m below its present level (Dawson 1992); this would have left a land connection from high, wet Puerto Rico eastward to the amalgamated Virgin Islands of Great Guania that would have amounted to a classic filter bridge: It let some species make the traverse, but others were stopped by ecological factors beyond their adaptive abilities. The fan color difference between the subspecies of *A. cristatellus* may have arisen autecologically through sexual selection. Or, as I have suggested above for *A. cooki*, red may be a pleiotropic effect of a genotype for water-loss resistance. Other less visible differences, most probably physiological characteristics involving water-loss resistance, probably evolved too: synecological characteristics. If water-loss resistance is costly in energetic terms or expensive in terms of enzymes coded to produce it, it would be selected against in high-precipitation regimes where it is not needed. At the western end of the long eastern peninsula of Great Guania there would always be an abrupt transition zone from arid to mesic conditions; the genetic constitution for high water-loss resistance would meet the proto–Puerto Rican genotype that does not need such a hereditary burden. In a small zone, the intermediate genotype, or a mix of genotypes, might long prevail: a zone of intergradation would be selected for and subspecies status maintained.

Over multiple seasons and in several papers, it has been demonstrated that the *A. c. wileyae* of Guana Island not only resists water loss better than any other crested anoles of their size yet tested, but the difference is apparently not simply the result of phenotypic plasticity (Dmi'el et al. 1997; Perry et al. 1999, 2000). During drought years this resistance would be a real advantage, but in the rainy years since 1997 the genetic constitution providing this resistance could be excess baggage. If water-loss resistance is advantageous to crested anoles why has not the Guana genotype spread? Why did we not, after a decade of drought, find the Guana-type of crested anoles on Tortola? Surely, if the genetic ability was there it would have rapidly swept through the population, especially at the coastal site, The Bridge, East End, where we got our Tortola lowland animals for experiments: The turnover in anole populations is potentially annual or better in the face of strong selection.

My hypothesis for Guana crested anoles is that the genetic ability to resist water loss appeared by mutation in one individual; his or her descendants replaced the whole population at some past time of severe drought when they had the advantage over ordinary anoles. Subsequently, mutations undoing the resistance genetic code have not occurred; so Guana anoles are stuck with their hereditary baggage. A prediction of this notion is that in times of plenty of precipitation these high-resistance anoles are at a disadvantage. The saddled anole, *A. stratulus*, and the grass anole, *A. pulchellus*, lack water-loss resistance—at least to any special extent and at least in all the populations checked so far. Although there is good modal niche segregation between the three species, there is plenty of overlap. By 2000, after more than three wet years, we noticed that saddled and grass anoles had both become much more common than they had been in the drought years and crested anoles were relatively scarce; in the drought years crested anoles had outnumbered the sum of the other two by a factor of at least two.

Under very arid conditions, crested anoles can outcompete both saddled and grass anoles simply because they are bigger: The mass to surface-area ratio is greater, so they simply evaporate less water. On a small cay without considerable habitat diversity, this means that *A. cristatellus* will push *A. stratulus* and *A. pulchellus* to extinction. On a barely vegetated rock, broiling in the tropical sun, not even *A. cristatellus* can survive without the kinds of ecogenetic adaptations that lead to speciation. There are no crested anoles on Watson Rock or Carval Rock in the BVI, or on Frenchcap in the USVI. The only surviving reptiles on these rocks are sphaeros (genus *Sphaerodactylus*). Apparently the mutations providing water-saving large size that appeared in some ancestral *A. ernestwilliamsi* on Carrot Rock never appeared in the other rock populations.

One more crested anole population seems to have moved from the species' norm: that on West Dog. These grow big—to 70-mm snout-to-vent length (SVL) regularly—and have short snouts; they seem to have larger than average scales. To date, we have only a half-dozen specimens, divided between the Museum of Comparative Zoology and the U.S. National Museum: too few for statistical tests of significance. West Dog is small and parched, with a rugged coastline that defies gene-exchanging colonization. Until we get more specimens and more data we cannot evaluate this population. If the differences turn out to be modal only, and contained within the overall variation of crested anoles from the rest of the range of *A. c. wileyae*, then the West Dog population could be no more than the product of founder (or survivor) effect. If means are outside the range of other *wileyae* variation then at least genetic drift is indicated. Obviously, we must try to determine the West Dog population's water-loss resistance.

The geographic pattern of evolutionary differentiation in *Mabuya* skinks is uncannily congruent with that of crested anoles (Mayer and Lazell 2000). There is a Puerto Rican mainland form, *Mabuya sloanii nitida*; the nominate form *M. s. sloanii* on many of the eastern islands of the Puerto Rico Bank from Culebra to Anegada; a zone of intergradation between the two subspecies involving extreme northeastern Puerto Rico and at least one close offshore island, Icacos; and an endemic derivative of *sloanii* on

Carrot Rock: *Mabuya macleani* (Mayer and Lazell 2000). The Carrot Rock skink is pallid and almost unpatterned, in contrast with the rich copper brown and black, boldly patterned *Mabuya s. sloanii* of Peter Island and many other Virgin Islands; its color and pattern departures from the ancestral stock are much more striking than are those of *A. ernestwilliamsi*. It is not differentiated in size; however, it does have smaller scales than its relatives. I cannot invoke the same explanation for small scales—increasing their number in a given space—that I floated for the anole, though: *Mabuya macleani* has no digital gripping pads.

How rapidly can evolution proceed on Carrot Rock? The habitable (for lizards) surface area of that islet is about 1.3 ha. For *A. ernestwilliamsi* we can reasonably assume a density comparable to that found by Rodda et al. (2001b) in their earliest seral stage habitat for *A. c. wileyae*: about 1,000 per hectare. So, about 1,300 anoles on Carrot Rock. We can also assume that *A. ernestwilliamsi* resembles Guana's *A. c. wileyae* in reproductive output, with females yolking up an egg about every other week, but not so often in very dry weather; let us say about 24 eggs per year. Female anoles reach sexual maturity in a year, so one year constitutes a generation. We will make two assumptions: a balanced sex ratio and one-half of the offspring of a female that has the beneficial gene inherit that gene. This means *x*—the number of female offspring per individual in the population carrying the beneficial gene—is 24 divided by two (sex ratio) divided by two again (carriers) equals six. Thus,

$$g = \log 1{,}300/\log 6$$
$$g = 3.11/0.78$$
$$g = 4.$$

In just four generations, potentially, a new genotype could sweep the Carrot Rock population. Of course, no real gene would be that overwhelmingly beneficial and real lizards would never be so totally successful at instantaneously replacing other lizards, but the point is that evolution can proceed rapidly indeed

(Bjornstad 2001). The problem, then, is not the speed of natural selection and speciation; the problem is having the right stuff—those novel, highly beneficial genes.

Let us stick with the same density and reproductive data for Guana Island and assume 250 of Guana's 300 ha are populated by crested anoles, making a total population of 250,000. Now:

$$g = \log 250{,}000/\log 6$$
$$g = 5.4/0.78$$
$$g = 7.$$

There becomes a crippling problem with my simplistic formula; the formula assumes no lizards with the beneficial genotype die, rather they all replace other lizards without that genotype, and those die, disappear, or are vaporized. Seven generations exceeds the likely longevity of real crested anoles, although not by much: We have had several marked as adults that remained on their territories for four more years, implying they were at least five. I think the most we can say for rapid evolution in this case is that the whole population could be replaced by anoles with the novel, beneficial genotype in less than a century; that may not be unreasonable.

I have earlier hypothesized that the occasionally extremely beneficial genetic constitution for water-loss resistance appeared in Guana Island *A. c. wileyae* and swept through the population. Further, in times of plenty of rainfall this genetic constitution is not beneficial; it is detrimental excess baggage. How could it have swept through so large a population, leaving not one individual left to take the population back toward normality in rainy times, like we have now? If the crested anole population of Guana Island is, as I suggested above, augmented by occasional waif dispersers from (for example) Tortola, why has the population not yet been genetically rescued as suggested by Ingvarsson (2001)? Perhaps the answers to both questions derive from the agricultural efforts of the eighteenth century and the Little Ice Age. Haug et al. (2001) show dramatically reduced precipitation in coastal Venezuela beginning about 275 ybp

(AD 1726) and continuing through 175 ybp (AD 1826). Curtis et al. (2001) show a contemporaneous drought at Miragoane, Haiti; this is one of the few times in the Holocene record that precipitation and temperature appear closely correlated; it is the Little Ice Age (Haug et al. 2001). The Quaker families, the Parkes and the Lakes occupied Guana Island at this time and cleared a large portion of the island for sugar cane production, horticulture, and livestock paddocks. I hypothesize that habitat destruction via agriculture reduced the crested anole population and the water-loss resistant genotype was able to propagate during this drought period. However, the 175 years since have been insufficient for a successful waif colonization to restore genetic diversity. Of course, that is just a theory.

Another kind of problem appears when we look at Carrot Rock's *Mabuya macleani*. A count of skinks was 12 in 200 m² —conservatively, in good times, indicating a total population, *N*, of about 500; the sex ratio appears balanced. These Antillean *Mabuya* are highly *K*-selected; we have no reproductive data for *Mabuya macleani*, but *Mabuya sloanii* on Guana can produce four babies per litter once a year; the big babies can grow to sexual maturity and reproduce in one year, so *g*, generation time, is that year. Let us assume that one-half of the offspring of the mutant female with a beneficial gene inherit that gene: two of her four. This means that *X*, the number of offspring per individual in the population that inherit the gene is one because half the individuals are males. Thus,

$$g = \log 500/\log 1.$$

Unfortunately, the log of one is zero, and we cannot divide by zero. My simplistic formula can only work when *X* is greater than one. At *X* = 1 we hit the division by zero problem and for any value of *X* less than one we are forced to divide by a negative number; neither will produce a number of generations, *g*. I have only two actual litters of *M. sloanii* known—both of four babies. Let us pretend I have somehow managed to count five litters, four of four and one of five, giving me an average litter size of 4.2 per female. Now I can halve that (sex ratio) and halve the result again (carriers) and then $X = 1.05$. Thus,

$$g = 2.69/0.2$$
$$g = 14.$$

All fractions of a generation are rounded up. It will take the strongly K-selected skinks, despite their smaller population, notably longer to make a genotypic changeover than it might take the anoles. It seems r-selective reproductive strategies can potentially dramatically speed up evolution and speciation. Of course, that makes sense.

We cannot extrapolate from *Mabuya macleani* on Carrot Rock to *Mabuya sloanii* on Guana. Talk about distinct evolutionary roles! This is exactly what Simpson (1961) had in mind when he wrote about distinct evolutionary species not sharing the same evolutionary role: I know; he and I talked long about that issue. *Mabuya macleani* is abundant but *Mabuya sloanii* is scarce throughout its range. The Carrot Rock skink occupies the niche occupied by the ground lizard or ameiva on Guana: an active, mobile, questing, largely terrestrial predator; Guana skinks are much more furtive and act more like ambush hunters. The Guana Island population is so sparse that Rodda et al. (2001b) got none, and, therefore, we have no population or density estimate. Evidence is, however, that the *Mabuya* skinks in the Antilles, descendants of a waif or waifs from Africa, constitute a greater radiation than has been published on to date (Mayer and Lazell 2000).

PART 5

The Cast

Standardized species accounts could only have been achieved on the bed of Procrustes.

JAMES LAZELL (1976)

How many living things are there on Guana Island? We have had 25 years, usually for one month (a "scientists month") each year, to count. (There is a second "scientists month" on Guana now, for marine biology; that can be the subject of someone else's book.)

The first part of the counting is collecting specimens, identifying them, and making a list. Most of the plants and animals we have collected belong to known species, but quite a few represent new species: species never before scientifically documented or described; species without names. Of course, one must be an expert on a particular group of organisms to spot a new species. To get the new one officially named and scientifically recognized, the expert must select a type specimen: one individual that best represents the new species and will serve as the onomatophore, or name bearer. This individual, preserved and placed in a permanent museum collection, will enable subsequent researchers to identify with certainty the kind of organism described, no matter how incomplete the description. Without a type specimen, we might never be absolutely certain what species had been described. Next, of course, the new species must have a type locality: the place the

type specimen came from. Guana Island is now the type locality for dozens of species, some of which have not yet been found anywhere else.

Once the describing expert has selected the type specimen—and ipso facto the type locality—he or she must publish, in the peer-reviewed, primary, scientific literature, a formal description that will, one hopes, enable other biologists to recognize members of the new species when they encounter them. This process of peer-reviewed publication can be drawn out and tedious because other experts— those peers—may be unconvinced of the validity of the proposed new species.

Every species must have, minimally, a two-part Latin or Latinized name: first, capitalized, a generic name telling what group of species it belongs to; second, lowercase, a species name that, in combination with the generic name, is unique within its kingdom of life forms. A plant might have the same name as an animal, but no two kinds of plants can have the same name. Our system of binomial nomenclature was invented by the eighteenth-century Swedish naturalist Carl von Linné, who Latinized his own name: Carolus Linnaeus. If an organism has three names, the third is the subspecies or geographic race. Some species, but certainly not all, divide into recognizable geographic variants that interbreed and grade into each other (i.e., intergrade) where their ranges meet. Botanists, but not zoologists, sometimes name conspecific varieties of plants, signified by the abbreviation var., or cultivar, cv.

Formal botanical nomenclature can be very complicated. Not only are there the two Latin or Latinized generic and specific names (and maybe a subspecies or varietal third name), but the name of the author of the original name, or at least an abbreviation, comes next. The lovely little fern that grows in crevices of Guana's walls is about the simplest possible case in point: *Pteris vittata* L. The "L." stands for Linnaeus, who named this virtually worldwide species. Linnaeus is, as far as I know, the only person who ever named a plant who has a one letter official abbreviation. Well-known botanists typically have standard name abbreviations of three or four letters; for the exoteric, these can be looked up in the half-dozen volumes of Staflew (1976–88), copies of which are housed at all major herbaria, or botanical museums. Short names or less-well-known names of species' authors are usually written out. If the author has a very common name, (e.g., Smith, Jones, Brown), initials may be used to identify him or her. If there has been a change in the rank or position of the specific name since its author described it, the author's name (or abbreviation) is put in parentheses and followed by the name(s) of the reviser(s). For example, many Linnaean species are now put in genera that did not exist in his day so the "L." now goes in parentheses and whoever made the change comes next. For example, Guana's fishtail fern is *Nephrolepis falcata* (Cav.) C. Chr. I have no idea who Cav. and C. Chr. are or were, and, frankly, it is not important to me so I will not even bother to go look. It could only be important if one needed to know for purposes of comparison or taxonomic revision.

But we are not done yet. We look again at the name of the fishtail fern and see that it is even longer: *Nephrolepis falcata* (Cav.) C. Chr. cv.

"Furcans." What is all that about? Well, as defined earlier, "cv." means *cultivar*: an artificially produced and cultivated variety called "Furcans." It may actually be patented. How about the very next fern on Guana's plant list: *N. multiflora* (Roxb.) Jarrett ex Morton. First, you can reduce a generic name to its initial in lists and texts when the full generic name has already been introduced; in this case, *Nephrolepis*. Then we know that somebody whose surname is abbreviated "Roxb." named the species *multiflora* but did not place it correctly by current standards. Perhaps he put it in the wrong genus or called it a subspecies or variety. Anyway Roxb. gets put in parentheses. Next, we know Jarrett is the fellow who actually published the name in the form we regard as correct today. But "ex Morton" means from Morton: Jarrett published the name we use, but he got it from someone named Morton who failed to get it published. (Perhaps poor Morton was either bored to death by all this or died of old age trying to get his paper past his peer-reviewers.)

Some zoologists and most entomologists maintain the convention of putting original authors' names after the species, but they only put them in parentheses to signal a generic shift (not merely a change in rank). Mycologists (who study fungi) and vertebrate zoologists normally do not bother with author's names unless there is some problem or potential for confusion.

All species must belong to a genus. Rarely, a species is so different from all others that it is placed in a monotypic—one-of-a-kind—genus all by itself. Genera (the plural of genus) are grouped in families; a family of organisms is often readily recognized. It can be very useful to know what family a species belongs to. So I have worked on getting our species, in the lists and text that follow, properly sorted and allocated. Families group in orders; orders in classes; classes in phyla (singular, phylum); and phyla in kingdoms. Most of these categories are of scant interest except to specialists. I have left most of them out. Even at the level of kingdom, there is controversy over how many to recognize and how to delimit them. There is even controversy over determining what is a living thing. Over a quarter of a century ago, I told the story of the manufacture of a virus from off-the-shelf chemicals and the resulting clamber to, therefore, have viruses removed from the list of living things (Lazell 1976). For many, it was philosophically unacceptable for humans to be able to create life, so life had to be redefined with the viruses defined out. I felt that was not a very philosophical way to deal with the problem.

The three most conspicuous kingdoms of Guana's living things are plants (Plantae, the flora), animals (Animalia, the fauna), and Fungi. Probably more numerous in species or kinds, and certainly so in individuals, are what used to be called Protista, including bacteria, spirochetes, protozoans, and other noncellular or single-celled organisms. I do not even have lists of these for Guana, but they are incredibly important because many cause disease in humans and in other organisms whose health we value. For a lot of Guana's organisms I have asked the relevant experts to contribute some text and at least a list of the known species, usually grouped in families. For some groups the lists are annotated with a common name, or a habitat, or some note about the natural history. For some groups enough is known to provide species accounts; for some species these accounts are fairly detailed. Alas, our all-too-human prejudices and passion for big things, colorful things, and even noisy things is all too apparent. There is definitely a wide-open slot for a virologist on our annual field team.

The foundation reference to the flora of the British Virgin Islands is Beard (1945). This paper was published in February 1945, in Trinidad. At that time German U-boats still lurked in the Caribbean; the French islands, at least, were still essentially in enemy hands; and supplies for printing and disseminating forestry bulletins were scant indeed. Very few copies of this work were apparently printed, and the only one to reach the United States seems to be that housed at the Yale Forestry Library (but I have my own copy of it now). A reference that borrows heavily from Beard (1945) is Little et al. (1976), describing the flora of Virgin Gorda. D'Arcy (1967) provides lists of plants for Tortola, many of which also occur on Guana. In 1989, George Proctor, then with the Department of Natural Resources in Puerto Rico, first came to botanize Guana. I had met George four decades before, in 1957, when he was at the Institute of Jamaica and I was on my first Antillean expedition (Lazell 2003). George has compiled the "Flora Guanae," part 5 totaling some 339 species to date, which follows my brief introduction. Pedro Acevedo-Rodriguez, of the Smithsonian, author of the definitive "Flora of St. John, U.S. Virgin Islands," came in 2000 especially to look for novelties and rarities such as Jarecki's wild pine, a bromeliad.

Flora Guanae

AN ANNOTATED LIST OF THE VASCULAR PLANTS

George R. Proctor

IN THE FOLLOWING CHECKLIST, all numbers following names are Proctor collection numbers. The first (and only complete) set of this material is deposited in the herbarium of the Department of Natural and Environmental Resources, San Juan, Puerto Rico. Partial duplicate sets have been sent to the U.S. National Herbarium, Washington, DC; the Institute of Jamaica, Kingston, Jamaica; and the New York Botanical Garden, Bronx, New York.

A limited amount of fieldwork has been carried out on several islands near Guana. Although information for these islands is far from complete, these records have been appended to the current list where relevant. Presence of a species on St. John is also indicated based on Acevedo-Rodriguez (1996). Species not found on Guana, but on the following 5 islands, are put in parentheses. The following symbols are used: GC (Great Camanoe), GD (George Dog), LC (Little Camanoe), SI (Scrub Island), SJ (St. John).

To date, 339 species of vascular plants have been found growing on Guana Island outside of cultivation. In addition, 53 cultivated species have been observed (and in some instances collected): All of these are included in the checklist. The noncultivated flora consists of both indigenous and introduced/naturalized species. These two categories have not been separately listed because of the uncertain status of many species with relation to the original hypothetical primeval flora before human disturbance. For this reason it is difficult to compare the Guana list with lists published for various other islands in which "indigenous" versus "introduced" plants are more sharply differentiated than I am able to do. Of course, many cases are clear, but others are not. In the table presented by Howard and Kellogg (1987) for various islands, figures are given for indigenous and "cultivated/introduced" plants, but they do not say which if any species can be considered naturalized. However, bearing this in mind, the flora of Guana (339 species) presents a concentrated diversity higher than any other West Indian island for which figures are available (see table 20). Thus Guana's relatively uniform habitat renders its concentrated diversity even more noteworthy.

In general, the vegetation of Guana is far better preserved than is that of most of the world's dry islands. Apparently, much of the steeper portions escaped clearing for agriculture in the heyday of the eighteenth century, and goats (if they were ever present) have been

TABLE 20
Areas and Numbers of Plant Species for Some Islands
in the Northern Caribbean

	AREA (KM^2)	PLANT SPECIES
Guana	3	338
Virgin Gorda	21	372
St. Barts	26	326
Anegada	36	198
Tortola	55	484
Anguilla	91	321
St. Martin	97	392
Vieques	133	781

SOURCE: Original data.

off the island since the 1930s. The good condition of the vegetation probably has a direct bearing on the diversity of animal species found on Guana.

The meticulous investigations of the flora of St. John by Acevedo-Rodriguez (1996) have resulted in that island (area ca. 12 mi.^{2} or 31 km^2) having one of the best-known floras in the West Indies. It is not surprising given the general similarity of habitat (except for lower elevation) that by far the greater part of the Guana flora also occurs on St. John. However, it is noteworthy that at least 29 species occurring on Guana have not been recorded for St. John. Of these, 23 are believed to be indigenous and six naturalized introductions. These are listed as follows:

Acalypha chamaedrifolia (Euphorbiaceae)
Alternanthera pungens (Amaranthaceae)
Bastardiopsis eggersii (Malvaceae; formerly known as *Sida eggersii*)
Bauhinia monandra (Leguminosae: Caesalpinioideae; naturalized)
Cardiospermum halicacabum var. *microcarpum* (Sapindaceae)
Cenchrus incertus (Gramineae)
Commelina diffusa (Commelinaceae)
Cordia obliqua (Boraginaceae; naturalized)
Cyperus confertus (Cyperaceae)
Cyperus unifolius (Cyperaceae)
Digitaria bicornis (Gramineae)
Digitaria eggersii (Gramineae)

Eugenia underwoodii (Myrtaceae)
Haematoxylum campechianum (Leguminosae: Caesalpinioideae; naturalized)
Hohenbergia antillana (Bromeliaceae; naturalized)
Kalanchoe tubiflora (Crassulaceae; naturalized)
Lippia nodiflora (Verbenaceae)
Opuntia dillenii × *repens* (= *Opuntia triacantha?*; Cactaceae)
Pappophorum pappiferum (Gramineae)
Paspalum pleostachyum (Gramineae)
Paspalum setaceum var. *ciliatifolium* (Gramineae)
Pilea microphylla var. *succulenta* (Urticaceae)
Pilea microphylla var. *trianthemoides* (Urticaceae; naturalized)
Pitcairnia new species (Bromeliaceae; endemic)
Plectranthus amboinicus (Labiatae; naturalized)
Portulaca pilosa (Portulacaceae)
Sabal causiarum (Palmae)
Tabebuia lepidota (Bignoniaceae)
Tephrosia noctiflora (Leguminosae: Faboideae)
Zephyranthes puertoricensis (Amaryllidaceae)

There is no particular conclusion to be drawn from the above list except that it highlights the lack of floristic uniformity characteristic of disjunct sites even of similar habitats, given the random distribution of plant propagules under natural conditions. The very low level of endemism among all of these islands can perhaps be explained by their relative lack of isolation over a long period of time.

PTERIDOPHYTA

POLYPODIACEAE (SENSU LATO)

Nephrolepis falcata (Cav.) C. Chr. cv. "Furcans." Sight, cultivated as the fishtail fern.
N. multiflora (Roxb.) Jarrett ex Morton. 42015, rock crevices, summit of Guana Peak. SJ. This native of India has become widely naturalized in

the West Indies. The Guana plants are evidently self-introduced.

Pteris vittata L. 42634, crevices of old walls in Guana Island Club. SJ. Native of southern China, widely naturalized.

GYMNOSPERMAE

No indigenous or naturalized gymnosperms are known from Guana Island. *Araucaria heterophylla* and *Cycas revoluta* are cultivated near Grenada House.

ANGIOSPERMAE: DICOTYLEDONAE

ACANTHACEAE

Asystasia gangetica (L.) T. Anders. 43402. Introduced from the Old World tropics, naturalized. SJ.
Blechum pyramidatum (Lam.) Urban. 43480. SJ.
Crossandra infundibuliformis (L.) Nees. 43892, cultivated. Native of Africa; alleged to be an aphrodisiac.
Justicia periplocifolia Jacq. 47247. GD, SJ.
Oplonia microphylla (Lam.) Stearn. 42014, 43703, 44898. GD, SJ.
O. spinosa (Jacq.) Raf. 43870, Monkey Point. GC, SJ.
Ruellia tuberosa L. 43430. GD, SJ.

AIZOACEAE

Aptenia cordifolia N. E. Brown. Sight, cultivated. D. Jarecki photo. Native of South Africa.
Sesuvium portulacastrum (L.) L. 43437. SI, SJ.
Trianthema portulacastrum L. 42326. GC, SJ.

AMARANTHACEAE

Achyranthes aspera L. 42645. SJ.
Alternanthera crucis (Moq.) Boldingh. 42598, seen only at south end of Muskmelon Bay Beach. SJ.
A. pungens Kunth. 42558.
Amaranthus crassipes Schlecht. 43479. SJ.
A. viridis L. 43478. SJ.
Celosia nitida Vahl. 43414. GD, SJ.

Iresine angustifolia Euphrasen. 42617, 43407. GD, SJ.

ANACARDIACEAE

Comocladia dodonaea (L.) Urban. 47245, 48403. SJ.
Mangifera indica L. Sight, introduced and long-established. SJ.
Spondias mombin L. 44896, Quail Dove Ghut only. SJ.

ANNONACEAE

Annona glabra L. 43682. SJ.
A. muricata L. 43683. SJ.
A. squamosa L. 43407. SJ.

APOCYNACEAE

Catharanthus roseus (L.) G. Don. 43889, cultivated, escaped. The Guana plants represent the pure white form called var. *"albus."* SJ.
Nerium oleander L. Sight, cultivated. SJ.
Pentalinon luteum (L.) Hansen & Wunderlin (formerly known as *Urechites lutea*). 42593. SJ.
Plumeria alba L. 42563. GC, GD, LC, SI, SJ.
P. rubra L. 48828, cultivated. GC.
Prestonia agglutinata (Jacq.) Woodson. 42586. SJ.
Rauvolfia viridis Willd. ex R. & S. 42003, 48824. GC, GD, LC, SJ.
Tabernaemontana divaricata (L.) R. Br. Sight, cultivated.

ASCLEPIADACEAE

Asclepias curassavica L. 42525. SJ.
Cryptostegia grandiflora R. Br. 43857–58, cultivated, naturalized. SJ.
Matelea maritima (Jacq.) Woodson. 46551. SJ.
Metastelma grisbachianum Schltr. in Urban. 43871, Monkey Point. GD, SJ.

BIGNONIACEAE

Crescentia cujete L. 42522. LC, SJ.
Macfadyena unguis-cati (L.) A. Gentry. 42001, 43470. GC, GD, SJ.
Tabebuia heterophylla (DC.) Britton. 42581. GC, GD, LC, SJ. (See plate 10.)
T. lepidota (Kunth) Britton. 48816. According to Gentry (1982), this species occurs in the

Bahamas, Cuba, and the Haitian island of Tortue and is represented by a single record from Anegada. The citations by Howard (1974–1989), from Anguilla, St. Martin, and Barbuda were not accepted by Gentry for this species.

Tecoma stans (L.) A. Juss. ex Kunth. 43861, cultivated, naturalized. SJ.

BORAGINACEAE

Argusia gnaphalodes (L.) Heine. 43400, North Beach. LC, SJ. (See plate 11.)
Bourreria succulenta Jacq. 42024. GC, GD, SJ.
Cordia alliodora (R. & P.) Oken. 47235, Quail Dove Ghut only. SJ.
C. collococca L. 43449, 43712. SJ.
C. laevigata Lam. 47233, Quail Dove Ghut only. SJ.
C. obliqua Willd. 47229, naturalized.
C. rickseckeri Millsp. 42631, 44897. SJ.
C. sebestena L. 43908, cultivated, Grenada House. D. Jarecki photo.
Heliotropium angiospermum Murray. 42632. SJ.
H. curassavicum L. 42631, in saline soils. GD, SI, SJ.
H. indicum L. Sight. D. Jarecki photo. SJ.
(*H. ternatum* Vahl. GD, SJ.)
Tournefortia microphylla Bertero ex Spreng. 43872, Monkey Point; 47236, Grand Ghut. GD, SI, SJ.

BURSERACEAE

Bursera simaruba (L.) Sarg. 42006. GC, GD, LC, SJ.

CACTACEAE

Hylocereus trigonus (Haw.) Saff. 43462. SJ.
Mammillaria nivosa Link ex Pfeiff. 43849, East side of Monkey Point. GD, SI, SJ.
Melocactus intortus (Mill.) Urban. 43456. GC, GD, LC, SI, SJ. (See plate 12.)
Opuntia dillenii (Ker-Gawl) Haw. 44879. GC, GD, LC, SI, SJ.
O. dillenii × *repens.* 47250, rocky slopes near Guana Head. Perhaps the same as *Opuntia triacantha* (Willd.) Sweet. GD, LC.

O. repens Bello. 43418. GC, GD, LC, SI, SJ.
O. rubescens Salm-Dyck ex DC. Sight, cultivated only. SJ.
Pilosocereus royenii (L.) Byles & Rowley. 42592. GC, GD, LC, SI, SJ.
Selenicereus near *grandiflorus* (L.) Briton & Rose. 43420 (sterile only), rocky hillside near Guana Island Club. SJ. This plant is not very similar to any of the diverse forms of *Selenicereus grandiflorus* occurring in Jamaica, the type locality of the species. Further taxonomic study of these plants is necessary. The Guana population was presumably introduced long ago and has become naturalized.

CAPPARACEAE

Capparis baducca L. 43464. SJ.
C. cynophallophora L. 43408. GC, GD, SI, SJ. (See plate 13.)
C. flexuosa (L.) L. 42619. GC, GD, SJ.
C. indica (L.) Faw. & Rendle. 43463, 43667. SJ.
Cleome viscosa L. 42550, 43886. GC, SJ.
Morisonia americana L. 42572, 48812. SJ.

CARICACEAE

Carica papaya L. Sight, cultivated. SJ.

CASUARINACEAE

Casuarina equisetifolia L. Sight, planted, one tree only.

CELASTRACEAE

Crossopetalum rhacoma Crantz. 42606. GC, SJ.
Elaeodendron xylocarpum (Vent.) P. DC. 42564. GC, LC, SI. This species is listed under the name *Cassine xylocarpa* Vent. in Acevedo-Rodriguez (1996, 166). My reason for not accepting *Cassine* for our plant is as follows: Robson (1965) in a taxonomic study of African Celastraceae restricted the genus *Cassine* to two South African species. Subsequent authors have varied widely in their treatment of this taxon. In the current checklist, I follow Airy Shaw's (1988) usage, in

which *Cassine* is a genus of 40 species occurring in South Africa, Madagascar, tropical Asia, and Pacific islands, whereas *Elaeodendron* is recognized as a taxon of 16–17 species of wide distribution, including in the neotropics. (See plate 14.)

Maytenus laevigata (Vahl) Griseb. ex Eggers. 42005, 42664, 43465, 43704. SJ.

Schaefferia frutescens Jacq. 42568, 43395, 43442. GC, SI, SJ.

COMBRETACEAE

Conocarpus erectus L. 43867. SJ.

Laguncularia racemosa (L.) Gaertn. 42630. SJ.

Quisqualis indica L. 43857, cultivated. SJ.

Terminalia catappa L. 48823, naturalized. Native of India. SJ.

COMPOSITAE (Asteraceae of Acevedo-Rodriguez 1996)

Bidens alba (L.) DC. var. *radiata* (Sch. Bip.) Ballard. 43713. SJ.

B. cynapiifolia Kunth. 42541. SJ.

Chromolaena corymbosa (Aubl.) King & H. Rob. 43399. GD, SJ. This and the following species have customarily been included in a broadly construed genus *Eupatorium*. The current treatment follows the usage in Acevedo-Rodriguez (1996).

C. sinuata (Lam.) King & H. Rob. 42574, ridge east of Muskmelon Bay; 42591, northernmost hill on Long Mans Point. SJ.

Conyza canadensis (L.) Cronq. 42642, in field near White Bay. The closely related *Chromolaena bonariensis* (L.) Cronq. is cited from SJ.

Cyanthillium cinereum (L.) H. Rob. 43425. SJ. This species has been known as *Vernoniacinerea* (L.) Less. in most recent floras.

Emilia fosbergii Nicolson. 43460. SJ.

Launea intybacea (Jacq.) Beauverd. 43847, rocky seashore near Pinguin Ghut. SJ.

(*Lepidaploa glabra* (Willd.) H. Rob. Formerly known as *Vernonia albicaulis* Willd. ex Pers. GC, SJ.)

(*Melanthera aspera* (Jacq.) L.C. Rich. Recorded from GD but not from SJ.).

(*Pectis linifolia* L. GC, SJ.)

Pluchea carolinensis (Jacq.) G. Don. 43453. SJ.

Synedrella nodiflora (L.) Gaertn. 43882. SJ.

Tridax procumbens L. 42542. SJ.

(*Wedelia fruticosa* Jacq. GD, SJ.)

CONVOLVULACEAE

Convolvulus nodiflorus Desr. 42621. SJ.

Cuscuta americana L. 42021. GD, SJ. This species is placed in a separate family, Cuscutaceae in Acevedo-Rodriguez (1996). I prefer to follow Howard's (1974–1989, vol. 6) treatment.

Ipomoea eggersii (House) D.F. Austin. 42604, wooded hillside northeast of the Guana Island Club. GC, GD, SJ.

I. pes-caprae (L.) R. Br. 42638, 43873. SJ.

I. triloba L. 43484. SJ.

I. violacea L. 42608. SJ.

(*Jacquemontia cumanesis* (Kunth) Ktze. GD, SJ.)

J. havanensis (Jacq.) Urban. 42607. SJ.

J. pentanthos (Jacq.) G. Don. 42651. GC, GD, LC, SJ.

J. solanifolia (L.) Hall. f. 42537, thickets behind North Beach. SJ.

Merremia quinquefolia (L.) Hall. 42633. SJ.

Stictocardia tiliifolia (Desr.) Hall. 43448. SJ.

CRASSULACEAE

Bryophyllum pinnatum (Lam.) Oken. Sight, becoming naturalized. This species is a potentially noxious weed. SJ.

Kalanchoe tubiflora (Harvey) Hamet. 43423, naturalized, hillside northeast of the Guana Island Club.

CRUCIFERAE (Brassicaceae of Acevedo-Rodriguez 1996)

Cakile lanceloata (Willd.) O.E. Schulz. 42650. SJ.

CUCURBITACEAE

Cayaponia americana (Lam.) Cogn. 43424. SJ.

Doyerea emetocathartica Grosourdy. Specimen at USDA, Lane, OK. This and other cucurbits of Guana are currently under study by Angela Davis and Rudy O'Reilly of the USDA.

Momordica charantia L. Sight, naturalized. SJ.

ERYTHROXYLACEAE

Erythroxylum brevipes DC. 43472. GC, GD, LC, SJ.

EUPHORBIACEAE

Acalypha chamaedrifolia (Lam.) Muell. Arg. 42556. LC.

Adelia ricinella L. 42658, 43389, 43390. SJ.

Argythamnia candicans Sw. 42510. SJ.

A. fasciculata (Vahl. ex A. Juss.) Muell. Arg. 43711, Grand Ghut; 43869, Monkey Point. SJ.

Chamaesyce articulata (Aubl.) Britton. 43868, Monkey Point. GD, SJ.

C. hirta (L.) Millsp. 42539. SJ.

C. hypericifolia (L.) Millsp. 42506. SJ.

C. hyssopifolia (L.) Small. Sight. SJ.

C. mesembrianthemifolia (Jacq.) Dugand. 42612. LC, SJ.

C. ophthalmica (Pers.) Burch. 42505. SJ.

C. prostrata (Ait.) Small. 42557. SJ.

Croton astroites Ait. 42535. GC, GD, SJ.

C. betulinus Vahl. 42540. GC, SJ.

C. fishlockii Britton. This species was introduced to Guana Island in 1992 from GC. It is otherwise known only from Anegada, Virgin Gorda, Tortola, and SJ, and in all these places is quite rare.

C. flavens L. var. *rigidus* Muell. Arg. 42521. GC, GD, LC, SJ.

C. lobatus L. 43436. SJ.

Dalechampia scandens L. 43485. GC, SJ.

Euphorbia heterophylla L. 42555. SJ.

E. lactea Haw. Sight, cultivated.

E. milii Ch. Des Moulins. Sight, cultivated.

E. neriifolia L. 43419, cultivated.

E. petiolaris Sims. 43398, common near summit of Guana Peak. SJ.

E. tirucalli L. 42588, cultivated.

Flueggea acidoton (L.) Webster. 43440, slopes behind Bigelow Beach. SJ.

Gymnanthes lucida Sw. 42594. GC, SJ.

Hippomane mancinella L. 43438, Bigelow Beach. Also seen at north end of North Beach. LC, SJ.

Jatropha multifida L. 43875, cultivated.

Pedilanthus tithymaloides (L.) Poit. subsp. *angustifolius* (Poit.) Dressler. 43845, Pinguin Ghut. SJ.

P. tithymaloides subsp. *parasiticus* (Kl. & Gcke.) Dressler. 43906, cultivated.

Phyllanthus amarus Schum. 43492. SJ.

Ricinus communis L. Sight; eradicated as a noxious weed. SJ.

Savia sessiliflora (Sw.) Willd. 43466, 43710, 48809. GC, SJ.

Tragia volubilis (L.) Poit. 42523. GC, GD, LC, SJ.

FLACOURTIACEAE

Samyda dodecandra Jacq. 42000, 43679. GC, SJ.

GOODENIACEAE

Scaevola plumieri (L.) Vahl. 42610, North Beach. SJ. (See plate 15.)

S. sericea Vahl. Cultivated. 48360, 48827, Guana Island Club.

GUTTIFERAE (Clusiaceae of Acevedo-rodriguez 1996)

Clusia rosea Jacq. Sight, summit area, Guana Peak; also cultivated near Grenada House. SJ.

LABIATAE (Lamiaceae of Acevedo-Rodriguez 1996)

Leonotis nepetifolia (L.) Ait. 43434. SJ.

Plectranthus amboinicus (Lam.) Spreng. 43427, naturalized near Guana Island Club but not seen flowering at any time.

Salvia serotina L. 42512, in sandy soil near White Bay. SJ.

LAURACEAE

Ocotea coriacea (Sw.) Britton. 43684, 44906, foothill southeast of White Bay. SJ.

LEGUMINOSAE (Fabaceae of Acevedo-Rodriguez 1996)

Here divided into three subfamilies:

1. Caesalpinioideae

Bauhinia monandra Kurz. 43897, 48834, becoming naturalized.

Caesalpinia bonduc (L.) Roxb. 43447, south end of White Bay Beach. SJ.

Caesalpinia pulcherrima (L.) Sw. 43859, naturalized. SJ.

Chamaecrista glandulosa (L.) Sw. var. *swartzii* (Wikstr.) Irwin & Barneby. 42551, 43855. SJ.

Haematoxylum campechianum L. 47330, naturalized.

Hymenaea courbaril L. Sight, Quail Dove Ghut. SJ.

Parkinsonia aculeata L. 48829, cultivated. SJ.

Senna bicapsularis (L.) Roxb. 42654. SJ.

S. occidentalis (L.) Link. 43692. SJ.

Tamarindus indica L. 42516, 48825, very large planted trees next to agricultural area, becoming naturalized. SJ.

2. Faboideae

Abrus precatorius L. 43446. GC, SJ.

Alysicarpus vaginalis (L.) DC. 42635. SJ.

Canavalia rosea (Sw.) DC. 42616. GD, SJ.

Centrosema virginianum (L.) Benth. 43393, 48359. GD, SJ.

Crotalaria incana L. 43405. SJ.

C. lotifolia L. 43852, 48362. SJ.

Desmodium glabrum (Mill.) DC. 46520. SJ.

D. incanum DC. 42589. SJ.

D. procumbens (Mill.) Hitchc. 42589. LC, SJ.

D. triflorum (L.) DC. 46552. SJ.

Erythrina variegata (L.) DC. var. *orientalis* (L.) Merrill. 48839, cultivated.

Galactia dubia DC. 42573, 43406. SJ.

G. eggersii Urban. 42529. Endemic to SJ, St. Thomas. Tortola, and Guana.

G. striata (Jacq.) Urban. 42538, 48387. SJ.

Gliricidia sepium (Jacq.) Kunth ex Walp. 42644, planted, becoming naturalized. SJ.

Indigofera suffruticosa Mill. 42649. SJ.

Pictetia aculeata (Vahl) Urban. 42013, 43474. GC, SJ. (See plate 16.)

Piscidia carthagenensis Jacq. 43392, 48815. GC, GD, SJ.

Poitea florida (Vahl) Lavin (formerly *Sabinea florida*). 43685, 44907, foothill southeast of White Bay; 43705, Grand Ghut; 48818, north slope of Pyramid Hill. SJ.

Rhynchosia minima (L.) DC. 42647. SJ.

R. reticulata (Sw.) DC. 42011. GD, SJ.

Stylosanthes hamata (L.) Taub. 42646. LC, SJ.

Tephrosia cinerea (L.) Pers. 43853, Monkey Point. GD, SJ.

T. noctiflora Bojer ex Baker. 46674, 48363, along road below Grenada House.

Teramnus labialis (L.f.) Spreng. 48386. SJ.

3. Mimosoideae

Acacia macracantha H. & B. ex Willd. 42641. LC, SJ. (See plate 17.)

A. muricata (L.) Willd. 42653, 43469, northwest ridge of Guana Peak. SJ.

A. retusa (Jacq.) Howard. 42653, 48810. GC, SJ.

Desmanthus virgatus (L.) Willd. 42648. GD, SJ. Very small or stunted examples of this species have been called *D. depressus* H. & B. ex Willd; such plants on Guana Island are represented by 42640.

Leucaena leucocephala (Lam.) de Wit. 43457. SJ. (See plate 18.)

Pithecellobium unguis-cati (L.) Benth. 42585. GD, SJ.

Samanea saman (Jacq.) Merrill. 43691, planted, becoming naturalized. SJ.

LOGANIACEAE

Spigelia anthelmia L. 43860. SJ.

LORANTHACEAE

Dendropemon caribaeus Krug & Urban. 42012, 42562. GC, SJ.

LYTHRACEAE

Ginoria rohrii (Vahl) Koehne. 43473, Palm Ghut. SJ.

MALPIGHIACEAE

Bunchosia glandulosa (Cav.) DC. 42023, 42562, 44912. SJ.

Heteropterys purpurea (L.) Kunth. 43397, 43467. GC, GD, SJ.

Malpighia emarginata DC. Sight, cultivated. SJ.

M. woodburyana Vivaldi. 47241, Grand Ghut. GD, SI, SJ.

Stigmaphyllon emarginatum (Cav.) A. Juss. 42588. GC, GD, LC, SJ.

MALVACEAE

Abutilon umbellatum (L.) Sweet. 43902, 44909. SJ.

Bastardia viscosa (L.) Kunth. 43426 (var. *viscosa*). SJ.

Bastardiopsis eggersii (Baker f.) Fuertes & Fryxell (formerly known as *Sida eggersii*). 43706, 44428, Grand Ghut (23 trees counted; subsequently others have been found elsewhere on the island). This rare species is endemic to Culebra, Tortola, Jost Van Dyke, Guana, and Ginger Islands. It is reported to be cultivated on SJ. On Guana, it has been the subject of a detailed ecological study by Kraus (2002).

Hibiscus rosa-sinensis L. Sight, cultivated. SI.

H. schizopetalus (Masters), Hook. Sight, cultivated. D. Jarecki photo.

Malvastrum americanum (L.) Torr. 43877, 48404. SJ.

M. corchorifolium (Desr.) Britton ex Small. 42627. SJ.

M. coromandelianum (L.) Garcke. 42579. SJ.

Sida acuta Burm. f. 44908. SJ.

S. ciliaris L. 42554. SJ.

S. glabra Mill. 43901. SJ.

(*S. glomerata* Cav. GC, GD, SJ.)

S. repens Dombey ex Cav. 42620. SJ.

Sidastrum multiflorum (Jacq.) Fryxell. 42570. GC, GD, LC, SJ.

Thespesia populnea (L.) Sol. ex Correa. 43851, beach area east of Monkey Point. SJ.

MELIACEAE

Azadirachta indica A. Juss. 48401, 48832, planted near Guana Island Club as "Neem." Reported to be cultivated on SJ.

Swietenia mahagoni (L.) Jacq. 45388, planted and well established at lower end of road to Guana Island Club. SJ.

MOLLUGINACEAE

Mollugo nudicaulis Lam. 42517, along track at base of Quail Dove Ghut, appearing only after rains. SJ. In many publications, *Mollugo* is included in Aizoaceae.

MORACEAE

Artocarpus altilis (Parkinson) Fosberg. Sight, cultivated as "Breadfruit." Reported to be cultivated on SJ.

Ficus citrifolia Mill. 42613, 43707. SJ.

MYOPORACEAE

Bontia daphnoides L. 42027, thickets behind North Beach. SJ.

MYRTACEAE

Eugenia axillaris (Sw.) Willd. 42524, 47231. SJ.

E. biflora (L.) DC. 42004, 42019, 43686, 47246, 48819. SJ.

E. cordata (Sw.) DC. 42527, 42561, 48820. GC, SJ.

E. ligustrina (Sw.) Willd. 42657, 43702, upper slopes of Guana Peak. SJ.

E. monticola (Sw.) DC. 42018, 47231, 48384, 48885. SJ.

E. procera (Sw.) Poir. 42662, 43687. SJ.

E. underwoodii Britton. 43401, Grand Ghut. A rare shrub otherwise known from a few collections in Puerto Rico.

Myrcianthes fragans (Sw.) McVaugh. 42016, west side of ridge saddle just south of summit of Guana Peak, perhaps the tallest tree on Guana Island, with smooth orange bark. SJ.

Myrciaria floribunda (West ex Willd.) Berg. 47243. SJ.

Psidium guajava L. 43893, cultivated. SJ.

NYCTAGINACEAE

Pisonia subcordata (L.). GC, GD, LC, SI, SJ.

OLACACEAE

Schoepfia schreberi J. F. Gmel. 42528, 43680. SJ.

OLEACEAE

Chionanthus compacta Sw. 47245, upper slopes of Guana Peak. SJ.

Forestiera eggersiana Krug & Urban. 42530, 42531, 43435. GC, GD, SJ.

Jasminum grandiflorum L. 43894, cultivated.

J. sambac (L.) Sol. in Ait. 43895, cultivated, Grenada House.

OXALIDACEAE

Oxalis corniculata L. 43864. SJ.

PAPAVERACEAE

Argemone mexicana L. 43403. SJ.

PASSIFLORACEAE

Passiflora edulis Sims. 46521, becoming naturalized. SJ.

P. suberosa L. 42509, 42514. GC, SJ.

PHYTOLACCACEAE

Petiveria alliacea L. 43450. SJ.

Rivina humilis L. 42663. GD, SJ.

Trichostigma octandrum (L.) H. Walter. 43688, foothill southeast of White Bay. SJ.

PIPERACEAE

Peperomia humilis A. Dietr. 42010, 42659, northwest ridge of Guana Peak. SJ.

P. magnoliifolia (Jacq.) A. Dietr. 42660, near summit, Guana Peak. SJ.

PLUMBAGINACEAE

Plumbago auriculata Lam. 43896, cultivated.

P. scandens L. 43444, ravine behind Bigelow Beach. SJ.

POLYGONACEAE

Antigonon leptopus Hook. & Arn. 43421, naturalized. SJ.

Coccoloba uvifera (L.) L. 43396. SJ.

PORTULACACEAE

Portulaca oleracea L. 43862. SJ.

P. pilosa L. 48817, near northwest point.

(*P. teretifolia* Kunth. GD.)

(*Talinum fruticosum* (L.) Juss. GD, SJ.)

PUNICACEAE

Punica granatum L. 48836, cultivated, escaping, "Pomegranate."

RHAMNACEAE

Colubrina arborescens (Mill.) Sarg. 42624. LC, SJ.

C. elliptica (Sw.) Briz. & Stern. 42578, 47239. SJ.

Gouania lupuloides (L.) Urban. 42584. SJ.

Krugiodendron ferreum (Vahl) Urban. 42587. GC, SJ.

Reynosia guama Urban. 43709, Grand Ghut. GC, SJ.

RHIZOPHORACEAE

Rhizophora mangle L. Sight, Salt Pond, rare. D. Jarecki photo. SJ.

RUBIACEAE

Chiococca alba (L.) Hitchc. 43708, 43866, Grand Ghut. These specimens are anomalous in having four-parted flowers. Normally in this species the flowers are five-parted. SJ.

Erithalis fruticosa L. 43475. SJ.

Exostema caribaeum (Jacq.) Schult. 41999. GC, GD, SJ.

Guettarda odorata (Jacq.) Lam. 43401, 43698. SJ.

G. scabra (L.) Vent. 47237, Grand Ghut. SJ.

Psychotria brownei Spreng. 47234, 48833. SJ.

P. microdon (DC.) Urban. 41999, northwest ridge of Guana Peak. GC, SJ.

Randia aculeata L. 43391. GC, GD, LC, SJ.

Rondeletia pilosa Sw. 42017, 42603. SJ.

Scolosanthus versicolor Vahl. 42560, hillside northeast of Guana Island Club; 43697, Palm Ghut. GC, SJ.

Spermacoce assurgens Ruiz & Pav. 42511, 42636, 43888. SJ.

RUTACEAE

Amyris elemifera L. 42009, 42532, 43888. GC, SJ.

Citrus aurantifolia (Christm.) Swingle. 43690, naturalized. SJ.

Zanthoxylum martinicense (Lam.) DC. Sight, Quail Dove Ghut. SJ.

SAPINDACEAE

Cardiospermum corindum L. (Kunth) Blume. 42520. SJ.

Melicoccus bijugatus Juss. 43695, planted old tree. LC, SJ.

Serjania polyphylla (L.) Radlk. 42622. GC, SJ.

SAPOTACEAE

Manilkara zapota (L.) van Royen. 48826, cultivated. Reported to be cultivated on SJ.
Sideroxylon foetidissimum Jacq. 47238, Grand Ghut; 48364, SJ.
S. obovatum Lam. 42597, 43433, 44911. GC, SJ.

SCROPHULARIACEAE

Capraria biflora L. 42625. SJ.
Russelia equisetiformis Schlect. & Cham. 43907, cultivated. D. Jarecki photo. GC.

SOLANACEAE

Capsicum frutescens L. 47232. GD, SJ.
Cestrum laurifolium L'Her. 42655. SJ.
C. nocturnum L. 48835, cultivated.
Physalis angulata L. 43486, 43487. SJ.
Solanum americanum Mill. 43455. SJ.
S. polygamum Vahl. 41997, 42533, 43443. GC, SJ.
S. racemosum Jacq. 42008, 42515, 44405. GC, GD, LC, SJ.
S. torvum Sw. 43681. SJ.

STERCULIACEAE

Ayenia insulicola Cristobal. 43428, 43471, 44880. GD, LC, SJ.
Helicteres jamaicensis L. 42569. SJ.
Melochia nodiflora Sw. 43476. SJ.
M. tomentosa L. 42626. GC, GD, LC, SJ.
Waltheria indica L. 42577. SJ.

SURIANACEAE

Suriana maritima L. 42611, North Beach. SJ.

THEOPHRASTACEAE

Jacquinia arborea Vahl. 42596, 42609. GD, LC, SJ.
J. berterii Spreng. 41998, 42026, 46524. GC, SJ.

TILIACEAE

Corchorus aestuans L. 43482. SJ.
C. hirsutus L. 42648. GC, GD, LC, SJ.
C. siliquosus L. 43483. SJ.

TURNERACEAE

Turnera ulmifolia L. 46568. SJ.

ULMACEAE

Celtis iguanea (Jacq.) Sarg. 47240, Grand Ghut. SJ.
Trema micranthum (L.) Blume. 43488, south end of plain, east of White Bay. SJ.

URTICACEAE

Pilea microphylla (L.) Liebm. 43898, garden weed, rare. SJ.
P. microphylla var. *succulenta* Griseb. 43850, east side of Monkey Point.
P. microphylla var. *trianthemoides* (Sw.) Griseb. 43874, cultivated, naturalized, Grenada House vicinity.
P. tenerrima Miq. 42651, north end of North Beach, sheltered among stones near sea. SJ.

VERBENACEAE

Citharexylum fruticosum L. 42582, 47249. GD, LC, SJ.
Clerodendrum aculeatum (L.) Schtdl. 46263. LC, SJ.
Lantana involucrata L. 42605. LC, SJ.
L. urticifolia Mill. 42583, 44899. GC, GD, LC, SJ.
Lippia nodiflora (L.) Michx. 42639, on low, moist ground northeast of White Bay.
Priva lappulacea (L.) Pers. 42628. SJ.
Stachytarpheta jamaicensis (L.) Vahl. 43481. SJ.

VITACEAE

Cissus trifoliata (L.) L. 43693. LC, SJ.
C. verticillata (L.) Nicolson & Jarvis. 42614, 43694, form with red and orange flowers; 43876, form with greenish cream flowers. GC, SI, SJ.

ZYGOPHYLLACEAE

Guaiacum officinale L. 43858, perhaps planted, five other possibly wild trees also seen. SJ.
Kallstroemia maxima (L.) Hook. & Arn. 43416. SJ.
K. pubescens (G. Don) Dandy. 43905. SJ.

ANGIOSPERMAE: MONOCOTYLEDONAE

AGAVACEAE

Circumscription of this family follows that of Howard (1974–1989, see 1979 volume).

Agave angustifolia Haw. 43890, cultivated. Probably native of Mexico.

A. beauleriana Jacobi. Sight, cultivated.

A. missionum Trel. 43413, 43699. GC, GD, SJ. This species is endemic to the Virgin Islands (except St. Croix) and Puerto Rico.

Dracaena fragrans (L.) Ker-Gawl. Sight, cultivated. Native of Africa.

Sansevieria trifasciata Prain. Sight, cultivated, naturalized. SJ.

Yucca aloifolia L. Sight, cultivated. Originally described from Jamaica; widely naturalized in the West Indies. SJ.

Y. guatemalensis Baker. Sight, cultivated. In Guatemala, the flowers are commonly eaten in a type of omelet.

AMARYLLIDACEAE

Hymenocallis expansa (Herb.) Herb. 43891, cultivated. The similar *H. caribaea* (L.) Herb. occurs on Anegada, GD, and SJ. In *H. caribaea*, the perianth tube is 4–6.5 cm long, shorter than the segments, whereas in *H. expansa*, the perianth tube is 8–11 cm long, with segments 9–15 cm long.

Zephyranthes puertoricensis Traub. Sight. Although native to the region, it is not known if this white-flowered "zephyr lily" was introduced to Guana Island, where it is very rare.

ARACEAE

No indigenous Araceae have been observed on Guana Island. *Anthurium crenatum* (L.) Kunth, *A. grandifolium* (Jacq.) Kunth, and *Epipremnum aureum* (Lindl. & Andre) Bunting have been seen under cultivation.

BROMELIACEAE

Bromelia pinguin L. 43846, Pinguin Ghut. SJ.
Hohenbergia antillana Mez in DC. 44881, cultivated, naturalized. GD, SI. The Guana Island plants were introduced many years ago from SI.

The species is abundant on GD but is otherwise endemic to mainland Puerto Rico.

Pitcairnia new species. 47242, ridge leading to Palm Point, elevation ca. 450–500 ft. (137–152 m), endemic to Guana, very rare, currently is being described.

Tillandsia fasciculata Sw. 42652, near summit of Guana Peak, very rare. SJ.

T. utriculata L. 42602, widespread and common. SJ.

COMMELINACEAE

Callisia fragrans (Lindl.) Woodson. 43417, cultivated, escaping, probably naturalized as on many other West Indian islands. Native of Mexico. GC, SJ.

C. repens (Jacq.) L. 42533. GC, SJ.

Commelina diffusa Burm. f. 43904.

C. erecta L. 42552. GC, GD, SI, SJ.

Tradescantia pallida (Rose) Hunt. 43422, cultivated, becoming naturalized.

T. spathacea Sw. Sight, cultivated SJ.

CYMODOCEACEAE

Syringodium filiforme Kutz. 44914. SJ.

CYPERACEAE

Cyperus confertus Sw. 43878.

C. nanus Willd. 42020, 42025, 42567. LC, SJ.

C. planifolius L.C. Rich. 42565, 43439. GD, LC, SJ. The closely related *C. brunneus* Sw. has been found on LC.

C. rotundus L. 42629, 43865. SJ.

C. unifolius Bocklr. 42513, 42566, 43700, 43848. Closely related to *C. filiformis* Sw. of the Greater Antilles but has much shorter spikelets.

Fimbristylis cymosa R. Br. 42519. LC, SJ.

(*Scleria lithosperma* (L.) Sw. has been collected on GC, SI, SJ.)

GRAMINAE (Poaceae of Acevedo-Rodriguez 1996)

Anthephora hermaphrodita (L.) Ktze. 42543. GC, SJ.

Bothriochloa pertusa (L.) A. Camus. 42544. SJ.

Bouteloua americana (L.f.) Scribn. 42508. LC, SJ.

Brachiaria adspersa (Trin.) Parodi (*Urochloa adspersa* in Acevedo-Rodriguez 1996). 42549, 43477, 43879, 43880. SJ.

B. fasciculata (Sw.) S. T. Blake (*Urochloa fasciculata* in Acevedo-Rodriguez 1996). 42543. SJ.

Cenchrus echinatus L. 43410. GC, LC, SJ.

C. incertus M. A. Curtis. 43409.

Chloris barbata Sw. (often known as *C. inflata* Link). 42547. GC, GD, SJ.

Cynodon dactylon (L.) Pers. 43451. SJ.

Dactyloctenium aegyptium (L.) Beauv. 42518. SJ.

Digitaria bicornis (LM.) R. & S. 43881. GC.

D. ciliaris (Retz.) Koeler. 42507. SJ.

D. eggersii (Hack.) Henr. 47251, 48361, 48814, 48822. Originally described from St. Thomas, this rare species is otherwise known only from Virgin Gorda and the Sierra Bermeja in southwestern Puerto Rico.

D. insularis (L.) Mez ex Ekman. 43404, 46522. SJ.

Eleusine indica (L.) Gaertn. 42559, 43883. SJ.

Eragrostis ciliaris (L.) R. Br. 42536. GC, SJ.

(*Lasiacis divaricata* (L.) Hitchc. GC, SJ; to be expected on Guana.)

Oplismenus hirtellus (L.) Beauv. 42661, upper slopes of Guana Peak. SJ.

Panicum maximum Jacq. (*Urochloa maxima* in Acevedo-Rodriguez 1996). 42546. SJ.

Pappophorum pappiferum (Lam.) Ktze. 42600, Muskmelon Bay Beach. Elsewhere in the Virgin Islands it is known only from Carrot Rock and a very old (1880–81) record from St. Thomas. It is also known from a single locality in western Puerto Rico and from Mona Island. Otherwise, this rare species has a broad range throughout almost all of the neotropics.

Paspalum laxum Lam. 43432. GC, GD, LC, SJ.

P. molle Poir. in Lam. 42545. SJ.

P. pleostachyum Doell. 42599, south end of Muskmelon Bay Beach.

(*P. plicatulum* Michx. LC.)

P. setaceum Michx. var. *ciliatifolium* (Michx.) Vasey. 43887, in sandy field near White Bay.

P. vaginatum Sw. 43412. SJ.

(*Setaria setosa* (Sw.) Beauv. GC, GD, LC, SJ.)

S. utowanaea (Scribn. ex Mills.) Pilg. 46523, along crest of ridge between Guana Island Club and Muskmelon Bay, in rocky woodland, rare. GC, LC, SJ.

Spartina patens (Ait.) Muhl. 43431, North Beach. SI, SJ. This species apparently never flowers in the West Indies.

Sporobolus indicus (L.) R. Br. 48813, hill east of Muskmelon Bay, in a gravelly clearing. GC, LC, SJ.

S. virginicus (L.) Kunth. 43696, 43854. LC, SJ.

Tragus berteronianus Schult. 42595. SJ.

HYDROCHARITACEAE

Thalassia testudinum Banks & Solander. 44900, North Beach, in shallow sandy sea-bottom. SJ.

LILIACEAE

Circumscription according to Howard (1974–1989, see 1979 volume); however, the two included genera are obviously not closely related.

Aloe vera (L.) Burm. f. Sight, cultivated. SJ.

Asparagus densiflorus (Kunth) Jessop. 43889, cultivated.

ORCHIDACEAE

Epidendrum ciliare L. Sight, near summit of Guana Peak; also cultivated at Guana Island Club. SJ.

Psychilis macconnelliae Sauleda. 42571. GC, SJ. Mostly restricted to the Virgin Islands and Vieques but also reported from St. Kitts.

(*Tetramicra canaliculata* (Aubl.) Urban has been found on GD, SJ, and several other Virgin Islands. It is also known from Florida, Hispaniola, Puerto Rico, and the Lesser Antilles.)

Tolumnia prionochila (Kraenzl.) Braem. (formerly *Oncidium prionochilum*). 43678. GC, SJ. Endemic to the Virgin Islands and Culebra.

PALMAE

This is Arecaceae in Acevedo-Rodriguez (1996). Several exotic species of palm have been introduced to Guana Island over time, including *Chrysalidocarpus lutescens* Wendl., *Cocos nucifera* L., and *Veitchia merrillii* (Beccari) H. E. Moore.

Coccothrinax alta (O. F. Cook) Beccari. 41996. SJ. This was erroneously reported as a species of *Thrinax* by Lazell (1983b). Endemic to the Virgin Islands and Puerto Rico.

Sabal causiarum (O. F. Cook) Beccari. 42022. Occurs in the thickets behind North Beach, also near the lower end of Grand Ghut and along the top of the seacliffs at the northernmost end of the island. It appears to be absolutely indigenous at all of these sites. This species occurs elsewhere in the Virgin Islands only on Anegada; otherwise its range includes western Puerto Rico and scattered localities in southern Hispaniola. Not cited from Guana by Zona (1990). This species has been planted at Beef Island Airport.

PANDANACEAE

Several cultivated species, not identified.

RUPPIACEAE

Ruppia maritima L. 48402, salt pond. SJ.

SMILACEAE

Smilax coriacea Spreng. 47244, ridge leading to Palm Point, very rare. SJ.

Fungi

D. Jean Lodge • Peter Roberts

THE TRUE FUNGI BELONG to a separate kingdom from plants and appear to share a common ancestor with animals, though their relationship is now very distant. Similar to animals, all fungi are heterotrophic: deriving their energy from other organisms, living or dead. Around 100,000 species have been named to date, but it is thought this figure may only represent a small minority of the true total.

Three fungal phyla have been recorded so far on Guana Island. The most ancient of these in evolutionary terms (Triassic, over 200 million years ago [mya]) are the *Glomales* (phylum Zygomycota), which form symbiotic relationships with living plant roots called *arbuscular endomycorrhizae*. The term *endo-* in endomycorrhiza means "inside" and refers to the fungal penetration of the host plant's root cells, and *arbuscular* means "treelike" or "arborescent" and refers to the highly branched organ formed by the fungus inside the root cells. The arbuscles are structures formed for nutrient exchange between the partners. A mycorrhiza is the product of a mutually beneficial partnership between a fungus and a plant, analogous to that found in lichens. Endomycorrhizal symbionts are ubiquitous, and about 95% of higher plant species are dependent

to some degree on species in the Zygomycota for the uptake of mineral nutrients and water from soil. The oldest fossils of these fungi that have been dated are more than 200 million years old. Hart et al. (2003) review the role of arbuscular mycorrhizae in facilitating the coexistence of a diversity of plant species in the same habitat. Plants that otherwise might be at a competitive disadvantage may gain significantly greater benefits from nutrients and water resources if they harbor relatively high densities of these fungi. This may keep them viable in the habitat and thus maintain high plant species numbers. Now that is a different way of looking at diversity. The earliest land plants other than algae were mosses, ferns, and their allies, and the earliest fossils of these plants contain fungi of the Zygomycota. Although mosses and ferns have rootlike structures called rhizoids, rather than true roots, they also form symbiotic relationships with fungi in the Zygomycota. It has been hypothesized that symbiotic relationships with fungi enabled plants to colonize land. The spores of endomycorrhizal fungi are too large to be dispersed by wind, but they are ubiquitous in all types of soil and can be transported to "new" islands in (for example) the mud on the feet of birds.

The next most ancient phylum, making up the majority of fungal species worldwide, is the Ascomycota. These typically have both sexual and asexual stages, the latter including common molds (such as *Penicillium*, one of the common bread molds and a source of antibiotics) and many of the yeasts. The sexual stages include cup fungi, dead man's fingers (Xylariaceae), many leaf-spotting plant parasites, and some hyperparasites on other fungi. Molds, yeasts, and leaf spots certainly exist on Guana Island, but the ascomycetes have not yet been systematically surveyed.

The third, and most conspicuous, phylum is the Basidiomycota (see Bonderteiner et al. 2004). This group includes the mushrooms, shelf fungi, jelly fungi, coral fungi, earthstars, earth fans, puffballs, and stinkhorns. The majority of species decompose wood, leaf litter, and organic matter in the soil, playing a critical role in releasing and recycling nutrients. This is important for animals as well as plants. For example, most termites, including *Nasutitermes* on Guana Island (see "Termites: Isoptera," part 5, below), and many other invertebrates need basidiomycetes to precondition the wood by softening and enriching it in order for it to be palatable. The remaining basidiomycetes on Guana are mostly ectomycorrhizal symbionts forming a mutually beneficial relationship with the roots of certain trees and shrubs (seagrape, *Coccoloba*, and blolly, *Pisonia* on Guana). The term *ecto-* in ectomycorrhiza means "outside" and refers to the fungi remaining outside of the host root's cells. Hyphae of ectomycorrhizal fungi form a sheath on the surface of the host's fine roots and also surround (but do not invade) the root's outermost cells.

Though fungal mycelium (in soil and rotten wood) is always present on the island, the fungal fruiting bodies only appear after heavy rain and may not come up at all in some years, making systematic surveys challenging. Nonetheless, around 120 species have been recorded from Guana to date and (if it were possible to make a complete inventory) it is likely that the total actually present on the island would be well in excess of 2,000 species. The majority of these would be widespread, having a cosmopolitan, pantropical, or neotropical distribution. However, some of the basidiomycetes, especially those forming ectomycorrhizal symbioses with woody plants, would be more restricted.

Two ectomycorrhizal mushrooms associated with *Coccoloba* (*Amanita arenicola* and *Lactarius coccolobae*) have recently been described from Guana Island as species new to science (O. K. Miller et al. 2000), though both are known to occur on other nearby islands (including Puerto Rico). Other ectomycorrhizal fungi that have been found fruiting under *Coccoloba* and *Pisonia* include various mushrooms (*Inocybe*, *Lactarius*, and *Russula* species), boletes (*Xerocomus*), earth fans (*Thelephora*), and earthballs (*Scleroderma*). Conspicuous wood-rotting fungi on logs and dead standing trees include a number of shelf fungi (notably the pantropical, orange to scarlet, *Pycnoporus sanguineus*), some jelly fungi (including the edible *Auricularia cornea*), and several mushroomlike species. One of the large shelf-fungus species, *Ganoderma nitidum*, was reported for the first time for the Caribbean region from Guana Island (Ryvarden 2000). Conspicuous wood-rotting fungi on Guana Island include stinkhorn relatives (*Clathrus crispus* and *Lysurus* cf. *gardneri*, cf. *Lysurus cruciatus*) whose foul-smelling fruiting bodies attract flies that disperse the spores.

In the list that follows we have had to editorially, and somewhat arbitrarily, resolve controversies of familial and generic names and assignments. In cases where we are convinced the name given will eventually be replaced, we have put that name in quotations marks. We use the abbreviation *cf.* (Latin, meaning "confer") to indicate that our specimens are comparable to, but possibly not quite the same as, members of the group named. We use *sp.* for one unknown species and *spp.* for two or more unknown species. Families and genera within them are arranged alphabetically.

PHYLUM ZYGOMYCOTA

GLOMACEAE
Sclerocystis sp.

PHYLUM ASCOMYCOTA

PATELLARIACEAE
Rhytidhysteron rufulum

SARCOSOMATACEAE
Plectania cf. *rhytidia*

XYLARIACEAE
Xylaria cf. *mellisii*

PHYLUM BASIDIOMYCOTA

AGARICACEAE
Agaricus cf. *arvensis*
A. cf. *campestris* (*edulis* group)
A. spp., at least four more
Chlorophyllum molybdites
Lepiota cf. *guatopoensis*
Leucocoprinus birnbaumii (= *L. lutea*)
Leucoagaricus sp.

AMANITACEAE
Amanita arenicola
Limacella sp.

AURICULARIACEAE
Auricularia cornea

AURISCALPIACEAE
Amylosporus campbellii

BOLETACEAE
Xerocomus coccolobae

BOTRYODONTACEAE
Botryodontia denticulata

BROOMEIACEAE
Diplocystis wrightii

CANTHARELLACEAE
Cantharellus cinnabarinus

CLATHRACEAE
Clathrus crispus
Lysurus cf. *cruciatus*

CLAVARIACEAE
Ramaria sp.

COPRINACEAE
Coprinus plicatilis
C. spp., at least three more
Psathyrella cf. *candolleana*

CORIOLACEAE
Ceriporia xylostromatoides
Dichomitus setulosus
Fomitopsis dichomia
Hexagonia hydnoides
Pycnoporus sanguineus
Trametes villosa
T. hirsuta

CORTINARIACEAE
Gymnopilus sp.
Inocybe xerophytica

ENTOLOMATACEAE
Clitopilus sp.
Rhodocybe luteocinnamomea

EXIDIACEAE
Eichleriella leveilliana

GANODERMATACEAE
Ganoderma nitidum
G. resinaceum
Humphreya coffeatum

GEASTRACEAE
Geastrum cf. *minimum*

GLOEOCYSTIDIELLACEA
Gloeocystidiellum laxum

GRAMMOTHELEACEAE
cf. *Grammothele* sp.

HYMENOCHAETACEAE
Phellinus contiguus
P. pectinatus
P. rimosus

HYPHODERMATACEAE
Hyphoderma argillaceum
Hyphodontia sambuci
Radulomyces confluens
Subulicystidium longisporum
S. meridense

LACHNOCLADIACEAE
Scytinostroma duriusculum

LENTINACEAE
Lentinus cf. *bertieri*

LYCOPERDACEAE
Calvatia cf. *rubroflava*
Calvatia sp.

MERULIACEAE
Mycoacia sp.

PENIOPHORACEAE
Peniophora albobadia

PLEUROTACEAE
Pleurtus flabellatus

PLUTEACEAE
Pluteus, two spp.
Volvariella cf. *cubensis*
V. diplasia
V. cf. *taylori*

POLYPORACEAE
Polyporus tricholoma

RUSSULACEAE
Lactarius coccolobae
L. nebulosus
L. cf. *putidus*

Russula cremeolilacina (= *littoralis*)
R. spp., at least three more

SCHIZOPHYLLACEAE
Schizophyllum commune

SCLERODERMATACEAE
Scleroderma bermudense Coker with the tree *Coccoloba uvifera*
Veligaster nitidum (Berk.) Guzmán & Tapia with the tree *Pisonia subcordata.*

SISTOTREMATACEAE
Trechispora cf. *farinacea*

STEREACEAE
Stereum cf. *ravenellii*

THELEPHORACEAE
Thelephora sp.

TREMELLACEAE
Heterochaete shearii
H. sp.

TRICHOLOMATACEAE
"*Collybia*," four spp.
Crepidotus sp.
Gerronema cf. *icterinum, tenuis*
Henningsomyces puber
Lactocollybia angiospermarum
Leucopaxillus gracillimus
Marasmiellus cf. *coilobasis*
M. spp., at least five more
Marasmius atrorubens
M. guyanensis
M. haematocephalus
M. spp., at least six more
Mycena, two spp.
Neonothopanus nambi (= *Nothopanus hygrophanous*)
Resupinatus sp.
Tetrapyrgos nigripes

The Frog *(Eleutherodactylus antillensis)*

Kristiina Ovaska

W HISTLING FROGS OR COQUIS (genus *Eleutherodactylus*) dominate the amphibian fauna of Puerto Rico and the Virgin Islands. This large group of frogs is widely distributed in Central America, northern South America, and the West Indies, where several centers of endemism occur (Hedges 1989). A total of about 500 species are recognized, and new species continue to be discovered (Joglar 1998). Of the 17 species of *Eleutherodactylus* endemic to the Puerto Rico Bank, three occur in BVI: *E. antillensis*, *E. cochranae*, and *E. schwartzi* (MacLean 1982). Only one species, *E. antillensis*, is present on Guana.

Whistling frogs are typically small (about 3 cm or less in snout-to-vent length, SVL), secretive, and inconspicuous when not calling. Males of most Puerto Rican Bank species, including *E. antillensis*, are highly vocal, and their loud, whistle-like calls often dominate the night-sounds. All breed on land and have direct development; there is no aquatic tadpole stage. Depending on species, whistling frogs can be found in a variety of forested, riparian, and disturbed habitats. *Eleutherodactylus antillensis* is one of the few species that can occupy xeric forests (Joglar 1998). Because of their local abundance, whistling frogs often play an important role in the food web of the ecosystem as prey for a variety of invertebrate and vertebrate predators, such as spiders, snakes, and birds (Stewart and Woolbright 1996). They eat insects and other small arthropods. (See plate 19.)

West Indian whistling frogs have proven useful as model organisms for various evolutionary and systematic studies, including studies addressing acoustic communication (Narins and Capranica 1976, 1978, 1980), community organization (Drewry and Rand 1983), population ecology (Stewart and Pough 1983; Stewart and Woolbright 1996), parental care (Townsend et al. 1984; Townsend 1996; Ovaska and Estrada 2003), and biogeography (Hedges 1989). The ecology of one species in particular, *E. coqui* (the common coqui) of Puerto Rico, has been investigated in great detail because of this species' abundance and relative ease of study. However, even basic ecological and life-history information is lacking for most other species. Regrettably, as with many amphibian populations worldwide, several species of *Eleutherodactylus* in Puerto Rico have declined drastically since the 1970s for reasons that are largely unknown; 3 of the 16 species known from the island are probably extinct, and an additional 7 appear to have declined (Joglar and Burrowes 1996). Even some populations of the ubiquitous *E. coqui* have experienced declines (Stewart 1995). Baseline data on distributions, abundance, and habitat requirements of these frogs are largely lacking for smaller islands on the Puerto Rico Bank, precluding evaluation of their status.

In 1993, Jeannine Caldbeck and I initiated ecological studies of *Eleutherodactylus* in the BVI, focusing on courtship behavior (Ovaska and Caldbeck 1997a, Ovaska et al. 1998), acoustic communication (Ovaska and Caldbeck 1997b), distribution and biogeography (Ovaska et al. 2000), habitat use, and population dynamics. This summary of the natural history and population biology of *E. antillensis* on Guana is based on fieldwork during October of eight years between 1993 and 2001. October is one of the wettest months of the year (average of 16.3 cm of precipitation per month from 1960–84 on adjacent Tortola), and conditions are usually favorable for frog activity.

Eleutherodactylus antillensis is a small, brownish frog with a slender shape and long legs. Similar to other members of the genus, these frogs possess expanded toepads (with T-shaped terminal phalanges) that facilitate climbing. *Eleutherodactylus antillensis* can be distinguished from related species by its granulated underside, dark mottling on the inner surface of thighs, and eye color—the upper half of the iris is brick red (MacLean 1982).

Eleutherodactylus antillensis is widespread in Puerto Rico and on smaller islands on the Puerto Rico Bank (Schwartz and Henderson 1991). In the BVI, the species is known from eight islands (Ovaska et al. 2000). Over its range it inhabits a variety of wooded areas and forest edges, including xeric forest (Schwartz and Henderson 1991; Joglar 1998); dense forest with closed canopy appears to be unsuitable (Joglar 1998).

In the BVI, Guana is among the most arid islands where *E. antillensis* is found. Its distribution on the island is patchy, probably as a result of a paucity of suitable moist retreats. The species occurs throughout the northwest portion of the island but, based on nightly calling by males, its presence in other parts of the island is restricted to small, widely separated patches. Habitat features that appear to be important for the frogs on Guana include the presence of bromeliads and other plants that hold moisture in their leaf axils, used for diurnal retreats, and a layer of leaf litter, used for oviposition (Ovaska and Estrada 2003). These two habitat parameters showed a significant but weak correlation with numbers of calling males during nighttime surveys of the concentration in the northwest of the island; in another, smaller, habitat patch, the frogs were always associated with bromeliads (Ovaska et al. 2000). Interestingly, this species is not associated with bromeliads in wetter habitats on other islands (Tortola and Virgin Gorda), and the presence of these plants appears to gain importance with increasing aridity of the habitat.

The frogs are largely arboreal, and at night males call from elevated perch sites in vegetation, typically at heights of 0.5–2.5 m. On Guana, calling sites include leaves and branches of trees and shrubs, agave plants, and bromeliads. During the day we have found *E. antillensis* within bromeliads and under woody debris on the forest floor, but microhabitats used for retreats remain poorly known. In other areas the species has been found during the day "under grass roots, loose bark of trees, rocks, logs, trash, [and] in tarantula burrows" (Schwartz and Henderson 1991, 27). (See plate 20.)

As with many other anurans, males call from stationary locations and females initiate courtship by approaching males. Spacing of males is accomplished by acoustic cues (Ovaska and Caldbeck 1997b), and we have never observed fighting between males. During courtship the male leads the female to a secluded oviposition site on the ground; the male produces a special, multinote courtship call that apparently induces the female to follow (Ovaska and Caldbeck 1997a, 1997b). We have found seven egg clutches on Guana; many additional courting pairs disappeared under the leaf litter on the ground but were either caught before oviposition or had moved by the following morning when we inspected the site. Six clutches were under a thin (ca. 2 cm deep) layer of leaf litter, and one was under a grass mat. In addition to providing cover from predators, a leaf layer provides a moist environment for embryonic development and moderates temperatures. We monitored the temperature under leaves near one egg mass and on the surface for five days in October 2001. Average maximum daytime temperatures were about 2°C lower under leaves than on litter surface (29.1°C under leaves; 31.2°C on surface), whereas nighttime minimum temperatures were similar in the two microhabitats (24.9°C under leaves; 25.2°C on surface). Oviposition on the ground is unusual in *Eleutherodactylus* (Ovaska and Estrada 2003); many species lay their eggs in trees, often in bromeliads. Thermal and humidity factors may restrict oviposition sites on Guana.

The average size of six newly laid clutches (located by following courting pairs) was 31 eggs (standard deviation [SD] = 5.8, range = 25–42); the remaining clutch (found by raking leaves) contained only 9 eggs. The eggs were round, opaque, and laid in a globular cluster. Individual eggs measured ca. 4.0–4.5 mm in diameter within 24 h of oviposition but expanded as they absorbed water. Few clutches of *E. antillensis* have ever been found, and, to our knowledge, ours are the first reported observations from the natural range of the species. Clutch size of 11–32 eggs has been reported for this species in captivity (Michael 1997: nine clutches; Joglar 1998: two clutches), and the clutch size for an introduced population within residential gardens in Panama City was 11–28 eggs (mean = 19 eggs; Castillo and Mayorga 1984). Development to hatching of one clutch monitored on Guana took 15 days, just as observed in Panama (mean = 16 days, range = 15–17 days; Castillo and Mayorga 1984). We have inspected clutches during several days and nights but never observed parental attendance. It appears that parental care, which is common among *Eleutherodactylus* (Townsend 1996), is absent in *E. antillensis*, at least on Guana. Ovaska and Estrada (2003) reported a male in attendance at a 24-egg clutch on Puerto Rico. Spatial separation of retreats and oviposition sites may account for the lack of parental care. The thermal environment under leaf litter on the ground might be optimal for embryonic development but might not provide sufficient cover and moisture for adult frogs. (See plate 21.)

Virtually nothing is known of the growth and behavior of juvenile *E. antillensis* (Joglar 1998). Curiously, we have never found juveniles on Guana, although we have frequently encountered them in moist habitats on other islands (Tortola and Virgin Gorda). We suspect that juveniles initially dwell within the litter layer and that their secretive habits and low densities on Guana partially account for our lack of detection. All our surveys on the island have taken place in October, during the rainy season. Oviposition may be confined to a short period during the wettest portion of the year because of the aridity of the island, resulting in relative synchronicity of development. Hence older juveniles may not be available for detection in October, as young from the previous year have attained adult size but young of the year are still hidden within the litter.

Juvenile frogs probably mature during the subsequent wet season after hatching. On Guana, calling males are typically fairly uniform in size (27- to 31-mm SVL). Males are larger, on the average, than on Virgin Gorda but similar in size to those on Tortola (Ovaska et al. 2000). Females attain a larger size than males and measure up to 45-mm SVL. Females can breed repeatedly within a season, and in captivity produce clutches at intervals of as short as eight days (Michael 1997). Limited evidence suggests that individuals are short lived: of 7 resident adult males marked with long-lasting fluorescent elastomers, injected subcutaneously, on a 15 × 15 m mark–recapture plot on Guana in 1998, none were recaptured the following year; of 23 males similarly marked on the same plot in 1999, none were recaptured in 2001. The elastomers cannot be shed off with skin and have lasted over three years on other amphibians.

Auditory strip surveys (see "Line Transects," part 2, above) are an effective way of obtaining relative abundance estimates of anurans, especially for those species that (1) are highly vocal,

(2) have a relatively long breeding period, and (3) are dispersed throughout the area of interest (Zimmerman 1994). *Eleutherodactylus antillensis* fits all these criteria. Since 1996 we have monitored trends in abundance of calling males along a transect 825 m long and 10 m wide that passes through a concentration of frogs in the northwest portion of the island. The transect followed a narrow path and was marked with a flag every 5 m to facilitate counting. We carried out surveys after dark (1820–2150 h), and counted calling frogs within a 5-m strip on either side of the transect center (measured outward from the edge of the path); we did not trace locations of individual frogs unless required to verify the source of particular calls near the boundaries of the strip. The temperature during surveys averaged 26.4°C (SD = 1.7°C). Each survey took an average of 70 minutes to complete.

We conducted 24 surveys over five years from 1996 to 2001, done in October (2–6 per year; no surveys took place in 2000). Counts of calling males were consistently lower on dry nights (with no rain less than 24 h before surveys); hence we omitted these nights ($n = 7$) to reduce variability among years due to weather. The numbers of calling males (figure 15) remained relatively constant from 1996 to 1998 (average of 40–55 frogs/survey), peaked in 1999 (average of 107 frogs/survey), and were the lowest in 2001 (average of 36 frogs/survey).

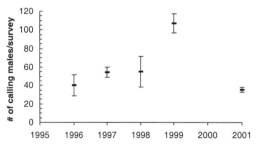

FIGURE 15. Number of calling males detected during surveys of a transect (825 m long, 10 m wide) on Guana done in the month of October over five years from 1996 to 2001. Only surveys within less than 24 h since rainfall are included. Horizontal bars, means; whiskers, standard errors. The number of surveys was 4, 2, 3, 5, and 3, respectively, by year.

To obtain density estimates for adult males, we first calculated density per hectare using the highest call-count for each annual survey period (Zimmerman 1994). We then adjusted this value upward based on the estimated proportion of males that were expected to call at any one time. We estimated this proportion based on mark–recapture data on a 15 × 15 m plot in the same habitat in October of both 1998 and 1999; we also inspected frogs for recaptures within a 3 m wide boundary zone surrounding the plot. A total of 9 males in 1998 and 26 males in 1999 used the plot. The percentage of calling males of those known to use the plot ranged from 58% to 78% during seven inspections on wet nights. Using these proportions, density estimates for adult males ranged from a low of 41–55 males/ha in 2001 to a high of 143–192 males/ha in 1999, an approximately threefold difference (table 21). Correspondingly, assuming an equal sex ratio, the estimated densities of all adults were twice these values, resulting in a total biomass of approximately 180–242 g/ha in 2001 and 629–845 g/ha in 1999, based on average weights of 1.7 g and 2.7 g for adult males and females, respectively (Ovaska et al. 2000). Densities obtained for 1999 are comparable to maximum estimated densities (ca. 400 individuals/ha) recorded for the species in Puerto Rico but much lower than average densities for adult *E. coqui* (3,265 frogs/ha; Stewart and Woolbright 1996). In Puerto Rico the total biomass of all *Eleutherodactylus* species in forest habitats was estimated to be around 3.7 kg/ha (3,700 g/ha). Our values must be interpreted with caution, because the actual sex ratio is unknown. An unknown number of juveniles are also missing from our estimates, precluding an estimation of total biomass.

Potential sources of bias in auditory strip counts include observer error and variability in the proportion of males that call at any one time owing to seasonal and diel patterns of activity, weather conditions (humidity, rainfall, temperature), and prevalence of alternative mating tactics that do not involve calling, such as satellite

TABLE 21
Estimated Density of Adult Males on Northwest Portion of Guana

	NUMBER OF SURVEYS	NUMBER OF CALLING ± SE	CALLING RANGE	OBSERVED DENSITY (HIGHEST COUNT/HECTARE)	ESTIMATED DENSITY (LOWER LIMIT/HECTARE)	ESTIMATED DENSITY (HIGHER LIMIT/HECTARE)
1996	4	40.5 ± 11.6	27–75	61.9	79.3	106.7
1997	2	54.5 ± 5.5	49–60	49.5	63.5	85.3
1998	3	55.0 ± 16.6	22–75	61.9	79.3	106.7
1999	5	107 ± 10.4	85–135	111.4	142.8	192.0
2001	3	35.7 ± 2.9	30–39	32.2	41.3	55.5

SOURCE: Original data. Based on surveys of an auditory strip transect (825 m long, 10 m wide). The percentage of males calling at any one time is based on the minimum (58%) and maximum (78%) values obtained from mark–recapture data in the same habitat in October 1998 and 1999. Surveys were conducted in October of each year, and only those surveys within less than 24 h of rainfall are included.

behavior (i.e., crouching silently by a rival calling male and intercepting approaching females). Thus, the proportion of males that call at any one time is often unknown. Observer bias was probably minimal in our study because the same observers always conducted the surveys, and the frogs were typically sufficiently spaced along the transect so that confusion was avoided. We attempted to minimize other sources of bias by carrying out surveys between dusk and midnight when calling activity was at its peak and by including only surveys on wet nights (with rain within fewer than 24 h) in the comparisons.

Fogarty and Vilella (2001) evaluated call-count methods for *E. coqui* and other species of *Eleutherodactylus* in Puerto Rico by comparisons with mark–recapture data. Call-count surveys provided accurate estimates of relative abundance in two different habitats (native forest and a plantation), and the authors concluded that this method provides an efficient way to obtain relative abundance estimates when detailed population information is not needed; more labor-intensive mark–recapture methods are required for the latter purpose. Fogarty and Vilella (2001) conducted their study during the dry season, whereas our study took place during the peak of the rainy season. The relatively low proportion of males calling in their study (mean of 25%) when compared to our study (mean of 68% on the mark–recapture plot) probably re-

flects this seasonal difference. We expect that auditory surveys result in more accurate density estimates for adult males during the rainy season, when calling activity is likely to be less influenced by weather.

Auditory strip transects are considered unsuitable for estimating total abundance of adults because they provide no information on females (Zimmerman 1994). This constraint applies to our density and biomass estimates, and we have presented these values as initial estimates only; their validity requires further investigation. Estimating abundance of female *E. antillensis* on Guana is problematic, as females are seldom seen except when courting. As an initial step to address this problem, more information is required on habitat use patterns and activity of females. Female *E. antillensis* are too small for conventional telemetry, but new approaches, such as harmonic direction finder methods (Engelstoft and Ovaska 1999), offer exciting possibilities for tracking movements of small animals and may be applicable to *E. antillensis*.

Worldwide declines of amphibian populations are usually considered as indicators of environmental problems, as semipermeable skin, requirements for moisture, and other attributes of amphibians render them particularly sensitive to pollutants and atmospheric changes (Blaustein 1994). In the Caribbean atmospheric pollutants appear as patchy, thin sheets (or lenses) and descend to the ground and vegetation with fog and rainfall (Stallard 2000). Dust and contaminants in these lenses are derived primarily through long-range atmospheric transport and include dust plumes resulting from fires and land clearing in Africa. The transport of pathogens in the dust is of particular concern, as epidemics of emerging diseases, particularly chytridiomycosis (a fungal skin disease), have been identified as the proximate source of death in declining amphibian populations on several continents (Daszak et al. 1999). It is unclear, however, whether disease is the actual cause of declines or whether other environmental stressors are involved that increase the vulnerability of amphibians to pathogens. Amphibian declines observed on Puerto Rico since the 1970s showed no obvious correlations with removal of forest cover, meteorological factors, or rainwater chemistry, but long-range transport of pollutants and microbial spores remains a concern (Stallard 2000). Several Puerto Rican species have not been seen alive for years and must be presumed extinct.

Unlike several other species of *Eleutherodactylus*, *E. antillensis* remains widespread and locally abundant in Puerto Rico and is considered secure there (Joglar 1998). Similarly, the species is widespread in the BVI in various habitats (Ovaska et al. 2000). Call-count data from Guana do not suggest a decline; low counts observed in 2001 were probably an anomaly as October was relatively dry that year. Our time series, however, is too short to depict multiyear trends with reliability.

Widespread, abundant species are not necessarily secure, however, as attested by precipitous declines and local extinctions of the northern leopard frog *(Rana pipiens)* over much of its range in western North America; this species was one of the most widespread and abundant amphibians on the continent (reviewed by Stebbins and Cohen 1995). Populations of such species may in fact be more suitable for monitoring programs than are small populations with a restricted distribution because larger sample sizes are possible and different geographic areas and habitats can be monitored. Populations of *E. antillensis* occupying xeric forest and other marginal habitats may be particularly vulnerable to any additional environmental stresses, so we hope to continue monitoring the population on Guana in the future.

The Iguana

The iguana is a sort of quadruped serpent, very frightful to look at but very good to eat. . . . There are few men who have seen it alive who dare to eat it, except those in that land who are used to that fright and ever greater ones.

OVIEDO (1522, IN STOUDEMIRE 1959)

I USED OVIEDO'S WORDS, above, to introduce the "Iguana" section of my doctoral dissertation more than three decades ago (it became the reference Lazell 1973), but Oviedo was writing from Puerto Rico. His iguanas were the last survivors of their species in that land, and a remnant of what is probably the largest species of iguana and, therefore, the largest New World lizard. That ponderous species is the stout iguana, today called *Cyclura pinguis*: the original iguanine occupant of the entire Puerto Rico Bank. By about 1930, the last surviving population of this species persisted on Anegada, BVI (Carey 1975; Pregill 1981)—a classic case of peripheral survival at the far edge of the original range (Lomolino and Channell 1995; Tynan et al. 2001). What happened? As their name implies, stout iguanas (*pinguis* means "stout") are notably more massive than green iguanas (*Iguana iguana* and *I. delicatissima*) or even other Greater Antillean rock iguanas (*Cyclura*). The heaviest individual I have ever actually weighed, a Guana Island male 51-cm snout-to-vent length (SVL), was 10.4 kg (23 lb). This was not a large individual. There are certainly individuals on Guana

today approaching 70-cm SVL and easily weighing twice that (Lazell 1997b). I have seen a photograph of a dead specimen on Anegada, taken decades ago, with people in the picture for scale, indicating a SVL of a meter or more. None like that survive on Anegada today. The relative mass of this species seems to render it both slower and less agile than its more gracile cousins. This led to its being easier prey for humans. (See plate 22.)

The bones in kitchen middens attest that large numbers were slaughtered for food by the Tainos, prior to European arrival. Wing (2001) calculated more than 2,000 kg were consumed at one site, Maisabel, on the north coast of Puerto Rico about 1,850 ybp, and a similar quantity at Tutu, on St. Thomas, USVI, about 560 ybp. The latter banquet was held scarcely half a century before Christopher Colombus came this way. By Oviedo's day, few stout iguanas survived on Puerto Rico and perhaps none in today's USVI. Stout iguanas do climb trees regularly (Goodyear and Lazell 1994; Lazell 1997b), but this affords little protection from humans armed with noose-poles or firearms. Mitchell

(2000a) suggests that the igneous rock islands such as St. Thomas, Tortola, and Guana provide few subterranean escape retreats, whereas limestone areas, such as Anegada, provide plenty. This probably explains survival on limestone Anegada in the face of predation from humans, dogs, cats, and mongooses (Mitchell 2000b).

Garcia (2001) suggested the igneous rock islands were "... probably not part of the historic range of the species." This is certainly not true. St. Thomas is largely igneous rock and the Tutu region notably so. Just next door, so to speak, on the Anguilla Bank, strongly terrestrial populations of *I. delicatissima* swarmed on igneous rock islets (Lazell 1973). Here, stout iguanas prefer plants native to the igneous rock islands over those found on karst (limestone) (Mitchell 1999, 2000a). Most convincingly, the explosive population growth and rapid individual growth trajectories of stout iguanas on Guana and Necker Islands indicate igneous rock substrates may provide nutritionally optimal habitats (Goodyear and Lazell 1994; Lazell 1997b, 2000a; Mitchell 1999, 2000a; Binns 2001, 2003; Perry and Mitchell 2003). It is imperative that stout iguanas be restored to every island on the Puerto Rico Bank that can sustain a population. We are losing potential island habitats because the introduced, exotic *I. iguana* is expanding its range on the Puerto Rico Bank rapidly and can predictably outcompete *Cyclura pinguis* (Lazell 2002c). Today, *C. pinguis* is probably the most critically endangered of all iguanas (Mitchell 2000a, 2000b).

The phylogenetic position of the stout iguana is remarkable because it appears to be the basal member of the entire *Cyclura* group (Malone et al. 2000). The stout iguana shares a broad parabasisphenoid (a floor bone in the skull), polycuspate lateral teeth, and the absence of caudal verticals (rings of enlarged scales on the tail) with members of *Iguana*. It agrees with other *Cyclura* only in lacking enlarged spikelike scales on the edge of the dewlap. In fact, despite the literature (de Quieroz 1987; Norell and de Quieroz 1991), the lack of dewlap spikes is the only anatomical characteristic that unites the genus *Cyclura* and separates it from *Iguana* (Lazell 1973, 1989b). I have long advocated uniting these two putative genera. Indeed, on the Anguilla Bank islands, *I. delicatissima*, which is highly terrestrial and apt to lose its green coloration with age, is arguably the closest species anatomically to *C. pinguis* among all iguanas: scarcely more than an intergrade between *C. pinguis* and *I. iguana*.

The evidence from mitochondrial DNA (mtDNA) contradicts anatomy (Sites et al. 1996). Wiens and Hollingsworth (2000) suggested that the mtDNA evidence, which allies both species of *Iguana* with lizards of the western North American deserts called chuckawallas, genus *Sauromalus*, was flawed. They strove for a consensus tree of relationships and got *Iguana* and *Cyclura* back together, separated from *Sauromalus*, but still as different genera. Malone et al. (2000) redid the mtDNA, confirmed the Sites et al. (1996) tree, and dismissed Wiens and Hollingsworth's argument against the mtDNA data. In my opinion, no consensus is advisable: the mtDNA data are just plain wrong. There is ample precedent for mtDNA to grossly mislead (Russo et al. 1996; Zardoya and Meyer 1996; Hagelberg 2003), and it is easily switched between lineages by occasional hybridizations.

My favorite mtDNA mess occurred with North American deer (Cronin 1991, 1992). Black-tailed deer are an intergrading subspecies of mule deer, but they have moose mtDNA, very different from that of any other mule deer. Mule deer and white-tailed deer are distinct species but can and do rarely crossbreed. As a result, two distinct kinds of mtDNA are found in both species; therefore, mtDNA has nothing to do with the evolution, biology, or other characteristics of the deer. It is easy enough to see how rare hybridizations could mix up the mule and white-tailed mtDNA, but how did those black-tails get moose mtDNA? A possible scenario would have a mating between a buck black-tail and a female moose—perhaps in the first contact between these two species after a few (or one) moose straggled across the Bering land bridge from Asia at some Pleistocene glacial maximum. Given no other mate choice, the female moose might have fallen in ungulate love with a deer (lonely moose in New England attach themselves to horses and cows, so far without issue). Anyway, the cross worked and the resultant hybrid fawn, a female, carried her moose mother's mtDNA. She, in turn, crossed back into the black-tailed deer population and her moose mtDNA—sequestered from and immune to natural selection (so they say)—spread through the subspecies.

Sound wildly far fetched? Well, consider the water striders—insects of order Hemiptera—studied by Sperling et al. (1997). There the mtDNA also contradicted biological relationships and an excellent fossil record—and was wrong as a molecular clock by an order of magnitude. And, there is the bizarre possibility of mtDNA lineage transfer from skuas to jaegers—seabirds of family Stercorariidae (Brook in Schreiber and Burger 2002). Sometimes mtDNA works fine, and sometimes it wholly misleads (Avise 1986, 1991, 1994; Russo et al. 1996; Zardoya and Meyer 1996). Hedges et al. (1992) cite immunological distance data indicating that *Iguana* and *Cyclura* split a mere 12 million years ago. If ancestral *I. iguana* was the stay-at-home South American member of the splitting pair and ancestral *C. pinguis* was the Antillean member, then perhaps ancestral *Iguana* dispersed north into tropical North America (Central America) at the Pliocene closure of the Isthmus of Panama and eventually contacted ancestral *Sauromalus*. A hybridization event may have passed *Sauromalus* mtDNA into the *I. iguana* population, and it may have spread through the species. In this scenario, *I. iguana* dispersing northward up the Lesser Antillean chain would have brought *Sauromalus* mtDNA to ancestral *I. delicatissima* stock on the Anguilla Bank, derived and weakly differentiated from its *C. pinguis* parent on the Puerto Rico Bank. Then the kind of iguana we call *I. delicatissima* really would be essentially a hybrid or intergrade between the species *I. iguana* and *C. pinguis*.

A great deal of misinformation about the morphology of iguanas is in circulation. Sites et al. (1996) and Norell and de Quieroz (1991) refer to "consistent" differences between *Iguana* and *Cyclura* that are really quite inconsistent. For example, most—maybe all—*Cyclura* have enlarged scales on the toes of the hind feet, called "comb" scales. These scales are often fused together. Most members of *Iguana* lack them but not all, making this difference is inconsistent. (See table 22.)

Most authors follow Zug (1971), who examined the arteries of two specimens of *C. carinata* and four *I. iguana*. Zug (1971) diagrams the arterial plumbing of a typical iguanine lizard and implies that *Cyclura* fits the depicted arrangement. He then says: "In all genera except *Iguana*, the celiac arises anterior to . . . the mesenterics. . . . In *Iguana*, the celiac may arise posterior to . . . or contiguous with . . ." one or both mesenterics. May arise. The very first *I. iguana* I checked, MCZ 69111 from St. Croix, USVI, looked just like Zug's diagram: The celiac arises well anterior to either mesenteric—just as in *Cyclura*. Day et al. (2000) state: "The genus *Iguana* is distinguished morphologically from *Cyclura* by the presence of gular spikes on the dewlap, and its continuous dorsal crest, which in *Cyclura* is divided into distinct nuchal, dorsal, and caudal regions." All right, I concede the dewlap spikes, although some *I. delicatissima* may have as few as four of these scales. But just look at the photographs on the cover and inside the very book Day et al. (2000) were published in. In the cover shot of *Cyclura nubila caymanensis*, there is not a hint of a break. In the photo of *I. delicatissima*, third down on the left on the second page of color plates, taken by Mark Day himself, there are nice big breaks.

The controversy about the relationships and taxonomic status of iguanas—notably the most morphologically generalized of all, our stout iguana—cannot be resolved by appeal to either of the data sets (mtDNA or morphology) currently available: They are utterly contradictory. One or the other is absolutely, unequivocally wrong. At least two more independent biochemical tests are needed. One of these should be a nuclear DNA sequence test of introns (neutral spacers between coding sites) or microsatellites. Other kinds of tests, such as albumin immunological distance or allozyme electrophoresis, would also be appropriate.

TABLE 22

Comb Scale Characteristics From the Right Hind Foot of Seven Iguanas in the Museum of Comparative Zoology (MCZ)

DIGIT	PHALANGE	NUMBER OF SCALES	FUSION	MCZ TAG NO.	TAXON
4	3	2	Yes	16155	*delicatissima*
4	2	2	Yes		
3	2	2	No	14137	*iguana*
2	1	4	No	31493	*rileyi*
3	1	4	No		
3	2	2	No		
2	1	2	No	11062	*inornata*
3	1	2	Yes		
3	2	3	No		
2	1	2	No	154071	*ricordi*
3	1	3	No		
3	2	3	?		
2	1	3	No	154072	*ricordi*
3	1	3	No		
3	2	3	Yes/no		
4	2	2	No		
2	1	3	Yes	12082	*pinguis*
3	1	3	?		
3	2	4	Yes/no		

The stout iguana is currently on the Red List of the International Union for the Conservation of Nature (IUCN) as "Critically Endangered." The population on Anegada "has suffered a steady decline over the past 49 years primarily due to introduced exotic predators (cats), competition and habitat degradation from feral grazing animals (goats, cattle, donkeys) and loss of habitat from development" (R. Hudson 2001). Mitchell (2000a) and Binns (2003) provide a detailed account of the situation. The most fundamental problem is the lack of a protected area on Anegada such as could be provided by a national park. An excellent plan exists but has not yet been implemented.

At present, the populations on Guana and Necker islands offer the best hope for the stout iguana's survival (see "Restoration," part 6, below). Following a goat eradication program completed in 2001 and a vegetation survey conducted by Numi Mitchell and John Binns, of the IUCN Iguana Specialists Group, in October 2002, we moved a dozen stout iguanas to Norman Island, BVI. The capturing effort began on Guana Island on 27 September 2003 and took until 2 October 2003. It was spearheaded by Mitchell and Guana's foreman Lynford Cooper, who, over the years, has learned more about Guana's iguanas (individuals, habitats, and behaviors) than anyone else. Wenhua Lu and I helped. We took a male 38-cm SVL, a female 41-cm SVL, and 10 hatchlings 78- to 115-mm SVL (average 103), to Bluff Bay on the south side of Norman. Everton Henry of the Norman staff got us safely to a landing in this rough and rocky, treacherous but scenic, embayment. "Us" on that day, 2 October 2003, included John Craven of the BBC and the video crew for the program *Sanctuary*. The attempt to restore *C. pinguis* to Norman Island was documented both on TV and in the news media (Henighan 2003).

On 27 June 2003, Guana Island manager Roger Miller e-mailed that he had gone down to North Bay Beach that morning and encountered at least eight, and up to a dozen, iguanas on one small area of the dune ridge—where the sand road crosses through the vegetation. He reported, "They were digging holes." This must have been an aggregation of nesting females. If eggs take about three months to hatch, these could produce babies in late September. We know some Anegada females are gravid in late July (Binns 2003). Hatchlings we regarded as "new" appear right through October, lending circumstantial evidence to both laying and hatching periods spread over at least one month. Hatchlings are immediately aggressive and territorial (Perry et al. 2003) and natural selection is as quickly imposed by raptors and snakes (LeVering and Perry 2003).

The Lizard Genus *Anolis*

Gad Perry

TRAVELING IN THE CARIBBEAN, it is almost impossible not to notice lizards of the genus *Anolis*. They are just about everywhere, and many species are brightly colored. The males frequently engage in courtship and territorial displays, which include attention-grabbing push-ups and the flashing of their colorful throat fans. In recent decades, many researchers have used Caribbean anoles in seminal ecological studies. So much work has been conducted on these animals that Perry and Garland (2002), who reviewed published data on the sizes of lizard home ranges, complained that other taxa were not getting enough attention. They pointed out that, worldwide, there were many more data for *Anolis* than for lizard families with twice as many species. Although many members of the genus are found in Central America and South America, most of the *Anolis* data that Perry and Garland (2002) found were for Caribbean species.

We have established, then, that anoles are fascinating creatures. In fact, there has been so much work on anoles that it would take a thick volume to review it all. Caribbean anoles have become model organisms for addressing a variety of evolutionary and ecological questions. It is impossible to ignore this vast literature completely. The following paragraphs very briefly,

and extremely selectively, attempt to provide some of the highlights by mentioning a few of the studies conducted by some of the leaders in the field. Earlier in this book in "Communities, Local Guilds, and Ensembles," part 3, Lazell described aspects of Guana's anoles including diet, perch heights, insolation, and populations.

Much of the work in the 1960s and early 1970s, such as that of Ernest Williams, Stanley Rand, and James Lazell has concentrated on two main topics: taxonomy and basic biology. Thus, many of the early studies sought to identify the species and subspecies found in the region and to describe their fundamental biological characteristics. One of the most enduring contributions has been the ecomorph concept, still extensively employed to describe Caribbean anoles (Rand 1964; E.E. Williams 1983; Irschick et al. 1997). Most of this work focused on among-island or island-group patterns, but studies of within-island patterns can also be found (e.g., Lazell 1972). Collectively, that work set the stage for more recent studies, using more sophisticated methodologies, by scientists such as Robin Andrews, Jonathan Losos, Joan Roughgarden, Thomas Schoener, and Judy Stamps.

In an early study, Andrews (1979) was one of the first to contrast the ecologies of anoles from

Caribbean islands with those of mainland species. One of the fascinating conclusions of such comparisons is that there are some profound differences between the two. For example, the ecomorph concept that is so prevalent in descriptions of Caribbean taxa is of much less value in characterizing most mainland species. Among Andrews's many important contributions since then have been multiple *Anolis* papers, including one of the most intriguing manipulations of tropical forest ever conducted (Chalcraft and Andrews 1999). In this study, leaf-litter moisture was manipulated in several plots on Barro Colorado Island in Panama, and the effects on the population of *A. limifrons* were carefully assessed.

Jonathan Losos began publishing on anoles in the mid-1980s and has contributed an impressive number of papers on Caribbean species starting in 1990. Many of these (e.g., Garland and Losos 1994) have dealt with ecomorphology, and especially locomotor performance. Most of these (e.g., Losos et al. 1998) have combined a strong evolutionary approach. These studies have greatly improved our understanding of the evolution of anole locomotor behavior and morphology, as well as some of the impacts on the ecology of the animals.

Joan Roughgarden has focused on ecological issues. She has brought a strong theoretical background to studies of community structure (e.g., Rummel and Roughgarden 1983) and feeding biology (e.g., Shafir and Roughgarden 1998).

Tom Schoener and his collaborators have mostly used anoles in studies of community structure. One of his early studies (Schoener 1974) focused on anoles in an influential attempt to provide a theoretical context for resource partitioning in ecological communities. Recent work (e.g., Schoener and Spiller 1996; Spiller and Schoener 1998) has used an experimental approach. Lizards were introduced to small islands that had previously lacked them, and their interactions with the biotic and abiotic environment was monitored over a period of years. Anoles appear to have a strong effect on

invertebrate populations in these small environments and are able to cause major changes in the abundance of prey such as spiders, even to the point of driving them locally extinct.

Judy Stamps and her co-workers (e.g., Stamps et al. 1994, 1997) used Caribbean anoles as model organisms for studying the development and ecological correlates of sexual size dimorphism. Stamps has also been carefully studying the factors that influence territorial behavior (e.g., Stamps and Krishnan 1998).

Given the wealth of knowledge available on anoles, one might expect one of two conditions to arise. So much might be known about the anoles inhabiting the Puerto Rico Bank, including Guana Island, that this chapter would have to be very long to encompass it. Alternatively, with so much previous work, there could be little left to say; this chapter would then be short, but rich in references to previous work. In fact, neither of these conditions pertains. Relatively little of the *Anolis* literature deals with the species found in the immediate area. Most of the existing information was collected on Puerto Rico itself, and little research has been conducted anywhere in the Virgin Islands. The main exception to this rule is Guana Island.

We should start by describing the cast. Taxonomists are forever describing new species and sinking old ones, so the number of taxa, especially in large and complex groups, is constantly changing. In their mammoth summary of the West Indian herpetofauna, Schwartz and Henderson (1991) list 128 species of *Anolis*. Of these, about 15 have been recorded from the Puerto Rico Bank (Schwartz and Henderson 1991; G. Mayer, unpublished data). Five *Anolis* species are currently known from the Puerto Rico Bank in the Virgin Islands. Four of those are native to the BVI. A fifth, *A. leachi*, has been introduced to St. Thomas in the USVI. A sixth species, the large *A. roosevelti*, is likely extinct (MacLean 1982; R. Thomas 1999).

Three species of *Anolis* are found on Guana Island: *A. cristatellus* (the crested anole), *A. pulchellus* (the grass anole), and *A. stratulus* (the saddled anole; Lazell 1983a). The three species are fairly closely related (Crother 1999) and are found throughout much of the Puerto Rico Bank. One likely scenario to explain this distributional pattern is that these species used to be found throughout the original landmass and that populations became separated as sea level rose some 10,000 years ago. This scenario makes today's populations relicts of the past distribution, and it suggests that islands lacking these lizards have lost them because of local extinctions. Alternatively, lizards dispersed overwater, following the rise in sea level. This scenario is also reasonable because anoles are known to be good dispersers. With some human help, *A. cristatellus* has succeeded, in historical times, in expanding its range to Costa Rica (Savage and Villa 1989) and Florida (L. D. Wilson and Porras 1983). Most likely, present-day distributions are the result of both scenarios.

Isolated populations, such as those of anoles on small Caribbean islands, separated by rising water, tend to diverge over time. This happens because of selection (to better adapt to local conditions) and genetic drift (a random evolutionary process that is especially important in small, isolated populations). As small changes accumulate, these processes lead to speciation, the emergence of new species. *Anolis ernestwilliamsi*, which is now found only on Carrot Rock, is an example of this. No new species have been described from Guana Island, but there is some evidence to suggest that the process of speciation may have already begun.

Below, I provide species accounts for the three anoles found on Guana. Much of this information is drawn from work at other locations, but it appears to be true for the animals throughout their range, including Guana. Wherever possible, however, I expand on this general account by providing more detail from work conducted on Guana itself. Because that work is ongoing, only some of it has previously been published.

Saddled anole, *Anolis stratulus*. This smallish anole is typically up to 40 mm in body length, with a tail about as long as its body, and a mass of about 2 g (see plate 23). Sexual dimorphism

is very weak (Butler et al. 2000). Males and females are mainly distinguishable by the presence of an orange dewlap, bordered by yellow, in the males. The body is a greenish gray in both sexes. In ecomorph terminology, the saddled anole is usually called a trunk-crown species, but on Guana, and wherever it occurs in the Virgin Islands that we have observed it, this species frequents the ground, especially in early morning. It is normally found perching relatively high on trees of various sizes. Although found at heights above 20 m, saddled anoles often come lower (Reagan 1992). On Guana, they are usually seen at heights of 1.5–3 m. Population densities in Puerto Rico can be very high (Reagan 1992), but the research on Guana Island shows much lower values (Rodda et al. 2001b). The diet primarily consists of small invertebrates (Lister 1981; Reagan 1996), but saddled anoles have also been observed drinking nectar (Perry and Lazell 1997a). Similar to other anoles, these lizards are sit-and-wait predators (Perry 1999).

When fighting, males will sometimes assume a position opposite one another, with blood-engorged orange tongues extended toward each other (Schwenk and Mayer 1991). Saddled anoles are typically remarkably unconcerned about the nearby presence of humans, allowing one to approach to less than a meter before running away.

Grass anole, *Anolis pulchellus*. Unlike the other two anoles found on Guana, this species typically inhabits a grassy or early successional habitat. The grass anole is relatively small, having an adult body size that is normally about 40 mm and a tail that is more than twice as long as the body (see plate 24). As with other members of this ecomorph, the morphology of *A. pulchellus* is quite different from that of most anoles: It is more slender and elongated and has relatively short legs. These characteristics, as well as its brown coloration and the yellow flank stripe, allow it to merge with its background—clumps of grass or the small stems of regrowing woody vegetation. Sexual dimorphism is intermediate, with the males being somewhat larger than the females (Butler et al. 2000). The dewlap is rose or crimson red, with yellow or white scales.

Home ranges of grass anoles tend to be relatively small, and numbers can be very high in favorable habitats (Schwartz and Henderson 1991). The food includes small invertebrates, typically captured from ambush (Perry 1999). These animals are active during the day. On hot sunny days, they become inactive during the middle of the day. At night, these lizards typically sleep in open areas, at heights of less than 1 m above the ground (Goto and Osborne 1989). When disturbed, grass anoles will "squirrel" themselves to the opposite side of a stem.

Crested anole, *Anolis cristatellus*. The crested anole is the largest of the three *Anolis* species found on Guana Island (see plate 25). Male crested anoles can reach a total length of over 20 cm, of which about two-thirds is tail. In unusual cases, an individuals mass can exceed 10 g. Most of the time, however, adult males are about 15 cm in total length and weigh about 7 g. The body color of crested anoles is tan or gray, and it can change rapidly in response to social and environmental cues. There appear to be some color differences between island populations within the BVI. For example, animals from Sage Mountain on Tortola tend to be greener than those on Guana. However, it is hard to say how much of this variation is genetic. Similarly, body size varies among islands. Sage Mountain animals tend to be smaller than those from Guana, but it is hard to know how much of this is because of phenotypic plasticity.

Only adult males possess the characteristic tail crest that gives the species both its common and its scientific names. Of 40 Caribbean *Anolis* species studied by Butler et al. (2000), the crested anole had the most pronounced sexual dimorphism. Females and juveniles are much smaller than males and might be confused with *A. stratulus*. However, the markings are somewhat different: The dark marks on the back form a series of saddles in *A. stratulus* but form longitudinal stripes in *A. cristatellus*. The larger size, crests, and behavior characteristic of the males have earned them the local name "man-lizard."

By far, the crested anole is the best-studied lizard on Guana Island. Not only has it received more notice in other parts of its range, but researchers working on Guana have also paid much more attention to it. Consequently, more is known about this species than about the other two. *Anolis cristatellus* is a classic trunk–ground species. The typical habitat is relatively exposed, with males being found relatively high (usually up to 2 m above the ground) and females and juveniles typically foraging lower, often on the ground. As is common for all anoles studied to date, crested anoles are sit-and-wait foragers, spending most of their time stationary and only infrequently darting out after their prey (Perry 1999). The diet mostly consists of small invertebrates but can include lizards (Schwartz and Henderson 1991) and even small fruits and flowers (Lazell and Perry 1997; Lazell and Mitchell 1998).

Male crested anoles are highly territorial, defending home ranges that average about 20 m². They spend much time displaying at other males, exposing a large dewlap with a greenish center, surrounded by a red margin as wide as the center spot. The dewlap strongly reflects ultraviolet (UV) light, producing a strong contrast with the typical environment inhabited by this species (Leal and Fleishman 2002). Males with relatively high endurance display more often in the field, and are more likely to win fights staged in captivity (Perry et al., 2004). Moreover, males with high endurance tend to control larger territories (G. Perry, K. LeVering, and I. Girard, unpublished data).

Crested anoles are normally active during the day, but around human settlements they may be seen foraging near lights at night (Perry and Lazell 2001). Crested anoles will bask, often maintaining a body temperature that is 1°C–2°C above ambient temperature in Puerto Rico (Hertz 1992) and on Guana (G. Perry, R. Dmi'el, and J. Lazell, unpublished data). Their relatively large size, compared to the other two species, allows them to be active in more exposed habitats and during hotter parts of the day (G. Perry, R. Dmi'el, and J. Lazell, unpublished data). These data help explain the observation of Thomas (1999), based on the work of Lazell (1983a), that *A. cristatellus* is the reptile most likely to be found on even the smallest of Puerto Rico Bank islands.

Crested anoles are preyed on by small raptors and by ground snakes *(Alsophis portoricensis)*, locally known as racers, and other snakes (Chandler and Tolson 1990). Leal and Rodríguez-Robles (1995, 1997) studied the response of the lizard to the presence of a snake, and described 13 different behaviors, ranging from immobility to biting. Similar to many lizards, anoles can autotomize their tails and then regenerate them. Many crested anoles on Guana have damaged tails (G. Perry and K. LeVering, unpublished data), but it is not only predators that can cause such damage.

The function of the tail crests has been the focus of study for several years. When only one sex possesses such a striking secondary sexual characteristic, it is reasonable to expect it to have a role in sexual selection. However, the relative size of the crest does not affect the outcome of agonistic male–male interactions, or the size of the home range, or the number of females associated with a male (G. Perry and K. LeVering, unpublished data). In contrast, manipulative experiments of free-ranging animals show that the presence of a tail crest is important in identifying an animal as an adult male (K. LeVering and G. Perry, unpublished data). Subadults, who are normally ignored, were immediately attacked by a territorial male when an artificial crest was glued to their previously crestless tails. In the absence of such a crest, both prior to attachment of an artificial crest and following its removal, the same subadult male was ignored. No physiological costs to having a large crest, either in thermoregulation (G. Perry and K. LeVering, unpublished data) or in water-loss rates (G. Perry, K. LeVering, and R. Dmi'el, unpublished data), have so far been identified. However, it is possible that maintaining a large crest is energetically expensive or increases the risk of predation.

Several studies have concentrated on water-loss rates in crested anoles from the BVI. These have consistently shown differences in water-loss rates among animals from various islands (Dmi'el et al. 1997; Perry et al. 1999, 2000), and even from different altitudes within Guana Island (G. Perry, R. Dmi'el, and J. Lazell, unpublished data). Populations from dry islands and low altitudes have lower water-loss rates than do populations from wetter habitats and islands. Interestingly, animals from Guana have consistently shown lower water-loss rates than would be expected, based on the perceived aridity of their environment (Dmi'el et al. 1997; Perry et al. 1999, 2000). To test whether this is a result of a misconception of conditions on Guana, a more comprehensive study, involving members of all three anole species widely found in the area, was conducted over several years. It showed that this phenomenon is unique to *A. cristatellus* and not shared by the other two *Anolis* species (G. Perry, R. Dmi'el, and J. Lazell, unpublished data). This strengthens the suggestion, first made by Dmi'el et al. (1997) and later reiterated by Perry et al. (2000), that the Guana population of *A. cristatellus* may be genetically diverging away from other isolated populations. If it were indeed embarking on a different evolutionary trajectory, the Guana population would merit special protection.

Other Reptiles

... despite the number of biologists working in the West Indies, especially in recent years, it is surprising how little is known about the natural history of about 95% of the herpetofauna.

ALBERT SCHWARTZ AND ROBERT W. HENDERSON (1991)

IN ADDITION TO the four iguanid lizards detailed earlier (the iguana and three anoles), eight or nine additional terrestrial species of reptiles inhabit Guana Island (Lazell 1991). Number nine, the boa, is questionable (not seen ever by me, or ever actually collected; see below). There are two species of sea turtles that regularly nest on Guana, and a third might be expected. Sea turtles crawl into this book when they come ashore to lay eggs, but are often observed off Guana loafing on the sea surface, or by divers and snorkelers among the reefs. On 27 June 2003, Guana Island manager Roger Miller e-mailed me that there were three fresh sea turtle crawls on North Bay beach that morning, presumably made the night before. He reported, "This now makes 16 sets of tracks over the last three weeks that I have seen." He did not attempt to sort out the species.

The green turtle, *Chelonia mydas,* is large— up to 230 kg (ca. 500 lb)—smooth shelled, and not at all green. It is the color of a wet paper bag. The common name comes from green turtle soup, which is made from the fatty tissue called "calipee" just inside the dorsal shell or carapace. This soup—the favorite of the late Sir Winston Churchill, it is said—is distinctly greenish. The channel between Beef Island and Guana Island is noted as an excellent place to see green turtles (K. Eckert et al. 1992). We have no records of these turtles nesting on White Bay, but this is surely an oversight. One green turtle nested on North Bay beach on 3 October 1992 (K. Eckert et al. 1992, 94) in the only year (1992) in which records have been kept and published. Greens haul themselves up the beach by swinging both front flippers forward together, in a rowing motion. Typically this results in a symmetrical track pattern with the two flipper stroke marks aligned opposite each other.

The hawksbill turtle, "*Eretmochelys*" *imbricata,* should in my opinion be regarded as congeneric with the green because the two species hybridize both in captivity and in the wild; the offspring, however maladapted, are viable. The two species are as closely related, therefore, as species can be and still qualify as real species (Simpson 1961).

Hawksbills are the smallest sea turtles, said to rarely exceed 80 kg (ca. 170 lb). As the name correctly implies, they have a rather pronounced, beaklike "overbite" (K. Eckert et al. 1992, 8), which is useful in prying sponges and other soft-bodied organisms from the reef. Their large, imbricate carapace scales (overlapping like roof shingles) are boldly patterned in shades of brown from reddish and yellow to near black. These scales are the basis for the "tortoise shell" pattern, so popular commercially prior to the development of plastics. The late ornithologist Erma J. Fisk recorded hawksbill turtles hatching from nests on White Bay beach in the 1950s and drew diagrams of their head squamation in her bird guide. Between 23 June and 26 October 1992 a dozen nests were found on North Bay beach. One of these, on 16 August, was hatching out (K. Eckert et al. 1992, 94). Hawksbills typically crawl across the beach by swinging their front flippers forward alternately, making a staggered pattern of stroke marks in the sand. This is the species one is most likely to see at Guana, either on the beach or in the water. Mehard (2001) reports hawksbills are declining, as green turtles increase, at St. Croix, USVI.

The leatherback or trunk turtle, *Dermochelys coriacea*, is known to nest on Tortola—at Josiah's Bay opposite Guana's White Bay beach. It is huge—it may exceed a ton—and possibly the largest living reptile (S.A. Eckert 1992). Leatherbacks appear virtually black in the water. They have five prominent keels (sharp ridges) on the scaleless, leathery shell. They just pass through our waters, occasionally stopping to unload eggs, on their vast migrations that can extend from the Arctic to the Antarctic oceans (Lazell 1976). Perhaps one will come up and nest some day on White Bay beach.

All sea turtles are endangered or threatened (Spotila 2004). Given good management and real law enforcement, their numbers would rebound and they could again be a part of sustainable commerce. Here are our terrestrial species; my order follows MacLean (1982).

TESTUDINIDAE

Red-legged tortoise, *Geochelone carbonaria*. This is a large, colorful land turtle reaching over half a

meter in carapace (dorsal shell) length (512 mm; Schwartz and Henderson 1991, 169). The carapace is dark—nearly black—with bright yellow to ocher scute centers; the plastron (ventral shell) is also yellow—bright to dull—with dark patterning. There are yellow markings on the face and also, especially in males, red facial markings and scarlet red scales on the forelimbs. The males are typically more brightly colored than the females and, as adults, have a more flared carapace edge and a deeply concave plastron. The concavity of the plastron accommodates the dome of the female's carapace during copulation. Males "cluck" when copulating and also sometimes when they are simply picked up. On several occasions I have induced clucking by picking up a male tortoise after holding him down elicited no vocal response. If the clucking is a stress or alarm vocalization, what purpose does it serve and how does that reconcile with its association with copulation? And, of course, what good does it do during copulation? It does serve very well to attract human turtle hunters, but that seems maladapted and—one would think—should be selected against.

Schwartz and Henderson (1991) note that this species is "highly esteemed as a food source by humans; distribution in the West Indies is probably the result of a combination of dispersal methods: natural, introduction by prehistoric Amerindians, introduction by early European settlers, and/or recent introduction as pets. . . ." I discuss our stocking of this species on Guana in the mid-1980s in "Restoration," part 6, below. Briefly, there are ample historical records of the presence of both *Geochelone carbonaria* in the Virgin Islands and pre-Columbian subfossils of a like-sized *Geochelone* from the neighboring Anguilla Bank (Censky 1988, Lazell 1993). Ernst and Leuteritz (1999) assumed these were actually *Geochelone carbonaria*, but I was more diffident and unsure. Back in 1985, I examined bones from the bat caves collected by Michael Gibbons (see "Human History," part 6, below). Among these was one I believed was a tortoise scapula, field tagged MCZ-19868, and a portion of a tortoise shell bridge, field tagged MCZ-19903. These have been misplaced or lost (M. Gibbons, pers.

comm.). It seems remarkable to me that Wing (2001) records no tortoise material from any Antillean Amerindian middens as there are a lot of *Geochelone* fossils known from all over the Caribbean (Auffenberg 1974).

GEKKONIDAE

House gecko, *Hemidactylus mabouia*. This is the pallid, rather flat, lizard conspicuous at night, especially at lights that attract insects. They have broadly expanded toe pads and can run across ceilings and even, sometimes, window-panes (I suspect only fairly dirty ones). They apparently are the same species as occurs right across Africa, in houses and outside on trees (Hedges 1996b). They are amazingly adaptable (Frankenberg and Werner 1981). On Guana, we frequently encounter these geckos on trees, too. I have described our pathetic efforts to quantify house gecko populations on Guana in "Mark," part 2, above. The lizards are common, even abundant, but we have no numbers as of yet. If these geckos are recent immigrants from Africa, as they certainly seem to be, then that implies the absence of a nocturnal, scansorial (climbing) gecko here originally: They have filled an empty niche that fairly overflows with species in the Old World and Pacific (Rodda et al. 2001b). Not only do we need population, life history, and ecological data on these lizards, but they also would provide a most interesting and informative study of dispersal, using molecular clock data, such as introns and mtDNA. It is remarkable that no one has taken this on yet.

Sphaero or ground gecko, *Sphaerodactylus macrolepis macrolepis*. (See plate 26.) The type locality (the place from which a species is scientifically described and named) of this species (and thus our nominate subspecies) is St. Croix—an island it may have been quite recently introduced to, and where it seems to be displacing the endemic *Sphaerodactylus beattyi* (MacLean and Holt 1979). The nominate subspecies occurs throughout the Puerto Rico Bank Virgin Islands, including some BVI as small as Watson Rock (Lazell 1994a), up to and including

Culebra, but not Vieques—which has its own subspecies. On Puerto Rico and its satellite cays there are eight more described subspecies (Schwartz and Henderson 1991). There are probably a hundred species of *Sphaerodactylus* in the West Indies, but only a few on South America and Central America, and one, at least, native to South Florida (Lazell 1989b). Thus, this group shows a classic Antillean radiation, probably derived from South America in mid-Tertiary times (Hedges 1996b). Sphaeros are typically crepuscular to diurnal, although sometimes they are about at night. They are densest in leaf litter in shady areas. In fact, they are not just the densest lizards on Guana Island but the densest nonaggregated vertebrate animal yet documented on Earth (Rodda et al. 2001b). In optimal habitat, such as seagrape woods on the flats, they may occur in numbers up to 67,600 per hectare. That is six or seven lizards per square meter. The geographically variable resistance to water loss in this species was documented by MacLean (1985, 1986).

SCINCIDAE

Slipperyback or Sloane's skink, *Mabuya sloanii sloanii*. This is a striking lizard, rich metallic brown over most of the dorsum with glossy, near-black anterior stripes setting off three light lines on the neck. These light lines may be silvery or golden. MacLean (1982) found these skinks "quite rare," but they are of regular occurrence on Guana and many other BVI. They are not common enough, however, for us to have ever managed a mark–recapture population estimate, and none turned up in the plot samples of Rodda et al. (2001b). Finding out about their density and ecology will be very difficult. Schwartz and Henderson (1991) list our form as a subspecies of *Mabuya mabouya*, but that is now regarded as an amalgamation of several distinct species (Mayer and Lazell 2000). These lizards bear live young, typically three to five per litter; Schwartz and Henderson (1991) noted a gravid female in December, but that may be for some other member of the species complex. The same caveat applies to other natural-history data that they list. There are probably now enough BVI specimens at MCZ to ontograph (sensu Lazell 2002a); this could at least be a stab at starting a life history. As with *A. cristatellus* and *Sphaerodactylus macrolepis*, the Virgin Islands geographic variant of this species is quite different from the mainland Puerto Rico form (or forms); there are intergradient populations on the small islets just east of the Puerto Rican mainland (Mayer and Lazell 2000).

TEIDAE

Ameiva or ground lizard, *Ameiva exsul*. This large lizard—males can be over 20 cm (about 8 in.) in snout-to-vent length—is active and conspicuous during the heat of the day, especially in open, sunny locations. Clearly abundant, it disappears under cover and down holes when the sun is not shining. Schwartz and Henderson (1991, 191) say it occurs from "sea level to about 500 ft" but it is much in evidence on Guana Peak at 805 ft (246 m) and, in open places trailside, on Sage Mountain, Tortola—the highest point in the Virgin Islands—1,720 ft (524 m). Ameivas on Sage Mountain are conspicuously darker than in the lowlands, perhaps resulting from selection for heat absorption. Schwartz and Henderson (1991) cite a lot of natural-history data for this species but almost all of it is for populations on Puerto Rico. We have never attempted a mark–recapture study of these lizards, but Rodda et al. (2001b) extrapolated about 200 per hectare in early seral-stage brush; they got none in seagrape stands. In October 2003, Robert (Bob) Powell, Avila University, attempted line-transect censusing of Guana's ameivas and got much lower numbers: about 34 per hectare in forest and 52 per hectare along White Bay beach. Bob estimated the total island population to exceed 8,000 individuals; he found the adult : juvenile ratio only 1.24:1 and suspected the season—rainy—was responsible for the high number of young recruits. He plans to return and attempt more censuses using various techniques.

There is striking geographic variation in this species and a number of subspecies have been named. Only those from the islands of Mona (*Ameiva exsul alboguttata*) and Desecheo (*Ameiva exsul desechensis*) are currently regarded as valid, but I suspect several others are too.

AMPHISBAENIDAE

Amphisbaena or worm lizard, *Amphisbaena fenestrata*. (See plate 27.) The lizard family Amphisbaenidae is often elevated to subordinal status—Amphisbaenia—within the order Squamata of class Reptilia. I agree that it is a group of weird reptiles, mostly legless except for several species of genus *Bipes* in Mexico that retain forelimbs (but not hind), but I think they are clearly lizards. I was glad to see that Pianka and Vitt (2003), in their definitive book on the world's lizards, agree and include amphisbaenas. All species of genus *Amphisbaena* lack legs. Their eyes are reduced to dark spots under their enlarged head shields and their body scales are arranged in annuli. This is the animal that I knew was on Guana Island for two years before I saw one because the late Oscar Chalwell told me "it has rings"—those annuli of scales—and "it bites." MacLean (1982) discusses its biting abilities, which are impressive for an animal that only reaches about 26 cm (>1 ft.) in total length. Amphisbaenas have a rather foxy face with bulging, muscular jowls, and sharp teeth. They have a median premaxillary fanglike tooth right up front; not many animals have dentition like that. Our species is very distinct from any on Puerto Rico or elsewhere in the Greater Antilles, where there are a dozen species. Indeed, *Amphisbaena fenestrata* is so distinctive that it was once put in its own genus. Gans and Alexander (1962) dispensed with that, and I agree. When they wrote, there were few specimens of the species known, none from Guana Island. Thomas (1966) added a few more specimens and some notes on what he found them under (stones) and where (mostly in shady woods). I was unable to add anything in my initial BVI fieldwork (Lazell 1980) and, even after

Oscar Chalwell clued me in, it would be 9 March 1984 before I got a specimen. Then along came Tom Sinclair in July 1988. He could find amphisbaenas at will. Although they might be almost anywhere on an island (he got one under a log just above the beach on Great Thatch Island; it is in MCZ), he found them concentrated at 160–180 m (525–590 ft.) on most islands and usually between 150–210 m (500–600 ft.) anywhere there were good woods. Armed with this knowledge, one could embark on a life-history study, or at least a mark–recapture population estimate; nothing like that has been done, as far as I know, for any member of this family. In October 2001, we caught several more specimens on Guana, including one regurgitated by a common snake, *Liophis portoricensis*, captured as part of a study by Arijana Barun (Barun and Perry 2003).

MacLean (1982) gives the range of our amphisbaena as "St. Thomas, St. John, St. Croix." The type locality is "St. Thomas and Santa Cruz" fide Schwartz and Henderson (1991, 558), but they do not include St. Croix in their mapped range. They do put a dot on St. John, but I have never seen a specimen from that island. Gans and Alexander (1962) tabulate striking scale count differences between populations but, admittedly, their sample sizes are much too small for statistical significance. It seems all aspects of this strange creature's life would repay study, from geographic distribution and variation to parasites and diet. Hedges (1996b) postulates an African origin for the entire Antillean amphisbaenid radiation, but whether this occurred directly or by way of South America cannot be determined with current evidence. This is one more example of an evolutionary and biogeographic conundrum amenable to molecular elucidation.

TYPHLOPIDAE

Typhlops or worm snake, *Typhlops richardi*. These very tough, shiny, plain brown little snakes are abundant but rarely seen by most people. Their tiny, black-dot eyes are covered by pigmented scales, so they are often called "blind snakes," but they can tell light from dark, and they avoid light (K. R. Thomas and Thomas 1978). They mostly live underground or at least under objects (e.g., stones). They eat termites and will climb trees to get them (Schwartz and Henderson 1991, 656). Our species belongs to a radiation of at least seven species on the Puerto Rico Bank and is confined to islands east of Puerto Rico and west to Tortola; there is also a population on St. Croix. A suite of species occurs on Puerto Rico itself, and on Virgin Gorda and Anegada *Typhlops richardi* is replaced by *Typhlops catapontus*. Details are provided by Hedges and Thomas (1991), but worm snakes are now known from several more small islands, such as Guana and Necker (Lazell 1995), not available to Hedges and Thomas when they did their work; they state (1991, 457): "A more exhaustive study of variation in the *Typhlops* of the islands east of Puerto Rico is needed." *Typhlops* sensu lato (e.g., Lazell 2002a) of one sort or another live on tropical and subtropical islands all over the world.

BOIDAE

Virgin Islands boa, *Epicrates monensis granti*. Described as exceedingly rare by MacLean (1982), I believe this is the most common snake on Tortola—where the three other snake species discussed here are common too. When MacLean wrote, he could locate only 11 actual specimens anywhere. Mayer and Lazell (1988) reported on several more specimens, and there are at least 20 salvaged road kills or electric-fence kills bottled at the H. L. Stoutt Community College in Tortola. Back in 1994, Greg Mayer and I took measurements and meristics on six of the community college specimens and tagged them BVI 1-6. (See table 23.)

BVI 6, a mangled road kill, has the highest ventral count ever recorded in this species and by far the highest saddle count. Overall, it is very dark and does not resemble others of the subspecies *granti*. All BVI specimens have generally high counts for their species, often the

TABLE 23
Some Characteristics of Tortola Boas

BVI	SVL	V	SC	D	SUP	SUB	SAD
1	910	265	—	36,48,28	13,14	15,15	57,—
2	584	265	88	34,44,26	13,13	15,15	58,18
3	474	268	87	37,46,26	13,13	14,15	57,20
4	652	268	86	35,46,24	13,13	15,15	60,20
5	673	261	87	38,46,26	12,13	14,15	65,24
6	765	274	—	—,46,30	—,13	—,13	71,—

SOURCE: Original data. Two boas, numbers 1 and 6, are females; the other four are males. Snout-to-vent length is SVL in millimeters. V is the ventral scale count including the anal plate. SC is the number of subcaudals (plates under the tail). D is the number of dorsal scale rows around the body at three points: neck just behind the head, midbody, and trunk just before the cloaca (vent). SUP is the number of supralabial (upper lip) scales, left and right. SUB is the number of sublabial (lower lip) scales, left and right. SAD is the number of dark dorsal saddles first on the trunk anterior to the cloaca, followed by the saddles (or bands) on the tail. A dash indicates that the specimen was damaged and the count could not be made.

highest recorded. The subcaudals are consistently higher than in USVI snakes. The fundamental systematics of the *Epicrates monensis* group are in dire need of review: Are the Puerto Rico Bank snakes really conspecific with those on Mona Island? I doubt it. What are the mainland Puerto Rican ones? They look very distinct from the *granti* subspecies to me. Are the St. Thomas specimens a different form, with lower scale counts, than those from Tortola?

And, for now, the biggest question of all: Does this boa occur on Guana Island? If not, did it ever? Grant (1932) stated that it occurred on Guana Island, but he provided no specimen. I have never seen one here. Mary Randall provided two excellent stories back in 1980. First, USVI biologist David Nellis (Nellis et al. 1983) visited Guana back in the 1970s. Mary said he excitedly reported seeing a boa in the stone wall at the northwest end of the North Bay woods. Second, and better, Mary kept a pair of caged finches in the office back then. One night a transversely banded snake got up to the finch cage, hanging from the ceiling, and swallowed both birds. The late Albert Penn was on hand, cut the snake in pieces, and threw the pieces away.

Next, in the mid-1980s, Elizabeth (Holly) Righter, the archeologist featured in "Human History," part 6, below, came back from visiting the Lake House ruins over on the southern shoulder of Guana Island. I cannot quote her

exactly, but she said roughly: "You know these common snakes you see all the time? Kind of skinny and striped? Well, I just saw a snake up at the ruins that was about the same length, but it was thick and had bands across its back, and it just sat there sort of in a ball." Miguel Garcia and Manuel Leal, old boa hands, came over from Puerto Rico one October. They could only stay a couple of days, and it poured rain each night, just when they would have liked to hunt. Of course, they did not see a boa, but they thought the sea grapes of North Bay woods were superb habitat.

Clive Petrovic, at the community college, has a colony of these boas. They thrive on *Anolis* lizards. They are nocturnal. We need a good hunter.

COLUBRIDAE

The small snake, *Liophis exigua exigua*. This snake has no vernacular name and is not distinguished from the young of the bigger snake, *L. portoricensis*, by most people. These are typically a warmer brown hue than their bigger relative and with much neater, more smoothly continuous dark lateral stripes. They are stockier at the same length, with a less narrowed neck (or flared jowls), and a notably shorter tail—if tails are undamaged. Ventral scale counts overlap slightly: 137–165 in *L. exigua* (Schwartz 1967) and 163–187 in *L. portoricensis* (Schwartz 1966). Dorsal counts in both species are usually either 17 or 19. Subcaudal counts are quite distinct—not over 102 in *L. exigua* or under 106 in *L. portoricensis*—but because many *L. portoricensis* (and no doubt some *L. exigua*) have stumped tails, that count is hardly worth making. The classic, generic-level characteristic that is supposed to separate these species, and their putative, erstwhile genera *"Arrhyton"* and *"Alsophis,"* is the presence of paired apical pits on the dorsal scales of *"Alsophis" portoricensis* (MacLean 1982: 41 and note the photograph facing p. 25). Unfortunately, young *L. portoricensis* at the sizes that wholly overlap the biggest *L. exigua* totally lack apical pits. This characteristic is perfectly useless.

We do not see these snakes regularly enough to do a mark–recapture guestimate. In the 2001 October field session, for example, I logged 38 *L. portoricensis* on Guana and only two *L. exigua*. We could make a guess at the total island population by saying 5% of all *Liophis* seen are *L. exigua*, so there might be 50 of the small species if there are 1,000 of the bigger ones. Do not believe that; I bet there are a lot more.

In the molecular study by Highton et al. (2002), our two species come out side-by-side, each other's closest relatives. They are widely separated from other members of the nominal genus *"Arrhyton."* This study is very controversial (references cited therein), but it clearly shows the chaos of current snake classification rendering it almost meaningless. Hedges (2001) provides biochemical evidence that all of these Antillean snakes—called xenodontines because of their unusual rear-fanged dentition—are very closely related. They are certainly as closely related phylogenetically as are members of the biogeographically corresponding lizard genus *Anolis,* and they are very similar in evolutionary degree of difference too: big ones and small ones, different colors and patterns and proportions.

A Guana Island female *Liophis exigua* laid three eggs on 16 July 1988. Those hatched 2 months and 18 or 19 days later on 3–4 October

FAITH, HOPE, AND SNAKES

Before 12 March 1984, no one ever suggested this small snake might live on Guana Island. At 7:45 a.m. on that day, I saw one just off the south side of the dining terrace, below the kitchen window. I truly believed I saw one. It just did not resemble the bigger common snake. Who would believe me? No one. Of course not. How many *Liophis exigua* had I ever seen? Maybe two or three, through the side of a bottle in a museum. How many babies of the common snake had I seen? Maybe one back in those days. All I was going by was the Gestalt or *je ne sais quoi* of appearance. Truly, I did not know. There is a real distinction between knowledge and belief. I believed. Just three days before I had gotten that first amphisbaena (see Amphisbaenidae, "Other Reptiles," part 5, above). I told everyone who would listen that I believed a second species of "common, ordinary" snake—like a baby of the one we all knew so well—lived here too. It is common on Tortola, St. Thomas, and St. John (MacLean 1982), so why not a few on Guana?

The years went by. On 25 July 1987, 3 years, 4 months, and 13 days after I saw my snake, Karen Koltes, ichthyologist with USNM, came up from the flat late in the afternoon. She said there was a little snake dead on the driveway, almost at the bottom. I could run downhill pretty well, and I got it: It is in MCZ today—a new Guana Island record then, and the last addition to the herpetofauna of Guana Island until we began bringing species back. Today we see these little snakes—up to "slightly over a foot" (MacLean 1982, 41)—regularly.

at MCZ. Kept at room temperature in Cambridge, Massachusetts (and no doubt air-conditioned at least through August), this incubation period was probably significantly prolonged. At hatching, the babies measured 86-, 90.5-, and 91-mm SVL and 125-, 129-, and 131-mm total length, respectively. The eggs, fresh laid, measured 21 × 8, 21 × 8.5, and 22 × 8 mm (I do not know which egg contained which baby). Schwartz and Henderson (1991) provide a lot of anecdotal natural-history data, but many of them apply to populations on Puerto Rico that belong to very different, notably larger subspecies (Schwartz 1967, under the generic name *"Dromicus"*). They are reported to eat *Eleutherodactylus* frogs and *Sphaerodactylus* lizards. A Guana captive regurgitated *Anolis* lizard eggshells, and Puente-Rolon (2001) reports one eating an *Anolis* egg on Puerto Rico.

The bigger snake, *Liophis portoricensis anegadae.* This is the common, conspicuous Guana Island snake, with no more specific vernacular name than its relative the small snake, above. I once proposed calling them "Anegada ground snake," but abandoned that when I met one (of several to come) a good 4 m up in a pillar cactus. Also, there is at least the suspicion that these may not really be the same as the Anegada population (Schwartz 1966). Guana specimens are typically dull gray brown, paler but well pigmented below, and have a ragged lateral stripe on each side produced by dark, near-black pigment on the fifth dorsal scale row. Schwartz (1966, Fig. 50) shows an essentially unstriped Virgin Gorda specimen, and subspecies *anegadae* is in part diagnosed by a "not especially conspicuous" stripe (Schwartz 1966, 210). Schwartz and Henderson (1991, 572) give the record snout-to-vent length as 923 mm; they do not give a tail or total length, presumably because large individuals are usually stump tailed. In October 2003, Robert (Bob) Henderson, Milwaukee Public Museum, began what we hope will be a long-term study of these snakes, which he refers to as genus *"Alsophis"*; I do not believe that is valid (as discussed above; see Highton et al. 2002). He paint-marked and microchipped 38 individuals; none were hatchlings or even very small. In fact, one caught by Fred Sibley (of bird and dragonfly renown) was the largest I have ever seen: a male 855-mm SVL (with a stumped tail 330 mm long) that weighed 308 g (11 oz.). Most individuals weigh about 50 g (< 2 oz.). Bob got too few recaptures to be willing to hazard a population guess (but I did; see part 2, "How to Count Snakes—and Other Things," above). One of his snakes traveled about 300 m. (See plate 28.)

These snakes often rear up and spread their necks laterally, just like a cobra. This is of great evolutionary interest to me because it would seem to be adaptively beneficial mimicry, but how can it possibly be? Who and where is the model? There are no hood-spreading cobras closer than Africa, and no evidence there ever were.

On Puerto Rico, *Liophis p. portoricensis* has been reported to cause severe envenomation in humans (Heatwole and Banuchi 1966, R. Thomas and Prieto Hernandez 1984). The enlarged rear fangs presumably vector the venom. On Guana, we have seen *L. p. anegadae* kill a baby iguana in just a couple of minutes with a single bite (LeVering and Perry 2003). However, we are frequently bitten by these snakes and none of us has ever developed symptoms. Schwartz and Henderson (1991, 574), reporting on the whole, geographically variable, species, report the diet as mostly lizards: anoles (62.5%), ameivas (15.6%), sphaeros (9.4%), and juvenile iguanas (3.1%); and then frogs (9.4%). That makes 100%, leaving no room for the house gecko, blind snake, and small snake *(L. exigua)* also listed as prey. I can confirm anoles, ameivas, and baby iguanas on Guana, and Barun and Perry (2003) got an *Amphisbaena* regurgitated by one, as noted earlier. On Guana, this normally diurnal snake occasionally hunts lizards at lights at night (Perry and Lazell 2001). Schwartz and Henderson (1991) report that the red-tailed hawk, *Buteo jamaicensis,* is a predator on these snakes; the next step in the night-light-niche hunting regime might involve red-tailed hawks hunting snakes at lights at night. This whole complex of snakes needs a thorough taxonomic review, including an attempt at molecular phylogeny, and both population and ecological (life-history) elucidation. Guana Island can provide the perfect starting point for all of this.

Bats

The order Chiroptera (from Greek, "foot wing") comprises the only truly flying mammals: bats. In all bats the forelimbs are greatly enlarged and support wings of skin. In all forms there are five fingers, the first—the pollex or thumb—very short and at least partially free of the flight membrane. The second digit supports the distal edge of the membrane and consists of the single metacarpal in advanced forms. The remaining three digits retain phalanges within the membrane. The wing membrane attaches to the ankle, but the foot bears at least vestiges of the five toes free of the membrane. In most forms there is an interfemoral membrane between the hindlimbs, supported by a sturdy strut, the calcar, extending from the heel, and often including some or all of the tail, if one is present. All bats have prominent ear pinnae, and most have a leathery, antenna-like structure within each pinna called the tragus.

JAMES LAZELL (1998a)

ON GUANA ISLAND, four species of bats are currently known: our only native mammals. Our bats each belong to very distinct genera and three different families, as follows:

Family Noctilionidae
    Genus *Noctilio*
        Species *Noctilio leporinus*, fishing bat
Family Phyllostomidae
    Genus *Brachyphylla*
        Species *Brachyphylla cavernarum*, cave bat
    Genus *Artibeus*
        Species *Artibeus jamaicensis*, fruit bat
Family Molossidae
    Genus *Molossus*
        Species *Molossus molossus*, mastiff bat

The most recent complete classification of bats was done by Nowak (1994). Humphrey and Bonaccorso (1979) state: "Bats are the numerically dominant group of mammals in the Neotropics"—even exceeding rodents in diversity and density. They continue to say that the phyllostomid bats, two of our four species, "exhibit great diversity. . . . The importance of this family in diversity and relative density suggests . . ." its "importance in tropical ecosystems." Since they wrote, Antillean bat community ecology has been given considerable attention (e.g., Rodriguez-Duran and Kunz 2001). I will take this up after the species accounts.

Fishing Bat, *Noctilio leporinus*. This species divides into two well-differentiated subspecies: the nominate *N. l. leporinus* of South America and its coastal islands such as Trinidad; and ours, *N. l. mastivus*, of the West Indies and Central America. Our form extends into the Bahamas at least to Great Inagua (Buden 1985). Our form is the largest member of the genus, attaining more than 13 cm in head-body length, with a forearm of over 9 cm and a wingspan of more than 61 cm (2 ft.). Males are bigger than females and said to be "bright orange rufous" (Nowak 1994). Our only Guana specimen, AMNH 256528, is a female and so gray brown; I have seen many live ones on Guana, but in the dark. In the night all bats are gray.

W. B. Davis (1973) separates our subspecies from the South American form not only based on larger size but also on the prominent, whitish middorsal stripe, weak or absent in *N. l. leporinus*. In my experience this bat is nowhere common. When we originally wrote up the bats of Guana (Lazell and Jarecki 1985), we had no specimen records for the entire BVI. There was a published sight record for Virgin Gorda (A. C. Johnson 1978) and Koopman (1975) had USVI specimens available from St. Thomas, St. John, and St. Croix and from the Puerto Rico Bank island of Vieques. Although the species occurs on Puerto Rico, Choate and Birney (1968) showed that the record for Mona Island was spurious, being based on specimens of *N. l. leporinus* from Monos Island, Trinidad.

Anatomically, these bats are remarkable; the pelvic girdle is unique among mammals: the ischia are fused at the ventral midline and dorsally to the urostyle, or tailbone. In males there is a pubic symphysis, but in females the pubic bones not only do not meet but also are parallel or divergent. The ischial ring of solid bone would seem to provide a foundation for movement of the enlarged hindlimbs, with their enormous, long-clawed toes. The wing membranes attach to the hindlimbs at the knee (thus departing from my introductory sketch, above), but the interfemoral membrane is very large and saclike, supported at its distal edge by enormous bony calcars. The tail comes about halfway down the interfemoral membrane and perforates it dorsally, so the tail-tip is free.

Facially, these big fellows live up to their occasional descriptive names of "bulldog" and "harelipped" bats. The lips are pendulous and fleshy and can stretch to accommodate a 7-cm (nearly 3 in.) fish; the lips are split to below the nostrils. The nose is protuberant, rather like that of a female proboscis monkey. There are no nose leaves and only a small chin pad. The large, pointed ears house a tuberculate or serrated tragus.

The whole family Noctilionidae consists of just the two species, ours and the before-mentioned smaller South American form. They are quite closely related to the leaf-nosed bats, family Phyllostomidae. R. J. Baker (1979) provided a detailed report on the karyology—number, size, and structure of the chromosomes—of the relevant forms. It is a tour de force on the subject and a must-read for those who may stray into thinking that karyotypic, or chromosomal, characteristics are somehow better than those of conventional morphology. R. J. Baker (1979), noting that fishing bats share a basically phyllostomid karyotype, says "... karyotypic changes become established in a species at such irregular intervals that one cannot depend on the rate of their establishment to indicate taxonomic position." He details examples of intraspecific karyotypic variation as well as many differences between species in the same genus. When I learned about genetics, people thought karyotypes were constant at the family level, at least. Molecular evidence from mitochondrial DNA, supported by some nuclear DNA, indicates that this species differentiated from the smaller, South American *N. albiventris* less than a million years ago, in the Pleistocene (Lewis-Oritt et al. 2001).

The late naturalist of St. Croix, Harry Beatty (1944b), notes of our fishing bat, at this, its type-locality: "... this is one of the most highly odorous of the bats," and one "can detect the pungent musk as the animal passes by even though it may be fifty feet away." The specimen we collected was not notably smelly, and I have often had them fly much closer to me than 50 ft. and not noticed their aroma. Their roost caves, however, out by the Guana Head, do truly reek. Although fishing bats seem to catch insects by

A SHOT IN THE DARK

We did not see a fishing bat on Guana until the Salt Pond was stabilized, for the benefit of the flamingos, so that it does not go dead dry. That was about 1987, and then they appeared. Ralph Rusher, then on the Guana staff, was the first to tell me about them. He would see a flock of about a dozen come into White Bay from around the Guana Head and, staying close inshore, pass just by or even over the dock, heading for the Salt Pond. They came, typically, a bit before midnight, say, 11 p.m. We had a large team of student assistants in July 1988, including Joe Walsh (now a PhD), then at Harvard. Lianna and I, authors of the Guana bat paper about to be eclipsed as soon as we proved a fourth species was present on Guana, were in residence; we all set off to catch a fishing bat on the night of 15 July 1988.

Joe Walsh directed activities. The idea was to stretch a 30-ft. mist net from the dock to a dinghy out in the bay. This proved to be rather akin to horizontal, aquatic bungee jumping, with a lot of stout men and nubile young ladies overboard at any given moment. It seemed as though the bats were likely to be safe that night, not that Joe had given up. He resprung and reset, and came back toward the dock as a yo-yo to its master's hand. Ralph was standing on the dock with me, watching the action and waiting for the bats. Ralph had a 45-caliber shot pistol loaded with number four goose shot.

At 11 p.m. sharp, here came the whole flock of bats. Here they came, among the happy throng of net-stretching bathers, and there they went. Ralph let them get well past dock, boat, net, Joe, and all, drew his side arm, and fired. Joe swam over toward the beach and retrieved his bat. Shooting one, of course, had no effect on the population—even if the entire population is just a dozen. They absorbed one death—happens all the time—and continued to prosper. The flock still comes, every night that I have gone, or sent a spy, to look.

mouth similar to other bats, they use their big feet and hooked claws to scoop up fish—at least far enough out of the water to grab the fish in the jaws. According to Nowak (1994), the interfemoral membrane may be used to carry fish back to the roost; he says they may eat up to 40 fish a night, finding them in the water by echolocation.

The female we collected, AMNH 256528, had a wingspread of 611 mm (61 cm or almost exactly 2 ft.). Head to tail-tip, she was about 12 cm with an 8.6-cm forearm. Dorsally, she was beige gray, but ventrally, she was bright yellow. She had an ashy middorsal stripe from shoulders to near her rump, where the stripe got bolder, broader, and yellow. She was very plump but not pregnant or nursing. Her stomach was full of fish scales with a few beetle parts. Seen in the light of day, the next morning, this was a spectacular animal.

Cave Bat, *Brachyphylla cavernarum*. As currently classified, this species is confined to the Puerto Rico Bank islands and the Lesser Antilles from St. Croix south to Barbados and St. Vincent. The only other member of the genus, which is rather isolated within the phyllostomid family, occupies Cuba, Grand Cayman, and Hispaniola (Nowak 1994). The nose leaf is reduced to a vestigial fringe; the ears are normal size for a bat, and well separated, with a rather short, broad tragus. The tail is a tiny rudiment, not extending far enough into the interfemoral membrane to be noticed in the living animal, so this bat appears tail-less.

There is weak geographic variation; Swanepoel and Genoways (1978, 1983) divided the species into three subspecies. In their 1978 paper they tabulated their raw data and provided several sorts of analyses. Jarecki and I reviewed it all and concluded: "Measurements . . . overlap so greatly as to preclude identification of most individuals to subspecies." We thought the putative subspecies unworthy of recognition unless and "until diagnoses can be framed which will permit identification of more individual specimens" (Lazell and Jarecki 1985). The only differences claimed are weak modalities in measurements.

When we wrote, our cave bats were known from a single specimen and several sight records. Our paper was published 12 June 1985, and on 5 July 1985 we got a second specimen, AMNH 256518, at the Bat Caves: a very fresh skull found lying on the cave floor by Michael Gibbons, anthropologist, who was excavating (see "Human History," part 6, below). When Fred Sibley and his ornithological team came along and began setting up mist nets, we had live cave bats caught and delivered to us: two on 21 October 1996 and two more on 19 October 1997; one of the latter is a specimen at the Yale Peabody Museum. There seems to be small colony on the Pyramid, probably roosting in a tuff cave out there.

Cave bats are omnivores, feeding on fruit, nectar, pollen, and insects. They are often seen visiting agaves in bloom at twilight and can be separated from fruit bats, at least up close and in profile, by their inconspicuous nose leaf. Nowak (1994) reports seasonal reproduction with births concentrated in late May and early June; scientists are rarely on Guana at that time.

The fondness of cave bats for the native agave or "century plant," *Agave missionum*, is especially worrisome. The weevil, *Scyphophorus acupunctatus*, has invaded the BVI and caused a devastating plague of the century plants (Osborne 2002; Woodruff and Pierce 1973). If it is merely fondness, not dependence, it may not be too bad, because some agaves have survived and should repopulate the islands.My concern is that during the dry "winter" months, December through March, when many agaves used to bloom, other food resources may be scant; there may be too few flowering survivors to sustain bats in the short-term future.

Fruit Bat, *Artibeus jamaicensis*. (See plate 29.) This species occurs from the Florida Keys and Mexico to Paraguay and Brazil and throughout the Greater and Lesser Antilles and the Bahamas north to the Caicos Islands, Inagua, and Mayaguana (Buden 1985; Lazell 1989b). Two subspecies have been recognized: nominate *Artibeus j. jamaicensis* from most of the range, including the Virgin Islands, and *Artibeus j.*

parvipes from Cuba. Buden (1985) provided a detailed analysis; the only difference is a modal one in size: the Bahamian populations are intermediate and individuals are not distinctive. I opine the subspecies are unworthy of recognition.

These bats really do eat mostly fruit; it is said to go through them so fast—in about 20 minutes—that a bacterial contribution to digestion could hardly take place; this brings up the possibility of special enzymes and an unknown digestive process. They also eat flowers, nectar, pollen, and insects; they can be seen feeding on agave flowers on Guana with cave bats and can be recognized, in profile, close-up, by their very large nose leaf. There is no external tail and the interfemoral membrane is reduced to two fringes bordering the hindlimbs, supported by the calcars, and almost disappearing in the body fur at the crotch.

Reproduction has been studied in Central America but not in the Virgin Islands. A basic pattern is a March–April birth peak, after which the females come into estrus (heat) and conceive again. The next birth peak is in July–August; this is followed by a second estrus, but the resulting embryos, although implanted in the uterus, remain dormant until about November, when embryological development begins again, leading to the next March–April births (Nowak 1994). On Guana, females occupy the Bat Caves as nurseries in March and April and can be found with their babies then. It was thought remarkable that females could or would carry their young in flight when we recorded them doing so (Lazell and Jarecki 1985), but Harry Beatty (1944b) had long ago reported that. For more about our fruit bats see "Parasitism," part 3, above.

Mastiff Bat, *Molossus molossus fortis*. This is our commonest bat—or at least the one regularly seen around the buildings. Nowak (1994) calls it the "velvety free-tailed bat," a good descriptive name. It occurs all over the West Indies and in much of Mexico, Central America, and northern South America south to Paraguay; it occurs in the Florida Keys as the subspecies *M. m. tropidorhynchus* (P. A. Frank 1997; Bowers 2003) but is absent from the Bahamas. The subspecies *M. m. fortis* is endemic to the islands of the Puerto Rico Bank: Puerto Rico and many of its coastal cays including Culebra, the BVI, and all of the USVI except St. Croix and its islets, on their own separate bank. The species breaks up into six separate, geographically replacing, subspecies in the Antilles (Hall 1981, 225), some of which are so distinctive that they are occasionally ranked as full species—similar to our form *fortis* (Krutzsch and Crichton 1985). It would be worthwhile to study the patterns of morphological variation with an eye toward accepting or refuting an overview of a variable evolutionary species in the Simpsonian sense elucidated by Lazell (1972). Biochemical evidence should also be developed.

These are little bats—a couple of inches in head-body length—but up to 66 mm; with the tail, which extends well out beyond the interfemoral membrane, the total length can exceed 10 cm (~4 in.). These bats have a very mastifflike face without leaves or frills; their broad, blunt ears meet at their bases on the forehead. They eat only insects as far as is known, and one can consume thousands of "no-see-ums" (tiny biting flies) in a night. They can roost under a corrugated iron roof, so their ability to withstand heat approaches the supernatural.

In Puerto Rico, Krutzsch and Crichton (1985) found mating occurred in February and March, with the single young born in June; this was followed by estrus and renewed mating, with the next young born in September. It is difficult to reconcile these data with our female of 8 July 1984, pregnant with an embryo 26 mm in crown-rump length—more than a third of her own size: Certainly she could not have held this baby for a September birth date. There remains a great deal to be learned about our bats.

One of the major points we made (Lazell and Jarecki 1985) was that the bat faunas of the other islands in the BVI are virtually unknown— notably the big islands of Tortola and Virgin Gorda, which predictably harbor several more

species, including some that are rare or even thought to be extirpated. More recent work on bat biogeography, ecology, and known extinctions greatly strengthens our case for the larger islands supporting at least six species, with obvious candidates being the red fig bat *Stenoderma rufum*, the free-tailed bat *Tadarida brasiliensis*, and a member of the genus *Monophyllus*—the last being long-nosed, leaf-nosed phyllostomid bats that seem to lack a common name.

McFarlane (1989, 1991) looked at Antillean bat faunas from the viewpoints of species packing and competitive exclusion. The notion is that a genus includes bats that are sufficiently similar in size, diet, and ecology so that two members of the same genus might not be able to coexist in a small space with limited resources. Thus a species/genus ratio might be high—say, three species per genus—on a large island with many ecological zones and a rich diversity of habitats, but would be expected to decline to one species per genus on small islands. You can see an emerging glitterality here, but the data look pretty good. One sets up the null hypothesis that bat faunas accumulate species at random from the available species pool; this will produce a species/genus ratio that says, in effect, that generic level similarities do not matter (see "A Unified Theory," Part 4, above). Next one compares the actual observed number of species per genus; in every case McFarlane got lower than predicted numbers of species per genus on the real islands. What this means is that a bat species has a better chance of surviving in a small place with limited resources if it is very different—anatomically, in size, in diet, and in ecology—from the other bat species present. Or, looked at the other way, if two species are very similar in all those ways, one will predictably exclude the other—*read* force it to extinction: Too much competition makes winners and losers, where the losers are lost forever.

Rodriguez-Duran and Kunz (2001) paint a broad-brush biogeography and do not consider the small islands of the Puerto Rico Bank, such

TABLE 24
Local Bat Species Sorted According to Roost Site, Body Size, and Diet

CODE		ROOST SITE	BODY SIZE	DIET
FOUND ON GUANA ISLAND				
Fishing	GLZ	Generalized	Large	Zoophagous
Cave	CLP	Cave dwelling	Large	Phytophagous
Fruit	GLP	Generalized	Large	Phytophagous
Mastiff	GSZ	Generalized	Small	Zoophagous
ABSENT OR LIKELY TO OCCUR OR TO HAVE OCCURRED ON TORTOLA				
Monophyllus	CSP	Cave dwelling	Small	Phytophagous
Tadarida brasiliensis	CSZ	Cave dwelling	Small	Zoophagous
Stenoderma rufum	TMP	Tree roosting	Medium	Phytophagous
Eptesicus fuscus	GMZ	Generalized	Medium	Zoophagous

SOURCE: Rodriguez-Duran and Kunz (2001). Codes are drawn from the first initial of the qualities in the three right-most columns: roost site, body size, and diet.

as the Virgin Islands. They believe a fundamental West Indian island bat fauna is—anything but random—a combination of six species, all in different genera: one *Monophyllus*, one *Brachyphylla*, *Artibeus jamaicensis*, *Noctilio leporinus*, *Tadarida brasiliensis*, and *Molossus molossus*. Puerto Rico has all of those and more, and Guana Island has four of them. Next, Rodriguez-Duran and Kunz assign every species a letter code depending on three factors: roost site, body size, and diet. Table 24 shows how our bats got sorted out; it also considers those absent bats that I think are likely to occur, or to have occurred, over on Tortola.

Why did I choose those four of the absent bats? *Monophyllus* is on the Rodriguez-Duran and Kunz (2001) list of basic bats, and two species are known from the Puerto Rico Bank: *Monophyllus redmani* still survives on Puerto Rico and *Monophyllus plethodon* is known there from subfossil bones (e.g., Choate and Birney 1968) and still survives just to the east in the Lesser Antilles; we must have once had a *Monophyllus*. *Tadarida brasiliensis* is on the Rodriguez-Duran and Kunz (2001) list and is known to occur as close by as St. John (Koopman 1975). *Stenoderma rufum*, although rare and not at all a basic bat, occurs right over on St. John (Koopman 1975). And *Eptesicus fuscus*, the big brown bat of eastern North America, lives on Puerto Rico and has very close, congeneric relatives over in the Lesser Antilles.

Now, in my little sketches of our species, above, we can see that the classifications, to quote Rodriguez-Duran and Kunz (2001) have some "blurry boundaries." All of our bats eat insects, but at least two—the big fishing and the little mastiff—eat (as far as we know) no plant matter at all. All of our bats roost in caves, at least sometimes; fruit bats do often take to the trees and mastiff bats will roost anywhere—even behind your mirror on occasion. How can we square this with the notion of competitive exclusion, or interspecies competition, when we have two such similar bats as our cave and fruit bats both living together in good numbers on Guana? If the fact that they are in different genera—a fact that depends on nose-leaf size, skull characters, and dentition—makes the difference, how? Would not one of the two potential *Monophyllus* species, both C, S, P, a combination unrepresented on Guana or in the BVI—do much better here? Or how about *Stenoderma rufum*, the red fig bat (the bat is red, not necessarily the figs)? It is a T, M, P: We have nothing like that.

For species that really do need caves, because of temperature and humidity, Guana is truly resource-limited. All of our caves are little things and fail to provide the length or depth needed to stabilize temperature or to hold high humidity. That may account for the lack of a *Monophyllus*. But *Tadarida brasiliensis* does not really need caves; it does fine in buildings and even under bridges—even in the arid southwestern United States. Morgan (2001) looked long and hard at the record of bat extinctions in the Antilles. It is doubtful that humans can be blamed for exterminating bats directly—at least prior to Columbus—because people probably did not eat many of our native bats: They are just too small to bother with. The circumstantial evidence points largely to two exterminating factors: first, postglacial sea-level rise, and second, deforestation—especially of the lowlands. When I was describing the wildlife and biogeography of the Florida Keys (Lazell 1989b), I mapped the entire Caribbean Basin to show how enormously land areas—and distances between them—changed with 130 m of sea-level rise in about 20,000 years: Land was reduced to about 25% of its glacial maximum extent, and distances between the major islands of the Greater Antilles and the mainlands of North America and Central America were increased by about a factor of three. What I did not consider was the geologic nature of the inundated terrain. Much of that drowned land was limestone, and within that limestone were undoubtedly huge cave systems. The simplest explanation for 75% of West Indian bat extinctions is that the last few survivors of once-vast populations of obligate cavernicolous bats died out in the paltry few and small caves still above sea level, where we can find their bones. They may have faced competitive exclusion away from their optimal (drowned) habitats in the habitats dominated by more generalized, less site-specific (often smaller, too) species.

As for the rest, they were mostly tree dwellers. For a bat to use a tree successfully as a roost site, the tree needs to be either rather tall—so the bat is unlikely to be eaten by a

A FIFTH BAT

In October 2004, while this book was in production, the frog-chronicling team of Kristiina Ovaska and Jeannine Caldbeck expanded their efforts to include bats, and their personnel to include Michelle Theberge. They caught many bats of all our previously known species, except the cave bat, and plan to report on these separately. Most remarkably, on 16 October 2004, Theberge netted a free-tailed bat, *Tadarida brasiliensis*. This species in family Molossidae is about the same size as our common mastiff bat, *Molossus molossus*, but has a strikingly corrugated upper lip, long ears that fold forward to reach the tip of the snout, and several distinctions of pelage and dentition. Because they also caught mastiff bats, we were able to compare the free-tailed bat to them directly and all of us could confirm the identification. Theberge prepared the specimen, a male, as skin, skull, partial skeleton, and tissues in ethanol and sent it to The Museum, Texas Tech University—nowadays the most appropriate repository for Caribbean bats. It bears my field-tag number Z-39340.

This fifth bat species was certainly not what I expected. It is rare in the Virgin Islands (only previously recorded on St. John, USVI by Koopman 1975). Its niche factors seem solidly similar to those of its common cousin, the mastiff bat, calling into question the ecological underpinnings of bat biogeography described above. I still expect the red fig bat to turn up on Guana, so maybe there are *six* species present.

I have not recast table 24 to include this fifth species.

snake or an iguana up in the leafy canopy—or hollow (or, even better, both). Tall trees and wolf trees—the latter hollow or broken—are the first to fall before the forester's ax. Of course, big trees and hollow trees did, sometimes, survive up in the mountains of the bigger islands. Why did not, then, more of those extinct bats? An explanation is the late Karl Koopman's "paramontane diversity rule"

(Koopman 1983): Elevation is far more influential in directly affecting species diversity than is area (contra MacArthur and Wilson 1967), but most of the increase in diversity associated with higher elevation *involves species that live in the lowlands.* Is that not fascinating?

Well, all right, then why are there so few bat species on Guana Island? Should we not be able to find at least a fifth? It cannot be an obligate big cave dweller and it cannot be a rain-forest species. We do have forest, with big, tall trees, and a lot of hollow ones, too: No one culls Guana's wolf trees. So what is it going to be? My bet is on *Stenoderma rufum*, the red fig bat, rare and colorful. It will take a real bat snatcher to bag it, but I am betting it is there.

Birds

Ornithology is not an exact science like mathematics. The work proceeds slowly, perseveringly, in an endless search through material for vestiges of order, and something new discovered about a species often proves to be as important as its original discovery.

MARY WICKHAM BOND (1971)

WRITTEN RECORDS of Guana's birds go back more than half a century to the initial efforts of Jonnie Fisk in 1952. There is a break from 1961 to 1976. The Cambridge Ornithological Expedition of 20 June to 19 August 1976 came over from England (Mirecki 1977), and I came first in 1980 with Robert Chipley, then of The Nature Conservancy (Lazell 1980). Since that time, records were kept for most years up to 1990 by Chipley (Chip) and Gregory Mayer (Mayer and Chipley 1992). In July 1987, the late Jonnie Fisk observed, listed, and banded birds while Chip concentrated on bridled quail-doves. See Mayer and Chipley (1992) for an analysis of land-bird presence and the remarkable lack of turnover in our avian fauna.

1993 was a very big birding year. The late Debbie Paul, David and Charlotte Hill, Charles Bartlett (of planthopper fame), and Kerry Sherred compiled detailed lists. Wayne Arendt, USDA Forest Service, Puerto Rico, came in 1994 and did an exhaustive survey. Also, 1994 was the first year Fred Sibley, then at Yale, brought his group and set up mist nets; then it became identification of birds in the hand, not just in the bush. 2003 was another big birding year. In October, Clint Boal and his wife Tracy, from Texas Tech University, began their studies. The Sibleys came, too, continuing their bananaquit studies and recording many other species. Angela Davis, Wenhua Lu, and I added sightings. Clint and Tracy caught two species of migrant wood warblers new to Guana, bringing Guana's total to 107 species. However, comprehensive lists have not been made for any year since 2000 (table 25, pp. 233–236).

In preparing table 25, an avian listing history, I edited Jonnie Fisk's early lists. For example, I can do nothing with "tern" or "plover" in 1952. That list is not exclusively from Guana, so I have questioned "Great Blue Heron" and left out "Northern" of the two listed kingbirds. I have converted all names here, and in all other lists, to those of Raffaele et al. (1998); for example, "Sparrow Hawk" to American Kestrel and "Honey Creeper" to Bananaquit. For 1953, I cannot allocate "Peeps North Beach" beyond some sandpiper of the genus *Calidris*, and I dare not

BOND, JAMES BOND

Many people are surprised to find out that James Bond had a wife, Mary, who often accompanied him in the field, and that he wrote books himself and did not just dominate those written by Ian Fleming. Indeed, Fleming met and knew the real James Bond, and Mary, only briefly, in Jamaica, where Bond was, of course, engaged in full-time espionage, unraveling the systematics, life histories, and distributions of birds (M. W. Bond 1966, 1980). I knew Bond a little better when I was a boy in Philadelphia, where he was curator of birds at the Academy of Natural Sciences. Beginning in 1957, I pestered him with questions about Caribbean birds and sent him occasional notes about birds I had spied on well into the 1980s. Anyone alive before about 1990 who was interested in West Indian birds had to know James Bond, at least through his writing: He literally wrote the book on the subject, *The Birds of the West Indies*—five times (the fifth edition came out in 1985).

I have both that colorful volume and a brown, battered, stained, sea-bleached first printing of the 1947 version. The old one was not, alas, the one I took with me on my first Antillean expedition of 1957 (I lost that one; the oldest one I have today of my own is 1971). The old book in my hand of 1947 is not really mine. It was given to The Conservation Agency by Erma J. Fisk—Jonnie to those who knew her—wife of financier Bradley Fisk. The Fisks were among the original cottage builders on Louis and Beth Bigelow's island—Guana Island—purchased in 1935. After World War II, when the Fisks and the Bigelows and all others could come back to Guana, Jonnie Fisk's interest in birds had grown to the point where she, too, was sending reports of her spying back to Bond in Philadelphia. Jonnie's notes—right in the old brown book—begin in 1952 and continue most years until 1961. Then there is a break in Jonnie's records until 1987. It took me that long to find Guana Island, to make the connection between Guana's pioneer birder and the woman I met in the Florida Everglades back in about 1970, to track her down, on Cape Cod, and to bring her back.

guess which species of "coot." Sad, because we have no other record of a coot. One species we see occasionally flying by, Masked Booby, cannot be tabulated because I lack a dated sighting.

Guana Island has proven critically important in our expanding knowledge of Neotropical migrants: birds that breed in North America and migrate south to the tropics after nesting. McNair et al. (2002) record many species of migratory birds that were previously unknown, or very little known, in the eastern Caribbean captured by the Sibleys and their co-workers on Guana (and even more are added below). That report only goes to 1998, at which time the only other banding station in the whole West Indies east of Puerto Rico was on Barbados. That one has since closed down. Feduccia (2003) shows two major migratory pathways from North America to Guana, and beyond: all the way to South America. One of these is the great transAtlantic jump off from maritime Canada and New England. Blackpoll Warblers do this; Guana may be the first landfall for many of them. The second seems less ambitious: move down the east coast to Florida, cross over to the Greater Antilles, and island hop east and south to the Virgin Islands and Lesser Antilles. This may be what our Swainson's thrushes do. Migratory birds can vector communicable diseases such as the West Nile virus (Malakoff 2003), so maintaining an active program of ornithological research on Guana is of vastly more than anecdotal importance.

For those interested in pursuing ornithology in greater depth, I recommend Terborgh and Faaborg (1980) and *Evolutionary Ecology of Birds* by Bennett and Owens (2002). This is a research monograph documenting the most topical issues in the field. As with this book, theirs builds on the foundation laid by the great ornithologist and biogeographer David Lack. Guana Island and the BVI would be ideal for conducting the kinds of ornithological studies their book describes and suggests.

Pied-billed Grebe, *Podilymbus podiceps.* This is a regular breeding resident species throughout the Greater Antilles and northern Lesser

Antilles, but it is scarce on small islands for lack of freshwater habitat (Raffaele et al. 1998, 216). Mirecki (1977) reported three or four on Tortola in June 1976. They will visit and feed on salt water occasionally. Jonnie Fisk got the only Guana Island record between 10 February and 10 March 1961. Rowan Roy (in litt. 27 December 1993) reports the occasional pied-bill in the regular Tortola Christmas bird count over the years. They swim low in the water and cannot abide staying on the surface long—pop! Down they go for a dive, only to pop up again, often surprisingly far away. Pied-bills are drab except for the bold black-and-white bill of the adult.

White-tailed Tropicbird, *Phaethon lepturus catesbyi*. Regularly seen in flight at Guana, this species was first documented nesting in the BVI on Carval Rock and Sandy Cay, and probably on Norman Island (Lazell 1980, 1981). On 13 March 1982, a pair were observed using a nest hole in the cliff on the north face of Long Mans Point on Guana for our first nesting record (Norton et al. 1989). To quote myself (Lazell 1981):

"Our two species are hard to tell apart until you get used to just what to look for. They are largely snow-white patterned in jet-black; both have long, white, streamer-like tails and red bills when adult, so the names are no help at all. Juvenile birds of both species have yellow bills and a fine barring of black on their white backs—like a Plymouth Rock chicken. I can only identity a juvenile when I see whom its parents are. But the adults are easy enough: Red-billed—bill blood red or scarlet; back and forewing patterned with fine black barring, like the young. White-tailed—bill yellow to orange-red; upper back and forewing plain white."

To the above, I would now add that in flight the white-tailed has a solid black band over the wing secondaries, where the red-billed (see the following account) is barred and appears gray at a distance.

In their account, van Halewyn and Norton (1984) note "3–4 sites" for nesting in the BVI versus "15–30" in the USVI (excluding St. Croix). Walsh-McGehee (2000) lists Guana as well as Fallen Jerusalem, Great Tobago, and Norman Island, with question marks for Peter Island, Round Rock, Virgin Gorda, and unspecified Dog Islands. She reckons 40 to 100 pairs total in the BVI.

Our subspecies is named for Mark Catesby, who chronicled American fauna and flora from 1712 to 1726 (Hanley 1977).

Red-billed Tropicbird, *Phaethon aethereus mesonauta*. This is the more common of our two resident species and regularly nests on the north face cliffs of Long Mans Point, Guana Island, from January to October. Jonnie Fisk first recorded red-bills nesting on Guana on 27 February 1957. We have found fledglings in a nest on Carrot Rock in October. For this species, the BVI is a stronghold; it is much less common, and an infrequent nester, as far west as Culebra and Puerto Rico, as well as eastward in the Lesser Antilles (Raffaele et al. 1998, 222). However, this is the species I found, and suspected was nesting, at Sombrero on 1 June 1963 (Lazell 1964). I saw no white-tails there at that time. Walsh-McGehee (2000) also records it as the more numerous species at Saba.

The van Halewyn and Norton account (1984, 180) says the red-bill is ". . . most numerous . . . in the Virgin Islands" and "probably cannot expand its range much farther westward in the Greater Antilles because it may be confined to waters with somewhat increased productivity." They chart only one nest at one site in the BVI. Because they were unaware of my records (Lazell 1980, 1981) and Mirecki (1977) reported no nesting, I do not know where they got this record. Walsh-McGehee (2000) does list "Great Tobago, Guana Island, Misc. other isles." in the BVI, with "fewer than 50" estimated breeding pairs, but sites no reference; apparently not Norton et al. (1989) because we did not list Great Tobago, but did list Broken Jerusalem, Carrot Rock, West Seal Dog, and Ginger Island—in addition to Guana. Fred Sibley found three nests with young on Great Tobago in October 1999 and suspected there were more. Walsh-McGehee (2000) notes "that the numbers of birds had not increased substantially in [a] fourteen-year period. Given the disparity of Saba's estimate and a quadrupling of sites, the 1998 estimate should have been dramatically higher if populations at other sites were not declining." She recommends that the status of this species be revised from "to be monitored" to "vulnerable," and says "this species deserves global conservation consideration and . . . the West Indies supports a substantial portion of the world's population." I note that the subspecies *Phaethon aethereus mesonauta* is confined to the Caribbean (van Halewyn and Norton 1984).

Masked Booby, *Sula dactylatra dactylatra.* "A very rare and local year-round resident in the West Indies" (Raffaele et al. 1998, 222), but never recorded nesting anywhere in the BVI. Occasional individuals are seen flying and diving at White Bay, Guana Island, but I lack detailed records or photographs. Mirecki (1977) did not record this species at all from the BVI, but in a typescript list received from Rowan Roy, the BVI's premier birder, 17 April 1980, this species—as "blue-faced booby"—is listed as seen between 1958 and 1978 offshore. The van Halewyn and Norton (1984, 181) account says: "With no more than twelve known nesting sites this is the scarcest booby in the Caribbean region." They give 30–76 estimated pairs at one to three nesting sites in the USVI. Schreiber (2000) tabulates 40–50 pairs at Cockroach and Sula Cays and Frenchcap Rock in the Puerto Rico Bank USVI (i.e., excluding St. Croix, where the population is reported as extirpated).

Brown Booby, *Sula leucogaster leucogaster.* An abundant resident, this species has been regularly and consistently present on Guana Island for as long as records have been kept. Nevertheless, it is a bird of controversy and mystery here. It is the third species on Jonnie Fisk's 1952 list, and she first recorded nesting on Guana between 19 February and 18 March 1955. Mirecki (1977) counted 120 birds in the Guana Island nesting colony between 20 June and 19 August 1976 and about the same number—60 nests—on Great Tobago. I found them nesting on Guana in March 1980 (Lazell 1980, 40) and also counted 180 nests on Great Tobago, about 200 nests on Little Tobago, and 4 nests on the

Indians, just west of Norman Island. Raffaele et al. (1998, 223) note that the "breeding season is prolonged and varies from year to year." In view of all this, it is amazing that Schreiber (2000) lists only Great Tobago and Little Tobago nesting colonies with 100–195 pairs total. Imagine my surprise when Fred Sibley (in litt. 9 April 2001) agreed with Schreiber: "She and I felt there was no evidence of nesting on Guana. . . . There are lots of adults and lots of immatures but never any downy young and no adults on nests. . . . " Therefore they ". . . would not list Brown Booby as a nester." Could Fisk, Mirecki, and I all be wrong? I cannot now ask the other two, and I did not doubt that I was seeing actual nesting but did not describe just what I did see in my field notes. It will be very important to check and—especially—to photograph downy young if any are seen.

Young brown boobies are brown all over, but the adults have snow white bellies—except on very bright, sunny days when flying over White Bay. Then they may have brilliant turquoise blue bellies, reflecting the sea over the sand. No bird guide informs the inexperienced birder to recognize the blue-bellied booby.

Brown Pelican, *Pelicanus occidentalis occidentalis.* This is an abundant resident breeder. Pelicans are extremely popular until they become a nuisance—for example, by perching and pooping on swimming platforms, docks, boats, and even rooftops (where they can contaminate cisterns). Still, most people love them and many are fascinated by their aerial acrobatics as they dive for fish and by their handling of the catch, afloat, after a successful dive. This is the very first species listed by Jonnie Fisk on her "Life List" of 1952 at Guana. Brown pelicans underwent a catastrophic crash along the Gulf and Atlantic coasts of North America (subspecies *Pelicanus occidentalis carolinensis*) beginning in the mid-1940s and continuing until the ban on DDT and other organochlorine pesticides in 1972. The Pacific coast form (*Pelicanus occidentalis californicus*) was not noticed to be dying off officially until 1969, and both forms have now substantially recovered. Guravich and Brown

SHARPSHOOTER

When Liao Wei-ping first came from Guangdong Province, China, to Guana Island in 1984, the brown booby was the only species of bird present that also occurred at his native Hainan Island. He yearned for a museum specimen to take home to his base, the Guangdong Institute of Entomology's Department of Zoology (how is that for getting the numerical priorities straight?), Academia Sinica. I fixed him up with a .22-caliber rifle, but the birds were always flying irretrievably out over the sea. Liao is not a strong swimmer, and, anyway, the best shots were from cliffs no one would willingly jump off of. Days went by with no success. Michael Gibbons, Anthropology Department, University of Massachusetts, Boston, was in residence, digging subfossil bones and Amerindian potsherds; he is not without a sense of humor.

Coincidentally, Liao was trying to catch a scaly-naped pigeon in his mist net, set across the top of the driveway parking area in the col between Anegada and Barbados Houses on Guana. The pigeons regularly rocketed through this col but always over Liao's net. Gibbons, walking by one day, observed this phenomenon and Liao's despair. "I'll get you one," said Gibbons, "you just have to call them down: Coo-ee, Cooo-eee. . . ." Swoosh! A fine pigeon hit the net—bagged instantly. But Liao's jubilance was short lived in the face of failing to get a booby. I knew of a rock, close inshore but just out of sight from the beach, where, some mornings, a booby usually perched. Gibbons and I hatched a plot.

We told Liao to stay in the temporary lab we annually set up in the common room of Anegada House, working on his notes and specimens. Gibbons would call in a booby and shoot it—in flight, of course—so that it would fall right at the Anegada House door. Liao chuckled and said ok, he could hold up his end of the deal if we could manage ours. Early the next morning, Gibbons and I set off for the chosen rock, which was plenty far enough away from Anegada House to be out of audible rifle shot. Sure enough, there perched a fine adult booby. I am by no means a crack shot; only by incredible luck could I have hit a flying booby with a rifle. However, this was a job I believed even I could do.

I lay down prone on the ground, rested the rifle solidly on a big rock, lined up the booby in the buckhorn sights, and gently squeezed off a round. The booby toppled into the shallow water, and we waded over and picked it up. Back up the hill we went; Gibbons took up his station just outside the south side window of the common room, just inside which Liao was working. He gave me a minute or two to lay the dead booby on the doormat and take up my station below the steps leading up to the main door.

"Booo-bee, Booo-bee, Booo-beee" called Gibbons. Bang! He fired the .22 into the air, and I slung my leather water flask over the porch, banging it into the door. Liao came running, yanked open the door, and beheld his prize, incredulous. He had his bird for direct comparison to his Chinese specimens back home. But that is not the end of the story. Liao immediately set about preparing his specimen. When it was skinned—carcass bared, skin wrong side out—Liao called us over. His increased incredulity was written all over his face.

All three of us carefully examined the skin and carcass of that booby. There was no sign of a bullet hole.

The BVI had a short, simple list of protected birds; boobies, scaly-naped pigeons, and most other common birds were not on it. It was a simple matter in those days to get a letter from the government stating Liao's specimens were not illegally obtained, and that was all one needed to do to get them into the United States. Once back at the Museum at Harvard, we packed those specimens up with a fine collection of specimens that had been legally salvaged by the Massachusetts Audubon Society but were languishing unattended in sanctuary freezers all over the state, until Liao prepared them. That collection included a common loon (another species in common with China) a glossy ibis, several egrets and herons, and a fish crow that—as far as I know—is even today the only extant voucher specimen for the

continued

continued

northernmost breeding population known, from the Boston suburbs. Although Liao has now moved to the BVI and Guana Island, his collection at the Guangdong Institute remains a permanent testimony to his diligent efforts—and one small mystery.

(1983) have chronicled the story of the fall and rise of North America's brown pelicans in pictures and prose. In the BVI, at least, there was apparently insufficient hard pesticide application to cause a post–World War II die off. Our pelicans have been here right through it all.

As noted by Raffaele et al. (1998, 225), these huge birds nest in colonies, typically in trees; breeding "may occur during any season and may vary from year to year." On Guana there are typically active nests from March to September, with an occasional nestling visible in October. Jonnie Fisk first recorded nesting on Guana, "halfway to Monkey Point," between 19 February and 18 March 1955. Recently, nesting has been concentrated at Pelican Ghut between Northeast East End and Southeast East End on Guana's far eastern shore. The nesting birds are best observed from a boat.

The extreme variability in nesting season and the broad spread of breeding activity among pairs in a single nesting colony or geographic area has made it virtually impossible to tightly correlate adult plumage with time of year or apparent nesting and breeding activity. For example, Raffaele et al. (1998, 224) state that the chestnut nape ("reddish-brown hind neck") is characteristic of the breeding adult, but that in some breeders the neck remains all white; typically, they say, the all-white neck is characteristic of the nonbreeding adult. Well, with birds likely to breed at any time of year, or at different times in different years, how can one tell? Schreiber et al. (1989) undertook a comprehensive study of museum specimens and living birds from both coasts of North America and some from Puerto Rico. Of over 1,275 museum specimens examined, 900 were from Florida. Needless to say, because of the enormous variation in the timing of breeding, data only begin to make sense when a particular population can be studied long term. For example, at Tarpon Key, Tampa Bay, Florida, Schreiber et al. (1989) found that breeding adults of both sexes began appearing in the "yellow head/dark neck" plumage associated with the onset of breeding as early as mid- January (1972) to as late as early February (1974). Most years, 1969–76, this boldest of plumages appeared right about the last days of January. Only in 1969 did 100% of breeding adults appear in this plumage at any one time, and that was in mid-March. In all other years some began losing their yellow while others still had it; birds retaining a dark nape but lacking yellow are believed to be incubating eggs or raising chicks. Beginning in January (1972), but normally not until March (five of the eight years), adults shifted to "white head/dark neck." This plumage was 100% of the Tarpon Key colony from the first of June through all of August in seven of the eight years (in 1971 a small percentage kept their yellow just past 1 June). The shift to "yellow head/white neck" began in late September or very early October and was completed on or before 1 December in every year. This is the nonbreeding "winter" plumage, at least at Tarpon Key, Florida.

Percentages have never been tallied on Guana Island, or anywhere in the BVI that I know of. I certainly have the impression that our birds do not fit the Florida pattern, but that would make sense because we are far to the south, in the less seasonal tropics. It is critically important not to confuse juvenile pelicans, which are all muddy brown of varying shades all over, with the three adult plumages. Iris color should also change, from plain brown in nonbreeding birds to "straw white or bluish white" in prebreeding and courting birds; as incubation begins the iris darkens again. Schreiber et al. (1989) describe in detail other changes in the colors of bills and feet and body feathers. At Guana, all of this would make a fascinating study.

Magnificent Frigatebird, *Fregata magnificens*. These spectacular birds, Jonnie's second Life List species, are common in the skies over Guana; year-round BVI residents, they are known to nest in the BVI only at Great Tobago, over to our west. Nesting is typically from August to April (van Halewyn and Norton 1984, 181; Raffaele et al. 1998). There were active nests on Great Tobago in every October that we looked. Mirecki (1977) found 25 nests on Great Tobago and 5–7 nests in mangroves off Anegada, in June to August 1976. Nesting reported to them on George Dog could not be confirmed, although roosting birds were present; I have looked and never found nests there. Lindsey et al. (2000) found none either and tabulated only the Great Tobago colony as surviving. I found 200 pairs nesting on Great Tobago in March 1980 (Lazell 1980); Fred Sibley saw more than 300 pairs in early October 1997 and 300–500 nests in October 1999. Lindsey et al. (2000) reckoned the Anegada population extirpated but gave 500–600 pairs for the BVI, presumably now only on Great Tobago. This is a threatened species noticeably declining worldwide. Fishing line is a ghastly killer of these birds.

Frigatebirds have long bills, hooked at the tip, and long swallowtails. Their wings are very long, and, according to Raffaele et al. (1998, 227), they have the greatest wing surface to weight ratio "of any birds." They weigh only about 250 g or a bit more than 8 oz.—even though they look huge in the sky. Frigatebirds are notorious robbers of other seabirds, obtaining most of their food through aerial piracy. They can catch their own fish, squid, or crustaceans at or near the surface with their long bills, but they cannot dive or swim like boobies, pelicans, or tropicbirds. They rarely catch their own food.

The adult males are black all over. The adult females have a white "vest" and are otherwise all black. Juveniles are white-headed and have extensive white below, including but not confined to the "vest" area. Males in breeding and courting condition can inflate a spectacular crimson to blood red throat pouch, studded with black feathers at the sides.

Great Blue Heron, *Ardea herodias*. This is a rather mysterious bird. Raffaele et al. (1998) state that it is a common nonbreeding resident in our area with breeding confirmed only "on Cuba and the Virgin Islands." Jonnie Fisk began recording them, questionably from Guana but surely in the BVI, in 1952. Mirecki (1977) recorded "the exceptional number of seven . . . on the southern promontory of Guana Island" (12 August 1976). We see the odd individual every year or so but have no evidence of nesting, on Guana or in the BVI generally.

There is considerable geographic variation in great blues, with a general tendency toward narrower black striping on the crown, and generally paler blue gray plumage, in the tropics. Juveniles lack crown patterning and are browner than adults. Closer attention should be paid to the few birds that come our way, and, especially, good photographs taken and dates and localities logged, in order to figure out who our birds are.

Great White Heron, *Ardea occidentalis*. This bird only deepens the great heron mystery. Many, including Raffaele et al. (1998, 228), consider this a color variant of the great blue. It is rare in the BVI, but Mirecki (1977) records it on Anegada on 14 July 1976 and Jonnie Fisk listed it on 24 February 1955. She questioned her own record because Bond (1947) did not record it as occurring east of Cuba. Jonnie noted the "large size, vivid yellow bill" but "legs not visible," which would have been the clincher. She concluded it might have been a mere great egret, which is much smaller and has black legs. Great whites are bigger than great blues and have yellow to red legs. The great egret (see that account next) is a common North American species that Jonnie must have known well; I bet she saw a real great white.

I made an in-depth study of the great heron mystery in the Florida Keys, where both forms occur together (Lazell 1989b, 89–93), and concluded that the great white and great blue are distinct species. They do, rarely, hybridize, and there are very rare all-white individuals of the great blue—but these white herons are anatomically distinct from real great whites. My opinion was largely based on field evidence and a 36-osteological-character analysis done by Karin Zachow at the University of Miami; her data were impressive. I postulated that the great white, a full species, *Ardea occidentalis*, "speciated in the eastern Greater Antilles in geographic isolation from great blues. Subsequently, it invaded westward, spreading into the range of Greater Antillean great blues (e.g., in Cuba), while they spread east into its range.... Hybridization between great whites and great blues took place, and still takes place, at recently colonized sites peripheral to the main range of the great white: Florida Keys and coastal keys of South America. In these peripheral areas the great white is a newcomer (probably at Wurm glacial maximum) and selection for character divergence is as yet incomplete."

My hypothesis would have Hispaniola and/or possibly the Puerto Rico Bank as the original home of the great white heron. This would necessitate another hypothesis: The great white has been extirpated, or nearly so, from its original home islands and largely survives at the periphery of its former range. Exactly so: That is what I believe. Great whites, nesting in coastal mangroves, would have been among the first large birds Amerindians would have eaten up. Their kitchen middens might provide us with the bones needed to tell the tale, but the relevant midden sites are probably under several meters of seawater today.

Great Egret, *Ardea alba*. This bird is uncommon in the Virgin Islands (Mirecki 1977; Raffaele et al. 1998) and not known to nest. It is white with a yellow bill but black legs. Unless Jonnie Fisk's "great white" was one of these, we have no definite Guana records until the mid-1990s. This species is nearly worldwide in distribution in tropical and temperate regions, and seems to be increasing in numbers, at least in the Antilles.

Snowy Egret, *Egretta thula*. This bird is said to be a common resident in the Virgin Islands (Raffaele et al. 1998), but it was not noticed on Guana until the big birding year of 1993. Inter-

estingly, Mirecki (1997) regarded it as a "rare visitor . . . formerly more abundant." This was perhaps the most sought-after of the birds hunted for their plumes in the millinery trade of the nineteenth and early twentieth centuries, and it was nearly extirpated from the southeastern United States. It was the slaughter of birds such as egrets, especially at their nesting colonies, for decorations for ladies' hats, that truly initiated the great conservation movement in America and the world (Hanley 1977).

Little Blue Heron, *Egretta caerulea*. Our most common resident heron, this species has not been recorded nesting. It was on Jonnie Fisk's first list of 1952 and on all of her subsequent lists; most any day one can find one or more at the Salt Pond. Mirecki (1977) states that "breeding confined to the larger mangrove swamps" has, unfortunately, been much reduced since he visited. Adults are slatey blue—almost black—with a purplish gloss on the neck. Juveniles are white with blue gray bills and dull greenish legs; with age, the dark feathers come in and the bird appears spotted or piebald. Youngsters are more often seen, at least in October, than adults. Little blues often come up around the buildings, especially at night, to eat insects, especially cockroaches—good for them. Sometimes their footsteps on the roof alarm the uninitiated as they can sound bigger than they are.

Cattle Egret, *Bubulcus ibis*. This is a frequent visitor but does not nest on Guana. The cattle egret spread from Africa to the New World, arriving on the hump of South America by about 1870. Remarkably, the first North American specimen was shot by Massachusetts Audubon Society collectors at Sudbury, Massachusetts, on 23 April 1952 (Veit and Petersen 1993). If they had not shot it and put it in the Museum of Comparative Zoology at Harvard, then no one would have believed them. By that time the species must have already colonized the West Indies. Cattle egrets are quite terrestrial and love to hang out with exotic ungulate livestock such as Guana's donkeys. They eat mostly insects, typically stirred up by the grazing animals. They are white with yellowish bills and legs that get orange or rosy in breeding season, which Raffaele et al. (1998) say is "primarily from April to July, but also in other seasons." They nest in groups, in trees, frequently mangroves. In full adult plumage, they have buffy crest and elongate back feathers. There is no account of this species in Bond's 1947 book, and Jonnie Fisk never recorded it. In my "First American Edition" (1961), Bond provides an account under the generic name *Ardeola* and says: "Widespread in the West Indies, but not recorded from this region prior to 1952," when some guys from Massachusetts notified folks to start looking.

Green Heron, *Butorides virescens*. These small herons are not green. They are slatey with variable amounts of brown, usually chestnut to mahogany, on their necks; they have black caps and white-striped throats. Their legs and feet are yellow. This is an occasional visitor to Guana recorded first by Jonnie Fisk in 1953. The clearing of the mangroves around the Salt Pond, noted by Fisk in 1954, destroyed its main habitat—and that of several other species (see "Human History," part 6, below). Mangrove recovery along the northeast side of the Salt Pond has facilitated the return of the odd individual visitor.

Raffaele et al. (1998, 233) use the scientific name given above but have the range wrong. Monroe and Browning (1992) exhaustively analyzed specimens in a taxonomic study and concluded that *Butorides virescens* stops at Tobago and southern Central America; they call the birds on Trinidad and all of South America *Butorides striatus*. I must say their evidence for full species status is very weak, and, if one regards the two as subspecies, then our birds are *Butorides striatus virescens*. In an exhaustive follow-up study, Hayes (2002) documented over again what appears (certainly to me) to be intergradation between mere subspecies. On his evidence, I would retreat to the previous taxonomy, but Hayes (2002) found some indication of "assortative mating"—where birds of a feather did mate together; he let the two remain binomially separate. A true resolution will be best achieved biochemically with DNA sequencing, combined with field study of breeding birds in the contact zone.

Yellow-crowned Night-Heron, *Nyctanassa violacea*. Although common regionally, this species is seen irregularly on Guana. Jonnie found it in 1952, and most years there are a few about. The immatures are drab, but the adult is very handsome, patterned in violet to blue gray with a boldly striped black-and-white head, long white crest plumes, and—in breeding plumage—a bright yellow crown. As the name implies, these birds are typically nocturnal; they often hunt insects around the buildings at night and sometimes surprise guests: They can look pretty big and scary on a walkway in the dark. A youngster took on insect control work at Anegada House for the month of October 1993 and could be seen regularly on duty as early as 3:30 p.m.

Greater Flamingo, *Phoenicopterus ruber*. The story of our flamingos, which came from the Bermuda Aquarium, Museum, and Zoo initially in 1987, is told in "Restoration," part 6, below. Actually, we hope the story is only just beginning; Caitlin O'Connell-Rodwell, of Stanford University, plans to attempt initiation of courtship and nesting, using artificial stimuli. This has worked very well in other flocks too small to begin reproductive activities on their own. The restored population on Anegada has increased from 18 birds in 1992 to 63 in 2001 and seems established (Lazell 2001). I was told (Rondel Smith, pers. comm.) that there were 85 in 2003.

White-cheeked Pintail, *Anas bahamensis*. These beautiful small ducks were once extremely abundant regionally. They were hunted to rarity, which probably accounts for their absence from all of Jonnie Fisk's lists, prior to 1987. With the end of hunting in the BVI, these birds rebounded; Mirecki (1977) found them "locally abundant," but not on Guana Island. Mary Randall, who first came to Guana in 1974, recorded the first nesting pair in 1979; they were nesting when Chip and I came in 1980 (Lazell 1980), and Chip recorded a female with three ducklings on 23 July 1986. They have nested and resided on Guana ever since. They restored themselves. One banded by the Sibley crew was recovered in the USVI, so they travel.

Blue-winged Teal, *Anas discors*. Regionally, this is an abundant winter visitor (Mirecki 1977; Raffaele et al. 1998), but we have recorded it on Guana only in 1996 and 1997 to date. Rather a drab little duck, its forewing is bright blue. Like the pintail, it is a shallow water dabbler in the Salt Pond.

Osprey, *Pandion haliaetus*. This is a worldwide species that Raffaele et al. (1998, 249) say does not nest in the Virgin Islands, despite being "resident." Mirecki (1977) reported that Anegadians described nesting and the nest on Anegada, but he could not confirm it. Ospreys—often called "fish hawks"—are regularly seen, typically flying over, at Guana in most years. Jonnie Fisk believed they were nesting on Guana in February 1959, with good evidence. Chip recorded one on Guana on 13 July 1984 (not tabulated).

Sharp-shinned Hawk, *Accipiter striatus*. This is described as a "vagrant" in the Virgin Islands by Raffaele et al. (1998, 252), and it is understandably absent from the summertime list of Mirecki (1977). One was spotted by Fred Sibley hunting around the dump in October 1998. This is a small-bird eater, and we have a lot of small birds that they can eat.

Red-tailed Hawk, *Buteo jamaicensis*. These magnificent big raptors frequent Guana's skies and seem, based on their behavior, to have nested here at least occasionally. They soar and scream—the latter is a harsh, metallic noise, "like a rusty hinge" (Raffaele et al. 1998, 254). They are very fond of eating snakes and other reptiles; our iguana restoration has benefited our resident red-tails greatly by providing hatchling iguanas late in each summer, no doubt edible (transportable) well through the dry season and into the nesting season, January to July.

American Kestrel, *Falco sparverius caribearum*. These beautiful little falcons are year-round nesting residents on Guana, and they are widespread over the western hemisphere. They perch high and conspicuously, often calling a high-pitched "killy, killy, killy"—resulting in the local common name "killy hawk" (not to be confused with "kittyhawk," a laughing gull). Females are larger than males and russet in color, and males have russet bodies but blue gray wings; both sexes have bold, nearly black-and-white facial patterning. Immatures look drabber than females and have weak facial patterning. Our subspecies is confined to the Puerto Rico Bank islands and the Lesser Antilles, including St. Croix. The bulk of the food is insects, but many of us have seen them take anole lizards. Most remarkably, LeVering and Perry (2003) observed (and showed me and others), photographed, and videotaped a female kestrel eating a baby iguana on 8 October 2002. On Guana in October 2003, the Boals caught four kestrels; one had been banded by the Sibleys in 2002. They banded the other three and subsequently saw a fifth unbanded bird.

Peregrine Falcon, *Falco peregrinus*. This is a near-worldwide species, famous because of its alarming decline in the DDT pesticide era. Once probably a frequent migrant and winter resident, peregrines are now rare. One appeared on Guana in October 1999. Winter, dry season visitors to Guana have a good chance of seeing this species, which is regaining its numbers. They are magnificent, nearly black-and-white falcons; they feed on ducks, terns, and like-sized birds using a dive-and-strike method of killing their prey.

Clapper Rail, *Rallus longirostris*. Locally called a "marsh hen," this long-billed, secretive wader nests in the BVI (Mirecki 1977) but has not been proven to do so on Guana. Although it was among Jonnie Fisk's earliest records (1953), no one spotted another until the big bird year of 1993. I suspect clapper rails could be found more frequently on Guana, around the better-vegetated Salt Pond edge, especially if one played a tape of their "kek, kek, kek" . . . grating voice. To quote Raffaele et al. (1998, 260), they are "far more often heard than seen." They are crepuscular: active at dawn and dusk. Fred Sibley believes that we would see more if they were of regular occurrence and that they are more likely rare vagrants.

Common Moorhen, *Gallinula chloropus.* Raffaele et al. (1998, 263) say: "Generally a common year-round resident throughout the West Indies. Occurs worldwide." Jonnie Fisk recorded six on Guana in February of 1953, and Chip recorded one present from 8 to 21 July 1987. Then there is a gap until 1993. These aquatic rails are large, and they are slatey blackish with white streaks on their flanks and a bright red bill and frontal shield; the bill has a yellow tip.

Black-bellied Plover, *Pluvialis squatarola.* Raffaele et al. (1998, 267) say: "Generally a common non-breeding resident in the West Indies from August to May," with some records through the rest of the year. Jonnie Fisk recorded them in the winters of 1957 and 1959, Chip and I recorded the species in 1980, there are October records from 1993 to 1995, and one was banded in 1998. We normally see single individuals at the Salt Pond.

Wilson's Plover, *Charadrius wilsonia.* Recorded first by Jonnie Fisk on 2 February 1958, and noted as "Resident." These conspicuous shorebirds come to nest from March to at least July. They make a scrape nest on the ground, usually near the Salt Pond, and have been in residence every year since at least 1980 (and Mary Randall said they were here when she came in 1974). Mirecki (1977) found them nesting at many BVI sites, but not on Guana in 1976—probably an oversight. Chip found three nests on the Flat near the Salt Pond on 14 July 1984; two had single eggs and one had three eggs. On 17 July the former three-egg nest had a chick in it. The Sibley crew netted and banded two pairs on Guana in October 1996, and both were still present, with some unbanded individuals, in 2000. In 2001 they recaptured a female banded in October 1999. These mid-sized plovers have a broad, near-black breast band and a heavy black bill—giving them the common name "thick-billed plover."

Semipalmated Plover, *Charadrius semipalmatus.* This plover is a regular migrant through the BVI in October and may be seen occasionally in almost any month. They are only irregularly seen on Guana, typically among the Wilson's plovers. They are smaller and have stubby bills, narrower breast bands, and orange legs.

Killdeer, *Charadrius vociferus.* These are big plovers with two dark breast bands. They are resident in the BVI, but prefer large, open fields, such as the lawns at the community college on Tortola. Jonnie Fisk recorded them on Guana in 1953 and 1955, and we spotted them again in 1993. Guana is not a very good habitat for this insectivorous bird of the grasslands.

American Oystercatcher, *Haematopus palliatus.* This is a resident species in the BVI, but I know of no nesting record—certainly not on Guana. Raffaele et al. (1998, 270) say: "Restricted exclusively to a relatively scarce habitat type occupied by no other bird in the West Indies except, perhaps, Ruddy Turnstone." This habitat is "stony beaches and rocky headlands of offshore islands and cays"—not my idea of a "scarce" habitat! Although our records are spotty over the years, I believe an oystercatcher could be found on Guana on almost any given day if one checked those rocky shores and headlands (e.g., Monkey Point) regularly. These are large, handsome, black-and-white, modified sandpipers with big, chisel-like, orange red bills, which they use for opening shellfish. They have pink legs. They are unmistakable.

Black-necked Stilt, *Himantopus mexicanus.* These elegant, slender, graceful shorebirds nest on Guana and have been present every year since at least 1980 (back to 1974, according to Mary Randall). Mirecki (1977) found them on many islands in the BVI in 1976 but not on Guana. Jonnie Fisk recorded them in 1955 and 1957, in the winter, out of breeding season. At any given time, there may be few or none at the Salt Pond or as many as two dozen. Chip found a hatchling on 24 July 1986. In October 1997 the Sibley crew banded seven on Guana and several months later got a recovery of one of those from St. Thomas. In October 2001 they recaptured one of the females banded in October 1997; she must have been at least five years old. Typically, we have two or three pairs. They nest on the ground near the Salt Pond and will defend their nest with pugnacity.

Greater Yellowlegs, *Tringa melanoleuca*. This large, brown sandpiper is both resident and, more commonly, a migrant throughout the West Indies. It does have bright yellow legs. Jonnie Fisk thought she had it on Guana in winter of 1953, and she was sure in 1957. We spotted them again in 1993, but these are far less frequent here than their smaller cousins, discussed next.

Lesser Yellowlegs, *Tringa flavipes*. This is our regular, large, brown sandpiper, present in small numbers almost every October and during most winters. It is difficult to identity with certainty unless seen together with the greater yellowlegs.

Solitary Sandpiper, *Tringa solitaria*. These birds are rare migrants or winter residents at Guana, recorded only as single individuals at the Salt Pond in October most years now that the Sibley and Boal teams have been checking. These sandpipers are smaller than lesser yellowlegs; they have a pale eye-ring and dull greenish legs.

Willet, *Catoptrophorus semipalmatus*. This very large shorebird is regarded as "an uncommon year-round resident in the Virgin Islands" (Raffaele et al. 1998, 274). Mirecki (1977) records nesting on Anegada in 1976. We have only a single undated sight record by the late Debbie Paul in October 1993, but she knew a willet when she saw one. In flight, the broad, flashing, near-white band through the dark wing is highly distinctive.

Spotted Sandpiper, *Actitis macularia*. This is a common resident but not known to nest anywhere in the West Indies. We normally see youngsters and nonbreeding adults that lack spots. These small, brown sandpipers typically are white below with a prominent zone of brown coming down from the neck in front of the wing. As they walk along the pond edge or shoreline, they bob their rumps and tails up and down in a teetering motion; this is distinctive.

Whimbrel, *Numenius phaeopus*. Although Mirecki (1977) found good numbers of these big curlews on Anegada, Beef Island, Prickly

Pear Cay, and Necker Island in the summer of 1976, it is generally rare in the BVI. We have only one undated sight record by Debbie Paul and Liao Wei-ping in October 1993. Though brown, the crown is boldly striped in light and dark; the bill is extremely long and curved downward.

Ruddy Turnstone, *Arenaria interpres*. This is a common resident shorebird of pond edge, beach, and rocky coastline, but it does not nest here. Most of ours are in their relatively drab winter plumage, but even so they are boldly marked in brown and dark gray on the head and neck. Small flocks of 5 to 14 are seen on Guana regularly; it seems remarkable that Jonnie Fisk never recorded it. One (of two) banded 17 October 1998 on Guana was picked up on 20 May 2003 at Fort Macon, North Carolina.

Semipalmated Sandpiper, *Calidris pusilla*. This could easily have been Jonnie's "peeps" at North Beach and Monkey Point in February 1953. This is a common resident regionally, but it does not nest in the West Indies. There are one or two, sometimes up to six, typically hanging around the Salt Pond in October. The small sandpipers called "peeps" are notoriously difficult to identify. You need to really check out their characteristics, preferably in comparison with each other. These are classic "LBBs" (little brown birds).

Western Sandpiper, *Calidris mauri*. This little brown bird, a classic "peep," is a nonbreeding resident, like the semipalmated, above, and the least sandpiper, below. We have a single sight record by David and Charlotte Hill on 26 October 1993, and probably the same bird was seen by Kerry Sherred on 27 October 1993. Only an expert can be sure of their identification.

Least Sandpiper, *Calidris minutilla*. Yet another little brown sandpiper, this "peep" is a common migrant regionally. Debbie Paul and I saw one at the Salt Pond on 20 October 1993. It was among semipalmated sandpipers, so its smaller size (tiny) and light greenish (not black) legs made it distinctive. A decade later, in October 2003, the Sibley and Boal teams found this species again, in the same place.

White-rumped Sandpiper, *Calidris fuscicollis*. Regarded as a rare migrant in the Antilles, this "peep" is probably often overlooked if simply seen working the shoreline with its little brown relatives. In flight the white rump is distinctive. We have one undated record on Kerry Sherred's October 1993 list. I too saw the bird, in flight, but failed to log the date. On 20 October 2003 the banding team got two: birds in the hand, indeed.

Stilt Sandpiper, *Calidris himantopus*. Mirecki (1977) saw "up to fifty" of this "peep" on Anegada in 1976, but it is otherwise very rare in the BVI. All agree with Raffaele et al. (1998, 281): it is "difficult to identify." Ah, but in the hand, out of the mist net, weighed, measured, and firmly banded on 17 October 1997, it has been identified on Guana Island by Fred Sibley and his team.

Common Snipe, *Gallinago gallinago*. Despite its name, this big, stocky "sandpiper" of woods and swamps is uncommon in the Virgin Islands (Raffaele et al. 1998, 283). Jonnie Fisk saw them regularly in the 1950s, but we did not record another until 10 October 1997, when Fred Sibley flushed one out of the woods at the southeast end of White Bay.

Rednecked Phalarope, *Phalaropus lobatus*. Rare vagrant. A breeding bird of the high arctic tundra, this species is usually pelagic when not nesting and lives far out at sea. On 7 October 2004, the Boals spotted one swimming and actively feeding in Guana's Salt Pond. The bold eyestripe, slender all-black bill, and streaked back made it unmistakable. Although phalaropes belong to the sandpiper and snipe family, they have lobed feet and swim. This record may be the first for the Virgin Islands; it came too late to include in table 25.

Laughing Gull, *Larus atricilla*. This is the common seagull of the southeastern United States, the Virgin Islands, and the West Indies generally, but it is rare in winter. They vanish in September and return in April most years but were hanging around in October 2003. This is the "kittyhawk" that serenaded Orville and Wilbur Wright with its plaintive mewing and

got its name noted in the history of the world. It nests in scattered colonies in the BVI, but has not yet been proven to nest on Guana. A careful search of the booby and tropicbird nesting cliffs on the north face of Long Mans Point in spring and summer (May to July) would likely get us nesting records. These gulls devil pelicans, often perching on a pelican's head while he floats, draining his pouch; the gull hopes to pester the pelican into dropping fish. It is hard for me to account for Guana's paucity of actual records in view of the general abundance of this species. I suspect the birders are so used to them that they simply pay little attention.

Caspian Tern, *Sterna caspia*. This is a rare nonbreeding resident regionally. It is difficult to tell from the very common royal tern, below. The bill is deep red, not orange; the cap in nonbreeding plumage is flecked with white, but the forehead is not all white; the tail is only slightly forked. Still, I would doubt sight records of this species. However, right there on the bird list, in my own handwriting, is the note "Caspian Terns—25 Oct '93." Debbie Paul backed me up, recording them on 26 October 1993.

Royal Tern, *Sterna maxima*. This is the regular, big, white tern we see over the water coming to and from Guana. It can only be confused with its Caspian cousin, above. Breeding in the BVI is only suspected on Anegada (Mirecki 1977).

Sandwich Tern, *Sterna sandvicensis*. This is an uncommon resident. It was suspected to nest on Anegada by Mirecki (1977), and Raffaele et al. (1998, 291) seem to confirm that. There is an interesting biological situation with these terns involving the subspecies *eurygnatha*, which has apparently recently spread north from South America. Otherwise similar, *Sterna eurygnatha* has an all yellow bill—not black with a yellow tip. The two kinds of terns usually nest together in colonies but seem not to have made up their tiny minds whether they want to be different species or not: Some hybridization does occur. Of course, this is often true of very closely related species (white-tailed and mule deer; wolves and coyotes; black ducks and mallards, etc.).

Time will tell, as either the hybrids will be more fit and prosper, leading to large-scale introgression, or the hybrids will be less fit, even defective (e.g., mules) and there will be ipso facto selection against the mixed matings. We see sandwich terns—so far, for me, only the regular black-with-yellow-tip bill—at White Bay, or traveling to and from Guana, but fail to consistently record them.

Roseate Tern, *Sterna dougallii*. This is a Guana Island special. Roseate terns—ranging widely over the world—are rare and very local as nesting birds. Mirecki (1977) found a colony of 30 pairs nesting on Guana in 1976, and only one other proven nesting colony—10 pairs on Cooper Island—in the BVI. This is despite the species being a common resident, with the largest population in the tropical Atlantic (up to 2,500 pairs) on the small islands of the Puerto Rico Bank (e.g., Guana). In the account of van Halewyn and Norton (1984) it is noted that Puerto Rico Bank birds may travel all the way to Guyana in their first year and that nest site fidelity "seems weak." van Halewyn and Norton explicitly plot the Guana colony, so it came as a cold shock to see that Saliva (2000) tabulated our birds as "extirpated." Saliva cited van Halewyn and Norton (1984) but not Mirecki (1977); he does not clearly cite a source for extirpation of the Guana Island colony.

Roseate terns are abundant around Monkey Point and along Guana's southeast coast in summer, where we presume they nest. The fact that I did not find nests in April 1980 (Lazell 1980) is meaningless because the nesting season is May to July (Raffaele et al. 1998). All authors suggest this species is declining and threatened worldwide, and in need of close study where it still occurs. Just such a place is Guana Island.

Common Tern, *Sterna hirundo*. Despite its name, this mostly white tern is scarce in the West Indies. There are no certain nesting records for the Virgin Islands, perhaps because this species is hard to tell from the commoner roseate tern. Even as winter residents or migrants, "common" terns are "uncommon to rare" (Raffaele et al. 1998, 292). Robert Chipley found several among a resting flock of roseates on 12 July 1984; they are best identified this way (by direct comparison). David and Charlotte Hill recorded these terns between 19 and 26 October 1993, and I have an undated record for October 1994.

Least Tern, *Sterna antillarum*. According to J. A. Jackson (2000), this "is still considered a common breeding species" regionally. He provides historical evidence of nesting on Anegada and Beef Island, as does Mirecki (1977). The Anegada colonies were still present in 1990, but J. A. Jackson (2000) reports the Beef Island colony as extirpated. However, a colony of about 30 pairs was nesting on Great Thatch Island, BVI, in 1997 (J. A. Jackson 2000). Both Liao Wei-ping and I observed a least tern up close, on the ground, on Guana Island in October 1993.

Bridled Tern, *Sterna anaethetus*. This is a regular offshore dark tern that nests on small islets such as Carrot Rock (Mirecki 1977; Lazell 1980). Several of us observed them on different occasions off White Bay in the big bird season, October of 1993.

Sooty Tern, *Sterna fuscata*. Although less common than the bridled tern, this species nests locally on Carval Rock (Mirecki 1977) and we observed it in flight off Guana in the big bird season, October of 1993. Both of these pelagic, dark terns—bridled and sooty—can be seen regularly if one concentrates on looking for them.

Brown Noddy, *Anous stolidus*. This is an abundant resident and nesting species in the BVI (Mirecki 1977; Raffaele et al. 1998, 295). It is pelagic and rarely seen close to shore. Its color pattern is the reverse of most terns, being very dark brown all over except for a white cap. Liao Wei-ping listed it for Guana, without date, in the big bird year of 1993. They are regularly seen on the crossing to and from Guana Island.

Rock Dove, *Columba livia*. This is the familiar street and park pigeon of the world, an introduced exotic from Europe. Fred Sibley reports (in litt. 13 August 2001): "In 1994 two banded racing pigeons arrived on the island, and left within a few days after being fed by the staff."

Scaly-naped Pigeon, *Columba squamosa*. This is currently Guana's common, big, dark pigeon. It nests high in the trees and booms in explosive flight. Scaly-napes actually just have soft, red neck feathers, but with black edges—giving a scaly appearance. Their flesh is not palatable: I know; I have tried eating them on several occasions, beginning in Haiti in 1957. I believe poor quality of meat accounts for its continued relative abundance while the other large, more edible pigeons were hunted out in many areas. I hope the restored white-crowned pigeons will eventually outnumber the scaly-napes.

White-crowned Pigeon, *Columba leucocephala*. This magnificent, big, slatey blue pigeon now is a nesting resident again. Mirecki (1977) says: "once a common resident. . . . Now it is extirpated as a breeding species and only a casual visitor to the islands." White-crowns are delicious, which got them in trouble too deep to get out of by themselves. We had to bring them back. That project was accomplished largely through the good offices of the Puerto Rico Department of Natural Resources and Carlos Ruiz and Pedro Ruiz (who are not related). Our four birds of 1996 nested on Guana in 1997, but unsuccessfully. In 1998, with 10 more youngsters, they nested successfully on Guana. By October 2000 pairs of snowy-crowned adults and unpaired ashy-crowned young-of-the-year could be readily found in Guana's forest. More details are provided by Lazell (2002c) and in "Restoration," part 6, below.

Zenaida Dove, *Zenaida aurita*. This is the *tortola* from which our large neighbor island takes its name. These doves resemble American mourning doves but have squared-off, not long and pointed, tails. They are abundant, resident, nesting birds on Guana. The Sibley crew recaptured in October 2001 two males that had been banded in October of both 1997 and 1998. The record was set in October 2003 when one at least five years old was recaptured.

Common Ground-Dove, *Columbina passerina*. These very common, tiny, brown doves waddle along the roads and trails. They are most easily observed out on the White Bay flat, but they are everywhere. Fred Sibley notes (in litt. 13 August 2001): there are "frequently nests in the forks of the tree cactus." They will also nest on the ground. Chip found three nests all with two eggs on 18 July 1986. In October 2001 the Sibley crew recaptured a female banded in October 1998.

Bridled Quail-Dove, *Geotrygon mystacea*. (See plate 30.) Guana Island is a stronghold for this beautiful species, confined to the islands of the Puerto Rico Bank and south in the Lesser Antilles to St. Lucia. It is absent from many Lesser Antillean islands and "extremely rare and local on Puerto Rico and Vieques" (details in Raffaele et al. 1998, 304). Jonnie Fisk began recording them on Guana in 1954. Mirecki (1977) found "small numbers" on Guana and Tortola and gives old literature records for Virgin Gorda, Jost Van Dyke, Great Thatch, and Norman, Peter, and Beef islands. At the same time, Philibosian and Yntema (1977) listed bridled quail-dove as endangered, "accidental" only on Puerto Rico, and known to nest only on St. Thomas and St. Croix in the USVI. Robertson (1962) had reported the species declining on St. John.

Chipley (1991) describes vocalizations, roosting, courtship, nesting, and predation on Guana during four Julys. Briefly, nests are built of twigs and located 2.5–6 m up in trees, bushes, vines, or stumps. Two buffy eggs are the usual clutch, but raising just one chick is difficult. Pearly-eyed thrashers are very active nest predators on bridled quail-doves; one of Chip's July 1987 nests was raided and three nests contained a fledgling, so the breeding season must begin earlier— probably with the onset of summer rains that will bring on fruit production. Seaman (1966) listed seven species of plants used by bridled quail-doves on St. Croix: royal palm (currently absent from Guana but scheduled for restoration); maiden apple; *Calophyllum antillanum* (no common name); birch berry or stopper; *Arthrostylidium capillifolium* (no common name); croton or marawn; and turkey berry, in the potato family. Chip added two more on Guana: torchwood *(Amyris elemifera)* and gumbo-limbo or

When I first visited Guana Island, 24 March 1980, Mary Randall confirmed the presence of the bridled quail-doves I had read about there in Mirecki (1977). I called Robert Chipley, then of The Nature Conservancy. Chip arrived in the BVI on the 10 April 1980, and the next day we headed over to Guana, arriving for lunch. Between 2:30 and 3 p.m., Chip located two to four quail-doves in Quail Dove Ghut, which had been named long before we got there, but whether by Jonnie Fisk or Mary Randall, or someone else, I do not know. Chip came back again, 1984 through 1987, in what by then had become "annual scientists' month," in July. He netted, radiotracked, color marked, and color banded quail-doves over that four-year span; many of his data were published (Chipley 1991), but a lot were not. Fortunately, those data were not lost, and Chip forwarded the material to me in August 2001 for inclusion here: see "How to Count Snakes—and Other Things," Part 2, above. I say "fortunately" because working up those data makes such a good story about how intractable population estimation can be: We do not know within an order of magnitude how many of these birds, demonstrably rare in the world, actually live on Guana. I suspect the transect-census data for Quail Dove Ghut are about right, with 6 to 10 individuals frequenting the 3,000-m² strip sampled. But, I believe, this figure cannot be extrapolated to all good quail-dove habitat (back in 1984, about 121 ha; today increased to at least 150 ha). I suspect the birds concentrate in a few small areas when berries are ripe and falling and are absent from most of the range most of the time. The reforestation efforts of Fred Kraus, now with the Bishop Museum, Hawaii; Lianna Jarecki; and Liao Wei-ping have expanded quail-dove habitat north and west from Chipley's (1991) map, wrapping around the hilltop where the buildings are. This makes seeing quail-doves on Guana much easier than it was a decade ago.

turpentine tree *(Bursera simaruba)*. By far the most important foods on Guana in July that Chip documented were berries of stoppers, genus *Eugenia;* we have several species, very abundant in the forest understory. Seaman (1966) noted hipproboscid flies as ectoparasites on at least 30% of his St. Croix bridled quail-doves.

Yellow-billed Cuckoo, *Coccyzus americanus.* This species is apt to be secretive and was not detected on Guana Island until the big bird year of 1993, but it has been seen almost every October since, and even by me. We tend to think of yellow-bills as migrants from North America, but Raffaele et al. (1998, 317) point out that they do nest "rarely" in the Virgin Islands, and so they might on Guana.

Mangrove Cuckoo, *Coccyzus minor.* This is our common, resident, breeding cuckoo on Guana. An inveterate skulker, it is typically hard to see but easy to hear. The call starts out as a rattling "kah-kah-kah . . ." and winds down to a trailing off "cow, cow, cow." The blackish ear patch and buffy abdomen (not white) separate this species from the yellow-bill.

Smooth-billed Ani, *Crotophaga ani.* This strange-looking member of the cuckoo family is a common nesting resident virtually throughout the West Indies. It prefers agricultural areas and savannah habitats but also likes woodland edges. We can find some on Guana almost every year, usually quite easily because of their loud, whining "wheep, wheep" calls. They are almost always in a small flock of up to a dozen, and that flock may be all we have. Chip found a chick on 19 July 1987. Anis were conspicuously absent from Guana in October 2003. I wonder what happened to them.

Anis are slate gray, almost black, all over and have a huge, parrotlike bill; they are largely insectivorous. Often several pairs will share a nest and work together raising the young (Raffaele et al. 1998). Anis are graceless, ungainly flyers and generally seem awkward and clumsy in trees; this makes them fun to watch.

Puerto Rican Screech-Owl, *Otus nudipes newtoni.* This is a bird of mystery, or perhaps history. These little owls were present on Guana when I arrived in 1980 and well known to Mary Randall. It was not unusual for one to perch and trill on the roof of Dominica House in the evening. However, Jonnie Fisk never recorded it at all. Mirecki (1977) heard it calling on Tortola in 1976 and regarded it as reliably reported from Virgin Gorda. We collected pellets from three caves on Guana Island during the 1980s. These contained mostly cockroach exoskeletal material, beetle elytra, and a few lizard bones. The pellets were stored at The Conservation Agency (TCA) in Rhode Island for several years (Norton et al. 1989), but they were transferred to Greg Mayer, then at Harvard, and Michael Gibbons, at University of Massachusetts, Boston, for specific bone identifications, and—along with a lot of bones from archeological sites—they subsequently disappeared.

At 6:37 p.m. on 4 July 1986, a screech-owl flew right across the Anegada House lawn, directly in front of me. I wrote, "plump, heavy head; classic owl. . . . Fluttered then glided." No one has ever seen or heard of a screech-owl on Guana Island since, and there has been no subsequent accumulation of pellets in any of the caves. I admit, as of 2003, that it has been years since anyone looked, and it is high time to look again.

Our screech-owl was presumably the Virgin Islands subspecies, *Otus nudipes newtoni,* but, without a specimen in hand, that could not be proved. The status of this owl on any of the Virgin Islands is unknown, but the subspecies would certainly seem to be rare and endangered.

Short-eared Owl, *Asio flammeus.* This mid-sized, usually terrestrial owl is regarded as a vagrant in the Virgin Islands and as an uncommon resident on Puerto Rico (Raffaele et al. 1998, 324). Mirecki (1977) had a probable report of this species on Anegada in the 1970s. On 26 October 1991, birders Christina Leahy and Cora Brayton, a veterinarian studying our flamingos, found one on Guana Island. Over the next several days, they were able to relocate this bird several times and collected feathers from its resting sites on the ground. Short-ears are crepuscular and fly low over open areas. I doubt one would find much to eat on Guana because they especially like mice, of which we fortunately have few or none.

Chuck-will's-widow, *Caprimulgus carolinensis.* This bird is a rare resident, September to May, in the Virgin Islands (Raffaele et al. 1998, 326). Mirecki (1977) had only old sight records. Occasional members of this family, Caprimulgidae, called "goatsuckers"—based on ancient superstition from Europe—or "nightjars," have been spotted on Guana over the years, typically at twilight, and not well enough to identify. However, on 26 October 1994 Fred Sibley flushed out a female in the thick woods south of White Bay, and, on 28 October 1994, a male. He could identify them to species, thus solving a mystery. Chuck-will's-widows are probably more regular here than our meager evidence indicates, but one would have to hunt for them to prove that.

Antillean Mango, *Anthracothorax dominicus.* Raffaele et al. (1998, 334) say: "It is increasingly rare among the Virgin Islands and has been extirpated from most of them." Mirecki (1977) could find none in the BVI in 1976. However, they were present on Beef Island in 1980 (Lazell 1980), and Rob Norton found them on Anegada 18 June 1988 (Norton et al. 1989). In March and April 1982 there was a pair on Guana, hanging out between Queen's Terrace and the laundry. I saw them regularly and pointed them out to many people. I only identified them because of the female, who has pale underparts and white-tipped outer tail feathers: No other hummingbird on Guana looks like that. The males are big hummers, similar to the abundant green-throated carib (below), and—as with both sexes of the carib—look plain black in most light conditions. My sight records did not make it into Norton et al. (1989), and, over the years, no ornithologist really believed me. Then, in 1994, along came Wayne Arendt, who did the most exhaustive land-bird survey ever attempted on Guana. He found one right where I had left them years before. Nobody doubts him.

It is believed that the Antillean mango is victimized and driven away by the carib hummers. Indeed, Wayne's bird was being harassed by caribs. The carib may be a more adaptable human commensal than the mango. If so, Guana's areas of forested terrain, unmodified by human edifices, may be an important refuge for this species. I have seen females from time to time in the vicinity of lower Quail Dove Ghut.

Green-throated Carib, *Eulampis holosericeus*. This is an abundant breeding resident. A big, essentially black, hummingbird with a long, decurved bill, both sexes show a flash of iridescent green on the lower throat in good light (good luck!). The range of this species extends east and south through the Lesser Antilles and west to Puerto Rico, where it gets rare once away from areas east of San Juan. Mirecki (1977) noted it on Guana, and Jonnie Fisk listed it every year.

In a series of papers, Bleiweiss et al. (1997) and Bleiweiss (1998a, 1998b, 1998c) used Guana Island caribs and crested hummingbirds (below) to elucidate the relationships within this spectacular family using molecular biology: DNA.

Antillean Crested Hummingbird, *Orthorhynchus cristatus*. This is an abundant breeding resident, a tiny fellow with a conspicuous crest and a straight bill. Mirecki (1977) listed it for Guana, and Jonnie Fisk recorded it every year. So have we.

Belted Kingfisher, *Ceryle alcyon*. This is a regular visitor to Guana, resident but not nesting in the West Indies. Recorded in most years, individuals hang out at the Salt Pond or most anywhere along the coast. Kingfishers are blue gray, crested, and banded across the chest; the male has one blue gray band, and the female has one of those and, below it, a russet brown band.

Caribbean Elaenia, *Elaenia martinica*. Although an inconspicuous little gray bird, this is an abundant breeding resident. Jonnie Fisk did not record it her first year, 1952, but she got it in every other year; we have never missed it. It is surprising that Mirecki (1977) did not record it on Guana, but we suspect that was an oversight (Mayer and Chipley 1992). Elaenias are small flycatchers, with gray with pale wing bars and a very slight crest; they can display a narrow, cream-colored crest stripe when agitated. One banded in October 1996 was recaptured in October 2001, but the record is the one caught

in October 2003, seven years after the original banding. Our species occurs throughout the Puerto Rico Bank and Lesser Antilles.

Gray Kingbird, *Tyrannus dominicensis*. This is an abundant breeding resident that makes itself highly conspicuous, perching and calling from high in the trees. Locally called "chickery," this is the "pi-pi-rit" of the French islands and the *"el Pitirre"* in Spanish (the former name of the Journal of Caribbean Ornithology). Those names commemorate its loud calls. Kingbirds, classic members of the tyrant flycatcher family Tyrannidae, attack hawks, cats, and anything else that bothers them. Their superb aerial maneuvering, honed by catching insects on the wing, makes them immune to reprisals.

Caribbean Martin, *Progne dominicensis*. Raffaele et al. (1998) say these are present from January to September. They are often overlooked; it might well nest on Guana. Martins have nested on Great Camanoe and unspecified Dogs (Mirecki 1977). These are members of the swallow family, Hirundinidae; they are almost always seen in flight, rather high up, and moving fast. They are larger than our other swallows and look plain black.

Tree Swallow, *Tachycineta bicolor*. This is an uncommon migrant in the Virgin Islands (Raffaele et al. 1998, 366). Mirecki (1977) did not list it for the BVI, and I can find no previously published record for these islands. On 22 October 1999, Fred Sibley and his crew spotted one on Guana. Because this species is dark above and white below and has a shallowly notched, not deeply forked, tail, it is unmistakable.

Barn Swallow, *Hirundo rustica*. This is a common migrant in the Antilles, with a nearly worldwide distribution. Our paucity of records is no doubt a reflection of lack of looking skyward for small, fast, high-flying birds with rusty throats, white bellies, and long, streaming swallowtails. Sibley and his crew spotted five on 25 October 1995 and two on 21 October 1999 on Guana Island. A number were seen on several occasions in October 2003.

Swainsons's Thrush, *Catharus ustulatus*. (See plate 31.) This is a rare migrant in the western

Caribbean, never previously recorded east of Cuba and the Bahamas (Raffaele et al. 1998, 376). On 9 October 2000 the Sibley team, aided by Stephen Durand from the Division of Forestry and Wildlife, Dominica, misted-netted, weighed, measured, photographed, and banded an adult Swainson's thrush on Guana. The Boals caught and banded one on 16 October 2003 and subsequently saw another, unbanded. This is a north temperate to boreal breeding species in North America. To date it is our most remarkable neotropical migrant range extension. The fact that two individuals were full adults, not youngsters of inexperience, argues for Guana being on the regular migratory route for some Swainson's thrushes.

This species, and its close relatives in the genus *Catharus,* has been the subject of cutting-edge studies in evolution and migration. Outlaw et al. (2003) developed a molecular phylogeny that indicates Swainson's thrush is the oldest migratory member of the complex, but only going back about four million years. Tropical nonmigratory resident species have evolved at least three times independently. I have often wondered how migratory species that use a magnetic-compass sense survive magnetic reversals, which have occurred many times in Earth history. They sense the magnetic field but cannot tell north from south (Stokstad 2004). All right, but how do they avoid hopeless directional confusion if and when they cross the equator? Cochran et al. (2004), working with Swainson's and gray-cheeked *(Catharus minimus)* thrushes, found the birds set their mental compasses each evening by the position of sunset. As they get closer to the equator this becomes ever closer to due west. They migrate at night, so they just have to avoid being confused by sunrise the next morning.

Northern Mockingbird, *Mimus polyglottos.* Once a nesting resident species, this conspicuous, familiar bird has disappeared from Guana Island. It is common over on Beef Island and Tortola. Mirecki (1977) recorded it in 1976 from Guana and another dozen of the BVI. However, it is a newcomer to these islands, first recorded in 1933 (Mirecki 1977). It is edificarian (a species dwelling around edifices or human constructions). Jonnie Fisk recorded it on Guana on her first list of 1952 and every year after. Most years there was only one pair, but from 19 February to 6 March 1954 there were a minimum of two pairs. When I arrived in 1980, the resident pair hung out on the flat around the Beach House and the garden; they seemed to especially like the grass tennis courts and mowed golf course. After hurricane Hugo in 1989 we could only locate one individual. There was still one (we can only presume the same one) present in the big bird year, October 1993. When Wayne Arendt, Fred Sibley, and the birding team arrived in October 1994 the last mockingbird was gone. The tennis courts are clay today; much of the White Bay flat has grown up in scrub jungle; the habitat may not be as hospitable to mockingbirds now, even if an occasional bird pops over from Beef Island or Tortola to consider the real estate. Mayer and Chipley (1992) wrote soon enough to avoid this possible example of turnover in our avifauna, but mockingbirds could not have been here for long anyway. They may come back.

Pearly-eyed Thrasher, *Margarops fuscatus.* This is certainly the most conspicuous—and probably the most abundant—bird on Guana Island. Pearly-eyes are bold and saucy, attacking not just human food and drink but other birds—especially eggs and nestlings—too. On 12 July 1984, Chip watched one pecking pieces out of the head of a mangrove cuckoo that it apparently killed: The cuckoo had thrasher feathers in its bill. On 17 July 1987, Chip watched a pearly-eye attack and kill an adult male saddled anole lizard. On 27 July 1987, one of three bridled quail-dove nests Chip had under observation was raided and its egg or eggs destroyed. Chip regards pearly-eyes as the major impediment to quail-dove reproductive success. The Sibley team has to constantly guard their nets against thrasher predation on captured birds. Around the dining area and in the garden, where they punch holes in the mangos and papayas, they can be a real nuisance. When David Hill, founder of the Rare Animal Relief

Effort (RARE), and his wife Charlotte came birding in 1993, he became concerned that the density of pearly-eyes might endanger other birds. We contacted Wayne Arendt, U.S. Forest Service, who is the authority on pearly-eyes in Puerto Rico, where they have attacked and killed nestling parrots. Wayne's study of 1–11 October 1994 is an ornithological landmark for Guana, but—despite the abundance of pearly-eyes—he found no evidence that other species were being adversely affected. Joe Wunderle, also of the U.S. Forest Service, repeated the survey in 2001 with similar results.

Hunt et al. (2001) studied the molecular biology of pearly-eyes and their relatives on other West Indian islands; although not involving Guana Island birds, this work elucidates the relationships among this classic insular radiation.

White-eyed Vireo, *Vireo griseus*. Described as "very rare" in the Virgin Islands (Raffaele et al. 1998, 388), this species was not listed by Mirecki (1977) or Jonnie Fisk. Our only record of this migrant or winter resident is an individual spotted by Charles Bartlett on 17 October 1993, and verified by Debbie Paul and Kerry Sherred. Roy (1996) tabulates these as rare in autumn and winter in the BVI. Closer attention paid to little gray birds might reveal a lot more. It has bold white wing bars, yellowish flanks, and a really flashing white eye accentuated by "spectacles" of white, and—for a vireo—a rather heavy bill.

Yellow-throated Vireo, *Vireo flavifrons*. This is an uncommon migrant and possible winter resident (Raffaele et al. 1998, 391), but it is easily overlooked because it likes to glean insects high in the forest tree canopy. Fred Sibley spotted one on Guana Peak, 12 October 1994, and Clint and Tracey Boal netted, weighed, measured, and banded another, 15 October 2003. This species has white wing bars, inconspicuous "spectacles," a drab iris, and—most notably—a bright yellow throat and upper breast.

Red-eyed Vireo, *Vireo olivaceus*. This neotropical migrant is a vagrant in the Virgin Islands. The Sibley crew, however, have mist-netted, weighed, measured, photographed, and banded

seven individuals on Guana Island spread through eight Octobers, 1994–2001, and the Boals banded one on 17 October 2003.

Black-whiskered Vireo, *Vireo altiloquus*. Today this is a rare breeding resident in the BVI. Jonnie Fisk logged it as "abundant" on Guana before the mangrove clearance of 1954. Mirecki (1977) found it only on Beef Island. On 14 July 1985, Robert Chipley found one on Guana, which several of us got to see (Norton et al. 1989). Rob Norton found it regularly on Anegada, 1980–88 (Norton et al. 1989). The Sibley team recorded three individuals on Guana, 12 October 1994. The black whisker marking identifies this vireo, which often perches high and conspicuously when singing. Chip's July bird was doing just that, but in October they have ceased singing for the year and are much harder to find.

Golden-winged Warbler, *Vermivora chrysoptera*. This is a neotropical migrant from North America that rarely seems to pass through the Virgin Islands. The Sibley team mist-netted, weighed, measured, photographed, and banded a gorgeous adult on 14 October 1997 and saw a juvenile on 18 October 1997. The broad, bright yellow wing bars are distinctive.

Nashville Warbler, *Vermivora ruficapilla*. A rare vagrant in migration and winter anywhere in the West Indies, one of these was associated with two prairie warblers on the west side of Guana Peak when spotted by Fred Sibley, 12 October 1994. The white eye-ring sets this species apart.

Northern Parula, *Parula americana*. This is a common migrant and winter resident in the Virgin Islands first recorded on Guana by Jonnie Fisk, 1953–55. None were noticed again until the Sibley team saw one on 28 October 1994 and subsequently caught and banded individuals on 19 and 22 October 1996. This is a dark warbler with a yellow throat and breast, wing bars, but an incomplete eye-ring. The West Indian habitat of this species is critical to its survival because it is only known to winter in these islands (Raffaele et al. 1998, 396).

Yellow Warbler, *Dendroica petechia*. Jonnie Fisk recorded this species annually, 1952–55. The mangroves were clear-cut and the previous swamp area sprayed—no doubt with DDT—in 1954; her 1955 record would be the last for many years. Remarkably, Mirecki (1977) did record it on Guana in 1976, but no one else could find it until 1993. Since then, it has been recorded in most years again, at least in October.

There are two very distinct kinds of yellow warblers on Guana and elsewhere in these islands. There is the resident, breeding form, *Dendroica petechia cruziana,* and the North American migrant, *Dendroica petechia petechia.* They are very difficult to separate in the field, but it is easy once they are in the hand: The resident has rounded wings with the first primary feather shorter than the second; the migrant has longer wings with the first primary feather the longest of all. The resident form loves mangroves, where it typically lives and nests. The migrant might be anywhere.

We believe Jonnie Fisk probably recorded resident breeding birds because she associated them with mangroves. We can hope Mirecki (1977) did too, but there were few (and small) mangroves in 1976. Even in 1980 I believed there was no suitable habitat despite considerable regrowth. The birds seen from 1993 to 1997 could not be identified to subspecies.

At last, on 12 October 1999, the Sibley crew netted, weighed, measured, photographed, and banded one: a beautiful adult of the short-winged resident form. The native had returned (see Mayer and Chipley 1992 for more on the significance of this). We still do not know if they are nesting, but the nesting season is March to July, and our mangroves should be watched then.

Chestnut-sided Warbler, *Dendroica pensylvanica.* This is a rare migrant or winter resident (Raffaele et al. 1998, 397). Fred Sibley saw one on 26 October 1994 and another on 18 October 1997. This species has a white eye-ring but yellowish wing bars; the underparts are gray.

Magnolia Warbler, *Dendroica magnolia.* This is a rare migrant in the Virgin Islands (Raffaele et al. 1998, 398). The Sibley team mist-netted, weighed, measured, and banded one on 20 October 1996. The Boals banded one on 12 October 2003. There are conspicuous white markings across the midsections of the tail feathers, and the rump is yellow.

Cape May Warbler, *Dendroica tigrina.* This is an uncommon migrant and winter resident in the Virgin Islands (Raffaele et al. 1998, 198), but Jonnie Fisk recorded it in 1954. It went unnoticed again for 40 years, until several were seen at the bird feeder on 16–17 October 1994. One of these was mist-netted and banded; it stayed at the feeder for four days, aggressively repelling hungry bananaquits. Cape May warblers have been studied on their wintering grounds in the Dominican Republic by Latta (2002). They occupied three vegetation zones; the preferred zone was pine forest at relatively high elevations where fruit was available all winter. Dry forest was less desirable unless individuals could find and defend trees with a lot of hemipteran insects that produce honeydew—a favorite food. Desert thorn scrub was "always suboptimal," lacking both fruit and honeydew, and thus necessitating insectivory. In this habitat, which corresponds to most of Guana Island, "conditions become increasingly difficult during the late-winter dry period." We cannot expect Cape May warblers to do well on Guana. The heavy breast striping and yellowish rump are an identifying combination.

Black-throated Blue Warbler, *Dendroica caerulescens*. This is a rare migrant or winter resident in the Virgin Islands (Raffaele et al. 1998, 399), previously recorded in the BVI on Anegada in 1970 (Mirecki 1977). Fred Sibley spotted one on Guana on 26 October 1994, and the team caught and banded two in October 1996. The Boals saw one in October 2003. The strong sexual dimorphism and great differences between adults and young birds makes all but adult males in breeding plumage difficult to identify without a bird guide at hand. Look for a white patch at the base of the primaries on the otherwise dark, unbarred wing. The black-throated blue has been the subject of novel, high-tech research with respect to its migration. Wester et al. (2002) provide a broad overview. Rubenstein et al. (2002) assayed stable isotopes of carbon and hydrogen in the feathers of about 700 of these warblers. The birds molt and grow new feathers at or near their breeding sites, which extend from Michigan to the Canadian Maritimes and south along the Appalachians to Georgia. The isotopic ratios in the feathers conform to those in the insects the birds ate, and these in turn correspond to the ratios in the plant tissues the insects ate. The hydrogen isotopic ratios in the plants vary geographically because they vary in rainfall. Hobson (2002) maps bands of relative abundance of deuterium (the common hydrogen isotope), but the data do not greatly strengthen the case put forward by Rubenstein et al. (2002). For example, the two southernmost bands that could include breeding black-throated blues are each based on a single station. The station that includes northern Georgia's putative band seems to be located somewhere around Dallas, Texas (the map is small, contains no state lines, and Hobson does not tell us). The station that includes the Carolina uplands is located at Cape Hatteras and is believed (imagined?) to extend in a grand curve west and south to the Big Bend. Within the entire remainder of the warbler's range as mapped by Rubenstein et al. (2002), Hobson shows only four other isotope-reporting stations. That is two fewer than the mapped departure points of the birds.

Anyway, Rubenstein et al. (2002) concluded that Black-throated Blues ". . . appear to segregate on the wintering grounds with respect to breeding latitude." They found that the northernmost breeders wintered the farthest west, mostly in Cuba. Their easternmost station was on Puerto Rico, well west of Guana Island. Will the Guana birds turn out to be the southernmost breeders, from north Georgia? The race is on: Will we get feathers from Guana before the high-tech team gets isotopic ratios from north Georgia—or at least some place more relevant than Texas?

Blackburnian Warbler, *Dendroica fusca*. This is a very rare migrant in the Virgin Islands (Raffaele et al. 1998, 400). Mirecki (1977) did not list it, and I can find no published record for the BVI. On 9 February 1959, Jonnie Fisk recorded it on Guana Island in sea grapes. The Boals saw one in October 2003. These warblers have broad, flashing-white wing bars and a pair of bold, light, dark-bordered stripes on their backs.

They have orange or yellow head markings and throats. They are unmistakable.

Prairie Warbler, *Dendroica discolor*. This is a fairly common migrant and winter resident in the Virgin Islands (Raffaele et al. 1998, 403; Mirecki 1977). Jonnie Fisk recorded it in 1954, Wayne Arendt found them right at the Queen's Terrace in October 1994, and the Sibley team also saw them again in October 1996. They have not caught one yet. This warbler wags its tail similar to a phoebe. It has bold, dark streaks on its yellow breast and is white under the tail, and whereas it has inconspicuous white spots in the midsection of the tail, it does not have white outer tail feather tips.

Palm Warbler, *Dendroica palmarum*. This is a rare migrant and winter resident in the Virgin Islands (Raffaele et al. 1998, 403) and was not recorded by Mirecki (1977) for the BVI. Rob Norton got the first BVI record on Tortola, 20 April 1980 (Norton et al. 1989). I spotted one on the road passing the garden, near the dump, on Guana on 21 October 1995. This is a walking warbler that flicks its tail. The underside of the tail—easily seen as it flicks up—is yellow; the breast streaking is weak; and the outer tail feathers flash white tips.

Blackpoll Warbler, *Dendroica striata*. This is a classic neotropical-passage migrant in the Virgin Islands (Raffaele et al. 1998, 404) but it is not recorded for the BVI by Mirecki (1977) or Roy (1996). Until very recently, this, our most common October migrant, was virtually unrecorded in the Virgin Islands; was this an oversight or has their migratory route shifted? Wayne Arendt first recorded small flocks of four to eight individuals in 1994; they pass through, apparently heading south. The Sibley crew began catching and banding them in 1995 and by 2000 had banded 135. They got 83 in 1998 but nary a one in 1999. Between 2000 and 2003 the crews banded 6–19, averaging 12, each year. A. K. Davis (2001) studied blackpolls at their migration departure area in Nova Scotia, checking especially for fat deposition. They cannot eat during overwater migration and must have enough fat—fuel, quite literally—to make the trip. We believe most take off from capes jutting into the Atlantic (e.g., Nova Scotia and Cape Cod) and fly all the way over the ocean. A. K. Davis (2001) found they had insufficient fat reserves to make the whole distance to South America. This means that stopovers in the Antilles, on islands such as Guana, may be critical to their survival as a species. This is a very difficult bird to identify in the field because the sexes are not so dimorphic and the immatures, which resemble females, also resemble several other species. There are white wing bars, but no other distinctive markings, so a process of elimination helps.

Black-and-white Warbler, *Mniotilta varia*. This is a common overwintering species, arriving as early as August and heading north again in April (Raffaele et al. 1998, 407). Mirecki (1977) listed it as a migrant in the BVI. Jonnie Fisk recorded wintering birds in 1954 and 1955. Wayne Arendt recorded one female on Guana 40 years later, in October 1994. The Sibley team has caught and banded six individuals over the span of 1994–98. This species is well named: It really is black and white; the crown is boldly striped.

American Redstart, *Setophaga ruticilla*. This is a common, nonbreeding bird in the Virgin Islands (Raffaele et al. 1998, 407), regarded as "transient" by Mirecki (1977). Jonnie Fisk recorded redstarts in each winter of 1953 through 1955. Wayne Arendt logged two in October 1994, and the Sibley team caught and banded one on 28 October 1998. Redstarts may be gray brown to nearly black above with a broad orange (male) or yellow (female) band through the otherwise dark wing along the bases of the primaries and secondaries. The midsection of the tail is similarly colored.

Worm-eating Warbler, *Helmintheros vermivorus*. Raffaele et al. (1998) regard this species as rare in the Virgin Islands generally. Roy (1996) tabulates it as rare and present only in winter and spring. Guana's only record to date is an adult netted, weighed, measured, photographed, and banded by the Boals on 18 October 2003. Look for this olive brown warbler with a rich buff breast and golden ocherous crown bearing four bold black stripes in the midstory of woods. It is likely to be a winter resident.

Ovenbird, *Seiurus aurocapillus*. This is an uncommon nonbreeding resident in the Virgin Islands (Raffaele et al. 1998, 407), listed as a migrant by Mirecki (1977). Fred Sibley saw the first one on Guana 28 October 1994, and the team caught and banded three in October 1998. This is a terrestrial warbler, olive in color above, with a white breast sporting big black spots. If you can see it, the crown has an orange, black-bordered, stripe.

Northern Waterthrush, *Seiurus noveboracensis*. This is a fairly common nonbreeding species that may be found in any month (Raffaele et al. 1998, 409; Mirecki 1977). Jonnie Fisk recorded it first on Guana in 1955. Wayne Arendt found some in North Bay Woods in October 1994. The Sibley team finds waterthrushes regularly down on the flat near the Salt Pond or over near the dump. Between 1994 and 2000, they caught and banded five. Waterthrushes, despite the name, are big terrestrial warblers. They are dark brown above and pale tan or buff below with a prominent, buff eyebrow stripe and dark streaks on the throat. They bob and teeter similar to spotted sandpipers and palm warblers.

Kentucky Warbler, *Oporornis formosus*. This is a very rare nonbreeding resident in the Virgin Islands (Raffaele et al. 1998, 410). On 20 October 1996 the Sibley team mist-netted, weighed, measured, and banded one on Guana Island. The Boals banded another on 20 October 2003. Roy (1996) tabulates it as rare in winter only. Adult males have black facial markings, but these are reduced or absent in juveniles and females. However, in any plumage, this is a dark olive green warbler with a yellow breast and a yellow eye-ring or "spectacle." There are no wing bars, tail spots, or other markings.

Hooded Warbler, *Wilsonia citrina*. This gorgeous little bird is rare in the Virgin Islands (Raffaele et al. 1998) and tabulated by Roy (1996) as found in winter only. Both Fred Sibley and I believe we have seen them on Guana, but I would not have included this species had the

Boals not mist-netted, weighed, measured, banded, and photographed an adult male on 17 October 2003. Only males develop the full black hood, but all hooded warblers are olive above and bright yellow below, have dark lores (the area between the eye and the bill), and have a lot of flashing white in the lateral tail feathers.

Bananaquit, *Coereba flaveola sanctithomae*. This abundant, breeding, year-round resident belongs to a complex of geographically replacing forms that span the Antilles and occupy parts of Central America and South America (Raffaele et al. 1998, 415). Biaggi (1983) differentiates our subspecies in the Virgin Islands, *Coereba flaveola sanctithomae*, from *Coereba flaveola portoricensis* of Puerto Rico: our birds have brighter yellow bellies contrasting with paler flanks. These are modified warblers with decurved bills that they use to probe and pierce flowers for nectar; they also eat insects and spiders and, especially, feed them to their nestlings.

Our bananaquits have been intensively studied by the Sibley team each October since 1994. Typically, the birds are netted at the dining area and down on the flat. They are weighed and measured and can be sexed, as adults, by wing length. Color bands provide a code for year of first capture and site. The first and most significant fact emergent from this study is a remarkable difference between years in terms of ages, sexes, feeder use, and general activity—apparently correlated with rainfall and the positive responses of vegetation and insects to the rainfall.

Fred and Margaret (Peggy) Sibley have worked up their data for the span of 1994 through 1997; these reveal the variations. In the first year, October 1994, a simple Lincoln index indicated 120–150 bananaquits came in to the feeder by Dominica House. This is not a population density estimate because we do not know how far they came in from, so we do not know the area sampled. The sex ratio was balanced. This was a very dry year, following nearly a decade of relative drought.

In 1995, they got the same mark–recapture estimate, 120–150. The sex ratio remained

BEWARE THE BANANAQUIT

The bird feeder by the dining terrace is typically stocked with sugar, which bananaquits and several other species (including the occasional anole lizard), come to eat. Adorable as these pretty little birds are most of the time, they can be a nuisance around the dining area, hopping in food (especially cut fruit) and perching on the rims of glasses full of fruity drinks. The problem is not so much that they imbibe. The problem arises when, satiated, they reverse direction on the rim preparatory to departure and decide to jettison a bit of excess weight before flying off.

balanced, and there was about a 25% survival rate from 1994. The long drought broke this year and conditions were relatively wet during October. However, this would not have immediately affected the population.

In 1996, relatively few birds came to the feeder. Conditions remained wet and lush, and there had been time for these to affect the population. Two-thirds of the 42 new birds banded were the young of the year, and the sex ratio of the few adults that could be sexed was 3:4, males to females. The sample size was, however, too small to significantly differ from 1:1.

In 1997, conditions continued moist and lush, and about 90 birds were marked. This was close to the mark–recapture estimator of 90–100 total. The sex ratio was more than 3:1 males over females, and most birds were adults (84 total). One bird remained from 1994, 16 from 1995, and five from 1996. If the majority of females were busy elsewhere and not coming to the feeder, then the total 1997 population might have been about 160–180 birds—up a little from the drought years but pretty close.

One can set up the following extended hypothesis: In hard times, the sugar feeder enables the survival of about 120–150 bananaquits. When the drought broke, and the effects of its break worked through the population in 1996, most of the adult birds were busy getting in-

sects to feed nestlings and paid little attention to the sugar feeder. Most of the birds that did come to the feeder were immatures. By 1997, the second full year of good times, females may have been concentrating on catching insects for nestlings, but males could afford to lollygag at the sugar feeder. A new population level, at 160–180 total, may have become established. In 1997, 63 new birds were banded. It will be fascinating to see the 1998–2003 data because the good times have continued. Fred Sibley reports (in litt. 23 January 2004) that the predicted pattern is holding up.

Population turnover is high, with most birds not surviving more than one year. The basic 25% survival from one year to the next could still be true of 1997, if we assume that there were a lot of females (as many as males) but they were not coming to the feeder. Soon we will know more. In the meantime, the Sibley team banded 31 bananaquits down on the flat over the four years. One of these was subsequently caught up at the feeder, but no bird banded at the feeder was ever caught anywhere else on Guana Island. Most of our bananaquits seem to be permanent guests at the hotel. In October 2001, the Sibley crew recaptured a male bananaquit banded in October 1995 (that is six years!), two males and one female banded in 1997, six males and two females banded in 1998, and nine males banded in 2000—all dates in their respective Octobers. In October 2003 they got another banded six years before, also a male.

Rose-breasted Grosbeak, *Pheucticus ludovicianus*. This is a migrant and winter resident in the Virgin Islands, regarded as rare (Raffaele et al. 1998, 422). Fred Sibley located two individuals on Guana Peak trail on 22 October 1994. Exactly a year later, 22 October 1995, the Sibley team mist-netted, weighed, measured, photographed, and banded a lovely male near the Salt Pond. Another individual was present near the dump on 21 October 1998, and another was banded on 17 October 2003. These are big finches with massive, conical bills. The male is largely black above and white below, and it has a bright rose pink triangle on the breast. The female is boldly striped and streaked from crown to tail.

Blue Grosbeak, *Guiraca caerulea*. This is a very rare migrant or winter resident in the Virgin Islands (Raffaele et al. 1998, 422). Fred Sibley located an immature male on Guana, at the garden, on 10 October 1997, and presumably the same bird on 12 October 1997. Males are blue, and females are brown; the clincher is the heavy, conical, finch bill in combination with reddish brown wing bars in any plumage.

Indigo Bunting, *Passerina cyanea*. This is a common migrant and winter resident regionally (Raffaele et al. 1998, 423). On Guana Island a young male was netted and banded on 22 October 1995, and others were seen in October 2003. One was banded on 16 October 2003. Adult males in the all-blue breeding plumage might be seen in their northward, spring migration, but in fall and winter buntings here will be basically brown; the males have some leftover blue in wings and tail.

Black-faced Grassquit, *Tiaris bicolor*. This is an extremely abundant, nesting, year-round resident on Guana Island. Look for them perched on the rungs between chair legs in the dining area—or most everywhere else. Both sexes are basically little brown birds—sparrows to most of us—but males do have sooty faces. Grassquits often get caught in the lowest span of mist nets set for other birds and can attract our big native predators: iguanas and snakes. I have watched iguanas gulp down unfortunate grassquits on several occasions and recorded, on 13 October 2000, one of our common snakes trying to eat one out of a net. The results were a mess. The snake killed the bird almost immediately, indicating that the snake produces a saliva venomous to at least birds (and baby iguanas: see above), but the snake could neither extricate the dead bird from the mesh nor swallow it mesh and all. Eventually snake, bird, and net got so wound up with each other that it took us 20 minutes to untangle them all, and in the process the mist net was damaged. In October 2001, the Sibley crew recaptured a female grassquit banded in October 1997 and two males and one female banded in October 2000. New age records in October 2003 were of two males banded five years previously, thus making them at least six years old.

Lesser Antillean Bullfinch, *Loxigilla noctis*. According to Raffaele et al. (1998, 428), this black finch "expanded its range westward to St. John and St. Croix in the Virgin Islands in the 1970s where it is now locally common." Mirecki (1977) provides details of its colonization of the BVI: in 1972 "it was found on Norman and Peter Islands." However, in 1976 none could be located, indicating to Mirecki (1977) that "those found were migrants or that the colonization was short-lived." I know of no BVI records again until 20 October 1998 when the Sibley team netted and banded a female on the flat. This finch is black (male) or brown (female and immature), pretty much all over. Males have an orange-brown chin spot and all others have orange undertail coverts. Look for this bird around the buildings and the bird feeder: It likes human company.

Bobolink, *Dolichonyx oryzivorus*. This is a classic neotropical-passage migrant, passing through typically in small flocks from August to December (Raffaele et al. 1998, 433). A bird of open fields, it is likely to be present only on the White Bay flat, where birds were seen on several occasions—even by me—in October of 1995, 1997, 2000, and 2003. Bobolinks out of breeding plumage are buffy birds with sharp-pointed tail feathers.

There are two ornithological mysteries from Guana. First, Jonnie Fisk recorded two species of grassquits on the island in 1952 and 1953: the black-faced (no mystery at all) and the blue-black. The blue-back grassquit is a very small finch confined, in the West Indies, to the far southern Lesser Antillean island of Grenada (Bond 1947; Raffaele et al. 1998, 424). The males are all black; the females all brown. At least the male could not be mistaken for a black-faced grassquit. What could she have been seeing? The only essentially plain, all-black bird that might fit the bill is the larger

TABLE 25

A Chronicle of Bird Species Listed in Order of First Recorded Sighting on Guana Island

	1952	1953	1954	1955	1957	1958	1959	1961	1976	1980	1982	1985	1986	1987	1988	1989	1990	1991	1993	1994	1995	1996	1997	1998	1999	2000
1 Brown Pelican	×	×	×	×	×	×	×	×	×	×				×					×	×	×	×	×	×	×	×
2 Brown Booby	×	×	×	×	×	×	×	×	×	×				×					×	×	×	×	×	×	×	×
3 Bananaquit	×	×	×	×	×	×	×	×	×	×				×					×	×	×	×	×	×	×	×
4 American Kestrel	×	×	×	×	×	×	×	×	×	×				×					×	×	×	×	×	×	×	×
5 Black-faced Grassquit	×	×	×	×	×	×	×	×	×	×				×					×	×	×	×	×	×	×	×
6 Pearly-eyed Thrasher	×	×	×	×	×	×	×	×	×	×				×					×	×	×	×	×	×	×	×
7 Green-Throated Carib	×	×	×	×	×	×	×	×	×	×				×					×	×	×	×	×	×	×	×
8 Antillean Crested Hummingbird	×	×	×	×	×	×	×	×	×	×				×					×	×	×	×	×	×	×	×
9 Scaly-naped Pigeon	×	×	×	×	×	×	×	×	×	×				×					×	×	×	×	×	×	×	×
10 Zenaida Dove	×	×	×	×	×	×	×	×	×	×				×					×	×	×	×	×	×	×	×
11 Common Ground Dove	×	×	×	×	×	×	×		×	×				×					×	×	×	×	×	×	×	×
12 Little Blue Heron	×	×	×	×	×	×	×			×				×					×	×	×	×	×	×	×	×
13 Magnificent Frigate Bird	×	×	×	×	×	×	×			×				×					×	×	×	×	×	×	×	×
14 Smooth-billed Ani	×	×	×	×	×	×	×			×				×					×	×	×	×	×		×	×
15 Gray Kingbird	×		×	×	×		×		×	×				×					×	×	×	×	×	×	×	×
16 Spotted Sandpiper			×	×			×							×						×	×	×	×	×	×	×
17 Red-tailed Hawk	×		×		×					×				×					×	×	×	×	×	×		×
18 Belted Kingfisher		×		×	×		×							×					×	×	×	×	×	×	×	
19 Yellow Warbler	×	×	×	×					×															×	×	
20 Red-billed Tropicbird	?	?	?	×	×		×			×				×					×	×	×	×	×	×	×	×
21 Great Blue Heron	?	?	?	×	×		×			×									×	×	×	×	×	×	×	×
22 Yellow-Crowned Night Heron	×									×				×					×	×		×	×	×		
23 American Oystercatcher	×	×		×	×					×				×					×	×			×			

(continued)

TABLE 25 (continued)

	1952	1953	1954	1955	1957	1958	1959	1961	1976	1980	1982	1985	1986	1987	1988	1989	1990	1991	1993	1994	1995	1996	1997	1998	1999	2000
24 Caribbean Martin	X	X								X										X						
25 Northern Mockingbird	X	X					X		X	X				X					X							
26 Caribbean Elaenia		X	X	X	X				X	X				X					X	X	X	X	X	X	X	X
27 Royal Tern		X	X	X					X	X				X					X	X	X	X	X	X	X	X
28 Lesser Yellowlegs		X	X		X														X	X	X	X	X	X	X	X
29 Parula Warbler		X	X	X					X											X		X				
30 American Redstart		X	X																	X				X		
31 Common Snipe		X	X	X	X		X																X			
32 Kildeer		X		X															X							
33 Green Heron		X								X									X	X		X			X	
34 Greater Yellowlegs		?			X														X							X
35 Common Moorhen		X																	X							
36 Clapper Rail		X																	X							
37 Bridled Quail-Dove			X	X					X	X				X					X	X	X	X	X	X	X	X
38 Black-and-white Warbler			X	X															X	X	X		X	X	X	X
39 Cape May Warbler			X																	X	X					
40 Mangrove Cuckoo				X	X	X	X		X	X				X					X	X	X	X	X	X	X	X
41 Black-necked Stilt				X	X					X				X					X	X	X	X	X	X	X	X
42 Northern Waterthrush				X															X	X	X	X	X	X		
43 Osprey				X	X		X												X			X			X	
44 Prairie Warbler				X																X	X	X			X	
45 Great White Heron				?																						
46 Black-bellied Plover					X		X			X										X	X					
47 Sandwich Tern										X				X					X		X					
48 Wilson's Plover						X				X				X					X	X		X	X	X	X	X
49 Blackburnian Warbler							X																			
50 Ruddy Turnstone								X		X									X	X	X	X	X	X	X	X
51 Pied-billed Grebe								X																		

No.	Species
52	Cattle Egret
53	Black-whisked Vireo
54	White Cheeked Pintail
55	Roseate Tern
56	White-tailed Tropicbird
57	Puerto Rican Screech Owl
58	Antillean Mango
59	Common Tern
60	Greater Flamingo
61	Yellow-billed Cuckoo
62	White-crowned Pigeon
63	Semipalmated Sandpiper
64	Bobolink
65	Barn Swallow
66	Laughing Gull
67	Snowy Egret
68	Semipalmated Plover
69	Yellow-throated Vireo
70	White-eyed Vireo
71	White-rumped Sandpiper
72	Western Sandpiper
73	Least Sandpiper
74	Caspian Tern
75	Whimbrel
76	Willet
77	Brown Noddy
78	Least Tern
79	Sooty Tern

(continued)

TABLE 25 (continued)

	1952	1953	1954	1955	1957	1958	1959	1961	1976	1980	1982	1985	1986	1987	1988	1989	1990	1991	1993	1994	1995	1996	1997	1998	1999	2000
80 Bridled Tern																			X							
81 Blackpoll Warbler																				X	X	X	X	X	X	X
82 Rose-breasted Grosbeak																				X	X	X	X	X		
83 Chestnut-sided Warbler																				X			X			
84 Black-throated Blue Warbler																				X		X				
85 Ovenbird																				X	X					
86 Indigo Bunting																				X	X					
87 Rock Dove																				X						
88 Nashville Warbler																				X						
89 Chuck-Will's Widow																				X						
90 Red-eyed Vireo																					X	X	X	X		
91 Palm Warbler																					X					
92 Solitary Sandpiper																						X	X	X		X
93 Blue-winged Teal																						X	X			
94 Great Egret																						X	X			
95 Magnolia Warbler																						X				
96 Kentucky Warbler																						X				
97 Stilt Sandpiper																							X			
98 Golden-winged Warbler																							X			
99 Blue Grosbeak																							X			
100 Lesser Antillean Bullfinch																								X		
101 Sharp-shinned Hawk																								X		
102 Peregrine Falcon																									X	
103 Tree Swallow																									X	
104 Swainson's Thrush																										X

SOURCE: Original data. Those that have the longest span with the most yearly records are listed before those first recorded in the same year. Note that some, such as the Blackpoll Warbler and Solitary Sandpiper, have been of regular occurrence since first—and very recently—recorded. This indicates that either things have changed or that earlier observers missed regularly occurring species. Detailed lists, begun in 1952, were not kept after 2000.

Lesser Antillean bullfinch. These bullfinches appeared in the BVI in 1972 and 1998. Could there be a pattern of bullfinch range expansion every 20–26 years or so, bringing birds from the Lesser Antilles east and north into our area?

Second, on 28 July 1984, Mary Randall came back from a walk out toward Muskmelon Bay and excitedly reported seeing a very different sort of bird. It was apparently a cuckoo, but huge—much larger than the ordinary mangrove cuckoo she knew well—and darker. It had bright red skin around the eye. She was describing a *Saurothera* species: a lizard-cuckoo. The next day Liao Wei-ping and I were working in Anegada House when I heard a loud avian voice that brought me right up out of my chair. It was the classic "kah-kah-kah . . ." of a cuckoo, but loud and booming, getting faster and louder until it abruptly stopped. I am a lizard hunter by profession, of course, and I had come to know the competition—lizard-cuckoos—over on Puerto Rico and Hispaniola to the west. I was sure I was hearing a *Saurothera*. Liao and I piled out into the shrubbery. We heard the bird several more times, and Liao got a look at it. Checking the bird guide, Liao was convinced it was a lizard-cuckoo. I was incredulous. Imagine my surprise, however, to discover years later that the Puerto Rican lizard-cuckoo, *Saurothera vieilloti,* is known from St. Thomas—which you can see from Guana Island—from an actual specimen in the Natural History Museum, London (Wetmore 1927, 425; Raffaele et al. 1998, 318).

Ornithology in the eastern Caribbean is in its infancy. We obviously do not yet fully know what species occur where. With the deeper questions of demographics, inter- and intraspecific turnover, life histories, and fundamental questions of ecology (e.g., diet), we have only a few tentative hints for a handful of species. It has only recently been demonstrated that ecosystem degradation in the tropics has deleterious effects on North American migrants far away on their breeding grounds (Kaiser 2003). There is a lot to do.

ONE MORE BIRD

On 7 and 8 June, 2005, David and Charlotte Hill revisited Guana after over a decade. They spotted groups of two to six white-winged doves (*Zenaida asiatica*), and showed them to Wenhua Lu and me. This species, common in the Greater Antilles and American Southwest, has been expanding its range eastward in recent years. Raffaele et al. (1998) reported it from the USVI and Clive Petrovic (pers. comm., 8 June 2005) says it reached Tortola, BVI, "about 10 years ago." It is very conspicuous: about the size of the zenaida dove but with a broad white band across the wing boldly visible both in flight and when perched. It was certainly not present in October 2004.

An Introduction to Invertebrates

Barry D. Valentine

PHYLUM PLATYHELMINTHES

Flat worms, flukes, and tapeworms. These are the most primitive animals with bilateral symmetry and a recognizable head end with a brain and, usually, eyes. The two parasitic classes containing the flukes and tapeworms are surely present in various mollusk and vertebrate hosts, but they have been studied on Guana only in lizards (Goldberg et al. 1998; see "Parasitism," part 3, above). The third class, Turbellaria, includes the planarians and relatives that are mostly free living or commensals. There are about a dozen orders of turbellarians, one of which, order Tricladida, includes the true planarians, which are featured in every general biology class. There are about 800 species of triclads: About 70 of these are marine, and the rest are divided between freshwater and terrestrial habitats. We have found terrestrial planarians in leaf litter from Sage Mountain on Tortola but have not yet found them on Guana. They should be present.

PHYLUM NEMATODA

Round worms and thread worms. Thousands of species of nematodes have been described and estimates of the final total reach half a million.

This diversity is compounded by reports of extraordinary abundance of individuals (e.g., 90,000 in one rotting apple). The surface of nematodes consists of a continuous firm cuticle that is flexible but not elastic, thus they can wiggle from side to side but cannot lengthen or shorten their bodies. Hoards of species are internal parasites of plants and animals, including Guana's anole lizards (Goldberg et al. 1998; see "Parasitism," part 3, above). Additional legions are free-living predators or scavengers in sand, soil, leaf litter, and any organic debris. Our largest Guana species, about 40 mm long, emerged from the body cavity of a roach; and uncountable microscopic species occur in everything from beach sand to rotten *Agave* stems. Few are identified.

PHYLUM MOLLUSCA

Bivalves, snails, cephalopods. Guana terrestrial snails (class Gastropoda) are being studied by Adam Baldinger, Museum of Comparative Zoology, Harvard, who has found 4 orders, 15 families, 24 genera, and 28 species. Most of these species are terrestrial, but a few are on rocky marine coastlines or are aquatic in brackish water. Fred Kraus, Bishop Museum, provides an account of the strictly terrestrial species below.

Slugs, which are snails with the shell internal or absent, have not been seen on Guana.

PHYLUM ANNELIDA

Segmented worms. Earthworms (class Oligochaeta) are present in beach sand, soil samples, and rotting vegetation. The smaller aquatic oligochaetes have not been investigated, but our two big terrestrial species have. Both belong to the family Megascolecidae. Sam James, Maharishi International University, came and studied our wonderful worms and included them in his work on the Greater Puerto Rico Bank (James 1991). One species, *Trichogaster intermedia,* is incredibly elastic; an individual can expand its length by a factor of four. These worms form individually into firm subspherical balls in dry weather, protected by a shellaclike glue. In rainy weather, they crawl and sprawl all over the better-vegetated portions of Guana Island. Our *Trichogaster intermedia* is endemic to the Virgin Islands of the Puerto Rico Bank, all east of Puerto Rico, similar to our "worm" lizard, *Amphisbaena fenestrata* ("Other Reptiles," part 5, above), which occupies the same habitat. They are abundant but have a very restricted, small geographic range in the world. They can attain 40 cm (16 in.) in length. Our second species, *Pontodrilus bermudensis,* is scarce, small, and bright pink. It barely attains 7 cm (<3 in.). Sam James found *Pontodrilus bermudensis* only in Quail Dove Ghut, but it probably also occurs in beach sand, its favored habitat all over the warmer, coastal regions of at least the Atlantic New World. These worms blithely violate the classic abundance : range relationship.

PHYLUM TARDIGRADA

Water bears. Tardigrades are microscopic animals with four short, fat body segments and four pairs of soft legs tipped with claws. Most of the several hundred species occur in mosses and lichens where they are active when the substrate is wet. During dry periods, they shrivel into a desiccated anabiotic state (suspended animation) that can survive for years. When water returns, they plump up and return to normal activity. In the desiccated state they can be blown about like dust particles, so it is impossible to predict where they will turn up. Tardigrades are almost certainly on Guana, but a good microscope (they are <1 mm long) and a lot of patience are needed to find them.

PHYLUM ONYCHOPHORA

Onychophorans, peripatus, walking worms. Thanks to Lianna Jarecki and Scott Miller, we have seen live onychophorans from Tortola and Virgin Gorda, respectively. Onychophorans may occur on Guana in constantly humid sites that are free from desiccation and flooding. These segmented soft-bodied creatures resemble elongate, velvety caterpillars, but they have a prominent pair of fleshy antennae and too many pairs of fleshy legs (more than 20). They are predators, capturing their prey with a sticky liquid that is squirted from a pair of papillae lateral to the mouth.

PHYLUM ARTHROPODA

Arthropods. This, the largest of all phyla, includes arachnids, crustaceans, myriapods, and insects. All have an inert external skeleton that covers a segmented body and jointed legs.

CLASS ARACHNIDA

Arachnids have four pairs of legs (three in some juveniles), a pair of pedipalps that resemble legs in some groups, a pair of pincherlike or fanglike jaws called chelicerae, no antennae, and the head and thorax fused to form a cephalothorax. Most have exclusively liquid diets aided by extraoral digestion and a pumping pharynx. Most are predators, but some mites feed on plant juices. The more familiar groups include scorpions, spiders, daddy longlegs, and ticks and mites. Of 11 orders, 7 occur on Guana, and 2 more are expected.

ORDER SCORPIONIDA

Scorpions have pedipalps that are larger than the legs and end with pinchers; the flexible abdomen has seven broad segments followed by six slender segments, the last with a sting and two venom glands. About 67 of the 800 world species are considered dangerously venomous, the rest are painful but not life threatening. Very toxic species can, in rare cases, cause death in from 1.5 to 42 h, so a victim should be monitored constantly for 48 h. Antivenin is available for some especially toxic species. Atropine and oxygen help overcome respiratory paralysis, and intravenous calcium gluconate (20 ml) helps to reduce convulsions. Some pain medications (morphine, Demerol) appear to increase the toxic effects of some venom. Adult scorpion lengths range from 12 to 180 mm— about 0.5 to over 7 in. There are three species of native scorpions in the Virgin Islands (Franck and Sisson 1980), and all occur on Guana. The largest, *Centruroides griseus,* attains about 80 mm (3 in.); it is ocherous brown with darker, horn brown dorsolateral stripes. It is agile and climbs well, and it is often encountered in trees. Its sting really does hurt, but it is not life threatening. Some *Centruroides,* however, are deadly. A somber little species, *Heteronebo yntemai,* attains about 25 mm (1 in.). It has big, boxlike pinchers and its sting rarely penetrates human skin (and does not hurt much). The smallest species, *Microtityus waeringi,* is dusky gray brown and has dark, slender pinchers. It reaches about 17 mm (0.75 in.) and is incapable of stinging humans. Specimens of all three of Guana's species have gone to AMNH, where they are being studied biochemically by Lorenzo Prendini.

ORDER PSEUDOSCORPIONIDA

Pseudoscorpions look like tiny scorpions except that they have very short, fat abdomens and no sting. On Guana the largest is less than 4 mm long. They occur in cryptic sites such as leaf litter and under loose bark, and they occasionally cling to other animals using the pinchers (chelae) of the large pedipalps, presumably as a means of dispersal. The chelicerae have silk-spinning glands, and the big pinchers of the pedipalps usually have one or two venom glands that help subdue prey; because of their small size, they are harmless to humans. About

2,000 species have been described, 20 of these from St. John, USVI. We have about 10 from Guana, but they have not been studied.

ORDER SOLPUGIDA

Almost 1,000 species of these fast, agile nocturnal predators are known. They occur in most of the warmer parts of the world, but at present only one 4-mm specimen is known from Guana. The single species known from St. John and Puerto Rico is less than 20 mm long. An African species gets up to 75 mm long. Solpugids have huge chelate chelicerae, and leglike pedipalps that are longer or thicker than the first three pairs of true legs. The chelicerae are capable of biting viciously but do not appear to contain venom glands; despite this, there are reports of delayed healing, perhaps owing to secondary infections.

ORDER ARANEIDA

Spiders are the most familiar arachnids on Guana, including four species of tarantula, a trap-door species that may use a modified abdominal pad to close its silk-lined burrow, and dozens of hunting and web-spinning species. Lazell provides an account of our species, below, which really illustrates how little we now know. The chelicerae of spiders are fanglike, not pinchers, and venomous; the pedipalps are smaller than the legs, and the abdomen lacks external segmentation and has three pairs of apical or ventral silk-producing spinnerets. All spiders are venomous, but most do not have the strength to pierce human skin. The few that do (the widows and recluses) inflict a painful bite that in very rare cases is lethal. There are no effective treatments for spider bites. Although all spiders share a common body plan, anatomical details, ecology, and especially behaviors such as web-spinning, prey capture, and courtship are astonishingly diverse. All spiders are liquid-diet carnivores, and females are usually larger than males and often cannibalistic.

ORDER AMBLYPYGIDA

Many people find the tail-less whip scorpions among the most repulsive of all arthropods, despite the fact that they are not venomous and are either incapable of or unwilling to bite. The nocturnal activity, flat body, extraordinary long and spiny pedipalps, whiplike front legs, and sudden, lightning-fast sideways locomotion send cold chills down many spines. There are at least two species on Guana. Our giant species is as yet unknown taxonomically, but might be a member of *Heterophrynus*. Guana's single specimen was on the wall of the main house at night. Our most common form is a member of *Phrynus*. There seems to be an even smaller kind, either the juvenile of *Phrynus* or, more likely, a third species. Specimens of Guana's amblypygids are being studied by Lorenzo Prendini at AMNH. The smaller forms have body lengths up to 15 mm and anterior front legs about 50 mm long. The giant species has a body length of about 50 mm and a front-leg length approaching 200 mm. The small kinds live under stones that are not deeply embedded in the ground.

ORDER OPILIONES OR PHALANGIDA

Daddy longlegs or harvestmen are familiar, harmless, and poorly represented on Guana; we have seen only three species, one with the familiar small, rounded body and long slender legs, and two with flatter, crablike bodies, and shorter, heavier legs; the former on vegetation and sides of buildings, the latter hidden in thick leaf litter. All can be distinguished from spiders because the two body regions, unlike spiders, are broadly attached without a narrow constriction between, and the abdominal region is clearly segmented. Like spiders, the chelicerae and pedipalps are small, but the former are pincherlike, not fanglike.

ORDER ACARI

A few ticks and many mites are present on Guana, but their study is very incomplete. All are small to minute and have lost external segmentation, the body lacks clearly defined regions, chelicerae have pinchers or are modified for piercing, and pedipalps are leglike or have pinchers. Ticks are external bloodsuckers of ver-

tebrates; our iguana tick, *Amblyomma antillarum,* is discussed (see "Parasitism," part 3, above). Mites are free-living predators or parasitic on or in a huge variety of animals and plants. Ticks and mites include the only disease-carrying arachnids, and some mites are the only arachnids that are successful in freshwater habitats. (One European spider is aquatic, and several can enter the water, but none spend their entire life submerged.) Free-living mites can be extraordinarily diverse and abundant in soil and leaf litter, but our inventory has just begun. Without any special effort we have recognized 20 families and about 50 species, and we know that this must be only a small fraction of the final totals. The world fauna has over 300 families.

Four orders of arachnids have not been found on Guana. Two of these, Schizomida (micro whip scorpions) and Palpigradida (palpigrads) are known from St. John. These are obscure, minute, white, soil-inhabiting organisms that we expect to find. The order Uropygida (large whip scorpions) is less likely as the known West Indian species are all from Hispaniola and Cuba (Lazell 2000c). The small order Ricinulei (ricinuleids) is unknown in the Antilles; about 30 species are recorded from Texas to Brazil.

CLASS CRUSTACEA

Pill bugs, beach fleas, crabs, etc. Most of the 40,000 species of crustaceans are marine, but Guana has three orders that include a few terrestrial species. All have five or more pairs of legs and two pairs of antennae (in isopods—pill bugs—one pair of antennae is vestigial), and at some stage of their life history all have some appendages with a basal branch. Each appendage (antenna, mouthpart, leg, swimmeret) can have branches that are similar or very dissimilar, depending on location and function. Although a 45-lb lobster is on record and a Japanese crab has leg spans up to 3.6 m (~12 ft.), the vast majority of crustaceans are less than 3 mm long. The terrestrial species on Guana have body lengths from about 2–100 mm, at least 4 in.

ORDER ISOPODA

Pill bugs, sow bugs, roly-polies. The body is depressed (flattened) and capable of rolling into a compact ball with the seven pairs of legs and other appendages completely hidden. The head is distinct and not fused with the rest of the body, which is clearly segmented and flexible, and the eyes are not on stalks. Pill bugs appear to be scavengers, feeding on decaying leaf litter and other organic debris. Guana has at least six species taken in Berlese funnels, extracted from beach sand, and walking around at night.

ORDER AMPHIPODA

Beach fleas, sideswimmers, scuds. Amphipods share most features with isopods except the body is compressed from side to side. Although the abdomen is often curled, they are incapable of rolling into a tight ball. Our species are all associated with marine shorelines, where they occur under drift debris and in the sand and vegetation above high tide line. Whereas locomotion in terrestrial isopods is confined to walking, amphipods can swim, leap suddenly, and run backward, and they are much more agile and difficult to capture.

ORDER DECAPODA

Crabs, lobsters, shrimp, etc. Crabs have five pairs of legs (usually the first pair are enlarged and with pinchers), and unlike isopods and amphipods, they have lost body flexibility because the head and entire body are covered with a single shieldlike plate, the carapace. In addition, decapods have stalked eyes rather than sessile. Lazell provides an account of Guana's land crabs, below. Some species return to the sea to reproduce, others complete their life cycle on land.

Four classes of arthropods collectively called "Myriapoda" occur on Guana. Myriapoda is not a formal taxonomic category; it is a diverse group of distantly related animals, conveniently possessing more legs than insects, arachnids, and most crustaceans. The minimum leg count for an adult is nine pairs, but some hatchlings start with as few as three pairs. The four classes are aligned in two groups, centipedes with sym- phylans and millipedes with pauropods, but each class has fundamental differences from its partner.

CLASS CHILOPODA

Centipedes. These are active predators with 15–191 pairs of legs in adults (15–103 pairs on Guana). Those with 23 pairs or less are rapid runners on exposed substrates or in unimpeded crawl spaces. Those with 31 pairs or more are usually in tighter crawl spaces as in soil, sand, rotten wood, packed leaf litter, etc., where the ability to squeeze through is more useful than speed. A unique recognition feature of all centipedes is the pair of enlarged poison fangs called prehensors beneath the head; these are multipurpose, highly modified legs of the first body segment, used to grasp, kill, and macerate prey, to groom antennae, and to excavate retreats in softer substrates. Because of the prehensors, all centipedes are venomous, but only the larger species can pierce human skin.

Two orders on Guana have individuals with constant leg counts of 21 or 23 pairs, or 31 or more pairs. The other two orders add legs as they mature, hatching with 4, 6, or 7 pairs and attaining 15 pairs when adult. Details about each order follow.

ORDER SCOLOPENDRIDA

This order has 21 or 23 pairs of legs and includes two families and two species on Guana. The giant (200 mm or 8 in.) tropical centipedes, genus *Scolopendra,* belong here. Guana specimens, not identified to species, have been exhibited in the "insect zoo" at the Smithsonian.

ORDER GEOPHILIDA

This order has 31–191 pairs of legs (the number is constant throughout life) and is our most diverse group, with three families and three or four species, and (at present) leg counts between 53 and 103 pairs.

ORDER SCUTIGERIDA

This order includes the long-legged, worldwide house centipede, but the Guana species may not be the same. Muchmore (1993) lists the St. John scutigerid as *Scutigera linceci* (Wood), not the cosmopolitan *Scutigera coleoptrata* (Linnaeus), so we suspect the Guana specimens are the former.

ORDER LITHOBIIDA

This order should occur on Guana but we have not found it. They hatch with 6 or 7 pairs of legs and, as with house centipedes, end up with 15. These centipedes have alternating long and short body segments. Because all four centipede orders are essentially worldwide, the absence of lithobiids from Guana and other Virgin Islands is an unexpected anomaly.

CLASS SYMPHYLA

Symphylans, garden "centipedes." Symphylans are small, soft-bodied, blind, white myriapods, 1–8 mm long, with 6 or 7 pairs of legs at hatching and 12 pairs as an adult. Prehensors (poison fangs) are absent, so food is mostly decaying plant tissues, fungi, and, for some, living roots and underground tubers. Fewer than 200 species are known, two of these on Guana—as yet unidentified—where they live in the soil, in leaf litter, or under stones or dead bark. Individuals can be abundant, but small size, secretive habits, and active, agile locomotion make them one of the less familiar groups of arthropods.

CLASS DIPLOPODA

Millipedes. Although this is the largest class of myriapods, it is also the most incompletely studied. Probably 10,000 species are known to specialists, and thousands more are expected. Millipede diversity on Guana appears to be low; we have found only 4 of 15 orders, 5 of 115 families, and perhaps 9 species. Most occur in humid sites—soil, leaf litter, rotten wood—and most feed on decaying plant materials. One large Guana species is arboreal, resting on tree branches during the day, and probably feeding on bark or lichens at night. Millipede adults have from 13 to 375 pairs of legs, whereas some juveniles hatch with as few as three pairs. Guana specimens of unknown ages range from 3 to 97 leg pairs. A distinctive feature of millipedes is that body segments are strongly fused in pairs (forming diplosegments) so most body rings have two pairs of legs. The first five diplosegments behind the head have lost some legs, permitting that end to coil up more compactly and thus increasing protection for the head. At the posterior end, one to six diplosegments are legless.

Millipedes do not bite or sting; all can roll into a flat coil or sphere when disturbed. Most have chemical defenses and produce odorous or irritating compounds; hydrogen cyanide and various quinones being among the more infamous. These chemicals ooze from paired pores along the body sides, and a few West Indian species can squirt them up to a meter from the body.

ORDER POLYXENIDA

These are the only myriapods in which the soft, flexible exoskeleton is covered with flat, barbed setae arranged in transverse rows, lateral clusters, and posterior tufts. They are so odd that they may not even be millipedes. Adults are less than 4 mm long, occur commonly in leaf litter and under bark, and have 11–13 body segments and 13–17 pairs of legs. We suggest Guana has more than one species, but the material has not been studied carefully.

ORDER SPIROBOLIDA

On Guana, this includes the large, cylindrical, dark brown millipedes usually seen on tree trunks and branches. Some West Indies species squirt their defense secretions, but this has not been observed on Guana. Our species does exude a dark, toxic liquid if disturbed. Diplosegment counts range from 32 to 84; counts for Guana species are unknown because the specimens are disarticulated and have not been reassembled.

ORDER JULIDA

Julids are smaller and more slender than spirobolids, have different mouthparts, and are brown, creamy, or dark with a pale middorsal stripe, or pale spots. Diplosegment counts increase with age and range from 25 to 121; on Guana a few that are still intact range from 39 to 45. Several unidentified species occur on Guana.

ORDER SIPHONOPHORIDA
Beaked millipedes. Not yet taken on Guana but present on Tortola and Anegada. We mention them because of the unique mouthparts that are conical and form a beaklike projection that appears capable of ingesting only liquids. Individuals are small, cylindrical, and very slender, and some have the highest leg counts for any millipede. On Tortola, leg counts are from 71 to 97.

TABLE 26
Guana Island Nonmarine Invertebrate Survey

PHYLUM	CLASS	ORDER	GUANA STATUS	SPECIES
Platyhelminthes	Turbellaria	Tricladida	Probable	0
Platyhelminthes	Trematoda	?	Probable	0
Platyhelminthes	Cestoda	?	Probable	0
Nematoda		Mermithida	Present	1
Nematoda		Rhabdita	Present	?
Mollusca	Gastropoda	Archaeogastropoda	Present	2
Mollusca	Gastropoda	Mesogastropoda	Present	2
Mollusca	Gastropoda	Stylommatophora	Present	20
Mollusca	Gastropoda	Basommatophora	Present	4
Annelida	Oligochaeta	Opisthopora	Present	2
Annelida	Hirudinea	?	Possible	0
Tardigrada	?	?	Probable	0
Onychophora			Possible	0
Arthropoda	Arachnida	Scorpionida	Present	3
Arthropoda	Arachnida	Pseudoscorpionida	Present	Many
Arthropoda	Arachnida	Solpugida	Present	1
Arthropoda	Arachnida	Araneida	Present	Many
Arthropoda	Arachnida	Amblypygida	Present	2 or 3
Arthropoda	Arachnida	Phalangida	Present	3
Arthropoda	Arachnida	Acari	Present	Many
Arthropoda	Arachnida	Uropygi	Unlikely	0
Arthropoda	Arachnida	Schizomida	Possible	0
Arthropoda	Arachnida	Palpigradi	Possible	0
Arthropoda	Crustacea	Isopoda	Present	Several
Arthropoda	Crustacea	Decapoda	Present	12
Arthropoda	Chilopoda	Scolopendrida	Present	2
Arthropoda	Chilopoda	Geophilida	Present	2
Arthropoda	Chilopoda	Scutigerida	Present	1
Arthropoda	Symphyla		Present	2
Arthropoda	Diplopoda	Polyxenida	Present	1
Arthropoda	Diplopoda	Spirobolida	Present	1
Arthropoda	Diplopoda	Julida	Present	3
Arthropoda	Diplopoda	Siphonophorida	Probable	0
Arthropoda	Diplopoda	Polydesmida	Present	2
Arthropoda	Pauropoda		Present	1
Arthropoda	Insecta	(see table 27)		1,069+

SOURCE: Original data. Here + indicates survey relatively incomplete, ? present but not classified, and 0 not present.

Flat-backed millipedes. Although the internal form is cylindrical, the diplosegments have lateral projections that make the animals look flattened. This order has thousands of species, with diplosegment counts from 18 to 22 and leg pairs from 30 to 38. At present, only one species has been found on Guana.

CLASS PAUROPODA

Pauropods. Most biologists have never seen a pauropod because minute size (0.4–2 mm), rapid locomotion, and cryptic habits (soil and leaf litter) make them difficult to recognize and easy to overlook in the field. Hatchlings have 3 pairs of legs, which increase to 9, 10, or 11 in adults, and all have the antennal apex distinctively three-branched. Pauropods have extreme structural diversity. The majority, including the Virgin Island species, are slender, agile, fast running, soft bodied, white or cream colored, and disappear rapidly when exposed, but one family (not yet known from the Antilles) is sluggish, heavily sclerotized, rusty red, shaped like a

TABLE 27
Guana Island Insect Survey

ORDER	COMMON NAME	FAMILIES	GENERA	SPECIES
Protura	Proturans	1	1	1
Collembola	Springtails	6	?	12+
Diplura	Diplurans	2	2	2
Microcoryphia	Jumping bristletails	1	1	1
Thysanura	Silverfish	1	2	2
Ephemeroptera	Mayflies	0	0	0
Odonata	Dragon and damselflies	2	8	10
Blattaria	Roaches	4	13	15
Isoptera	Termites	2	8	9
Phasmida	Walking sticks	1	2	2
Mantodea	Mantids	1	1	1
Orthoptera	Grasshoppers and crickets	4	8	17
Grylloblatodea	Rock crawlers	0	0	0
Dermaptera	Earwigs	1	1	1
Embioptera	Web spinners	1	1	1
Plecoptera	Stoneflies	0	0	0
Zoraptera	Zorapterans	0	0	0
Psocoptera	Bark and book lice	?	?	10+
Mallophaga	Biting lice	?	?	?
Anopleura	Sucking lice	?	?	?
Hemiptera	hoppers, aphids, scales, bugs	36+	?	74+
Thysanoptera	Thrips	2	?	6
Megaloptera	Dobsonflies and alderflies	0	0	0
Neuroptera	Lacewing flies	5	5	6
Coleoptera	Beetles and weevils	57	240+	410+
Trichoptera	Caddisflies	0	0	0
Lepidoptera	Butterflies, skippers, moths	?	?	380+
Mecoptera	Scorpion flies	0	0	0
Diptera	True flies	33+	?	50+
Siphonaptera	Fleas	?	?	?
Hymenoptera	Wasps, ants, bees	34+	85+	125+

SOURCE: Original data. Here + indicates survey relatively incomplete, ? present but not classified, and 0 not present.

miniature pill bug and capable of rolling into a ball. We have found the latter group (in the eastern United States) only in damp sites under the bark of dead trees near or on the ground; their presence in the BVI is expected. The biology of pauropods is poorly known but reports suggest most feed on fungi and semiliquid products of decaying plants; others appear to eat solid foods and include some predators. Scheller and Muchmore (1989) describe 11 species of pauropods from St. John, USVI.

For our Guana Island nonmarine invertebrate survey, see table 26.

CLASS INSECTA

Insects are by far the most diverse animal group on Guana, with over 1,000 species recorded (table 27) and hundreds more expected. Of 32 world orders, 22 are known to be present and 3 more, all parasites, although not recorded, are almost certainly present. Among the other seven missing orders, four have aquatic larvae, one lives at the edges of snowfields and glaciers on high mountains, and two (zorapterans and scorpion flies) have not been found.

The insect surveys of St. Croix by Beatty (1944a) and Miskimen and Bond (1970) provide the only comparable data from the Virgin Islands. Together they total 18 orders and 1,220 species. St. Croix is a much larger island with a land area of about 218 km² (84 sq. mi.). Its maximum dimensions are 9.6 × 3.7 km (6 × 23 mi.) and maximum elevation is 350 m (1,165 feet). Rainfall averages between 15 and 50 in. per year (380–1,235 mm) depending on the site. Despite its size and diversity, St. Croix appears to have a smaller insect fauna, suggesting either less adequate sampling, more degraded habitats, or lack of land connections historically to Great Guania, or the Puerto Rico Bank. Our current, still incomplete, knowledge of Guana insect diversity is summarized in table 27.

Land Snails

Fred Kraus

Scientists remain woefully ignorant of the ecology of most individual mollusk species and the ecological role these species play in ecosystem processes.

CHARLES LYDEARD ET AL. (2004)

L AND SNAILS ARE among the most diverse groups of terrestrial animals but are easily overlooked because most species are small (< 1 cm), secretive, and require special attention to discover. Most species live in leaf litter or under logs because of their need to keep moist, but the most conspicuous land snails to humans are those that are arboreal. Most land snails feed on detritus or fungi, but a few groups are predators on other snails.

Land snail species diversity is greatest in the tropics but still poorly understood because the ranges of many land snail species are very small, owing to their exacting physiological requirements and limited powers of dispersal, and researchers on this group are few. Diversity and abundance is normally highest in habitats that have exposed limestone because the amount of environmental calcium available for shell building typically limits land snail population size; this is, of course, not a limiting factor when limestone (calcium carbonate) is available on the surface. However, in the BVI this habitat

is limited. There are only a few, scattered limestone beds on Guana Island. The only island in the BVI largely made of limestone is Anegada. Approximately 40–50 species of land snails occur across the Virgin Islands, with about half of these found on Guana Island.

The nonmarine (terrestrial and freshwater) snail fauna of Guana Island has changed considerably since the 1930s. At that time Clench (1939) provided a list of snails collected on Guana by two visitors. Of 13 species listed in Clench's report, I have been unable to locate 6 and some of these are certainly extinct. For example, one freshwater species has disappeared because of the loss of standing freshwater habitat from the island. As well, some of the snails now found on the island are alien species that have been introduced since Clench's time. This includes *Bulimulus guadalupensis,* which, although native to the Puerto Rican Bank, was apparently absent from Guana Island during the 1930s but is now the most abundant large snail on the island, occurring by the tens of thousands on the flat. Most of these alien snails were probably brought in as unintentional hitchhikers on the horticultural plants that have been widely used to landscape around the cottages and the flat. And, indeed, S. E. Miller (1994) lists four snail species as verified hitchhikers on potted plants imported to Guana from Florida in 1990. Of these, three (*Polygyra* cf. *cereolus, Praticolella griseola, Succinea* cf. *luteola floridana*) were not present on Guana at the time of the shipment; it is possible that they have become established since then, although I have no evidence for that yet.

Below I provide a list of land snails currently known to occupy Guana Island based on studies conducted by me from March–October 1991, July 1992, and July 1993. Similar to Jacobson (1968), I exclude those species inhabiting the nearshore zone (e.g., *Littorina, Melampus*) that are sometimes included in other lists of land snails (*e.g.,* Muchmore 1993). I provide notes only for those species large enough or active enough above ground to possibly be noticed by nonspecialists. Mention of populations elsewhere in the Virgin Islands is based on my own field observations. The remaining species are either small (< 1 cm) or can only be observed by rummaging through leaf litter and, hence, are unlikely to be observed unless specifically sought. Asterisks (*) denote clearly introduced exotic species. I am indebted to John Slapcinsky for critically reviewing my manuscript and to Kurt Auffenberg for identifying many of the specimens.

ANNULARIIDAE

Chondropoma tortolense. This Virgin Islands endemic is a high-spired shell of approximately 2 cm, bearing a regular network of both axial and spiral ridges, giving the shell a roughened appearance. Furthermore, the animal has an operculum, or trap door, with which it can close the aperture of the shell, sealing the animal inside to protect it from desiccation or predators. It is generally distributed around Guana Island, living under rocks and in leaf litter, although it may sometimes be observed attached low to the bark of trees or crawling on rocks in the daytime.

BULIMULIDAE

Bulimulus guadalupensis. This is the most common large snail on the island, occurring by the tens of thousands in the flat, but may also be observed around the cottages and in the woods behind North Beach. It lives in leaf litter and on low plants. It is a cylindrical shell of dirty white or white and brown, with the brown most commonly arrayed as one or two spiral bands. It reaches a length of approximately 2 cm. This species is widely distributed from Jamaica and Hispaniola through the Puerto Rican Bank and Lesser Antilles (Breure 1974) but, as noted above, appears to be recently introduced to Guana Island.

Drymaeus virgulatus. This species has a similar shape to *Bulimulus guadalupensis* but is larger and with a more pointed spire. It is white with brown axial flamules instead of spiral bands. This species is rare on Guana Island although appropriate habitat is widespread. Only three old shells have been found at widely scattered locations around the peak. This species is widely distributed from the Puerto Rican Bank through the Lesser Antilles to Aruba, Bonaire, and Curacao (H. B. Baker 1924) and may readily be observed on Anegada Island, where specimens estivate in the hundreds attached to shrub trunks.

CAMAENIDAE

*Zachrysia provisoria.** This is the largest land snail on Guana Island, attaining a length of 2.5 cm. The shell is round in appearance, with numerous axial ridges and a uniform straw yellow color. The species is native to Cuba (Pilsbry 1928) and was no doubt inadvertently introduced to Guana Island via the horticultural trade, as it has been to numerous other localities around the Caribbean. Currently, this species appears to be restricted to the area around the cottages, where they are most often encountered (and crushed underfoot) on the pathways feeding on dead vegetation after rain.

HELICINIDAE

Alcadia foveata. This small (~8 mm) snail has a low helical shell with a slightly elevated spire, and the bottom of the shell lacks the opening (umbilicus) seen in all other land snails on Guana Island. Furthermore, it is the only land snail on Guana besides *Chondropoma tortolense* to have an operculum. This species is endemic to the Virgin Islands and is generally distributed around Guana Island. It most often lives under rocks and in leaf litter but may sometimes be seen crawling on rocks or low tree trunks during the day.

SUBULINIDAE

*Subulina octona.** This human commensal is cosmopolitan throughout the tropics because of the ease with which it is transported in agricultural and horticultural materials. It is a translucent and narrowly cylindrical smooth shell with rounded whorls and reaches a length of approximately 2 cm. It is most readily observed around the cottages or on the flat, where it lives under stones and in leaf litter, but may be seen feeding on the surface at night after rain.

XANTHONYCHIDAE

Hemitrochus nemoralina. This arboreal snail is one of the more conspicuous native species and is distinguished by its low, helical shell, yellow ground color with a dark brown spiral band, and pink aperture. The shell is approximately 1 cm in width. As Lazell (1983b) noted, these snails are most readily seen on the *Coccothrinax alta* palms in Palm Ghut (which he misidentified as *Thrinax*), but live snails can be found in any of the moister, shaded environments in the island's higher elevations. It is found throughout the Virgin Islands and is common on most islands, occurring from sea level to the top of Sage Mountain on Tortola. In addition to *Coccothrinax*, I have found *Hemitrochus nemoralina* living on *Acacia muricata, Agave missionum, Eugenia biflora, Eugenia ligustrina, Guapira fragrans, Krugiodendron ferreum, Pilosocereus royenii, Pisonia subcordata, Plumeria alba, Psychotria microdon, Rhizophora mangle, Rondeletia pilosa,* and *Tabebuia heterophylla.*

Plagioptycha musicola. This species has a similar size and shape to *Hemitrochus nemoralina* but is solid brown with axial grooves and a somewhat angulate shoulder to the last whorl of the shell. It lives among rocks on the slopes and in leaf litter on the flat, but it can sometimes be seen crawling about on rocks or low on tree trunks. It is generally distributed throughout the island and is common throughout the Virgin Islands and Puerto Rico.

Additional land snails from Guana Island of a more secretive nature and unlikely to be seen by the casual observer are listed here:

FERUSSACIIDAE

Cecilioides consobrinus

OLEACINIDAE

Varicella terebraeformis

PUPILLIDAE

Gastrocopta pellucida
Gastrocopta servilis
Pupisoma dioscoricola

SAGDIDAE

Hyalosagda subaquila
Lacteoluna selenina

STREPTAXIDAE

*Gulella bicolor**

SUBULINIDAE

Allopeas gracile
Allopeas micrum
Beckianum beckianum
Opeas hannense

SUCCINEIDAE

Succinea hyalina

ZONITIDAE

Hawaiia minuscula

Land Crabs: Crustacea, Decapoda

THE DIVERSITY AND ABUNDANCE of our terrestrial crabs often amaze visitors to the West Indies who come from temperate climes. They may be found from the splash zone of rocky shores to the tops of mountains. There are species inhabiting beaches, swamps, and forests. Some attain a very large size. There are a dozen species known from Guana. Specimens have been deposited in the U.S. National Museum and Museum of Comparative Zoology (USNM and MCZ). I am indebted to Fenner Chace, at USNM, and Ardis Johnston, at MCZ, for identifications and correspondence. I have arranged them alphabetically by family name and then by genus and species. The basic reference is Chace and Hobbs (1969).

COENOBITIDAE

Soldier Crab, *Coenobita clypeatus*. Although quite terrestrial and even arboreal, these shell dwellers remind many northerners of the related marine species called "hermit" crabs. They like to live in snail shells, especially those called "topshells" or, locally in the BVI, "welks." They will reside in most anything that fits, at least in the absence of choice, and may be found hauling around bottles and even beer cans. Outside their acquired shell (or other armor) they resemble asymmetrical, curled lobsters or prawns. One pincer is greatly enlarged and usually brightly colored in red, bluish, or purple. The curved, naked body is unappealing in appearance but cooks up very well—similar to a lobster tail. Soldier crabs can be extremely abundant; for example, in the North Bay woods on Guana. They account for one good reason that many people wear pith helmets in coconut groves: the crabs climb up the trees and are apt to fall out (coconuts themselves are another good reason). One of the more remarkable characteristics of our soldier crabs is the loud chirping or squeaking noise some individuals make when they have been picked up. About one-third of the North Bay woods soldier crabs chirp. It seems that this is a distress call analogous to the "fear scream" of birds (Hogstedt 1983), but what adaptive purpose can it serve? No other soldier crabs ever come hustling to the rescue. Many organisms—including heliconiid butterfly pupae (see Butterflies: Lepidoptera: Rhopalocera, part 5 below), male tortoises, and long-horn beetles—make apparent distress calls when apprehended or manhandled, but the benefits are not always obvious.

GECARCINIDAE

Beige Land Crab, *Cardisoma guanhumi*. A common name for this species elsewhere is "blue" land crab, but ours have precious little blue and are about the color of a manila folder. They get huge—to carapace (upper shell) lengths of over 100 mm (4 in.). They are excellent for eating. They make big burrows in swampy coastal woods, as along the road between the dock and the Salt Pond on Guana. To many visitors, their burrows look like woodchuck holes. To observe them best, go at night with a flashlight.

Black-back Land Crab, *Gecarcinus lateralis*. A big, colorful, woodland species—up to 80 mm in shell length. The central area of the back is near-black, but the edges are usually bright red. Regarded as a choice item on the Antillean menu, these crabs are best hunted at night. Herpetologists often find them in the daytime under rocks and logs.

Purple Land Crab, *Gecarcinus ruricola*. This highly colored woodland species also gets quite large; as with its congener above, it is highly desirable food. These land crabs are steamed (after a few days of purging). The meat is picked from the shell and legs and packed in the back shell (carapace). Starchy material, such as rice or breadfruit, and peppers, or other chopped vegetables, and spices are added. A dollop of butter is placed on the exposed surface (outside of shell is down), and the crab back is baked or broiled until browned: Outstanding.

GRAPSIDAE

Globose Shore Crab, *Cyclograpsus integer*. This small species only attains about 13 mm (half an inch) in shell length but is quite pretty: orange tan with banded legs. They occur in the intertidal or splash zone of rocky coasts. I have only infrequently found one.

Variegate Shore Crab, *Geograpsus lividus*. (See plate 32.) Our most colorful terrestrial crab, this one attains 26 mm (just over an inch) in shell length. The record specimen, in USNM, was collected on Necker Island (Lazell 1995). On Guana, I have only found it just up from the cobble beach at Pinguin Ghut. It is boldly patterned in turquoise gray blue and near-black. The leg hairs are orange. It is decidedly uncommon.

Sally Light-foot, *Grapsus grapsus*. This is perhaps the most conspicuous of our crabs because it scampers and scuttles over big boulders and coastal cliffs in daytime. It also likes docks and pilings. This species attains about 70 mm (nearly 3 in.) and is rich olive spotted and marbled with near-black. When individuals molt, they often leave their old, shed shells perched high up on the rocks in the sun; these then turn rich orange red and black, giving rise to another common name: calico crab.

OCYPODIDAE

Ghost Crab, *Ocypode quadrata*. This is our beach crab. Pallid as White Bay sand, it trots from hole to hole in broad daylight and bright sun. Reprobate thieves, ghost crabs will steal pens, pencils, sunglasses—anything small enough to carry—zoom away, and pop down a hole with their ill-gotten gains. Fortunately they are rather small—about 35 mm (1.5 in.)—and cannot carry books, purses, beach chairs, or bottles.

Mangrove Crab, *Sesarma roberti*. Restricted on Guana Island pretty much to the mangroves on the north side of the Salt Pond, this species climbs up the roots and trunks. It attains about 27 mm (a bit over an inch) in shell length and is patterned in brown and cream color.

Salt Pan Fiddler, *Uca burgersi*. These little fellows—barely 20 mm ($<$ 1 in.) in shell length—fairly swarm on the mud among the salt grass stems around the Salt Pond. The females have pincers of about equal size, but males have one pincer enormously enlarged—almost as big as the rest of the crab. This hypertrophied member is used in complex, stylized combat choreography by rival males. The larvae of these crabs are a mainstay of the flamingo flock. These crabs may oviposit in burrows adjacent to landlocked ponds (Jarecki 2003), such as Guana's. This implies population isolation and a metapopulation

PORCELLANIDAE

Porcelain Crab, *Petrolisthes quadratus*. This is a small, drab species, typically about 1 cm (about a half an inch) in shell length. The shell is very hard and dense, as befits its habitat in the wave-washed, rock-rubble, intertidal zone. It lives under rocks, often clinging to the underside when the rock is turned over. It is not seen exposed, at least in daytime. This is our only land crab that has been studied in terms of activity and density. In July 1988, Christine (Chris) Henderson marked individuals in a 48-m² grid along North Bay beach. She never found one actually in the water or in a place that was not at least damp with seawater. She found them from the water's edge to 340-cm inland, at the edge of the damp zone. Chris did a triple catch, marking a total of 28 porcelain crabs, but the point estimator for the population was 81—with a 95% confidence interval of 34–96. Travel distance on marked crabs was very great: 65–212 cm (average 139 cm). Thus the area sampled was actually 128 m², extending far beyond the grid strip. Chris thought the marking method—colored thread tied on the pincers—invalidated the results because the number of individuals found in the second and third rounds that were missing pincers was so high. She suspected tying thread might cause a crab to drop off a pincer. Someone needs to test this hypothesis using a second marking method. The density estimate for our population came to ~0.63/m², or about two crabs in every 3 m². This accords well with unquantified observations.

XANTHIDAE

Mystery Crab, *Ozius reticulatus*. Fenner Chace (in litt. 18 October 1988) identified this species among those collected on Guana by Christine Henderson in July 1988. I do not remember these crabs, although I must have pickled them. Apparently, they were too small to tag. Ardis Johnston can find no account of them in the literature, and I have been unable to recontact Chace (he retired many years ago). If anyone reading this knows anything about *Ozius reticulatus*, please let me know.

Spiders: Araneida

There are more than 32,000 known species of spider in the world. . . . All spiders are carnivorous, ravenous eaters who feed on massive quantities of protein, in liquid form, usually the juices of their prey.

PETER DAVID (2002)

JAMES ORTIZ OF the Natural History Museum, Orange County, California, and his wife Carol, came to Guana in 1994 and made a preliminary list of our spiders. Richard Bradley, Ohio State University, identified more in 2003 from litter samples (in this list indicated by "RB" in parentheses). I have added some more along with a few notes (in this list indicated by "JL" in parentheses). Otherwise the annotations are from Ortiz's list dated 12 April 1995. In many cases, we have not been able to identify our spiders to species or even genera yet.

MYGALOMORPHA

THERAPHOSIDAE

Cyrtopholis bartholomei—large "tarantula" in burrows in open areas. (See plate 33.) Common. Smaller than above, dark, low carapace, under logs in damp areas. Common. Arboreal, hirsute carapace, obvious double tarsal pads called scopulae. Rare. Small, chevrons on back of abdomen. Rare.

CTENIZIDAE

Ummidia sp.—large, black trap-door spider. Specimens in MCZ number 10299 (JL) and Valentine collection. Occasional.

DIPLURIDAE

At least two unidentified species (RB). Rare.

ARANEOMORPHA

ARANEIDAE

Argiope argentata—smaller than most neotropical specimens, lateral projections of abdomen not prominent. Common.

Gasteracantha tetracantha—Four spines on abdomen; two color phases on island. Common.

Neoscona moreli—sweeping tall grass and shrubs in plantation area. Common.

Eustala sp.—in tall shrubs and trees. Common.

Cyclosa walckenaeri—in low shrubs; appears as debris in web. Common.

Nephila clavipes—large; able to catch and eat anole lizards. MCZ number 44910 (JL). Occasional.

Metepeira labyrinthea—in shrubs and low trees, also on docks; communal, labyrinthia. Common.

Ocrepeira serrallesi. Occasional.

Leucauge argyra—in horizontal webs, sheltered areas. Common.

Tetragnatha sp.—immature. Rare.

Mastophora P.—tubercles on carapace. Rare.

Acacesia hamata. Rare.

ULOBORIDAE

Miagrammopes sp.—Rare.

SCYTODIDAE

Scytodes sp.—in termite nests and under rocks. Rare.

LYCOSIDAE

Trochosa sp.—immature. Rare.

SEGESTRIIDAE

A single unidentified species (RB). Rare.

MIMETIDAE

Mimetus sp. Rare.

CLUBIONIDAE

Trachelas borinquensis. Rare.

Micaria sp. Occasional.

CORINNIDAE

Three unidentified species (RB). Rare.

THERIDIIDAE

Argyrodes sp.—a kleptoparasite. Occasional.

Achaearanea sp. Rare.

CAPONIIDAE

Orthonops sp. Rare.

OECOBIIDAE

Oecobius concinnus. Rare.

DEINOPIDAE

Deinopis sp. Rare.

GNAPHOSIDAE

Drassyllus sp. Rare.

A second unidentified species (RB). Rare.

ANYPHAENIDAE

Aysha sp.—near *Aysha velox*; female only. Rare.

Hibana tenuis—see Brescovit (1993) (JL). Occasional.

A third unidentified species (RB). Rare.

LINYPHIIDAE

Erigoninae—Three species. Occasional.

OONOPIDAE

Oonops sp. Rare.

A second unidentified species (RB). Rare.

LOXOSCELIDAE

Loxosceles virgo—in buildings and under rocks that have hollow area below. This species is toxic to humans. Common.

SELENOPIDAE

Selenops lindborgi—in buildings and on *Agave* leaves at night. Common.

PHOLCIDAE

One species. Rare.

Two species. Common.

OXYOPIDAE

Hamataliwa sp.—Known only from Great Camanoe at this time. Rare.

Oxyopes salticus—in open areas on grass and low shrubs. Common.

HETEROPODIDAE (SPARASSIDAE)

Olios sp.—large. Occasional.

Pseudosparianthis antiguensis. Occasional.

SALTICIDAE

(1) *Metaphidippus* sp.—green iridescence. Common. (See plate 34.)

(2) *Metaphidippus* sp.—chevrons on abdomen. Common.

(3) *Metaphidippus* sp.—similar to (2). Rare.

Hentzia sp. Occasional.

Planthoppers and Other
True Bugs: Hemiptera

... perhaps the most detrimental order of insects. ...

ROSS ARNETT (1985)

THE ORDER HEMIPTERA includes 40 families of true bugs and all the insects once put in the now-defunct order "Homoptera," the cicadas (which Americans are apt to call "locusts"), aphids, adelgids, scale insects, whiteflies, and the various tree-, leaf-, and planthoppers. They have piercing–sucking mouthparts, composed of four stylets modified from mandibles and maxillae, sheathed by a segmented labium. These form a sort of beak or rostrum that arises at the lower back of the head in planthoppers (in true bugs the mouthparts are up front). Planthoppers use their beaks to penetrate plants, from which they suck liquid sap that is typically rich in carbohydrates like sugar. Arnett (1985, 207) says they occasionally mistake us for plants and inadvertently pierce us, which hurts, even if they do not suck blood. The broader evolutionary relationships within Hemiptera are not well known. Morphological and biochemical analyses do not agree, and some molecular studies are contradictory, or at least fail to convincingly resolve the lineages. The order is extremely diverse: There are many very large, noisy, brightly colored forms (e.g., some tropical cicadas); bizarrely shaped creatures (e.g., some leafhoppers); and—at the other extreme—scale insects and adelgids that hardly appear to be insects at all but rather just yucky stuff stuck on their host plants.

The true bugs, suborder Heteroptera, with mouth parts up front where we think of them as normally being, include the Coleorrhyncha that are very primitive, resembling some Paleozoic fossils of a half-billion years ago. There are only about 20 species of those, confined to Australia, New Zealand, and South America. The Sternorrhyncha are the ones that bring most of the disrepute to their order: aphids, adelgids, whiteflies, scale insects, and jumping plant lice. Many of these are major agricultural pests. A lot of these live on Guana Island, but they have yet to be studied. Their name literally means "sternum rostrum," indicating their snouts or beaks have been shifted posteriorly onto their chests. The last suborder, Auchenorrhyncha, includes cicadas, leafhoppers, treehoppers, planthoppers, and spittle bugs (the latter have never been found on Guana). Their name "aucheno-" refers to the ventroposterior part of the head and "rhyncha" to the beak or rostrum.

Charles Bartlett, University of Delaware, is currently working on our new species and an overview of their relationships based, in part, on DNA sequencing. In the meantime, he has produced an excellent preliminary summary (Bartlett 2000). There is only one known cicada (family Cicadidae): *Proarna hilaris,* a small, silvery gray species. There is also only one known treehopper (family Membracidae): *Deiroderes inermis.* There are 25–30 species of leafhoppers (family Cicadellidae) on Guana, but they have not yet been sorted and identified even though a good collection has been made. The remaining 27 species are all fulgoroids: members of superfamily Fulgoroidea, or planthoppers.

Bartlett (2000) collected about 1,300 specimens on Guana from 1993 to 1997. He made a "discovery curve" or "species accumulation curve" by treating every specimen as an observation, randomizing the sequence of observations (with a computer), and plotting the number of accumulating specimens against the number of species found. Bartlett adds a most welcome bit of sophistication to his analysis by calculating Chao's (1984) estimator of species richness (see "Species Richness," part 1, above). For the first couple hundred specimens the curve zooms up steeply to 23 species. It took about 600 more specimens to flatten the curve at 27 species, and, after 800 specimens, no more species have been added. That is a run of twice as many specimens as it took to add any of the 27 species (about 500 vs. about 250 specimens needed to add the last species).

Has Bartlett gotten them all? All he says about that is "a high percentage of Guana Island planthopper species were detected," but that it "would be optimistic to claim that all species were detected." It would be fun to know the specimen number of Guana's rarest species, *Tangella schaumi,* represented by only one specimen out of the 1,300 (it might have been the very first one he caught). So I checked up on this specimen. It has a remarkable history. Beetle expert Mike Ivie caught it right out on North Beach, at night, on 10 July 1994. The date of col-

lection puts it near the midpoint of Bartlett's specimen accumulation.

For planthoppers, the observed number of species is 27, only one (*Tangella schaumi*) is known from a singleton and four (*Oliarus viequensis,* a new species of *Neopunana, Flatoidinus spinosus,* and a *Prosotropis* that might be *marmorata*) are known from two specimens. Using Chao's (1984) formula, we add to the observed 27 the number of singletons squared (one squared is still one) divided by twice the number of doubles: two times four is eight. Eight divides into one 0.125 times, so the estimate 27.125 is rounded to the nearest species: 27—very comforting.

In table 28, I have arranged our species in order of the number of specimens reported by Bartlett. At least one species, *Toya venilia,* is now represented by many more specimens as a result of the work done by Denno et al. (2001), described below. This species was probably underrepresented in Bartlett's collections because it occupies a very restricted habitat that is also very poor hunting ground for other species: salt grass.

The most wide ranging of all of our species is *Saccharosydne saccharivora,* literally the "sugar eater." It was originally described from the southernmost isle of the Lesser Antilles, Grenada, and is known from all over the West Indies, the southeastern United States, Central America, South America, and even Hawaii—to which we can assume it was introduced. It is a pest of sugar cane (Anonymous 1833) and has traveled with that crop, its beloved host. It can survive on other grasses, too, but none seem to suit it as well as sugar cane. I am surprised that it is not even more widespread in the Pacific and all the way to Australia. It is rare on Guana, being represented by only five specimens. The explanation for this rarity, of course, is that only a few ornamental sugar canes grow on Guana. *Saccharosydne saccharivora* is supposed to be all too abundant in places where cane sugar is grown as a major crop. As a crop pest, *Saccharosydne saccharivora* is a human commensal. This begs the question: to what extent is the range : abundance relationship an artifact? How often are abundance and wide range positively correlated

TABLE 28

Fulgoroid Hemiptera (Planthoppers) of Guana Island Listed by Sample Size

SPECIMENS COLLECTED	SPECIES	RANGE
1	*Tangella schaumi*	PRB
2	*Oliarus viequensis*	PRB, Hispaniola, Antigua, FL
2	*Flatoidinus spinosus*	VI
2	*Prosotropis* sp.	Guana, St. John
2	*Neopunana* sp. 1	Guana
3	*Neopunana* sp. 2	Guana
3	*Neopunana* sp.	Guana
3	*Colpoptera maculifrons flavifrons*	Mona, PRB, St. Croix, Antigua
5	*Saccharosydne saccharivora*	GA, LA, SA, CA, SEUSA, HI
13	*Anchidelphax havanensis*	GA, LA
13	*Neomegamelanus elongatus reductus*	PRB
17	*Pseudoflatoides albus*	PRB
18	*Bothriocera eborea*	PRB, LA
24	*Sayiana viequensis*	Vieques, Guana, Anegada
27	*Catonia* sp.	Antigua, Guana, Tortola
28	*Toya venilia*	PRB, LA to Montserrat
55	*Melormenis basalis*	Mona, PRB, LA to Montserrat, HI
64	*Catonia cinerea*	PRB, St. Croix
79	*Acanalonia depressa*	VI, St. Croix
83	*Pintalia alta*	PRB
86	*Oliarus slossonae*	PRB, Cuba, FL
92	*Thionia argo*	PRB, St. Croix
97	*Tangia viridis*	PRB, St. Croix
131	*Quilessa fasciata*	PRB
139	*Catonia arida*	PRB
168	*Neopunana caribbensis*	PRB
268	*Petrusa epilepsis*	GA, LA, SA

SOURCE: Bartlett (2000). A very general range is tabulated. Abbreviations: PRB, Puerto Rico Bank; VI, Virgin Islands on PRB and not including St. Croix; LA, Lesser Antilles; GA, Greater Antilles; FL, Florida; HI, Hawaii; SA, South America; CA, Central America; SEUSA, Southeastern United States.

to each other because the organisms benefit directly from human efforts or indirectly from habitat modifications brought on by humans? Interestingly, Bartlett (2000) did not record *Saccharosydne saccharivora* from any other Virgin Island except Guana, despite the facts that sugar cane is grown on several of them as a crop and he has planthopper records from at least 14 (not counting St. Croix). Six of those fourteen (Tortola, Jost Van Dyke, St. Thomas, St. John, Virgin Gorda, and Anegada) have sugar cane crops (as, of course, does St. Croix).

The most abundant Guana species, *Petrusa epilepsis*, with 268 specimens, is also widespread: virtually throughout the West Indies, south in South America to Brazil, and west to Columbia. Alas, I have been able to find nothing of its natural history.

Of our species, 59% (16 of 27) are only known from the Puerto Rico Bank. This includes our rarest species, *Tangella schaumi*, but it is widespread on the Bank, known from Puerto Rico, its coastal cay of Caja de Muertos, Vieques, St. Thomas, St. John, and Jost Van

Dyke. That is essentially the same range as three of our four most common species, *Quilessa fasciata, Catonia arida,* and *Neopunana caribbensis*—all represented by over 130 specimens. However, *Oliarus viequensis,* tied at two specimens for second rarest, is quite widespread: west to Cuba and north to Florida. At least three, possibly four, species to date are known only from Guana.

Species : area data for West Indian fulgoroids are scarce and probably quite incomplete—except maybe for Guana Island. Bartlett (in litt. 11 April 2001) has provided the data used in figure 16. Nevertheless, the Puerto Rican fauna is "relatively well known" (Bartlett 2000, based on Maldonado Capriles 1996) and a lot has been published on the adjacent Banks of St. Croix and Mona. Vieques, more than twice the size of Mona, has fewer known species, but it has the same number as St. Croix, which is more than one-third larger. Vieques, on the Puerto Rico Bank, is expected to have a large fauna because species were stranded there by sea-level rise and did not have to colonize across water. Even so, Guana has almost one-third more species and is only 2% of Vieques' area: As of now, Guana is a far outlier for this group, as it is with so many others.

Bartlett's work continues. He will describe the new species of *Neopunana* at least, even with only two or three specimens of each, because they belong to family Delphacidae, which is his particular specialty. He would especially like to sample more islands to test the notion that Guana is truly different. I have suggested two in particular: Necker, which is very small, one-tenth of Guana's size at 30 ha, but in good ecological condition; and Norman, closely comparable to Guana in size, but a devastated goatscape in poor ecological condition. We would predict that each would have fewer species than Guana: Necker because it is so small; Norman for habitat destruction. Alas, even if true (almost a certainty), smaller numbers will not "prove" anything: We have no replicates; we cannot reject a null hypothesis. Maybe no man is an island; but every island is—unique unto itself.

Meanwhile, Robert Denno and his colleagues Barbara Thorne (of termite fame, see "Termites: Isoptera," part 5, below) and David Hawthorne, all from University of Maryland, have done an indepth investigation of our salt grass hopper *Toya venilia* (Denno et al. 2001).

One of the most frequent questions, or set of questions, I am asked concerns flightlessness in island species. I have looked into the case histories of several vertebrate species, notably flying lizards (actually gliders, genus *Draco*) in the Philippines and flightless rails on Tasmania and various Pacific Islands and as an extinct species in the Virgin Islands. I agree with Charles Darwin that, for these species, it is the windiness of small islands that selects against flying or gliding. Strong winds can overwhelm weak fliers or gliders, smash them into trees and rocks, and even carry them out to sea. It is not that the wings are simply useless on an island with no predators, it is that they are detrimental: a death trap.

Evolution, of course, does not proceed by use or disuse of structures such as wings and flight muscles. Instead there must be a selection pressure that favors wing reduction or absence. The windthrow theory, even if it is sometimes certainly correct, cannot explain all—or perhaps even most—examples of

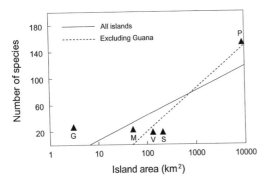

FIGURE 16. Islands on or adjacent to the Puerto Rico Bank with reasonably well-known planthopper faunas. P, Puerto Rico; S, St. Croix; V, Vieques; M, Mona; G, Guana. Dotted line, regression line without Guana data. Solid line, with Guana data included. *x*-axis, island area in log scale. *y* axis, number of species. Data from Charles Bartlett (in litt.).

winglessness. Therefore, controversy has ensued and is neatly described by Denno et al. (2001), complete with references to the various points of view.

On Guana Island they found two populations of the salt grass hopper, *Toya venilia*: one in the essentially undisturbed meadow at the Salt Pond, and the other in the windward side dunes at North Beach. They have been collecting salt grass hoppers around the BVI since 1996, but nowhere have they found two populations so strikingly different on any one island. Guana's Salt Pond meadow population has about 5% winged males and less than 1% winged females. Although we do not know the genetic mechanism, wings in salt grass hoppers must have a very simple genetic switch: They have either big, good wings or no wings at all. Guana's North Beach dune population has 69% winged males but no winged females. Dave Hawthorne (in Denno et al. 2001) did a DNA analysis and determined that Guana's two populations are genetically significantly different, another example of intraisland geographic variation, and the only one so far proven to be heritable.

On Anegada, intraisland variation was 75%–78% winged males and 0%–25% winged females. On Virgin Gorda the percentages were 46%–76% and 0%, respectively. On Beef Island it was 2%–27% and 0%, whereas on Tortola no winged individuals were found in any population. The two Guana populations are only 500 m apart, separated by the main northwest-to-southeast ridge of the island, about 20–100 m high for its relevant length. The trade wind would tend to blow things from North Beach toward the Salt Pond if they could make it over the ridge. Salt grass hoppers apparently make the trip rarely but maybe often enough to generate that 5% winged males. Of course, the North Beach dune population lives in the windiest place on Guana Island, as windy a site as any in the West Indies. Therefore, right away, the windthrow theory that there is a selective disadvantage to wings fails the test.

What are the advantages of having functional wings? Well, we all immediately think of escaping predators. Denno et al. (2001) noted three more: searching for resources such as food (planthoppers live on sap, essentially, and hardly need to search for water); getting around in complex habitats (e.g., they hate to walk on the ground); and finding mates.

What are the disadvantages of having wings? Well, if you are not really at risk of being smashed into a tree or blowing out to sea, you still must incur the enormous expense of making wings and their muscles. A veritable cascade of enzymes for catalyzing the biochemical steps involved in growing and using wings (all encoded for in DNA) as well as metabolic energy are needed. Denno et al. (2001) cite compelling evidence that wingless females (called brachypters, or literally, "short-wing") produce more eggs and offspring than winged ones (called macropters, or literally, "big-wing"). This would seem to explain the overall relative rarity of female macropters in all habitats: 0 in 13 of 16 populations sampled. Although it is difficult to quantity a benefit to males as simply as counting eggs per female, we can assume energy saved by brachyptery is used for other beneficial purposes by the males, too.

On Guana, where the two populations are so different with respect to the percent of macropterous males (5% vs. 69%), there must be a real advantage to males having wings in the windward side dunes. Denno et al. (2001) considered the behavior of the hoppers and the structure of the salt grass patches. Both sexes of hoppers "call" by vibrating their abdomens. The vibrations are transferred to the grass stems on which the hoppers perch through their legs or beaklike mouthparts. Females stay put, but males actively travel around in search of them. A female's vibrations can be picked up a meter away by a male if the grass stems are in contact, but he cannot hear her if he is on a separated plant. In the dune patch, the salt grass stands are sparse and individual plants are often isolated. A male who can fly is thus able to land on several disjunct grass plants, listening at each one for a female to strike up a duet with. Dueting leads to courtship and mating. Thus male macropters have an advantage in this habitat but no advantage in the dense, matted salt grass meadows on flats such as the one at Guana's Salt Pond. In dense meadows, neither sex benefits from wings.

The only place the group found a fair number of female macropters was in the (occasionally) mowed lawn in the Settlement on Anegada: 25%. One can imagine that the whirring blades of the lawn mower may spell doom to a female who cannot flit away to resettle on a still-living grass stem. Thus, two factors select for macroptery in island salt grass hoppers: (1) Sparse, scattered grass plants select for male macropters but not female; and (2) severe disturbance (e.g., lawn mowing), selects for both male and female macropters, but for females this is mitigated by the severe toll taken on reproductive output by making and using wings and wing musculature.

Denno et al. (2001) found no difference at all in the percentage of macropters in our island species, *Toya venilia*, and the set of related species using similar habitats on the North American continent. Here, flightlessness has nothing to do with getting blown into things or out to sea. Instead, habitat persistence promotes the evolution of flightlessness. If the habitat is stable and remains continuous for a long time (hundreds of years), then energy expended on flight apparatus and its biochemistry can be allocated to reproduction without penalty. This is obvious for females, but even flightless males sire more offspring than flyers. Denno et al. (2001) cite evidence that even though flightless males may mate with the same number of females as flyers, they carry higher sperm loads. Flying for these planthoppers is necessary in temporary and very patchy habitats (e.g., dunes and mowed lawns) where flying facilitates resource tracking and mate locating, or simply escape.

John Medler, Bernice P. Bishop Museum, Honolulu, was the first person to publish on a Guana Island planthopper, *Melormenis basalis* (F. Walker 1851). Medler has been whittling away at a complete revision of the planthopper family Flatidae, whose name literally celebrates the fact that, relative to the usual tectiform (rooflike) construction of most fulgoroids, they are usually quite flat. Medler's is no small task because many planthoppers were described well over a century ago (e.g., *Melormenis basalis*), and the descriptions are so vague that one can scarcely tell what they refer to (i.e., Walker's). All Walker provided for a type locality was "West Indies," and his description said nothing of the male genitalia, regarded as fundamental to most species identifications. So Medler had to examine F. Walker's actual specimens—not just of *Melormenis basalis,* but also of a host of flatid species. Those specimens were supposed to be in the Natural History Museum, London. F. Walker, back in 1851, had noted two specimens, male and female of *Melormenis basalis.* Medler could find only the female, which he designated the official onomatophore, or name bearer of the species. However, this female was not really satisfactory for systematic purposes: no male genitalia. In cases such as this, taxonomists have to describe the species all over again, making their best judgment as to the identity of the original, and select a specimen that future workers can examine to verify what the "new" describer was seeing. Fortunately, Scott Miller had collected specimens on Guana and deposited them in the Bishop Museum. So Medler (1990) selected, dissected, and described one of those males, designating it the plesiotype of the species *Melormenis basalis,* now a resident in a drawer in Hawaii.

This species is middling common on Guana (55 specimens) and middling widespread: It occurs on both Mona and Puerto Rico banks, St. Croix, and down the Lesser Antilles to at least Montserrat. It is somber brown with darker spots. In flight it is quite mothlike, but at rest it holds its wings in an unflatid, typically fulgoroid tectiform position.

The team of R. F. Denno, B. L. Thorne, and D. J. Hawthorne continue their work. Their larger project has been to compare sympatric populations of three unrelated species of insects morphologically, ecologically, and molecularly at the level of differentiation, first within Guana and then on other islands in the BVI. The three species were selected because they can be found in close proximity at virtually every site selected, but they have very different dispersal abilities. In addition to the planthopper, they are considering the termite *Nasutitermes acajutlae,* and a cicadellid leafhopper, *Tideltellus marinus,* with excellent potential dispersal ability.

Bartlett (in litt. 17 May 1995) added five species of Hemiptera to Guana's list; Scott Miller (in litt. 18 December 2001) added a sixth; and Bob Blinn, North Carolina State Museum (in litt. 1 July 2003), added a seventh: *Oncopeltus aulicus* (Fab.) (plate 35).

Lygaeidae
Ozophora divaricata Barber
Ozophora quinquemaculata Barber
Pseudopachybrachis vinctus (Say)
Nabidae
Arachnocoris berytoides (Uhler)
Pentatomidae
Chinavia sp.

Tingidae
Corythaica carinata Uhler

We have two more hemipterans of note. One of these is the treehopper, *Deiroderes inermis,* family Membracidae, collected on Guana, Tortola, and Virgin Gorda by Jason Cryan (Cryan and Dietz 2002). They also record the species from St. John, USVI, and Puerto Rico—from whence it was first described in 1945. Most treehoppers have fancy "horns" or apparently ornamental spikes and spines, but *Deiroderes inermis* (the name means "unarmed") lacks these. Cryan et al. (2000) sequenced nuclear DNA of a Guana Island specimen and found *Deiroderes inermis* to be a basal member of the membracid subfamily Stegaspidinae, but relationships to other genera remain enigmatic.

Another hemipteran ("The Water Boatman," part 5, next) is often spectacularly abundant and of major ecological importance. I will delegate its description to the expert who has studied it on Guana, and throughout the BVI, intensively.

The Water Boatman

Lianna Jarecki

THE WATER BOATMAN, *Trichocorixa reticulata* Guerin Meneville (Hemiptera: Corixidae), occurs from the Americas (including the entire continental United States, the Caribbean, and South America) to China (Hungerford 1948). Corixids are excellent dispersers and can fly 80–100 km in search of new habitats (G. A. Cole 1968). *Trichocorixa reticulata* is one of the most abundant animals in salt ponds of the BVI. Cullen (1994) showed that corixids represented the main food item captured by black-necked stilts *(Himantopus mexicanus)* in Fraternidad Lagoon, Puerto Rico.

Corixids have modified scoop-shaped forelegs and oarlike hind legs with swimming hairs. They surface periodically to capture air, which they store in a physical gill called a plastron.

Trichocorixa reticulata is not a predator, as is its close relative, *Trichocorixa verticalis*, which feeds on brine shrimp *(Artemia)* in the Great Salt Lake. Instead, *Trichocorixa reticulata* feeds on microbes and detritus from the benthos (Jang and Tullis 1980; Balling and Resh 1984; Hungerford 1948; Pennak 1978). *Trichocorixa reticulata* can regulate ions and osmotic pressure of its hemolymph (bloodlike body fluid) to 100 parts per thousand of salt (ppt) (Jang and Tullis 1980), three times the salinity of sea water. This species is found in all BVI salt ponds when salinities do not exceed 100 ppt, and it represents a major food resource for resident and migrating waterbirds in the BVI (Jarecki 1991, 2003). It is a staple in the diet of Guana's flamingos.

Termites: Isoptera

... temperate and tropical forest and grassland communities depend heavily on termite activity for their continued existence.

MARGARET COLLINS (1991)

IN JULY 1986 THE LATE Margaret S. Collins (1922–96) first came to Guana Island. In her first report (in litt. 11 January 1987) she noted sampling 75 colonies and collecting seven genera, stating "Guana has the richest fauna of the islands sampled [5]. . . . The most intriguing question . . . remains identification of the factors involved in the great diversity that obtains on such a small island as Guana."

Guana, with eight documented species, has a more varied termite fauna than has been found on nearby islands thus far and is unusually rich in species for an island of its size. For the species: area relationship for termites of islands of the Puerto Rico Bank (see fig. 17). Su et al. (2003) recorded 11 species on the closest other island bank, St. Croix, which has a land area today that is about 100 times larger than Guana. Three types of termites are found on Guana: nasutes, family Termitidae, are the most specialized with glue-squirting soldiers; drywood termites, family Kalotermitidae, living in sound but usually deadwood, sometimes occurring in buildings; and subterranean termites, family Rhinotermitidae, including pest species that enter structural timbers from the soil and can cause extensive damage.

Termites play an important role in communities as natural recyclers, breaking down the tough cellulose cell walls in wood or other plant material and returning the contained nutrients to the soil to be taken in by growing vegetation. This recycling activity is widely recognized as essential to the survival of forest, desert, and grassland plant communities in temperate and tropical areas all over the world. Some termites contain bacteria that fix nitrogen—an ability rare in the world. We are familiar with the nitrogen-fixing abilities of the bacteria inhabiting root nodules of legumes; these are important, as the only other sources for usable nitrogen are lightning and manures or commercial fertilizers. The bacteria in the termites break down cellulose and provide nitrogen used for growth; any excess is excreted. Termite excrement is called "frass"; for a fascinating description of its derivation and various definitions, as well as its several uses and values, check Berenbaum (2003). Our termites use frass in nest construction. Termite nests, broken into small pieces, are used as fertilizer mulch in some areas.

Termites have complex societies characterized by cooperative care of juveniles, specialized

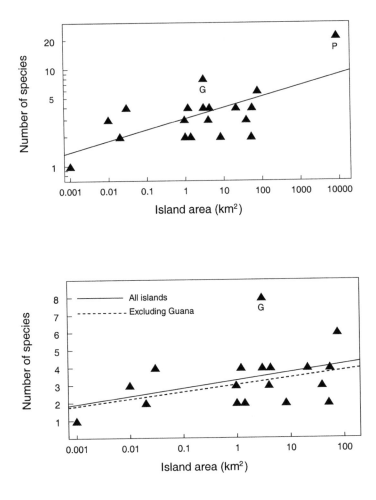

FIGURE 17. Species : area relationship for termites on islands of the Puerto Rico Bank. G, Guana; P, Puerto Rico. All islands: Anegada, Beef Island, Great Camanoe, George Dog, Eustatia, Great Thatch, Ginger Island, Jost Van Dyke, Carrot Rock, Little Tobago, St. Thomas, Necker Island, Cooper Island, Peter Island, Scrub Island, Tortola, Virgin Gorda, St. John. Both *x*- and *y*-axis are in log scale with a regression line included. Below, the same islands but without Puerto Rico and with *y*-axis in numerical scale. Dotted line, regression without Guana data included. Solid line, regression with Guana data included.

division of labor, and reproduction by only one pair or a limited number of males and females. Individual termite behavior is much more rigid than mammalian behavior. The colony as a unit, however, has great flexibility and has been compared to a complex organism in its ability to adjust to changes and environmental challenges. Termites provide many research questions of interest as well as essential recycling activities permitting the survival of plants in their communities (Merbach et al. 2002). Most species of termites are beneficial, but those that feed on the wood in buildings can cause significant structural damage. Nonpest species live in dead wood, or forest litter, leaves, or humus, and process plant material in and around their nest site.

After several years of growth, termite colonies have periodic (usually annual) dispersal flights. These nuptial flights release hundreds, thousands, or even tens or hundreds of thousands of fertile males and females from single colonies. Most of these winged termites will be captured by hungry birds, bats, lizards, snakes, ants, or other predators, but those that survive will become the founding kings and queens of new termite colonies. After a reasonably short flight, the dispersing termites land and lose their wings. A female releases a chemical signal to attract a male, causing the male to follow the female closely, and the pair will seek a place to hide and begin a new colony. Members of a colony are specialized morphologically for different roles in life. These distinct forms are called "castes". The castes of higher termites, for example, family Termitidae, typically have a "royal couple," the king and the queen, who reproduce, and a large population of nonreproductive workers, along with one or two forms of soldiers. As long as the queen lives, she produces chemicals that prevent any of the workers from developing into reproductives. Typically, one individual in the queen's first brood of eggs will become a soldier. As the colony grows, more soldiers are produced and will eventually comprise up to 20% of the total population, depending on the species. Soldiers produce a chemical that prevents other individuals from differentiating into soldiers, but this inhibition is density dependent. If the density of soldiers in the colony falls below a certain level, some immatures will metamorphose into soldiers. Once soldier density goes up enough to restore the chemical suppression levels, immatures will become workers. If there are two soldier castes, they are often derived from metamorphosis of different instars or developmental stages.

All termites produce olefins and other waxy hydrocarbons with chains 23–45 carbon atoms long per molecule. Any one species may produce a mixture of about 60 kinds of these molecules in the cuticle of the exoskeleton. The composition of the mix typically correlates with habitat moisture and the water-loss characteristics of each individual species (Collins 1991). In a series of papers, Haverty et al. (1991, 1996, 1997) analyzed cuticular hydrocarbon mixtures of our termites (and some others) and developed protocols for their collection. They found that workers can have qualitatively different mixes than soldiers, and—even within Guana—different colonies may have quantitatively different sets. The hydrocarbons are best extracted from a sample of at least 100 termites (200 for small species) that are frozen, thawed, and then soaked in hexane for 10 minutes. The hexane is drained off and the extract condensed; it is then run through a gas chromatograph mass spectrometer, which produces a profile of the molecules present and their relative quantities. For the best results, each mixture is subjected to a temperature program beginning at 20°C and increasing at 3°C per minute to 320°C, followed by a 16-minute hold, making (by my arithmetic) a 116-minute total run. The different hydrocar-

bon molecules elute from the mix at specific temperatures; typically, the longer the carbon chain of the molecule the higher its eluting temperature. Each species has a very distinctive hydrocarbon profile. Usually, those that are most abundant in the driest habitats have a set of late eluting, or high boiling point, long-chain molecules that are believed to be the best waxes for water-loss resistance.

Cuticular hydrocarbons have played an increasing role in the taxonomic study of Guana's termites and their relatives (Haverty et al. 1992; Thorne et al. 1994, 1996c; Collins et al. 1997). Comfortingly, the species recognized by their conventional morphological characteristics also sort out by their cuticular hydrocarbons. Thus, hydrocarbon chemistry can be used to identity termite species even in the absence of morphologically distinctive specimens. The hydrocarbons can even be determined from the solid excrement (frass) in the nest and galleries in the absence of living termites (Haverty et al. in press).

The most abundant and conspicuous species in the BVI is the nasute, *Nasutitermes acajutlae*. It builds covered runways or galleries on trees and walls as well as big nests in trees. You can detect a turpentine-like odor if you break into the galleries or nest and may even get brown stains on your hands. This material is a chemically unique glue that the soldiers squirt out of their nozzle-shaped heads when disturbed. It can stop an attacking ant in its tracks, gumming up antennae and legs. The soldiers are blind; it is unknown how they manage to aim so well, but we suspect they react to air currents and chemical cues. These "Bazooka-headed" termites can defend the nest against an ant attack until the damage is repaired. Workers, the fat and brown individuals that make up the majority of the colony population, mend breaks in the nest or in the dark, covered galleries or runways leading from the nest to the ground or to other trees. Workers construct all the new galleries, too.

Striking features on Guana are the large, dark brown nests on trees, built by nasute termites. The nest is composed of chewed wood fragments and secretions from the termites, a material called "carton". If mosquitoes are a problem, they can be repelled easily, pleasantly, and safely by simply taking a chunk of this carton, knocking all the termites out of it, and then lighting it as one would burn incense. The fragrant smoke is a potent, nontoxic mosquito repellent.

If you want to collect nasute soldiers, shave off the surface of a nest and place a sheet of moist cardboard on the exposed surface. The workers will flee to the interior of the nest and soldiers will swarm out onto the foreign cardboard. If you leave the cardboard in place for an hour, the soldiers will disperse and workers will return to begin incorporating the cardboard into a new nest surface. Termites collected from the cardboard at this time will be mostly workers, with only the usual small number of patrolling soldiers.

Just how Guana's termites recognize the members of their own colony and "know" to avoid or fight members of other colonies is not yet known. Of course, all colony members are very closely related, as they are the offspring of one or a few royal couples. Cuticular hydrocarbons could be the answer, but in the subterranean termite *Reticulitermes speratus* (Rhinotermitidae) of Japan, studied by Matsuura (2001), symbiotic gut bacteria produce the chemical cues for nestmate recognition. Apparently, each colony of termites has its own clones of bacteria. Chemicals from the bacteria are concentrated in and secreted by an abdominal gland wherever the termite goes. Thus its kin can recognize and follow it; termites can tell their familial galleries and trails from those of strangers and keep their work effort "all in the family."

Nasutitermes acajutlae is known from the Puerto Rico Bank southward to the South American continental shelf island of Trinidad and on the mainland in Guyana. With its suite of high-eluting hydrocarbons, it is vastly more abundant than other species, occurring even on such broiled, bleak outposts as Carrot Rock. Its much scarcer (in the BVI) congener *Nasutitermes costalis* lacks high-eluting hydrocarbons and

only lives in relatively moist habitats. *Nasutitermes costalis* occupies the entire range of *Nasutitermes acajutlae* and also the Greater Antilles west of Puerto Rico. It is a very common termite on other Caribbean islands.

Nasutitermes acajutlae has two remarkable traits that help in hard times when water or food are scarce. First, its workers build their galleries to and into tank bromeliads, the pineapple family epiphytes that hold water in their leaf axils (Thorne et al. 1996b). Sometimes the galleries are only extended to bromeliad roots, but usually they are constructed up into the leaves, even if the workers have to make a suspended gallery tunnel up to 12 cm long. In building these bromeliad galleries, the workers first lay down a trail of fecal deposits and other secretions from their nest to the target bromeliad. Then they begin building the tube from the bromeliad back to the nest. This way they have a water supply while they work. Tank bromeliads (the technical term is "*phytotelms*") are a major source of stored water in arid environments, used by many insects (e.g., mosquitoes) and even vertebrates such as lizards and birds. On Guana in a drought year the nasute termites accessed bromeliads on their nest trees in more than 97% of potential opportunities, and as many as eight bromeliads on a single tree. The nasute workers have an abdominal water sac and can transport water from the bromeliads back to their nests. Thorne et al. (1996b) report similar bromeliad tapping behavior in other species of termites from dry areas in southern Brazil. In a bizarre twist on the phytotelm story, Merbach et al. (2002) report that a pitcher plant from Borneo attracts nasute termites with edible white hairs on its water-filled trap lip, leading thousands of these termites per pitcher to slip to a watery death.

The second remarkable trait of *Nasutitermes acajutlae* is nodule production in their nests. This is apparently very rare in the world, reported only in a very few other species of nasutes in Panama and Jamaica and in a distantly related species in Java. The nodules made by our nasutes are a woodlike product, paler than nest carton, smooth and rounded, and fairly dense. They are arrayed in some nests, at least 2 cm in from the surface, in a zone surrounding the hard core of the nest. Only two of dozens of nests checked had nodules. Some nodules were solid, but others were being hollowed out—eaten—by larval termites. As far back as 1877, H.G. Hubbard reported finding nodules in nasute nests in Jamaica; he suggested that they are a food storage mechanism. Thorne et al. (1996a) analyzed nodules and regular nest carton (which is also sometimes eaten by hungry termites). The nodules were similar to nest carton in many respects, indicating that they are made from raw wood in the same way: chewed and glued, similar to Masonite. However, nodules are relatively rich in cellulose, the primary energy source for termites, with as much as double the amount as in normal nest carton, and nodules had only about one-third as much cutin. Cutin is indigestible to most organisms, including, we suspect—but do not yet know—termites. In the laboratory, various termites including *Nasutitermes acajutlae* eat nodules preferentially and with seeming gusto, even while ignoring nest carton. Thus, after 120 years, Hubbard's observations were verified and we know of yet another remarkable adaptation of our nasutes to the stress and vicissitudes of an arid climate.

The longest-running study of termite nest survival and development, spanning nine years (the next longest, two years in India), is reported by Thorne and Haverty (2000), done on Guana Island. Of the 35% of nests that survived the nine years, the biggest ones had the highest survival rate. Seventeen nests were monitored. Five were over 150,000 cm³, and four of these survived. Of the 12 smaller nests, only 2 survived; 11 were less than 100,000 cm³, and 9 of these succumbed. Following are three ways that new nests were generated from old ones. (1) Budding occurs when a satellite nest is built in direct contact, by covered trails, with the parent nest. A new royal couple may mature in a budded nest. (2) Moving also occurs; the colony departs the old nest and relocates to the newly built one.

Even the royal couple moves, including the physogastric queen (physogastry is the condition of being mostly an enormous abdomen containing egg-laying machinery). (3) Colonies may resprout after destruction of the original nest (e.g., by hurricane force winds). If the queen is killed, then some of the workers or developing nymphs are released from their slavery, which had been maintained by the antidevelopment hormone the queen secretes. Without her hormonal inhibition, workers can mature into ergatoids: sexually capable, wingless workers. Or nymphs (developing alates) may remain in their nest, never develop membranous wings, and become functional reproductives.

New colonies normally arise when an existing colony produces imagoes or sexually functional adults called "alates". Alates of *Nasutitermes acajutlae* have been found often in October on Guana. They are chestnut brown, except for their heads, unlike pallid workers and soldiers. A mated pair of alates will seek a knothole or crevice in a tree and take up residence. If all goes well, the new population will increase, but the young colony does not initially build a nest. The smallest carton nests yet documented were a little over 7,000 cm³ (the largest can be over 1,500,000 cm³—that is 1.5 m³!) The new population leaves its old home concealed within wood and builds a real nest when its numbers have grown to the point at which the workers can make a nest bigger than 7,000 cm³. But how many is that? We do not know. Because only a fraction of the individual termites in a colony are in the nest at any one time, counting them will not answer the question (Thorne and Haverty 2000 counted 10,000 individuals in a chunk of nest 10 cm in diameter). One cannot find and count the members of the colony that are not in their nest. There is still a lot to learn about our most common nasute.

The second nasute species, *Nasutitermes costalis*, builds much smaller, darker, and bumpy nests in Guana's moist ravines. The termites are smaller and darker than *Nasutitermes acajutlae*; alates are dark brown to nearly black and have charcoal gray wings. The soldiers are darker, too, and have reduced pilosity (that is, a coat of long, fine hairs) on their head capsules. *Nasutitermes costalis* is common on the heights of Tortola; for example, in the aridulate rain forest at Sage Mountain. The alates are said to be day fliers (Hubbard 1877).

The remaining termitid on the Guana Island is *Parvitermes wolcotti*, subject of taxonomic studies by Roisin et al. (1996) and Scheffrahn and Roisin (1995), who named a close relative on Hispaniola for Margaret Collins: *Parvitermes collinsae*. On Guana, *Parvitermes wolcotti* is so far known only from limited collections under rocks in Pinguin Ghut. It feeds on grounded dead wood, is very small, and notably fast moving. The alates have large, protruding, compound eyes and accessory ocelli on projections. Like *Nasutitermes acajutlae* and unlike *Nasutitermes costalis*, *Parvitermes wolcotti* has a fine set of late-eluting cuticular hydrocarbons, so it should be well adapted to very dry conditions. Perhaps its small size and highly mobile activity render it susceptible enough to water loss that it needs increased desiccation resistance even in relatively moist microhabitats.

Guana has a single representative of Rhinotermitidae, a possibly new species of *Heterotermes*. It is, or very close relatives are, found throughout the Caribbean. The soldiers have relatively narrow, pointed mandibles that curve at the tips. The heads are light and the mandibles dark. The pronotum—the body segment just behind the head—is narrow, giving these termites a "neck." The alates are small, light brown, night flyers. Most of the cuticular hydrocarbons of this *Heterotermes* are in the midrange length: 26–31 carbons to the chain. It lacks both very early and late-eluting hydrocarbons. Collins et al. (1997) suggested that *Heterotermes* might avoid water loss by living in the soil and foraging in carton-lined galleries that preserve high humidity.

The drywood termites, family Kalotermitidae, are represented on Guana by four species. They have proven highly informative on the evolution of termite social structure, life histories, and behaviors (Thorne and Traniello 2003). Guana's well-studied species are exemplary. The most common is a member of *Incisitermes*. At first it was called *Incisitermes snyderi*, then for a while we thought there was a second species, *Incisitermes incisus*, but now the consensus is that there is only one species with two very different soldier morphs. All colonies have the short-headed, *snyderi*-type soldiers, but only older colonies produce long-headed *incisus*-type soldiers. Are the two "species" one everywhere? Or does ecological release on a one-species island allow development of a soldier morph that resembles *Incisitermes incisus*? We do not know. Our *Incisitermes* is a terrible pest. It does not confine itself to consuming sound deadwood out in the forest; it insists on eating what the pest-control people call (I love the phrase!) "wood in service"—you know, like the chair I am sitting on. One relevant *Incisitermes* or the other occurs in North America, Central America, the Greater Antilles generally, and the Bahamas. Our *Incisitermes* has a fine set of early eluting hydrocarbons, a gap in the midrange, and the strongest of all sets of late eluters. These even beat out *Nasutitermes acajutlae*. It must need terrific water-loss resistance to live in dead, drywood indoors, out of the rain. None of our experts seems to have described more than the two soldier morphs or recorded a nuptial flight.

Procryptotermes corniceps is scarce on Guana. It gets its species name (literally, "horn head") from the extremely elongate, curved, and pointed mandibles of the soldier caste. The heads are sloped frontally and the anterolateral margin also has short horns. The alates are yellow brown and nocturnal. This species is known from Jamaica, the Puerto Rico Bank, the Turks and Caicos islands, and southward down the Lesser Antilles to Guadeloupe. It is strong on early eluting hydrocarbons and has a reasonably good set of late eluters, too, with a midrange gap.

The largest of all our termites, and the largest Antillean member of its family, is *Neotermes mona*. It attains a length of more than 1.5 cm in the soldier caste and alates with wings can be 2.2 cm overall. There are two soldier morphs, best separated on head capsule length:

the smaller with heads up to 3.4 mm, the larger with heads no shorter than 3.8 mm. Both morphs are usually present in mature colonies. The alates have distinctively colored rust red heads. Krecek et al. (2000) have redescribed this species, long thought to be unique to Mona Island, between Puerto Rico and Hispaniola. Margaret Collins discovered it on Guana, and Jan Krecek came to study it here and was rewarded with nuptial flights in October 2000 between 1 and 2 a.m. The species is now known to be widespread on Hispaniola and Puerto Rico, in the Turks and Caicos islands, and from Vieques and St. John. *Neotermes mona* makes only very early eluting cuticular hydrocarbons, causing Collins et al. (1997) to rank it as probably very poor at water retention. Krecek et al. (2000) found it in very dry habitats, however, and argued that it has "substantially lower moisture requirements than sympatric termite species in the families Rhinotermitidae and Termitidae that have soil access." Perhaps I should refrain from entering the fray, but it seems to me that Krecek et al. (2000) torpedo their own argument by pointing out that *Neotermes mona* galleries "occasionally extend into the xylem elements of living trees . . . possibly as moisture refugia during drought."

Guana's eighth and by far rarest species, *Cryptotermes brevis*, has the largest geographic range of all, vastly larger than most any species of anything except people. It occurs all over the tropical and most of the subtropical world. Of course, we can explain away its deviance from the cherished glitterality: It is a human commensal, exotic over most of its range (as on Guana). *Cryptotermes brevis* lives in buildings. You can tell when you have an infestation because of the frass piles under holes in the wood surface and the heads of dead soldiers—called "phragmatic" heads—jammed like grotesque little corks in the openings. They will render a support beam a paper-thin ghost of its former self. *Cryptotermes brevis* is rare because whenever it appears somebody calls the exterminator. To include it in the hydrocarbon study, Collins et al. (1997) had to use specimens from Hawaii, where it is much too abundant. In the BVI, *Cryptotermes brevis* has been found only in structural timbers and furniture and was typically missed in the biogeographic survey, reportedly because they did not look inside houses.

Dragonflies: Odonata

Dragonflies have been around for over 250 million years, and it's about time they got the recognition they deserve. Like birds and butterflies, they are large and colorful, diurnal and diverse.

DENNIS R. PAULSON (2001)

THE INSECT ORDER ODONATA includes both dragonflies and damselflies—which most of us erroneously think of as dragonflies. Odonate means toothy, and refers to the well-armed mouthparts. All odonates are voracious predators of other insects. Odonata is divided into three suborders: damselflies, the Zygoptera; dragonflies, the Anisoptera; and a couple of intermediate Asian species put in the Anisozygoptera. The anisopteran dragonflies have broader hind wings than forewings and hold their wings flat out, like an airplane, when perched. The zygopteran damselflies have nearly identical hind wings and forewings, which they hold vertically relative to the body when perched. There are other differences, too, that are less conspicuous, detailed by Dunkle (2000).

As with most insects, odonate bodies are divided into three major parts: (1) the head, with enormous, bulging, compound eyes and toothy mouthparts; (2) the thorax, a boxlike section to which wings and legs articulate; and (3) the abdomen, which is long, conspicuously segmented, and resembles a tail. All odonates have aquatic larvae, called "nymphs" or "naiads"; most live only in freshwater. Guana has no permanent freshwater, but pools persist in the gluts and low spots on the White Bay flat in the rainy season (July through November). Only limited reproduction of odonates can occur on Guana and no species has a permanently established population. Our odonates are gypsies: They come and go. Some move in front of, and along with, storms and may invade by the thousands. Ornithologist Fred Sibley has assiduously collected odonates on Guana over the years and has also made comparative collections from Tortola, Anegada, and Beef islands in the BVI. Sibley has published three papers on our odonates (Sibley 1999, 2000, 2002) and provided much of the information presented here. In addition, Dunkle's (1989, 1990, 2000) books provide more details on most of our species and lists of additional references. We have 10 odonate species.

Rambur's Forktail, *Ischnura ramburii*. At present, this is Guana's only known damselfly; if things stayed that way it would be easy to recognize as such, but another species—the rainpool spreadwing, *Lestes forficula*—is known from Tortola, so we suspect it is only a matter of time before it turns up on Guana. Medium-sized for a damselfly: 2.6–3.5 cm (an inch to nearly an inch-and-a-half). The males are blackish with bright green sides on the thorax and anterior abdominal segments. The eighth abdominal segment, near the end of the "tail," is often blue. Some females resemble males, but some are olive-drab, with the dorsum of the thorax tawny, and others are largely orange-red. The latter, called "red form," may become olive with age. The behavior of these forktails is very complex, involving up to seven hours of mating, male mimicry, courtship refusal displays, and cannibalism. Dunkle (1990) provides details and references and states that this species is named for French entomologist M. P. Rambur. This forktail occurs from Maine and Missouri to Oklahoma, and on Bermuda, the Bahamas and throughout the Antilles, Central America, and most of South America down to Chile. It is introduced in Hawaii. It is often abundant on West Indian islands, such as Puerto Rico, and even Tortola and Anegada.

Roseate Skimmer, *Orthemis ferruginea*. This species is about 5 cm (2 in.) head-body length. Males are either bright red or pink; if pink, then they are pruinose, which means covered with a waxy powder called "pruinescence". Females and juveniles are brown, rusty on the abdomen, and have white stripes on the thorax. Roseate skimmers range from Delaware to California and south to Costa Rica; there are subspecies or geographically replacing species throughout the West Indies. This species has expanded its range dramatically northward from extreme south Florida since about 1930. It has been introduced to Hawaii. It can breed in brackish water.

Band-winged Dragonlet, *Erythrodiplax umbrata*. This species is about 4.5 cm (1.7 in.) and extremely variable in color and pattern. Males are usually plain, black bodied, and have a broad black band through each wing, at least as adults. Females may be similar, or they may be plain brown with brown wing tips. This is a largely tropical species, occurring from Texas and Oklahoma and the Florida Keys and Bahamas throughout the West Indies, and south in South America to Argentina.

Slough Amberwing, *Perithemis domitia*. This is a small species, a little over 2 cm (0.9 in.). It has a yellow green thorax with brown stripes and a dull yellow abdomen with dorsolateral brown stripes. The wings are indeed amber. Dunkle (2000, 210) says this species typically perches in the shade, which is unusual among our dragonflies. This amberwing's distribution extends from the U.S. border area in Arizona and Texas south to Brazil. It is widespread but very local in the Greater Antilles; Guana Island and Tortola may be the easternmost known records.

Great Pondhawk, *Erythemis vesiculosa*. This is our largest species, about 6 cm (2.4 in.). It is bright green with dark bands on the abdomen. Described as "most ferocious" (Dunkle 2000, 212), these dragonflies are beloved by field biologists for their ability to wipe out whole swarms of mosquitoes. They can tolerate brackish water. They range from Arizona, Kansas, and Florida throughout the West Indies and Bahamas, and south to Argentina.

Wandering Glider, *Pantala flavescens*. Rather small, this species is about 4.7 cm (1.9 in.) and mostly ochre yellow. As males age, their yellow faces develop a rosy tint and their abdomens become oranger. Their gliding flight is also distinctive. Well named, this dragonfly is essentially cosmopolitan, occurring even on remote oceanic islands and all continents except Antarctica (actually, I suspect it even gets there occasionally). "Over the ocean they fly day and night for thousands of miles" (Dunkle 2000, 216). It can breed in brackish water.

Spot-winged Glider, *Pantala hymenaea*. Similar to the wandering glider, about 4.7 cm, this species instead has a mottled gray brown abdomen and a spot at the base of the hind wing. It is less common, generally, ranging from southern Canada south to Argentina and Chile, including Bermuda, the Bahamas, the Greater Antilles, and even the Galápagos Islands in the eastern Pacific. It was one of the species that invaded in numbers in the El Niño event of

October 1997 (Sibley 1999). It was far more abundant on Guana than its much more widespread wandering congener *Pantala flavescens*, but Sibley (in litt. 20 September 2001) could find no previous Antillean records east of the Dominican Republic. It almost undoubtedly reached Puerto Rico in October 1997, but as yet we have seen no record of it there. Sibley (1999) mentions *Pantala hymenaea* on Anegada in 1997 but does not list it for any island other than Guana in the 15 May 2001 report he sent to me. As of 28 February 2002, Sibley (in litt.) doubts the Anegada sight record. It tolerates brackish water.

Striped Saddlebags, *Tramea calverti*. This is another species about 4.7 cm (1.9 in.) in length. The name "saddlebags" derives from the dark band or blotch on the bases of the hind wings, "which make the body look larger than it really is" (Dunkle 2000, 217). This species is distinctive in having wide light gray or nearly white stripes on the brown thorax. The range extends from south Texas to Argentina, with vagrants occurring from California and Wisconsin to New England. It occurs in the Galápagos Islands and is widespread in the West Indies. Surprisingly, it appears not to have been recorded from Puerto Rico, similar to the spot-winged glider. Also like that species, it invaded Guana in great numbers in the El Niño event of October 1997. It seems to be migratory, with Cuban and Central American swarms reaching Venezuela in October (Dunkle 1989, 115). Individuals attempting that passage in 1997 could easily have been the source of the Guana invaders, carried northeast by the freak southwesterlies.

Vermillion Saddlebags, *Tramea abdominalis*. As with the preceding species, this one is about 4.7 cm long. This species is a rich red (females have a yellow forehead) with a narrow black middorsal stripe on the posterior abdomen. Its range extends from south Florida—with rare vagrants northward along the U.S. east coast—to Argentina, and includes Bermuda, the Bahamas, and the West Indies. It has been introduced to Hawaii.

Tawny Pennant, *Brachymesia herbida*. This is yet another 4.7 cm (1.9 in.) species. It appears largely near-black with dull ochre abdomen sides; the wings are tinted amber brown and unspotted. This is the most recent addition to the Guana dragonfly fauna, known from a single specimen collected in October 2001 by Fred Sibley. He has

TABLE 29
Known Dragonfly and Damselfly Occurrences in The BVI

		TORTOLA	ANEGADA	GUANA	BEEF ISLAND
1	Rainpool Spreadwing	×	×		×
2	Rambur's Forktail	×	×	×	×
3	Roseate Skimmer	×	×	×	×
4	Band-winged Dragonlet	×	×	×	×
5	Seaside Dragonlet	×	×		×
6	Slough Amberwing	×		×	
7	Great Pondhawk	×		×	×
8	Wandering Glider	×	×	×	×
9	Spot-winged Glider			×	×
10	Striped Saddlebags		×	×	
11	Vermilion Saddlebags	×	×	×	×
12	Red-tailed Pennant	×			
13	Tawny Pennant	×		×	×
14	Tropical Dasher	×			

SOURCE: Fred Sibley, pers. comm.

collected increasing numbers each year on Tortola and Beef islands, so a spillover to Guana was not unexpected. It is probably more common on Guana than a singleton indicates, but it closely resembles the females and juveniles of other species and may thus be overlooked. Tawny pennants occur from Texas and Florida south to Argentina; they are widespread in the West Indies and occur in the Galápagos.

I have discussed Guana's odonate fauna in terms of species richness above. For known dragonfly and damselfly occurrences in the BVI, see table 29. Anegada, at 3,872 ha—an order of magnitude larger than Guana, and with extensive freshwater and brackish water wetlands—has eight or nine known species. Tortola, at 5,494 ha—18 times the size of Guana Island—has 13 species recorded. Any or all of these can be expected to appear, sooner or later, if only briefly, on Guana. To date, *Pantala hymenaea*, the spot-winged glider, has not been found on Tortola, but it has occurred in big numbers on Guana. The four species known from the BVI, but not from Guana, are: the rainpool spreadwing *(Lestes forficula)*, a second species of damselfly; the seaside dragonlet *(Erythrodiplax berenice)*; the red-tailed pennant *(Brachymesia furcata)*; and the Caribbean dasher *(Micrathyria dissocians)*. The last has not been recorded in the United States, either.

Beetles: Coleoptera

Barry D. Valentine • *Michael A. Ivie*

I N HIS EARLY DESCRIPTION of Guana Island's fauna, Lazell (1989a) guessed there might be as many as 200 species of beetles present. Many regarded this figure as a preposterous example of his overweening ebullience. A very well-studied, nearby island such as St. Croix has about 200 beetle species (Beatty 1944a; Miskimen and Bond 1970), and it is about 100 times the size of Guana. Stubbornly, Lazell (1996b) reiterated his opinion. Of course, he was wrong. Previously, only 12 species were actually recorded from Guana in the scientific literature. We here increase this to 411 species. The true number of beetles living on Guana is still unknown and could rise above 500. All species are new records for Guana, except as specifically noted below.

Assistance in identification of Guana Island or Virgin Islands specimens used directly or to determine our material was provided by: Robert Anderson (Curculionidae), Fred Andrews (Latridiidae), Ross Arnett (Oedemeridae), George Ball (Carabidae), William Barr (Cleridae), Richard Beal (Dermestidae), Edward Becker (Elateridae), Chris Carlton (Pselaphidae), H. Franz (Scydmaenidae), Robert Gordon (Coccinellidae), Robert Hamilton (Attelabidae), Anne Howden (Curculionidae), John Kingsolver

(Dermestidae, Bruchinae), Alexander Kirejtshuk (Nitidulidae), David Kissinger (Apioninae), Peter Kovarik (Histeridae), John Lawrence (Ciidae), Ivan Lobl (Scaphidiinae), Richard S. Miller (Lycidae), Charles O'Brien (Curculionidae), James Pakaluk (Corylophidae), T. Keith Philips (Ptininae), S. Adam Slipinski (Cerylonidae), Ales Smetana (Hydrophilidae), Warren Steiner (Dytiscidae, Tenebrionidae), Michael Thomas (Silvanidae, Laemophloeidae), Charles Triplehorn (Tenebrionidae), Ruppert Wenzel (Histeridae), Richard White (Anobiidae), and Walter Wittmer (Cantharidae).

Responsibility for any errors in identification in these groups rests with us. In some cases, all records for a group are based on the unpublished work of a specialist, who is acknowledged next to the family, subfamily, or tribe name. Subfamilies are provided where it will assist understanding or clarify placement of groups sometimes considered families. The classification follows Lawrence and Newton (1995).

The specific location of at least one voucher for each Guana Island species is indicated by a word or letters to the right of the species name. Most vouchers are in the West Indian Beetle Fauna Project at Montana State University,

Bozeman, and/or the Barry D. Valentine Collection in Columbus, Ohio. For those cases where a voucher is not present in either of these collections, an acronym indicates the voucher repository as follows: BPBM, Bernice P. Bishop Museum, Honolulu, HI; NMNH, National Museum of Natural History, Washington, DC; NCSU, Department of Entomology, North Carolina State University, Raleigh.

An asterisk (*) indicates either a transient nonbreeding or nonestablished species or an introduced species suspected of no longer occurring on Guana. When an author of a name is put in parentheses this means the species was not placed in the genus regarded as correct today. In zoology (as opposed to botany) we do not follow this with the revisor's name.

CARABIDAE

CICINDELINAE
Cicindela suturalis Fabricius. Bozeman, Valentine.
Cicindela trifasciata Fabricius. Valentine.
Megacephala sobrina Dejean. Bozeman, Valentine.

TRECHINAE
Stylulus nasutus Schaufuss. Bozeman, Valentine.
Tachys ensenada Mutchler. Bozeman, Valentine.
Tachys pallidus Chaudoir. Bozeman, Valentine.

HARPALINAE
Aephnidius sp. Valentine.
Apenes iviei Ball & Shpeley. Bozeman.
Discoderus beauvoisi Dejean. Bozeman.
Pentagonica flavipes Leconte. Bozeman, Valentine.
Perigona nigriceps Dejean. Bozeman.
Plochionus amandus Newman. Bozeman, Valentine.
Pseudaptinus dorsalis Brulle. NMNH
Selenophorus alternans Dejean. Bozeman.
Selenophorus discopunctatus Dejean. Bozeman, Valentine.
Selenophorus latior Darlington. NMNH
Selenophorus propinquus Putzeys. NMNH
Selenophorus sinuatus Gyllenhal. Valentine.

DYTISCIDAE
Eretes occidentalis (Erichson).* NMNH
Copelatus sp.* NCSU

HYDROPHILIDAE

HYDROPHILINAE
Berosus metalliceps Sharp. Bozeman, Valentine.
Hydrophilus insularis Fabricius.* Bozeman, Valentine.
Enochrus bartletti Short. Bozeman.

SPHAERIDIINAE
Dactylosternum advectum Horn. Valentine.
Omicrus ?palmarum (Schwarz). Valentine.
Oosternum costatum Sharp. Valentine.
Phaenonotum extriatum (Leconte). Valentine.

HISTERIDAE

DENDROPHILINAE
Carcinops sp. Bozeman.

TRIBALINAE
Epeirus sp. near *antillarum* Marseul. Bozeman.

SAPRININAE
Hypocaccus sp. Bozeman.

HISTERINAE
Hister servus Leconte. Bozeman.
Idolia laevissima Leconte. Bozeman.
Omalodes laevigatus Quensel. Bozeman, Valentine.

PTILIIDAE
Three undetermined species. Valentine.

LEIODIDAE
Aglyptinus puertoricensis Peck. Bozeman.
Zeadolopus puertoricensis (Peck). Bozeman, Valentine.

STAPHYLINIDAE

PSELAPHINAE
Balega elegans (Reitter). Valentine.
Ephima simoni Reitter. Valentine.

SCAPHIDIINAE
Baeocera unicolor Pic. Bozeman.
Baeocera, four undetermined species. Bozeman, Valentine (two only).
Scaphisoma, two undetermined species. Bozeman.

Thirty or more species remain to be identified. Bozeman, Valentine.

SCYDMAENIDAE

Euconnus coralinus Reitter. Valentine.
Euconnus testaceus (Schaum). Valentine.

TROGIDAE

Omorgus suberosus (Fabricius). Bozeman, Valentine.

CERATOCANTHIDAE

Ceratocanthus pyritosus (Erichson). Bozeman.
Nesopalla iviei Paulien & Howden. Bozeman, Valentine.

SCARABAEIDAE

APHODIINAE
Aphodius lividus Olivier. Bozeman.
Aphodius nigrita Fabricius. NMNH
Ataenius beattyi Chapin. Bozeman, Valentine.
Ataenius michelii Chalumeau. Valentine.
Ataenius scutellaris Harold. NMNH

DYNASTINAE
Ligyrus cuniculus (Fabricius). Bozeman, Valentine.
Strategus talpa (Fabricius). Bozeman, Valentine.

MELOLONTHINAE
Phyllophaga iviei Chalumeau. Bozeman, Valentine.
Phyllophaga microphylla (Moser). Bozeman, Valentine.
Phyllophaga pleei (Blanchard). Bozeman, Valentine.

BUPRESTIDAE

BUPRESTINAE
Chrysobothris thoracica (Fabricius). Bozeman, Valentine.
Chrysobothris tranquebarica (Gmelin). Valentine.

POLYCESTINAE
Acmaeodera gundlachi Fisher. Bozeman.
Micrasta ornata Fisher. Bozeman, Valentine.
Micrasta sp. Bozeman, Valentine.

Micrasta uniformis (Waterhouse). Bozeman.
Polycesta porcata (Fabricius). Bozeman.

CHELONARIIDAE

Chelonarium punctatum Fabricius. Bozeman, Valentine.

THROSCIDAE

Aulonothroscus sp. Bozeman, Valentine.

ELATERIDAE

AGRYPNINAE
Aeolus sp. near *discollis* Candeze. Bozeman, Valentine.
Aeolus sp. near *granulatus* Candeze. Bozeman, Valentine.
Conoderus castaneus (Fabricius). Bozeman, Valentine.
Conoderus sticturus (Candeze). Bozeman, Valentine.
Lacon subcostatus (Candeze). Valentine.

ELATERINAE
Anchastus sp. Bozeman, Valentine.
Ischiodontus sp. Bozeman, Valentine.

CARDIOPHORINAE
Esthesopus poedicus Candeze. Bozeman, Valentine.

LYCIDAE

Leptolycus sp. (R.S. Miller 1991). Bozeman, Valentine.

CANTHARIDAE

Caccodes iviei Wittmer. Bozeman, Valentine.
Tylocerus barberi Leng & Mutchler. Valentine.
Tytthonyx discolor Leng & Mutchler. Bozeman, Valentine.
Tytthonyx guanaensis Wittmer (Wittmer 1992). Bozeman, Valentine.
Tytthonyx sp. Bozeman, Valentine.

PHENGODIDAE

One undetermined species. Valentine.

DERMESTIDAE

Cryptorhopalum preschi Beal. Bozeman, Valentine.

Cryptorhopalum quadrihamatum Beal. Bozeman, Valentine.
Cryptorhopalum sp. Bozeman, Valentine.
Trogoderma ornatum (Say). Bozeman, Valentine.

BOSTRICHIDAE

POLYCAONINAE
Melalgus femoralis (Fabricius). Bozeman, Valentine.

BOSTRICHINAE
Amphicerus cornutus (Pallas). Bozeman, Valentine.
Rhyzopertha dominica (Fabricius)*. Valentine.
Xylomeira tridens (Fabricius). Bozeman, Valentine.

ANOBIIDAE

ERNOBIINAE
Ozognathus sp. near *floridanus* Leconte. Bozeman, Valentine.

ANOBIINAE
Trichodesma, two undetermined species. Bozeman, Valentine (one only).

XYLETININAE
Megorama sp. Valentine.

DORCATOMINAE
Ascutotheca sp. Bozeman, Valentine.
Caenocara maculata Fisher. Bozeman, Valentine.
Petalium, three undetermined species. Bozeman, Valentine.
Protheca, four undetermined species. Bozeman, Valentine.

MESOCOELOPODINAE
Cryptorama carinatum White. Bozeman, Valentine.
Cryptorama ?densipunctatum White. Valentine.
Cryptorama impunctatum White. Bozeman, Valentine.
Cryptorama megalops White. Bozeman, Valentine.
Cryptorama ?sericeum White. Valentine.
Cryptorama ?tortolensis White. Bozeman, Valentine.

Cryptorama sp. near. *antillensis* White. Bozeman, Valentine.
Neosothes sp. Valentine.
Tricorynus insulicola (Fisher). Bozeman, Valentine.
Tricorynus, three undetermined species. Bozeman (one only), Valentine.

PTININAE
Lachnoniptus lindae Philips (Philips 1998). Bozeman, Valentine.
Ptinus strangulatus Fall. Valentine.
Pitnus antillanus Bell. Bozeman, Valentine.

TROGOSSITIDAE

Colydobius sp. Bozeman, Valentine.
Temnochila sp. Bozeman, Valentine.
Tenebroides transversicollis J. du Val. Bozeman, Valentine.

CLERIDAE

Neorthopleura murina (Klug). Bozeman, Valentine.
Phlogistosternus sp. Bozeman, Valentine.

MELYRIDAE

Ablechroides, two undetermined species. Bozeman, Valentine (one only).
Melyrodes, two undetermined species. Bozeman, Valentine.

NITIDULIDAE

Carpophilus mutilatus Erichson. Valentine.
Conotelus sp. Bozeman.
Epuraea (Haptoncus) luteolus (Erichson). BPBM
Stelidota ruderata Erichson. Bozeman, Valentine.

MICRIPIDAE

Smicrips sp. Bozeman, Valentine.

MONOTOMIDAE

Europs maculatus Grouvelle. Valentine.

SILVANIDAE

Ahasverus plagiatus (Grouvelle). Bozeman, Valentine.

Cathartus quadricollis (Guerin-Meneville). Bozeman, Valentine.

LAEMOPHLOEIDAE

Lathropus sp. Bozeman, Valentine.
Parandrita permixtus Grouvelle. Valentine.

PHALACRIDAE

Acylomus sp. Bozeman, Valentine.
Ochrolitus tristriatus Casey. Bozeman, Valentine.
Stilbus sp. Bozeman.

LANGURIIDAE

Cryptophilus integer (Heer). Bozeman, Valentine.
Hapalips filum Reitter. Bozeman.
Loberus testaceus Reitter. Bozeman, Valentine.
Telmatoscius sp. Bozeman, Valentine.
Toramus sp. Bozeman, Valentine.

BOTHRIDERIDAE

Bothrideres sp. Valentine.

CERYLONIDAE

Euxestus globosus Arrow. Valentine.

ENDOMYCHIDAE

One unidentified species. Bozeman, Valentine.

COCCINELLIDAE

STICHOLOTIDINAE

Delphastus nebulosus Chopin. Bozeman, Valentine.
Neaptera purpurea Gordon. Bozeman, Valentine.

SCYMNINAE

Diomus ochroderus (Mulsant). Bozeman, Valentine.
Diomus roseicollis (Mulsant). Bozeman.
Diomus, two unidentified species. Bozeman, Valentine (one only).
Nephus sp. Valentine.
Scymnus (Pullus) phloeus Mulsant. Bozeman, Valentine.
Stethorus caribus Gordon & Chapin. Bozeman.
Zagloba sp. Bozeman, Valentine.
Zilus sp. Bozeman.

COCCINELLINAE

Cycloneda sanguinea (Linnaeus). Valentine.
Psyllobora lineola (Fabricius). Bozeman, Valentine.

COCCIDULINAE

Pseudoazya trinitatis (Marshall). Valentine.

CORYLOPHIDAE

Hoplicnema sallaei Matthew. Bozeman, Valentine.

LATRIDIIDAE

Caserus sp. Bozeman, Valentine.
Metophthalmus muchmorei Andrews. Bozeman, Valentine.

MYCETOPHAGIDAE

Berginus sp. Valentine.
Typhaea stercorea (Linnaeus). Bozeman, Valentine.

CIIDAE

Ceracis furcatus (Bosc). Bozeman, Valentine.
Cis creberrimus Mellie. Bozeman, Valentine.
Cis hirsutus Casey. Valentine.
Cis melliei Coquerel. Bozeman, Valentine.
Malacocis sp. Bozeman.
One unidentified species. Bozeman.

MORDELLIDAE

Lu and Ivie (1999) recorded the species below from Guana.
Falsomordellistena danforthi (Ray). Bozeman, Valentine.
Glipostenoda guana Lu & Ivie. Bozeman, Valentine.
Mordella atrata Melsheimer. Bozeman, Valentine.
Mordella summermanae Ray. Bozeman.
Mordellistena lineata Ray. Bozeman, Valentine.
Tolidomordella basifulva (Quedenfeldt). Bozeman.
Tolidomordella leucocephala (Quedenfeldt). Bozeman, Valentine.

COLYDIIDAE

Endeitoma granulata Say. Bozeman, Valentine.
Monoedus sp. NCSU
Paha sp. Bozeman.

ZOPHERIDAE

Antillemonomma delkeskampi Freude. Bozeman, Valentine.

Aspathines aeneus (Thomson). Bozeman, Valentine.

Hyporrhagus marginatus (Fabricius). Bozeman, Valentine.

TENEBRIONIDAE

LAGRIINAE

Adelonia sp. Valentine.

PIMELIINAE

Trientoma puertoricensis Marcuzzi. Bozeman, Valentine.

BOLITOPHAGINAE

Rhipidandrus cornutus (Arrow). Bozeman.

DIAPERINAE

Adelina mystax Triplehorn & Ivie. Bozeman.

Adelina pici (Ardoin). Bozeman, Valentine.

Cryptozoon sp. Valentine.

Gondwanocrypticus sp. Valentine.

Gnathocerus curvicornis (Champion). Bozeman, Valentine.

Iccius rufotestaceous Champion. Bozeman, Valentine.

Phaleria picipes Say. Bozeman.

Phaleria punctipes Leconte. Bozeman.

Phaleria testacea Say. Bozeman.

Phaleria thinophila Watrous & Triplehorn. Bozeman.

Platydema micans Zimmerman. Bozeman, Valentine.

Trachyscelis flavipes Melsheimer. Bozeman, Valentine.

Ulomoides ocularis (Casey). Valentine.

OPATRINAE

Blapstinus opacus Mulsant & Rey. Bozeman, Valentine.

Blapstinus punctatus (Fabricius). Bozeman, Valentine.

Diastolinus clavatus Mulsant & Rey. Bozeman, Valentine.

Platylus dilatatus (Fabricius). Bozeman, Valentine.

Sellio tibidens (Quensel). Bozeman, Valentine.

TENEBRIONINAE

Nautes sp. Bozeman, Valentine.

ALLECULINAE

Allecula ramosi Campbell. Bozeman, Valentine.

Hymenorus wolcotti Campbell. Bozeman, Valentine.

Lobopoda thomasensis Campbell. Bozeman, Valentine.

COELOMETOPINAE

Strongylium paddai Ivie & Triplehorn. Bozeman, Valentine.

OEDEMERIDAE

Hypasclera nesoites (Arnett). Bozeman, Valentine.

Hypasclera simplex (Waterhouse). Bozeman, Valentine.

Oxacis laeta (Waterhouse). Bozeman, Valentine.

Oxycopis desecheonis (Wolcott). Bozeman, Valentine.

Oxycopis tenella (Waterhouse). Bozeman, Valentine.

Oxycopis vittata (Fabricius). Bozeman, Valentine.

MELOIDAE

Pseudozonitis marginata (Fabricius). Valentine.

Pseudozonitis obscuricornis (Chevrolat). Bozeman.

MYCTERIDAE

Physcius fasciatus Pic (Pollock 1995). Bozeman, Valentine.

SALPINGIDAE

Inopeplus praeustus Chevrolat. Bozeman.

ANTHICIDAE

Anthicus ?*antilleorum* Werner (damaged). Valentine.

ADERIDAE

Aderus brunnipennis (Leconte). Bozeman, Valentine.

Ganascus, two unidentified species. Bozeman, Valentine (one only).

Gymnoganascus sp. Bozeman, Valentine.
Pseudariotus sp. Bozeman, Valentine.
Zonates sp. Bozeman.

SCRAPTIIDAE

Naucles sp. Bozeman.

CERAMBYCIDAE

PRIONINAE

Solenoptera bilineata (Fabricius). Bozeman.

CERAMBYCINAE

Anelaphus nanus (Fabricius). Bozeman, Valentine.
Curtomerus flavus (Fabricius). Bozeman, Valentine.
Eburia quadrimaculata (Linnaeus). Bozeman, Valentine.
Elaphidion conspersum Newman. Valentine.
Elaphidion irroratum (Linnaeus). Bozeman, Valentine.
Elaphidion pseudonomon Ivie. Bozeman, Valentine.
Merostenus attenuatus Chevrolat. Bozeman, Valentine.
Methia necydalea (Fabricius) (Philips & Ivie 1998). Bozeman, Valentine.
Neocompsa cylindricollis (Fabricius). Bozeman, Valentine.
Nesostizocera vanzwaluwenburgi Fisher. Bozeman, Valentine.
Nesanoplium sp. Bozeman, Valentine.
Plectomerus sp. Bozeman.

LAMIINAE

Amniscus similis (Gahan). Bozeman, Valentine.
Ataxia alboscutellata Fisher. Bozeman, Valentine.
Batocera rufomaculata (DeGeer)*. Bozeman.
Ecyrus hirtipes Gahan. Valentine.
Lagocheirus araneiformis (Linnaeus). Bozeman, Valentine.
Leptostylopsis gundlachi (Fisher). Bozeman, Valentine.
Leptostylopsis sp. Bozeman.
Spalacopsis filum (Klug). Valentine.

Urgleptes puertoricensis Gilmour. Bozeman, Valentine.
Urgleptes sandersoni Gilmour. Bozeman, Valentine.

CHRYSOMELIDAE

These records are from the unpublished work on the Chrysomelidae of the Virgin Islands by Ivie and Shawn M. Clark.

BRUCHINAE

Acanthoscelides flavescens (Fahraeus). Bozeman, Valentine.
Acanthoscelides johnique Johnson. Bozeman, Valentine.
Acanthoscelides, four spp. Bozeman, (three only), Valentine (three only), NCSU (one only).
Amblycerus schwarzi Kingsolver. Bozeman, Valentine.
Caryedon serratus (Olivier). Bozeman, Valentine.
Ctenocolum crotonae (Fahraeus). Bozeman, Valentine.
Megacerus sp. Valentine.
Mimosestes mimosae (Fabricius). Bozeman, Valentine.
Stator chalcodermus Kingsolver. Valentine.

CRIOCERINAE

Neolema dorsalis (Olivier). Bozeman.
Neolema sp. Bozeman, Valentine.

HISPINAE

Chalepus sanguinicollis (Linnaeus). Bozeman, Valentine.

GALERUCINAE

Chaetocnema brunnescens (Horn). Bozeman, Valentine.
Cyrsylus volkameriae (Fabricius). Bozeman.
Epitrix sp. Valentine.
Homoschema nigriventre Blake. Bozeman, Valentine.
Homoschema obesum Blake. Bozeman, Valentine.
Longitarsis chlanidotus Blake. Valentine.
Longitarsis oakleyi Blake. NMNH
Lysathia occidentalis (Suffrian). Bozeman.

Megistops bryanti Blake. Bozeman, Valentine.
Syphrea sanctaecrucis (Fabricius). Bozeman, Valentine.

EUMOLPINAE

Chalcosicya crotonis (Fabricius). Bozeman, Valentine.

CRYPTOCEPHALINAE

Cryptocephalus krugi Weise. Bozeman, Valentine.
Cryptocephalus perspicax Weise. NMNH
Cryptocephalus stolidus Weise. Bozeman, Valentine.
Diachus nothus (Weise). Bozeman, Valentine.
Exema sp. Bozeman, Valentine.
Pachybrachis mendicus (Weise). Bozeman, Valentine.
Pachybrachis sp. Bozeman, Valentine.

ANTHRIBIDAE

These records are from Valentine (2003).

CHORAGINAE

Acaromimus new species Bozeman, Valentine.
New genus, new species. Valentine.

ANTHRIBINAE

Ormiscus new species Bozeman, Valentine.
Toxonotus new species Bozeman, Valentine.

ATTELABIDAE

Euscelus coccolobae Wolcott. Bozeman, Valentine.
Euscelus sexmaculatus (Chevrolat). Valentine.

BRENTIDAE

APIONINAE

Apion metum Kissinger. Bozeman, Valentine.
Apion sp. Bozeman, Valentine.

BRENTINAE

Nemocephalus monilis (Fabricius). Bozeman, Valentine.
Stereodermus exilis Suffrian. Valentine.

CURCULIONIDAE

These records are from or based on the ongoing, unpublished work of Wayne Clark (Anthonominae), Donald Bright (Scolytinae), and Charles O'Brien (remaining groups). We are grateful to them for letting us use their records here.

DRYOPHTHORINAE

Scyphophorus acupunctatus Gyllenhal. Bozeman, Valentine.
Sitophilus linearis (Herbst). Bozeman, Valentine.

CURCULIONINAE

Anthonomus alboannulatus Boheman. Bozeman, Valentine.
Anthonomus concinnus Dietz. Bozeman, Valentine.
Anthonomus convexifrons Hustache. Bozeman, Valentine.
Anthonomus sp. near *homunculus* Gyllenhal. Bozeman, Valentine.
Anthonomus macromalus Gyllenhal. Bozeman, Valentine.
Anthonomus pusio Gyllenhal. Bozeman, Valentine.
Anthonomus sisyphus Clark. Valentine.
Huaca apian Clark. Bozeman.
Huaca ayacho Clark. Bozeman, Valentine.
Mecinus pascuorum (Gyllenhal). NMNH
Neomastix numerus Clark. Bozeman, Valentine.
Smicronyx sp. Valentine.

BARIDINAE

Apinocis sp. Bozeman, Valentine.
Zygobaris sp. Bozeman.
Baridini sp. 4. Bozeman.
Baridini sp. 5. Valentine.

COSSONINAE

Acamptus sp. Bozeman.
Cossonus impressus Boheman. Bozeman.
Dryotribus mimeticus Horn. Bozeman, Valentine.
Himatium? sp. Bozeman.
Macrancylus linearis Leconte. Bozeman, Valentine.
Pseudopentarthrum sp. 1. Bozeman, Valentine.
Stenancylus sp. Bozeman, Valentine.
Stenomimus sp. 1. Valentine.
Stenotrupis sp. Valentine.

Wasps, Ants, and Bees: Aculeate Hymenoptera

Roy R. Snelling

OF THE 72 PRESUMABLY native species, 6 (8%) are currently known only from Guana Island. That they are not known from any of the adjacent islands is assumed to be the result of inadequate sampling on those islands: Guana Island is the most completely surveyed of all the BVI. The name of this order derives from Greek *hymenos* for "union" and *ptera* for "wing"; the hind wings (when present) typically bear tiny hooks called hamuli on the forward edges that unite them with the forewings. *Aculeate* means "with a stinger": an ovipositor, or egg-laying syringelike structure, modified to inject venom. Males lack stingers, and these structures are undeveloped in many species. But, as everyone knows, some wasps, ants, and bees have very well developed stingers, and a few can actually be lethal to humans. The Hymenoptera fauna of the Virgin Islands is very poorly known. The USVI have received sporadic attention, and their fauna is a little better studied, especially on the two larger islands, St. John and St. Thomas. St. Croix, although politically a part of the USVI, is faunistically a part of the Lesser Antilles rather than the Greater Antilles. By contrast, the BVI are virtually unknown in terms of their Hymenoptera fauna.

Only the fauna of Guana Island has been systematically collected, beginning with my first visit to the island in October of 1991 (Snelling 1992a, b). Several subsequent trips in following years (Snelling 1993a, b) added to the faunal list, and the results are presented below.

Eighty-one species of Hymenoptera are currently known from Guana Island. Nine of these are introduced; all except one, *Apis mellifera* Linné, as accidental intrusions. The remaining 72 species consist mostly of widely distributed West Indian species. Twenty-seven of these species (37.5%) are apparently limited to the Greater Antilles; the remainder consist of generally distributed West Indian species. Many of these are present on the mainlands of Central America and South America, as well. These data are tentative, largely because the Hymenoptera of the Caribbean remain known inadequately. Indeed, the Caribbean islands, as a whole, are the least well known area of the neotropics.

Most of the specimens collected during these visits are deposited in the collections of the Natural History Museum of Los Angeles County (LACM). Additional material of some species is also deposited in other collections, especially

The Natural History Museum (London), Museum of Comparative Zoology (MCZ), Harvard, and the U. S. National Museum of Natural History. Most specimens were hand collected: flying bees and wasps with a net, and ants with an aspirator. Several Malaise traps were also deployed at various sites on the island. Honey-water baits were also used but with limited success because of the rapidity with which birds discovered and consumed the bait. Sifted leaf litter (primarily for ants) was not especially successful; the litter was too thin and too dry and produced no species not already collected by hand. In the following lists, species currently known only from Guana Island are marked by an exclamation mark in parentheses; all are expected to be found on adjacent islands. Adventive (introduced) species are marked by an asterisk. As with beetles, above, authors' names in parentheses mean the generic assignment is different here than in the original description.

BETHYLIDAE

Bethylidae are small wasps (usually <3 mm long) that, as far as known, are exclusively predaceous on larvae of small beetles. Some are predators on larvae of beetles associated with stored grain products. The adult wasp paralyzes the beetle larva with its sting, prepares a cell in which to enclose it, and then lays one of its own eggs upon the larva. Females of many species in the subfamilies Bethylinae and Pristocerinae are wingless; males of these subfamilies do possess wings. Both sexes of Epyrinae are winged, as far as known. Bethylids are mostly dull black or dark brown, but others, such as some species of *Epyris* and *Anisepyris* may be metallic greenish or bluish, sometimes with the last two or three abdominal segments bright red. Sixteen species of Bethylidae are currently known from Guana Island; seven are known only from the Virgin Islands exclusive of St. Croix. Of the remaining nine species, most are probably generally distributed through much of the Antilles, but a few are known only from the Greater Antilles (Snelling 1996). The several species of

Dissomphalus and *Pseudisobrachium* remain unidentified. Between the two genera there are probably four or five species.

SUBFAMILY BETHYLINAE

Goniozus crassifemur Evans.
Parasierola rivularis Evans.

SUBFAMILY EPYRINAE

Anisepyris arawak Snelling(!).
Anisepyris aurichalceus (Westwood).
Anisepyris chupah Snelling.
Epyris guana Snelling(!).
Epyris jareckii Snelling(!).
Epyris karli Snelling(!).
Holepyris incertus (Ashmead).
Holepyris skip Snelling(!).
Rhabdepyris maboya Snelling.
Rhabdepyris versicolor Evans.
Scleroderma wilsoni Evans.

SUBFAMILY PRISTOCERINAE

Dissomphalus spp.
Pseudisobrachium spp.

DRYINIDAE

Dryinids are small wasps, mostly predaceous on various small Hemiptera, such as leafhoppers (Cicadellidae) and treehoppers (Membracidae). Females of many species, especially those that are wingless, are bizarre insects with peculiarly modified raptorial (talon-bearing) front tarsi. Only the following species has been collected on Guana Island; it is also known from Puerto Rico.
Thaumadodryinus snellingi Olmi (1993).

TIPHIIDAE

Of the two species found on Guana, little is known. *Tiphia* is a worldwide genus of external parasitoids, as far as is now known, exclusively on larvae of scarab beetles. The Caribbean species are not well studied, and our one species remains unidentified. *Myzinum* is a genus of brightly colored wasps; males are quite slender and are usually mostly yellow with sharply defined black markings. Females are often largely black with contrasting yellow marks, sometimes reddish also. Nothing is known of the biology of our single species, which ranges widely throughout the Caribbean and in the American tropics.

Myzinum haemorrhoidale (Fabricius).

Tiphia sp.

MUTILLIDAE

Female mutillids are wingless and antlike in appearance. Our species is only about 3 mm long and is rarely seen as it scurries through forest leaf litter; it is very alert and difficult to capture. Males are winged and occasionally turn up in Malaise trap samples. The one species known from Guana Island ranges from Cuba to the Virgin Islands.

Pseudomethoca argyrocephala (Gerstäcker).

SCOLIIDAE

Campsomeris is worldwide in distribution but especially diverse in the Old World tropics and subtropics. In the New World, species range from the United States to Argentina, with most of the species in South America. Our two species are common over much of the Antilles; *Campsomeris trifasciata* has also been collected in the United States (Florida). *Campsomeris* species are external parasitoids, presumably exclusively on scarab beetle larvae. A few species may exceed 20 mm in length. Although a few are largely or entirely black, others are richly marked in black with contrasting yellow or reddish marks, especially on the metasoma.

Campsomeris dorsata (Fabricius).

Campsomeris trifasciata (Fabricius).

FORMICIDAE

The ants are far and away our most diverse family of aculeates, even when we discount the adventive species (marked below by an asterisk). Most of the native species are widely distributed through the Greater Antilles and some range into the Lesser Antilles and even into Central America and South America.

SUBFAMILY DOLICHODERINAE

Dorymyrmex antillana Snelling, new status. This species was originally described by Auguste Forel (1911) as *Dorymyrmex pyramicus* subsp. *brunneus* var. *antillana*, and thus is an unavailable quadrinomial. The type locality is St. Vincent in the Lesser Antilles. This was treated as a synonym of *Dorymyrmex insanus* (Buckley) by Snelling (1973), an erroneous assignment, and later removed from that synonymy by Snelling (1995a). Workers differ from those of *Dorymyrmex insanus* by the posteriorly declivitous mesonotum (meaning the middle segment of the thorax slopes downward). The gyne, which is the reproductive female, differs from that of *Dorymyrmex insanus* (and all known North American species) by the presence of numerous short, stiff erect hairs on the head and mesonotum. Forel (1911) originally described this as a variety of the Argentinean *Dorymyrmex brunneus*, which it closely resembles. Workers, especially, are very similar but the mesosomal profile is subtly different. Again, the gyne differs in its extreme hairiness; those of *Dorymyrmex brunneus* do possess some erect hairs on the head and mesosomal dorsum, but not to the extent seen in *Dorymyrmex antillana*.

Tapinoma melanocephalum (Fabricius).* This is a common, cosmopolitan tramp species, often transported in potted plant material. It is now widely distributed in tropical regions throughout the world and occurs sporadically in subtropical regions (such as Florida) as well. Because it does tend and solicit aphids on garden plants it may sometimes be a minor pest.

SUBFAMILY FORMICINAE

Camponotus kaura Snelling & Torres. This species was recently described (Snelling and Torres et al, 1998) from Puerto Rico, where it is common. In the past it has been confused with *Camponotus ustus* Forel and most records under that name in the Greater Antilles actually refer to *Camponotus kaura*. This is our largest ant (large workers can be up to about 11 mm long). The workers are yellow in color and forage at night. Similar to many other species in the genus, this is a so-called "carpenter ant," nesting in wood. I have found it nesting most commonly in old termite galleries and beetle burrows in dead branches.

Camponotus sexguttatus (Fabricius). Another carpenter ant but diurnal. It differs immediately from the above species in its dark brownish body with contrasting whitish to yellowish spots on the posterior abdomen. It, too, nests in deadwood.

Paratrechina longicornis (Latreille).* A cosmopolitan tramp species, *Paratrechina longicornis* is a small, quite slender ant that runs very rapidly and erratically. Because of this erratic running, it is commonly known as "crazy ant." It is a household and garden pest, more a nuisance than a true problem species.

Paratrechina steinheili (Forel). A native species, common in forests and widely distributed in both Greater and Lesser Antilles. The body is brown and more robust than in the above species; sparse, long stiff hairs are present. Nests are usually in soil, often under a covering object, but may also be in hollow wood.

SUBFAMILY MYRMICINAE

Cardiocondyla emeryi Forel.*

Cardiocondyla obscurior Wheeler.* *Cardiocondyla* is an Old World genus and several species are common tramp ants, now widely distributed in tropical and temperate regions. The ants are minute and their taxonomy is currently very confused. On the authority of B. Seifert (pers. comm.) *Cardiocondyla obscurior* is here used for specimens that I had originally thought to be *Cardiocondyla wroughtoni* Forel; *Cardiocondyla obscurior* was originally described as a variety of that species. As far as is known, these species have no adverse economic or ecological impact in areas where they have become established.

Crematogaster steinheili Forel. Because of their habit of elevating the heart-shaped abdomen over the thorax when disturbed, these ants are sometimes called "cock-tail" ants. This small yellow species nests in plant cavities; colonies are usually very populous and several queens may be present. This native species is widely distributed in the West Indies.

Cyphomyrmex minutus Mayr. This occurs throughout the Caribbean, Central America, and northern South America; it is also found in Florida and other southeastern U.S. states. It has often been treated as a subspecies of the South American *Cyphomyrmex rimosus* (Spinola), which it closely resembles, but has been recognized as a separate species by Snelling and Longino (1992). Small and inconspicuous, it nests in soil under covering objects, in damp rotten wood, and occasionally in epiphytes. Colonies may possess a number of functional queens that may be dispersed over several satellite subcolonies. Larval food consists solely of the fungus that these ants rear in the nest on a substrate of insect droppings, usually of caterpillars. Workers forage singly and, when disturbed, feign death.

Monomorium ebininum Forel. This tiny (about 1.5 mm long) shiny black ant is common over much of the Caribbean. Colonies may number into the tens of thousands, with many queens and several satellite colonies. Nests are commonly in dead stems and branches of living trees; a colony may effectively occupy all the available cavities in a single tree. It has a tolerance for high temperatures (Torres 1981). The workers are diurnal and the feeding habits are similar to those of *Monomorium floricola* (see below). On Puerto Rico it was found to be an important control agent of the sugarcane borer—a moth, *Diatraea saccharalis* (Lepidoptera: Pyralidae)—by feeding on the egg clusters of this moth (Wolcott and Martorell 1937). Colonies are polygynous—have more than one queen—and the queens are ergatoid—meaning derived from workers. Because queens of both *Monomorium ebininum* and *Monomorium floricola* are wingless ergatoids, dispersal and establishment of new colonies is accomplished by budding or fission. Complex organic chemicals, dialkylpyrrolidines, produced in the poison gland, are used to repel other ant species in competitive interactions at food resources (Jones et al. 1982). This species is an aggressive competitor, occasionally being able to displace even *Solenopsis geminata*, one of the most aggressive ants on the island.

Monomorium floricola (Jerdon).* Native to southeastern Asia, *Monomorium floricola* is now widely distributed throughout the world tropical regions but usually limited to coastal and near-coastal areas. It may occasionally also become locally established in temperate areas. As with its congener *Monomorium ebininum*, this is a tiny ant with dark head and posterior abdomen, but with the thorax paler, often whitish in large part. It, too, has polygynous colonies that may become very populous. Liquid food, consisting of plant exudates, animal fluids, or honeydew from aphids and their relatives, is transported back to the nest in the crops of the foraging workers. In Puerto Rico, *Monomorium floricola* was the most abundant ant tending pineapple mealybugs (Plank and Smith 1940).

Mycetophylax conformis (Mayr). This seems to be generally distributed throughout the Caribbean islands and northern South America. Unlike *Cyphomyrmex minutus*, which is a quite compact ant, this species is both larger (up to about 3.5 mm long) and decidedly "spindly" with long slender legs. The few nests that I have found on Guana Island have been in sandy soil near White Beach. These ants cultivate the fungi that are used as food for the larvae. Some species apparently normally use freshly cut pieces of grass blades as a substrate. On Puerto Rico, workers were seen gathering fallen needles of the Australian pine *(Casaurina equisetifolia)* (J.A. Torres, pers. comm.). The fungal mass is fragile, according to Weber (1972).

Pheidole fallax Mayr.

Pheidole moerens Wheeler.

Pheidole sculptior Forel. The three species of *Pheidole* found on Guana are all characterized by the presence of two distinctly different subcastes of workers: larger "soldiers" with quite large, massive heads and much smaller "workers" that have normal-sized heads and slender bodies. Of the three, *Pheidole fallax* is the largest, about equal in size to the two species of fire ants (see below), but differs from these by the relatively dull body (distinctly shiny in the *Solenopsis* species). Unlike the fire ants, *Pheidole* lack a sting; both *Pheidole moerens* and *Pheidole sculptior* are timid and inclined to attempt flight rather than fight. Colonies of *Pheidole fallax* are much larger and the workers much more aggressive. Even without a sting they can certainly be annoying when they attack in defense of the nest. Our species of *Pheidole*, as with so many others in the genus, feed on seeds; although, *Pheidole fallax* is also an aggressive predator/scavenger and the diurnally foraging workers do bring in many insects and other small arthropods as food for the larvae. Harvard's famous biologist E.O. Wilson (2003) has recently monographed the New World members of the enormous genus *Pheidole*—624 species—to critical acclaim (Schultz 2003).

Solenopsis geminata (Fabricius). Because of their fiery sting this and the following species are known as "fire ants." On Cuba *Solenopsis geminata* is called "*la hormiga brava*." Colonies contain multiple queens and populations may run into the hundreds of thousands of workers. Workers are polymorphic, running the gamut from quite small workers (about 2.5 mm long) to relative giants (about 6 mm long) with greatly enlarged heads; the smaller workers are especially pugnacious. When the colony is disturbed, hundreds of workers rush out, ready to attack any intruder. Fire ants are often the dominant ants in an area and may be of some concern not only because of their stings but because they often attack nestling birds, immature mammals, and hatching reptiles. They also tend aphids and other insects damaging to crops. Such disservice is at least somewhat mitigated by the large numbers of other insects that they attack and kill. This species ranges from the southeastern United States west to Texas and from there south to northern South America, including western Amazon Basin and coastal areas of Peru and is presumably introduced into the Galápagos Islands. It is introduced in southeast Asia to India, south to northern Australia.

Solenopsis invicta Buren.* This important pest species was initially reported from Guana and elsewhere in the Greater Antilles by L. R. Davis et al. (2001). Its potential for inflicting pain and doing damage is enormous (Wojcik et al. 2001). These ants sting like wasps and anaphylactic shock kills up to one in a hundred human victims (Williams et al. 2003). Originally native to South America, *Solenopsis invicta* is now widespread, and spreading, in the southern United States; it is established on Puerto Rico and according to Williams et al. (2003), who report on efforts at biological control, also on Mona, Vieques, and Culebra.

Solenopsis pygmaea Forel. Unlike the two fire ant species listed above, this and the following species are minute (ca. 1.5 mm long) yellow ants that are seldom seen. They most often nest in soil, commonly under a covering object. Similar to their larger cousins, they are predaceous; some species, at least, are known to be predaceous on the brood of larger ant species, including the two fire ant species. Their tunnels are often closely intermingled with those of larger ants; from these tunnels they gain access to the brood chambers of the species on which they prey. Because they are commonly attracted to greasy or oily foods, some species may become household pests; they are called "thief ants" or "grease ants." Nests may be dispersed into satellite colonies; both species are polygynous. Of the two species known to be present on Guana, *Solenopsis pygmaea* is known to range from Puerto Rico (including Mona Island) to the Virgin Islands.

Solenopsis torresi Snelling (2001). This is known currently only from Puerto Rico proper and Guana Island but is likely more widespread.

Wasmannia auropunctata (Roger). The common name for this ant is "little fire ant" for, despite its minute size (about 2 mm long), it can deliver a surprisingly potent sting, especially on tender areas of the body. On Puerto Rico, it is an important agricultural pest. It is not closely related to the fire ants. The natural distribution in the New World is uncertain; currently, it extends from the southeastern United States (Florida) throughout the Caribbean and tropical lowlands of Central America to the Amazon Basin, and it is introduced into the Galápagos Islands. This species has also been introduced into various tropical regions in the Old World. Colonies are very populous and polygynous; the queens are much larger and bulkier than their workers. As with so many polygynous species, a colony may be divided into satellites. The workers are aggressive predatory foragers that often attack nests of other ant species, sometimes destroying the entire nest. Colonies may be in soil, in leaf litter, in rotten logs, or in hollow twigs and branches.

SUBFAMILY PONERINAE

Anochetus kempfi Brown. This slender, yellowish red ant is one of our more bizarre species. The head is elongate, and the mandibles are almost as long as the head. When hunting for prey, the ant opens its mandibles to a gape of 180°; trigger hairs arising from the mouthparts cause the mandibles to snap shut, impaling the prey, which is then stung (if necessary) before being carried back to the nest. Workers forage nocturnally and may sometimes be seen on tree trunks hunting for insect prey. Nests are in soil or damp rotten wood and several queens may be present. Because the queens are permanently wingless, new colonies are created when a newly fertilized queen and some workers depart to establish a colony elsewhere. Mature colonies contain about 100 workers. This species is tolerant of both dry and wet forest conditions. In the latter, workers may hold packets of eggs in their mandibles rather than place them on the floor of the brood chamber. This is possibly a mechanism to avoid molds that might kill the eggs or minute larvae (Torres et al. 2000).

Leptogenys pubiceps Emery. This species ranges through the Caribbean and Central America to Venezuela and Colombia. It may be found under rocks or other covering objects. Workers have slender, sickle-shaped mandibles adapted to piercing their prey; they forage at night (Deyrup et al. 1998) and are very timid.

Queens are workerlike (ergatogynes) and permanently without wings; males are winged.

VESPIDAE

SUBFAMILY ODYNERINAE

Pachodynerus atratus (Fabricius). A common species, it ranges widely in the Greater and Lesser Antilles (Menke 1986). It is easily recognized by the combination of shiny black body and the longitudinally folded wings when at rest. Males have a large yellowish area on the lower face (the clypeus). The multicellular mud nests are usually concealed within a plant cavity, including the interior of bamboo stems. Each cell is provisioned with several caterpillars, an egg is deposited within, and then the cell is sealed.

Stenodynerus sp. Little is known of this solitary vespid, which may prove to be an undescribed species. The few Antillean species of *Stenodynerus* are not well studied. It is richly marked black and yellow. Only a few specimens were collected on one occasion, flying around a bush. As with other *Stenodynerus*, the female probably nests in soil or in a plant cavity, such as a hollow stem. Unlike the *Pachodynerus*, the several cells within such a nest are constructed in a linear series, each separated by a mud partition. Each cell is provisioned with paralyzed caterpillars and sealed, and then a new cell is provisioned above it.

SUBFAMILY POLISTINAE

Mischocyttarus phthisicus (Fabricius). This wasp occurs throughout the Greater Antilles and on many islands of the Lesser Antilles. Although this social species is common, it is inconspicuous and seldom seen. The nests, constructed of plant fibers, are usually placed on a branch in a shrub and are often difficult to locate. The wasps are timid and often fly away from the nest rather than attack the intruder. The sting is mild. There are usually fewer than 20 adult wasps comprising a colony. Prey fed to the larvae consists of masticated smooth-skinned caterpillars. As is true of the following species, the nest population consists of an egg-laying "queen" and her daughters who normally remain unfertilized and are functional workers. Males are produced near the end of the life of the colony; they mate with newly emerged females who then depart to initiate new colonies.

Polistes crinitus (Felton). (See plate 36.) The common name for this social wasp on many islands is "Jack Spaniard." The range extends from Hispaniola to the Virgin Islands and from there into the Lesser Antilles. Distinctive color forms have been named from some of the islands but all seem to apply to a single morphological species. Nests are usually constructed in protected sites such as rock overhangs, under leaves, or, occasionally under the eaves of buildings. They may also be placed in dense shrubs, attached to the underside of a branch. On Guana, there are usually several nests in the bat cave. Colonies are initiated by a single original female, the "queen"; her progeny, upon maturing, become workers. The workers have the task of adding cells to the nest, within which the queen deposits additional eggs. Although all individuals in the colony forage for prey to feed the larvae, this is primarily the duty of the workers. Smooth-skinned caterpillars are the primary prey of these wasps. Colony size usually does not exceed about 50 adult wasps. The queen and workers of a colony can deliver a fairly painful sting, but the wasps are generally not very aggressive until the nest is seriously disturbed. J. Lazell (pers. comm.) reports that if a nest on a leaf or branch is cut off cleanly and falls straight to the ground, the wasps will stay with it and not attack the cutter. Good luck.

POMPILIDAE

The several species of Pompilidae on Guana are all predators on spiders. Interestingly, most of our species are specialized predators on burrowing spiders. The Pompilinae prey on terrestrial spiders.

SUBFAMILY PEPSINAE

Pepsis rubra (Drury). The two species of *Pepsis* are the largest spider wasps on Guana Island.

Females of both species are commonly seen as they forage in the forest litter for burrows of their prey. Males are more abundant and may sometimes be seen as they visit the flowers of various shrubs and trees; they are especially abundant at blossoms of *Coccoloba uvifera* when that tree is in bloom. *Pepsis rubra* is the model for the wasp-mimicking moth, *Empyreuma pugione* (Linné) (Lepidopera: Arctiidae). The presumed prey for both this species and the following is the local tarantula, *Cyrtopholis bartholomei* (Latreille). This spider is common and lives in ground burrows. Although I have followed searching females of this wasp, I have yet to witness an actual capture.

Pepsis ruficornis (Fabricius). Although less common than *Pepsis rubra*, this species is often seen on Guana. The two species are similar in size but are easily distinguished from one another. In *Pepsis rubra*, the wings are largely bright reddish and the antennae are black; in *Pepsis ruficornis*, the wings are uniformly blue and in females the antennae are mostly red, whereas in males often only the middle segments are red.

SUBFAMILY POMPILINAE

TRIBE APORINI

Aporus prolixus Bradley. This species was described from a female collected at Christiansted, St. Croix, USVI. Its presence on Guana Island suggests that it is likely to be found throughout the Virgin Islands. It has also been collected on Puerto Rico. Nothing is known of the biology of this species. As far as is currently known, all Aporini are predaceous on various species of trap-door spiders (Ctenizidae); an unidentified species of the genus *Ummidia* is known from Guana Island. Given the relatively small size of the wasp, the females probably seek out immature spiders.

Psorthaspis gloria Snelling (1995b). This is certainly the most attractive of all the wasps found on Guana Island, and it is evidently related to the Central American species *Psorthaspis variegata* (F. Smith). In addition to Guana (the type locality), it has been collected on St. Croix and

Puerto Rico; it probably occurs throughout the islands of the Puerto Rico Bank. This species is considerably larger than *Aporus prolixus* and, therefore, is likely predaceous on the adults of the trap-door spiders.

TRIBE POMPILINI

Poecilopompilus flavopictus (F. Smith). This common species is distributed throughout the Caribbean and in Central America and northern South America. Several "subspecies" (*hookeri* Rohwer, *mundiformis* Rohwer, and *mundus* Cresson) are described from the islands of the Greater Antilles. However, the variation within populations is such that probably a single name would suffice for all of these; the oldest available name would be *mundus*, described from Cuba. Because I have not studied the Central American populations in great detail, I am uncertain how to distinguish the island populations from those of the mainland and so have adopted a conservative course and not assigned subspecific names to our wasps. Species of *Poecilopompilus* are apparently exclusively predaceous on spiders of the family Araneidae, of which there are several known on Guana.

SPHECIDAE

SUBFAMILY CRABRONINAE

Ectemnius craesus (Lepeletier & Brullé). This is a common species over much of the West Indies. Although the color pattern is somewhat variable, no localized insular forms have been described. *Ectemnius* species mostly nest in pithy stems or in decaying wood. Individual cells are provisioned with the adults of flies of various families. *Ectemnius taino* Pate. This appears to be a new subspecies. The nominate form of *Ectemnius taino* is based on specimens from Hispaniola. The Guana Island form differs from the Hispaniolan in the more restricted yellow markings.

SUBFAMILY LARRINAE

Liris labiata (Fabricius). This is a widely distributed West Indian species; it has been recorded from Puerto Rico and other islands as *Liris ig-nipennis* (F. Smith), a synonym. Wasps of this genus nest in soil, provisioning one to several cells within the burrow. The usual prey consists of crickets (Orthoptera: Gryllidae). The identities of two of our species are uncertain pending a thorough taxonomic review of the genus:

Liris cf. *luctuosus* (F. Smith).

Liris sp.

Tachysphex alayoi Pulawski. This species was based on material from Cuba but is known to range over much of the Greater Antilles. Its preferred prey is unknown. As far as is known, all *Tachysphex* species construct shallow multicellular nests in soil. Various species use Orthoptera of several families to provision the cells.

Tachytes chrysopyga (Spinola).

Tachytes tricinctus (Fabricius). *Tachytes* is another cosmopolitan genus; most of the species occur in arid and semiarid habitats. Even in more forested and humid areas, the wasps tend to be found in clearings and other more open areas. The female wasps construct their multicellular nests in the ground; prey consists of grasshoppers (Orthoptera: Acrididae). Both sexes are often found at flowers.

SUBFAMILY NYSSONINAE

Bembix americana Fabricius.

Bicyrtes spinosa (Fabricius).

Stictia signata (Linné). These three genera are cosmopolitan, and each of our species is common throughout the Antilles. The above three species are commonly called sand-wasps because nests are located in fine, loose sand. Nesting females form large aggregations or colonies, but each female tends her own nest. Adult flies of various families are the usual prey; on Guana Island, females of *Stictia signata* are often seen capturing the tabanid flies that feed on the donkeys.

Epinysson borinquinensis Pate. This genus is known only from the New World; there are about 24 species described. As the name implies, this wasp was originally described from Borinquino, Puerto Rico; it is known only from islands of the Puerto Rico Bank. *Epinysson*

species are nest parasites of wasps of the genus *Hoplisoides*, and, although there are no observations of the host of *Epinysson borinquinensis*, it likely feeds on the following species.

Hoplisoides ater (Gmelin). This is another genus with species over much of the world; there are no known Australian species. Although not common, *Hoplisoides ater* is found over much of the Greater Antilles. Nests are in sandy soil, with the entrance often concealed beneath a covering object, such as a leaf or a small stone. Prey consists of Hemiptera of various families (Cicadellidae, Fulgoridae, Membracidae, etc.); each *Hoplisoides* species apparently is a specialist on a single hemipteran family. The preferred prey of our species is unknown.

SUBFAMILY PHILANTHINAE

Cerceris cf. *margaretella* Rohwer. The genus is another cosmopolitan one, with about 800 described species, most of them in the Old World. *Cerceris margaretella* was described from Puerto Rico, and the few specimens collected on Guana are probably conspecific. American *Cerceris* species provision their nests with adult beetles of various families, including Bruchidae, Chrysomelidae, Curculionidae, and Tenebrionidae. Females usually nest in aggregations on level, hard-packed soil. The burrows tend to be vertical, with the cells placed horizontally; up to 20 prey may be placed in each cell.

SUBFAMILY SPHECINAE

Prionyx thomae (Fabricius). This and the two following species are "thread-waisted" wasps; the first abdominal segment forms an elongate, very narrow petiole between the thorax and the remainder of the abdomen. Cosmopolitan, most species of *Prionyx* are found in the Old World, with about 20 species in the New World; *Prionyx thomae* ranges from the United States to Argentina. Females excavate burrows in the soil; a single cell is provisioned in each successive burrow. Prey consists of various species of grasshoppers. Both sexes are commonly seen at flowers; they are alert and quick to fly when threatened.

Sceliphron assimile (Dahlbom). This is another cosmopolitan genus with about 30 species worldwide; there are about 7 species in the New World. The species ranges from the United Staes (Texas) south to Panama and widely through the Caribbean. These are the common black and yellow mud-daubers that build multicellular nests of several elongate cells placed side by side, often in or on buildings. In nature, nests are often situated on rock overhangs or under the bark of trees. The cells are provisioned with small spiders.

Sphex ichneumoneus (Linné). This is another moderately sized, cosmopolitan genus; there are over 100 species worldwide, with about one-third of these found in the New World. This is perhaps the most wide ranging of our sphecids, common from the United States to South America, and through both the Greater and the Lesser Antilles. This is a large wasp, up to about 25 mm in length. The multicellular nests are in hard-packed soil and are provisioned with Orthoptera of several families (Gryllidae, Gryllacrididae, Tettigoniidae).

The following four families, Apidae, Colletidae, Halictidae and Megachilidae are bees. Bees differ from wasps in that the larvae feed on pollen and nectar from blossoms. Wasps, with few exceptions (in the family Vespidae), are predaceous. Bees have some or most of the body hairs branched or plumose; wasps generally have simple, unbranched hairs.

Of the 19 bee species found on Guana Island, only one, *Apis mellifera*, is truly social with three distinct castes within a colony: queen, worker, and drone (male). The remaining 18 species are mostly solitary, with the female (there are only two castes, females and males) of each species constructing her own nest, usually a simple burrow in soil, within which she constructs a number of cells, each provisioned with a paste of nectar and pollen. An egg is placed in each cell, and the cell is then sealed. The female then closes the burrow and may either die or begin a new burrow and series of cells. A few of our species may be subsocial or primitively social, all in the family Halictidae; several females

may cooperate in rearing larvae and providing food for them.

APIDAE

SUBFAMILY APINAE

Apis mellifera Linné* This is the common European honeybee, introduced into the New World many times during the colonial period. The honeybee is our only social bee species. As far as is now known, there are no introduced populations of the fiercely aggressive Africanized honeybee in the BVI yet, but they are present on Puerto Rico. *Apis mellifera* is an effective pollinator of introduced crops of commercial importance, but it is often less effective with many of the native New World flowering plants. However, because it is an aggressive forager, it may have serious adverse affects on native, nonsocial bee species by competing so aggressively that it drives the native species away. Although commonly encountered on Guana, it does not appear to be a dominant bee here.

Anthophora tricolor (Fabricius). A common species in the Greater Antilles, this ranges from Hispaniola to the Virgin Islands; it is also found in some islands of the Lesser Antilles. In both sexes, the hair on the thoracic dorsum is dark reddish, usually more conspicuously so in the females of this medium-sized (length is about 10–13 mm) bee. In both sexes there are white integumentary bands on the abdomen; narrow in the female (on segments 2–4) and broad in the male (segments 1–5). Nests are in hard-packed soil and usually several females nest in the same site, but each constructs her own burrow. The females visit a variety of flowers. On Guana, they have been taken at *Solanum persicifolium* and *Ipomoea pescapri*. They have also been collected at *Lantana involucrata*, which are visited for nectar only.

Centris decolorata Lepeletier. This is a common Caribbean species, present throughout the Greater and Lesser Antilles and along the Gulf and Caribbean coast from southern Texas to the Guianas; it does not extend into the mainland interior, nor has it been collected along the

Mesoamerican Pacific Coast. It is the largest of our *Centris* species, about 16–20 mm long. In the females, the abdomen is greenish and the thorax is covered with hairs that are buff-colored above and paler on the sides. Males are similar, but the abdomen is less obviously greenish and there are short yellow stripes at the sides of segments 2–5. Rarely, large males are encountered in which the abdominal stripes are broader and extend across the entire abdomen and the hind legs are noticeably stout; these are known as "metanders." On Guana Island males have been collected at flowers of *Melochia tomentosa*, *Canavalia rosea*, and *Caesalpinia bonduc*; there are no floral records for females here.

Centris haemorrhoidalis (Fabricius). This appears to be limited to the Greater Antilles and has been collected on the USVI (St. John), the BVI (Guana, Virgin Gorda), Hispaniola, Puerto Rico, and Mona Island. It is apparently absent from Cuba and is replaced on Jamaica by a similar species, *Centris dirrhoda* Moure, which ultimately may prove to be only a color form of *Centris haemorrhoidalis*. This is one of our more attractive bees: a robust species, about 15 mm long, with a metallic steel blue abdomen, and the last several segments contrastingly red; the pilosity (fine hair or fur) is black; the face is conspicuously marked with whitish (more extensive in the male), and males have a conspicuous white spot on each side of the second abdominal segment. It has been collected at flowers of *Solanum persicifolium*, *Stigmaphyllon periplocifolium*, *Malpighia emarginata*, *Melochia tomentosa*, *Ipomoea pescapri*, *Croton rigida*, and *Tamarindus indica* (males only were collected at the last four species).

Centris lanipes (Fabricius). This is another wide-ranging species, from Hispaniola and Puerto Rico through the Lesser Antilles to northern South America. It is our smallest species of *Centris*, usually about 10–12 mm long; the head (except limited yellow face marks) and thorax are black with buff-colored hair; the abdomen of both sexes is red. Unlike the other species of *Centris* here, all of which are ground-nesting bees, this one nests in deadwood and females

are often seen gathering mud with which to make the cells in the rotten wood. The megachilid bee, *Coelioxys abdominalis* Guérin, is a nest parasite of this species. Females have been collected on Guana at flowers of *Antigonon leptopus*, *Solanum persicifolium*, and *Caesalpinia bonduc*. Males have been taken on *Melochia tomentosa*.

Centris smithii Cresson. The range of this species extends from Puerto Rico (including Desecheo and Mona islands) into the Lesser Antilles. In this species, the abdomen is bluish with the last several segments bright yellow; males also have short yellow stripes on either side of segments 2 and 3; the remainder of the body is black with buff-colored hair; the face has yellow marks, more extensive in the male. A loose aggregation of nesting females was found on the trail to Long Man's Point; most nests were partially concealed under small stones. On Guana, females have been collected at flowers of *Stigmaphyllon periplocifolium* and *Caesalpinia bonduc*. Males have been collected at flowers of *Melochia tomentosa*.

Exomalopsis pulchella Cresson. Records for this species are from the southern United States (Florida) to St. Vincent in the Lesser Antilles; it has also been collected in Mexico. Females are small, about 5 mm long; the head and body are black, the head and thorax covered with mostly blackish hair; the abdomen appears polished, and the first two segments are almost entirely without hair. Males are 4–5 mm long, also black, but with pale buff-colored hair; there are weak stripes of pale hair across abdominal segments 2–4. Females have been collected at flowers of *Solanum persicifolium*, *Jacquemontia pentantha*, *Acacia macracantha*, *Cucurbita pepo*, and *Stylosanthes hamata*.

Melissodes nigroaenea (F. Smith). This species is known from Puerto Rico (including Mona Island) through the Lesser Antilles to Brazil (LaBerge 1956, 1962). Both sexes are about 10 mm long, and males are less robust than females; they are black with pale buff-colored hair on the head and thorax, and there are several well-defined hair bands on the abdomen; in the males the antennae are exceptionally elongate, almost as long as the entire body. On Guana, I have collected females only at blossoms of *Ipomoea pescapri*; males have been collected on that plant and on *Antigonon leptopus*.

SUBFAMILY XYLOCOPINAE

Xylocopa mordax F. Smith. The range of this species extends from Jamaica and Hispaniola to the Lesser Antilles. It is common on Guana, our largest bee, about 25 mm long, with a broad, somewhat flattened body. Females are entirely black with black wings, and males are entirely yellowish (but with obscure brownish bands on the abdomen) and with brownish, semitransparent wings. This is a carpenter bee; females construct their galleries in dry deadwood; several females may use a common entrance into the wood, but each constructs her own gallery of cells. G.C. Jackson and Woodbury (1976) and G.C. Jackson (1986) studied the nectar- and pollen-gathering activities and plants used as nesting sites. On Guana it has been taken at a large variety of flowers: *Caesalpinia bonduc*, *Canavalia rosea*, *Ipomoea pescapri*, *Coccoloba uvifera*, *Cardiospermum micrantha*, *Centrosemum virginianum*, *Jacquemontia solanifolia*, *Tabtebuia heterophylla*, and *Tecoma stans*.

COLLETIDAE

Hylaeus new species (!). Although this species is currently known only from Guana, it almost certainly is to be found on many of the adjacent islands. This is a small black bee, about 5 mm long, with only scattered hairs; both sexes have limited yellowish markings on the face and front part of the thorax. Most species of *Hylaeus* nest in plant stems or cavities in deadwood. Individual cells are lined with a cellophane-like material derived from the mandibular glands of the female. Females lack an external scopa—pollen-collecting apparatus—and instead ingest pollen to transport it to the nest. Both sexes have been collected on flowers of *Capparis cynophallophora*, *Schaefferia frutescens*, and *Cassine xylocarpum*.

Hylaeus wootoni (Cockerell)* A single male of this species was collected on 28 June 1993 at flowers of *Ipomoea pescapri*. This is a species of the western United States, ranging from the Pacific Coast to Montana, Colorado, and New Mexico. It is hard to imagine how it got to Guana Island.

HALICTIDAE

Augochlora buscki Cockerell. All records for this species are from the islands of the Puerto Rico Bank: Puerto Rico and the BVI. Both sexes of this small bee, about 10–12 mm long, are bright metallic green or, less commonly, bluish. These bees nest in moist deadwood, where several females may participate in mutual nest tending. Females collecting pollen have been collected only at flowers of *Ipomoea pescapri*; males have been collected on that plant as well as on *Melochia tomentosa*, and *Tecoma stans*.

Augochlora new species? The taxonomic status of the one specimen collected is uncertain. It does not appear to be any of the known Caribbean species, but until the genus is studied in detail, nothing further can be said of this bee. Similar to *Augochlora buscki*, it is bright metallic green and on casual inspection cannot be distinguished from that species.

Habralictellus rufopanticis Engel. Currently known only from Guana Island (the type locality) and Puerto Rico, but almost certainly present elsewhere on islands of the Puerto Rico Bank. This is a small bee, about 5 mm long, greenish, but with the abdomen largely reddish. Nothing is known of its habits, but females probably construct nests in the ground. Most of the specimens collected on Guana were at flowers of *Capparis cynophallophora*; other floral records here include *Schaefferia frutescens*, and *Cassine xylocarpa*.

Lasioglossum parvum (Cresson). The range of *Lasioglossum parvum* extends from Cuba and the Bahamas to the BVI. This species, and the following, are very similar in appearance and hardly distinguishable without a microscope. They are small (about 5 mm long), dull greenish bees. Females are sometimes attracted to perspiration and are commonly called "sweat bees." Nests are usually in soil and are often very deep. At least some species are semisocial, with two or more females participating in excavating the nest, foraging, and tending larvae; usually one female is dominant and is the "queen" of the colony. Most species of sweat bees are polylectic; that is, they have no discernible floral preferences. Floral records for this species include *Jacquemontia pentantha, Cakile lanceolatum, Capparis cynophallophora, Cassine xylocarpum, Ipomoea pescapri, Cardiospermum micranthum, Coccoloba uvifera*, and *Antigonon leptopus*.

Lasioglossum new species. At present this sweat bee is known only from Guana but likely occurs elsewhere in the Virgin Islands. Its habits are probably similar to those of *Lasioglossum parvum*. Specimens have been collected at flowers of *Schaefferia frutescens, Jacquemontia pentantha, Capparis cynophallophora*, and *Antigonon leptopus*.

MEGACHILIDAE

Coelioxys abdominalis Guérin. This is a continental species that ranges from Panama to Venezuela and Curacao, and from there through the Lesser Antilles to the Virgin Islands and Puerto Rico. The body length is about 10–12 mm, with head and thorax black and the abdomen bright red. In both sexes, but more obviously in the female, the abdomen is broadest at the base and tapers toward the tip. All species of *Coelioxys* are nest parasites of other species of solitary bees, usually members of the genus *Megachile*. However, *Coelioxys abdominalis* and related species are nest parasites of bees of the genus *Centris*, especially those related to *Coelioxys lanipes*, the presumed host on Guana. Specimens have been collected taking nectar from flowers of *Solanum persicifolium, Cardiospermum micrantha, Jacquemontia pentantha, Antigonon leptopus, Ipomoea pescapri*, and *Lantana involucrata*.

Megachile (Eutricharaea) concinna F. Smith*. The known distribution of this small leaf-cutting bee includes the BVI (Guana), Cuba, Dominica, Hispaniola, Jamaica, Mexico, Puerto Rico, Mona Island, and the United States. This species was first described from material from Hispaniola (Dominican Republic), but all species belonging to subgenus *Eutricharaea* are of Old World origin; *Megachile concinna* almost certainly was carried to Hispaniola during the colonial period. It may have originated in Africa, but has not been identified with certainty from anywhere in the Old World. This is a leaf-cutter bee. The female cuts small circular pieces from leaves and uses them to form the cells that are then provisioned with a nectar-pollen paste. Because the bees nest in almost any available cavity of suitable size, they are easily accidentally transported to new areas. Some species of subgenus *Eutricharaea* are commercially grown and distributed as pollinators for certain crop plants, such as alfalfa (lucerne), because they are more efficient than honey bees. I have only a single floral record for this species on Guana: *Ipomoea pescapri*.

Megachile (Pseudocentron) holosericea (Fabricius). This is known from the BVI (Guana), Puerto Rico, Mona Island, and the Lesser Antilles (Moure 1960). It is a larger species than *Megachile concinna* (up to about 15 mm) and, especially in the female, with the hair of the body slightly yellowish. The scopa of the female, on the underside of the abdomen, is yellowish red (white in *Megachile concinna*). The front legs of the male of *Megachile concinna* are simple and black in color; those of *Megachile holosericea* are brightly colored and expanded and modified. On Guana, females have been collected at flowers of *Solanum persicifolium, Jacquemontia pentantha, Melochia tomentosa*, and *Acacia macracantha*; males have also been collected at *Ipomoea pescapri* and *Antigonon leptopus*.

In summary, 81 species of aculeate Hymenoptera are recorded from Guana Island. Nine of these are adventive, not native. The only species known to be purposefully introduced is the European honeybee, *Apis mellifera*. Six of the introduced species are ants: *Cardiocondyla emeryi* Forel, *Cardiocondyla obscurior* Wheeler,

Monomorium floricola (Jerdon), *Solenopsis invicta* Buren, *Tapinoma melanocephalum* (Fabricius), and *Paratrechina longicornis* (Latreille). All of these ant species, except for the South American *Solenopsis invicta*, are common tramp species originating from the Old World.

ACKNOWLEDGMENTS

I wish to express my deep gratitude, first of all, to Henry Jarecki, without whose interest and support the work done on Guana Island by so many biologists would not have been possible. My own work there was facilitated by the enthusiastic interest of James "Skip" Lazell. For me, personally, none of this would have happened were it not for the suggestion first made to me more than a decade ago by that great facilitator, Scott E. Miller. My thanks also go to Mike Engel for his expertise in providing identifications for the halictid bees. Finally, I am indebted to my colleague at the LACM, Fred S. Truxal, for his comments on the manuscript.

Editor's Note: Three more species of Hymenoptera are known from Guana that are not aculeates. Two are tiny, parasitic wasps of the family Braconidae detailed by Paul Marsh (1988, 1993), who named one for Guana Island: *Coiba guanaensis* Marsh and *Holobracon limatus* (Cresson). The third is in the family Trichogrammatidae: *Brachista efferiae* (Pinto 1994). Its host is the robberfly *Efferia stylata* (Diptera: Asilidae).

Butterflies: Lepidoptera, Rhopalocera

Butterflies have long received more than their fair share of attention . . . , and this understandable inequity grows, indeed flourishes.

SMITH, MILLER, AND MILLER (1994)

THE ENORMOUS POPULARITY of butterflies with the biophilic public no doubt stems from their bright colors and conspicuous diurnal activity. For many of us growing up—especially in the dismal North—butterflies cavorting in spring and summer sunshine were irresistible. As children, we learned to catch them and kill them, pin them and mount them, and proudly exhibit them to admiring adults. Some children grow up and put away childish things. Some follow the desire to catch lizards, or beetles, or even elephants; a good few, however, remain set in their childhood love of these Lepidoptera, never to outgrow it. Indeed, second only to birds, butterflies are the most popular living creatures with the general public.

Butterflies are a miniscule fraction of the great order Lepidoptera. Most Lepidoptera are moths. Most moths—however big and beautiful—are nocturnal and do not flash in the sunshine. Actually, most moths are wee little things, and rather drab. On Guana, there are 31 species of butterflies known and more than two orders of magnitude—about 350—more moths. Becker and Miller (1992) long ago published work on our 31 butterflies; They also published a paper on the larger moths in 2002 and have further works in progress on the remaining moths.

The ordinal name Lepidoptera derives from Greek and means "scale winged." The wings are clothed in tiny scales, often brightly colored, that are rubbed off easily (to the great distress of budding young collectors), seem almost like dust, and leave the wings largely transparent. The name Rhopalocera also derives from Greek and refers to the antennae of butterflies that resemble elongate, slender clubs: a wiry stem with an expanded, oval tip. Only two of more than 20 superfamilies of Lepidoptera are thought of as butterflies: the skippers, in the Hesperioidea superfamily—in which the antennal clubs are typically slim and tapered to a point—and the rest are in the Papilionoidea superfamily. The myriad remaining families and superfamilies are all thought of as moths; they have numerous types of antennal shapes, from whiplike to feathery—the latter being very common—but never really clubiform (as the entomologists say).

All Lepidoptera undergo complete metamorphosis during a sedentary pupal stage. Larval butterflies—caterpillars—are generally rapacious herbivores and some are of great detrimental economic importance (but not on Guana Island). The metamorphosed adults, however, are major plant pollinators we could hardly live without.

The first comprehensive attempt to document the butterflies of the West Indies was Riley's (1975) *A Field Guide to the Butterflies of the West Indies*. This was followed by Smith, Miller, and Miller (1994), a beautiful "coffee table" volume covering the West Indies and South Florida. The foundation for the species : area relationship was based on West Indian butterflies as described by Munroe (1953); Davies and Smith (1998) expanded and updated this work, including Guana Island. Among all the islands considered, only Lignum Vitae Key, in the Florida Keys on the continental shelf of North America, is a more remote outlier than Guana in terms of high species count and small size. In the West Indies, Guana, at 300 ha, has as many species (31) as Tortola, at 5,444 ha, and nearly as many as Montserrat, at 200,000 ha (with 38).

Smith, Miller, and Miller (1994) report that there are about 350 species of butterflies in the West Indies, and the combined land area of all the islands is about 240,000 km² (or 90,000 mi.²)—about the same area as the United Kingdom. Further, the total butterfly fauna of the West Indies is more than half that of the entire North American continent (Smith, Miller, and Miller did not tell us, but that is 23,310,000 km², or 9,000,000 mi.², and 16 % of Earth's land area, according to the *Encyclopedia Britannica*). When looked at in this context, Riley's (1975, 11) lament that the fauna seems depauperate is not substantiated. Becker and Miller (1992) give the total Virgin Islands butterfly count as 41 species; remember, 31 are on Guana.

I have modified the text of Becker and Miller (1992) only very slightly, adding some information from Smith, Miller, and Miller (1994) in a few cases, some observations and opinions of my own, and some notes from entomologist Robert Denno, University of Maryland (see "Planthoppers and Other True Bugs: Hemiptera," part 5, above). I have deleted most of Becker and Miller's (1992) literature citations; check their original for their sources. I have kept the family name headings because these are very useful in locating more information should one want to explore further. The accounts below are intended to be used with a field guide such as Riley (1975) in hand; the technical terms are explained and illustrated there. Here are Guana's species.

Although butterflies are by far the most popular of insects to collect and systematically study, our knowledge of them is retarded by two problems. First, butterfly collectors—including professional entomologists—go for quality instead of quantity; they will discard a battered specimen as soon as they get a better one. Also they do not accumulate large series of specimens. This means that the specimens cannot be used for rarefaction analysis or Chao indices, or other quantitative approaches to diversity assessment; see, for a contrast, "Planthoppers and Other True Bugs: Hemiptera," part 5, above.

Second, butterfly biogeography—at least in the West Indies—has not been addressed in view of recent advances in historical geology. The picture presented by J. Y. Miller and Miller (2001) is totally out of sync with the configurations and timing of global plate tectonics as they are now known (A. A. Meyerhoff et al. 1996; Scotese 1997; Lazell 1999b) and not comparable to either of the prevailing views of Antillean geology and biogeography presented by Iturralde-Vinent and MacPhee (1999) or S. B. Hedges (2001).

With the exception of the V-mark skipper, none of Guana's butterflies is endemic at the species level even to the Puerto Rico Bank; they are all widespread. I cannot make a range : abundance comparison because abundances are not documented in most cases and are prone to major seasonal changes. Still, the wide-ranging nature of our species indicates vagile dispersers whose evolutionary patterns of speciation will not be heavily influenced by historical geology. Consider the case of our neighbor island in the BVI, Anegada, solidly united with us and Puerto Rico as recently as 8,000 ybp (Heatwole and MacKenzie 1967). Anegada has two endemic species (D. Smith et al. 1991) belonging to genera not represented on Guana. What are the factors that influence butterfly biogeography? It is a testimonial to the zeal and skill of Vitor Becker and Scott Miller that no one has been able to improve on their tally of 31 Guana species in over a decade. However, all of the gaps in our knowledge that they pointed out remain unfilled. There is thus a superb opportunity for a lepidopterist to jump in and write a truly definitive work on our butterflies in the context of the larger Antillean overview.

PAPILIONIDAE

Polydamas Swallowtail, *Battus polydamas thyamus*. Widespread in the New World, from the southern United States to Argentina, this is the only swallowtail in the British Virgin Islands. It is tail-less and black with yellow marks along margins. One specimen was seen (but not collected) gliding down the hill at North Bay, in July 1987. Doubtfully a resident as the larval host, *Aristolochia*, is not known to occur on Guana. However, Robert Denno has observed adults on several occasions since Becker and Miller's (1992) listing (pers. comm.).

PIERIDAE

Great Southern White, *Ascia monuste virginia*. A common species in the New World, ranging from the United States to Argentina. This and the following are the only two "whites" on Guana. This species is easily distinguished from female *Appias drusilla* by the wedge-shaped dark markings bordering the external margins. Presumably a resident as one of its food plants, *Cakile lanceolata* (Cruciferae), is common along the beach on North Bay.

Florida White, *Appias drusilla boydi*. This species ranges from Florida to southern Brazil. Males are plain white except for the finely black-edged forewing costa. Females have margins bordered in dark gray but not forming wedge-

shaped marks as in *Ascia monuste*. The larvae of *Appias drusilla* feed on various Capparidaceae, cultivated Cruciferae, and *Drypetes* (Euphorbiaceae). Common on the island and presumably a resident as several species of Capparidaceae are abundant.

False Barred Sulphur, *Eurema elathea*. This species ranges from Nicaragua south to Paraguay, and throughout the Antilles and Bahamas. Only the typical wet season form, common throughout the Caribbean, has been collected on Guana. It has the bar along the inner forewing margin fully developed. The dry season form, which has the bar greatly reduced and sandy underwings, also occurs in the area. Collecting in the dry season may reveal this form on Guana. Species of *Stylosanthes* have been recorded as the larval food plants and at least one, *Stylosanthes hamata*, occurs on Guana. Certainly a resident, this species is very common on the flat around the Salt Pond.

Little Sulphur, *Eurema lisa euterpe*. This deep yellow, black-bordered species occurs from the eastern United States to Costa Rica. This and the former are the only two species of *Eurema* recorded from the Virgin Islands. Males of the two are easily separated by the black bar along the inner margin of the forewing cell in *Eurema elathea*; females of *elathea* have hind wings with white ground color, whereas in *lisa* it is yellow as in the male. Its grass green downy larvae feed on a wide variety of leguminous plants such as clovers, peas, and beans: *Trifolium*, *Cassia*, *Mimosa*, etc. Three species of *Cassia* and several other leguminous plants are common on the island. Certainly a resident.

Cloudless Sulphur, *Phoebis sennae sennae*. Widespread in the New World, from the southern United States to Argentina, this is the only large bright yellow species seen on the island. Food plants are several species of *Cassia*, senna, clover, etc.

NYMPHALIDAE

Monarch, *Danaus plexippus megalippe*. This species ranges from Canada to Brazil, north of the Amazon. Four subspecies are recognized in the Caribbean. A female form called leucogyne with gray brown or pale gray upper side ground color often occurs in the area. A few specimens of *megalippe* were seen visiting *Asclepias* in the lowlands of Guana and several were collected. Milkweeds of the genera *Asclepias, Calotropis*, etc., are the common food plants. However, other milk plants such as *Apocynum* (Apocynaceae), *Ipomoea* (Convolvulaceae), and *Euphorbia* (Euphorbiaceae) have been reported.

Gulf Fritillary, *Agraulis vanillae insularis*. This orange butterfly with conspicuous silvery marks on the underwings is the only fritillary found on the island. Widely distributed in the New World, this species ranges from the United States south to Argentina. Food plants are various vines of the genus *Passiflora* (Passifloraceae).

Zebra, *Heliconius charitonius charitonius*. This black, yellow-barred species ranges from the southern United States to Venezuela and Peru. It is very common on the island. Food plants are various species of *Passiflora* (Passifloraceae). The pupae, hanging on the *Passiflora* vines, can make a loud squeak or distress call if handled; I wondered what good that did them? No one had come to their rescue while I was there, so I contacted Erika Deinert, Organization for Tropical Studies, Costa Rica, who wrote back on 6 September 2001 that, although the pupae do not stridulate when touched by predators such as ants, they do when contacted by fifth instar caterpillars of their own species. She notes that the caterpillars move on, so the vibrations may work to tell them they have found an already occupied, therefore unsuitable, pupation site.

Red Rim, *Biblis hyperia hyperia*. This beautiful black species, with red-bordered hind wings, ranges from Mexico to Paraguay. A few specimens were spotted and a couple collected near the orchard at White Bay. The species is presumably a resident as the reported food plant, the pine nettle, *Tragia volubilis* (Euphorbiaceae) occurs on the island.

Caribbean Buckeye, *Junonia evarete*. A few specimens of this widespread New World species were seen on the island. Two specimens were caught in a Malaise trap near North Beach. The only larval food plant in Jamaica is the black mangrove, *Avicennia germinans*, a plant not recorded from Guana. This species is very similar to, and easily confused with, the following.

Genoveva, *Junonia genoveva*. Two specimens were collected, one of them on flowers of *Stachytarpheta jamaicensis*, one of their larval food plants. *Junonia genoveva* is easily confused with the former species, but readily separated by the color of the antennae and by the broad oblique fascia on the ventral forewings. In *Junonia genoveva*, the antennae are pale cream or whitish with a dark club and the fascia runs to the external margin, whereas in *Junonia evarete* the antennae are dark tawny or brown with a black club and the fascia is closed by the marginal lines, not reaching the external margin.

Painted Lady, *Vanessa cardui*. This cosmopolitan species is "a notable migrant, a greater wanderer even than *Danaus plexippus*, liable to occur almost anywhere except in arctic conditions" (Riley, 1975). Two worn specimens were collected visiting flowers of oleander, *Nerium oleander* (Apocynaceae), on the top of the hill near the buildings. This is the first record from the Virgin Islands. Because the plant is an introduced exotic from the Mediterranean, these travelers were lucky to find it here.

LYCAENIDAE

Clench's Hairstreak, *Chlorostrymon maesites clenchi*. This beautiful species, with dark metallic blue dorsal wings and green ventral surfaces, ranges from Florida throughout the Caribbean Islands south to Dominica. It has been reared on *Albizia lebbeck* (Leguminosae) in the laboratory.

Fulvous Hairstreak, *Electrostrymon angelia boyeri*. This Caribbean species ranges from the Bahamas to the Virgin Islands. This and the following species are the only ones bearing two hairlike tails on each hind wing. *Electrostrymon angelia boyeri* is readily separated from the next species by the fuscous brown wing undersides and absence of oblique white fascia on the forewing undersides. The larvae of *Electrostrymon angelia* feed on *Schinus terebinthifolius* (Anacardiaceae).

Drury's Hairstreak, *Strymon acis mars*. This species, which ranges from Florida throughout the Antilles to Dominica, is the largest hairstreak on Guana Island. Easily distinguished by the two tails on the hind wings, the lead gray color, and the conspicuous white oblique band on the ventral wings. It is a resident despite its rarity on the island (Robert Denno, pers. comm.). According to Riley (1975) ". . . in Jamaica . . . it is attracted to a low-growing shrub, *Croton discolor* (Euphorbiaceae), which . . . may be the food plant of its unknown caterpillar." Plants of this genus, which are very common on the island, should be searched for eggs and larvae.

Bubastus Hairstreak, *Strymon bubastus ponce*. This is a widespread South American species, occurring through the Lesser Antilles north to Puerto Rico. Very similar to *Strymon columella* but tail-less. Distinguished from the other tail-less lycaenids on the island by the absence of a mark at the end of the discal cell on the ventral forewing. Despite its wide distribution, the early stages are unknown.

Hewitson's Hairstreak, *Strymon columella columella*. This is a very widespread species in the New World, ranging from the southern United States to Brazil. It is the only lycaenid on the island bearing a single tail on the hind wing. The ventral wing markings are very similar to *Strymon bubastus*. Its food plants are species of *Malva* and *Sida* (Malvaceae), two genera of plants common on Guana.

Hanno Blue, *Hemiargus hanno watsoni*. This species ranges from Costa Rica to Argentina and is very common on the island. We regard *Hemiargus hanno* as a species separate from *ceraunus*. The pattern on the ventral wing surfaces distinguishes it from the other three tail-less lycaenids on the island: *Strymon bubastus* lacks the mark at the end of the forewing discal cell; *Leptotes cassius* has more markings toward the wing bases; and *Hemiargus thomasi* has two black orbicular

marks on the hind wing margin near the tornus. Food plants include many species belonging to several genera of legumes such as *Crotalaria*, *Chamecrista*, and *Macroptilium*.

Thomas's Blue, *Hemiargus thomasi woodruffi*. This species ranges from southern Florida to St. Kitts, excluding Cuba and Jamaica, and is one of the four tail-less lycaenids on Guana. This and *Leptotes cassius* have two black orbicular marks on the ventral hind wings near the margin; however, *cassius* lacks other black markings. Certainly a resident on the island, the early stages are unknown.

Cassius Blue, *Leptotes cassius catilina*. This species ranges from Florida to Argentina and is the most common blue on the island. Smaller than *Hemiargus thomasi*, it is distinguished from the other tail-less lycaenids by the features discussed above. Food plants include several leguminous shrubs and bushes, such as *Galactia*, *Indigofera*, *Desmodium*, and *Crotalaria* as well as *Plumbago* (Plumbaginaceae). Several species belonging to all of these genera occur on Guana Island.

HESPERIIDAE

V-mark Skipper, *Choranthus vitellius*. This species is restricted to Puerto Rico and the Virgin Islands, but it was not seen on Guana Island by Smith, Miller, and Miller (1994, 226). It is our only species restricted to the Puerto Rico Bank. This bright yellow species is one of the most beautiful skippers on the island; easily distinguished from the others by its plain yellow underside. The larval host plants include sugar cane and other grasses.

Three Spot Skipper, *Cymaenes tripunctus*. This dull brown skipper ranges from the United States to Brazil. Common on the island, its larvae feed on Guinea grass.

Fiery Skipper, *Hylephila phyleus*. This species is also widespread in the New World, from southern New England to Argentina. Males are bright yellow but have dark marks scattered on the underside; females have yellow color generally reduced. Its larvae feed on various grasses.

Sugar Cane Skipper, *Panoquina sylvicola*. This species is easily recognized by its brown color, semitranslucent marks on the forewings, and the row of blue spots forming a line along the margin on the ventral surface of the hind margin. It ranges from the United States to Argentina. Food plants include various grasses such as sugar cane, bamboo, Johnson and Pimento grass.

Broken Dash Skipper, *Wallengrenia otho druryi*. This common species, the size of *Choranthus vitellius*, ranges from the United States to Argentina. It is almost all dark brown above, with reduced pale yellow markings; underside tinged rust orange. Food plants are coarse grasses, sugar cane, rice, etc.

Hairy Dusky Wing, *Ephyriades arcas philemon*. This species is confined to the Greater Antilles and the Lesser Antilles south to St. Kitts. The taxonomy of this large sexually dimorphic species is complex. Males are uniformly silky black with a costal fold from base to distal two-thirds of forewings; females are dull brown with translucent marks on forewings. Our specimens could be identified as *Ephyriades zephodes*. However, the traditional external characteristics are of limited value. Genitalia of one dissected male match *arcas*, so we assign the Guana specimens to *arcas*. As the species have been mixed, it is not clear to which species the recorded food plants belong. Larvae of one or both have been reared on *Stigmaphyllon periplocifolium* and *Malphigia fucata* (Malphigiaceae), and *Ceiba pentandra* (Bombacaceae). Considering this situation, more collecting, both of adults and larvae, is necessary to ascertain whether both species are present on the island and to determine the larval food plants.

Hammock Skipper, *Polygonus leo savigny*. This large species ranges from the southern United States to Argentina. It is commonly found in shady areas along trails resting upside-down under leaves. Its larvae feed on dogwood in the United States. It has been recorded feeding on *Lonchocarpus* (Leguminosae) on Cuba.

Tropical Checkered Skipper, *Pyrgus oileus oileus*. This species occurs from the southern United States to Argentina; it is certainly the most common butterfly on the island. It is found in open, sunny disturbed areas where its malvaceous weed hosts grow. It rests with wings wide open. Males are gray and females fuscous, spotted white.

Dorantes Skipper, *Urbanus dorantes cramptoni*. The two species of *Urbanus* present in the Antilles both occur on the island and are the only tailed skippers on Guana. *Urbanus dorantes* is distinguished from *Urbanus proteus* by the absence of metallic green. *Urbanus dorantes* is widespread in the New World, ranging from the southern United States to Argentina. Its larvae feed on various species of beans (Leguminosae).

Common Long-tail Skipper, *Urbanus proteus domingo*. This species ranges from the United States to Argentina. Easily distinguished from the former species by the iridescent green scales on the body and wing bases (in the former the wings are plain dull brown). Common on the island, the larvae feed on various species of beans as well other leguminous plants.

Moths: Lepidoptera

Vitor O. Becker • *Scott E. Miller*

T HE FOLLOWING LIST summarizes identifications of the so-called Macrolepidoptera and pyraloid families from Guana Island. Methods are detailed in Becker and Miller (2002). Data and illustrations for Macrolepidoptera are provided in Becker and Miller (2002). Data for Crambidae and Pyralidae will be provided in Becker and Miller (in preparation). General, but outdated, background information on Crambidae and Pyralidae are provided by Schaus (1940). Data for Pterophoridae are provided in Gielis (1992) and Landry and Gielis (1992). Author and date of description are given for each species name. Earlier dates were not always printed on publications; those in square brackets indicate that the year was determined from external sources not the publication itself. As in previous lists, authors' names are put in parentheses when their generic placement has been revised. Detailed acknowledgments are provided in Becker and Miller (2002), but, in addition, we are especially grateful to C. Gielis, E.G. Munroe, M. Shaffer, and M.A. Solis for assistance with identifications and to K. Darrow for assistance in all aspects of this project. Sampling on Guana Island was supported by The Conservation Agency, through grants from the Falconwood Corporation.

SPHINGIDAE

SPHINGINAE

Agrius cingulatus (Fabricius 1775). United States south to Argentina.

Cocytius antaeus (Drury 1773). Southern United States to Argentina.

Manduca sexta (Linnaeus 1763). Widespread in the New World.

Manduca rustica (Fabricius 1775). Widespread in the New World.

Manduca brontes (Drury 1773). Antilles north to Central Florida.

MACROGLOSSINAE

Pseudosphinx tetrio (Linnaeus 1771). (See plate 37.) United States through the Antilles to Argentina.

Erinnyis alope (Drury 1773). Widespread in the New World.

Erinnyis ello (Linnaeus 1758). Neotropical.

Erinnyis crameri (Schaus 1898). Southern United States through the Caribbean south to Brazil.

Erinnyis domingonis (Butler 1875). Southern United States through the Caribbean south to Brazil.

Erinnyis obscura (Fabricius 1775). Southern United States through the Caribbean south to Brazil.

Pachylia ficus (Linnaeus 1758). Widespread in the New World.

Callionima falcifera (Gehlen 1943). Southern United States to Argentina.

Perigonia lusca (Fabricius 1777). Southern Florida to Argentina.

Enyo lugubris (Linnaeus 1771). Southern United States, Antilles, south to Argentina.

Aellopos tantalus (Linnaeus 1758). New York south to Argentina.

Eumorpha vitis (Linnaeus 1758). United States through the Caribbean to Argentina.

Cauthetia noctuiformis (Walker 1856). Caribbean.

Xylophanes chiron (Drury 1770). Mexico through the Antilles to Argentina.

Xylophanes pluto (Fabricius 1777). Southern United States to Brazil.

Xylophanes tersa (Linnaeus 1771). Ontario, Antilles south to Argentina.

Hyles lineata (Fabricius 1775). Worldwide.

NOTODONTIDAE

Nystalea nyseus (Cramer 1775). Mexico through the Caribbean south to Brazil.

NOCTUIDAE

HELIOTHINAE

Heliothis subflexa (Guenée 1852). North America, Antilles, south to Argentina.

NOCTUINAE

Anicla infecta (Ochsenheimer 1816). Central United States south to Argentina.

HADENINAE

Leucania humidicola Guenée 1852. Antilles to Brazil.

Leucania dorsalis Walker 1856. Southern Florida through the Neotropics.

AMPHIPYRINAE

Neogalea sunia (Guenée 1852). Florida to Argentina.

Catabenoides terminellus (Grote 1883). Southern United States, Antilles.

Catabenoides lazelli Becker & Miller 2002. Guana, Anegada, St. Croix.

Spodoptera albulum (Walker 1857). United States, Antilles, south to Argentina.

Spodoptera frugiperda (J. E. Smith 1797). Widespread in the New World.

Spodoptera latifascia (Walker 1856). U.S. gulf states, Antilles south to Costa Rica.

Spodoptera pulchella (Herrich-Schäffer 1868). Florida, Greater Antilles.

Spodoptera dolichos (Fabricius 1794). U.S. gulf states, Antilles south to Costa Rica.

Magusa orbifera (Walker 1857). Widespread from Canada to Argentina.

Condica albigera (Guenée 1852). Mexico through the Antilles to Paraguay.

Condica mobilis (Walker [1857]). Southern United States through Antilles to Argentina.

Condica sutor (Guenée 1852). Southern United States, Antilles to Argentina.

Perigea gloria Becker & Miller 2002. Guana, Tortola.

Elaphria agrotina (Guenée 1852). Florida, Antilles south to Argentina.

Elaphria nucicolora (Guenée 1852). Neotropical.

Anateinoma affabilis Möschler 1890. Puerto Rico, Virgin Islands.

Micrathetis triplex (Walker 1857). Southern United States through South America.

AGARISTINAE

Caularis undulans Walker [1858]. Hispaniola, Jamaica, Puerto Rican Bank.

BAGISARINAE

Bagisara repanda (Fabricius 1793). Southeast United States to Paraguay.

ACONTIINAE

Amyna axis (Guenée 1852). Pantropical.

Ponometia exigua (Fabricius 1793). Southern United States, Antilles south to Brazil.

Cydosia nobilitella (Cramer 1779). Southern United States, Antilles south to Argentina.

Tripudia quadrifera (Zeller 1874). Southern United States, Antilles south to Brazil.

Tripudia balteata Smith 1900. Southern United States, Antilles south to Brazil.

Ommatochila mundula (Zeller 1872). Southern United States, Antilles south to Argentina.

Cobubatha metaspilaris Walker 1863. Antilles.

Eumicremma minima (Guenée 1852). Southern United States, Antilles south to Argentina.

Eublemma rectum (Guenée 1852). Southern United States, Antilles south to Argentina.

Spragueia margana (Fabricius 1794). Southern United States, the Neotropics.

Spragueia perstructana (Walker 1865). Southern United States, Antilles south to Costa Rica.

Thioptera aurifere (Walker [1858]). Southern United States, Antilles south to Brazil.

SARROTHRIPINAE

Characoma nilotica (Rogenhofer 1882). Pantropical.

Collomena filifera (Walker 1857). Florida, Antilles south to Brazil.

Motya abseuzalis Walker 1859. Florida, Antilles south to Brazil.

EUTELIINAE

Paectes obrotunda (Guenée 1852). Southern United States, Antilles south to Paraguay.

PLUSIINAE

Pseudoplusia includens (Walker [1858]). United States to northern Chile and Argentina.

CATOCALINAE

Ptichodis immunis (Guenée 1852). Mexico through the Antilles to Brazil.

Mocis latipes (Guenée 1852). Southern United States, Antilles south to Argentina.

Mocis antillesia Hampson 1913. Lesser Antilles, Bahamas.

Mocis repanda (Fabricius 1794). Antilles, Guatemala.

Ophisma tropicalis Guenée 1852. Southern United States, Antilles south to Argentina.

OPHIDERINAE

Azeta versicolor (Fabricius 1794). Neotropical.

Metallata absumens (Walker 1862). Southern United States, Antilles south to Brazil.

Plusiodonta thomae (Guenée 1852). Endemic to the Antilles.

Syllectra erycata (Cramer 1780). Florida, Antilles south to Brazil.

Litoprosopus puncticosta Hampson 1926. Haiti, Virgin Islands.

Diphthera festiva (Fabricius 1775). Neotropical.

Gonodonta bidens Geyer 1832. Florida, Antilles south to Argentina.

Melipotis acontioides (Guenée 1852). Florida, Antilles south to Brazil.

Melipotis fasciolaris (Hübner [1831]). Southern United States, Antilles south to Uruguay.

Melipotis contorta (Guenée 1852). Florida, Antilles.

Melipotis famelica (Guenée 1852). Southern United States, Antilles south to Venezuela.

Melipotis ochrodes (Guenée 1852). Mexico through the Antilles to Brazil.

Melipotis januaris (Guenée 1852). Southern United States, Antilles, northern South America.

Ascalapha odorata (Linnaeus 1758). Pantropical.

Epidromia lienaris (Hübner 1823). Neotropical.

Manbuta pyraliformis (Walker 1858). Florida, Antilles.

Ephyrodes cacata Guenée 1852. Southern United States, Antilles south to Colombia.

Concana mundissima Walker [1858]. Florida, Antilles south to Brazil.

Massala asema Hampson 1926. Antilles.

Lesmone formularis (Geyer 1837). Neotropical.

Lesmone hinna (Geyer 1837). Southern United States, Antilles south to Brazil.

Eulepidotis addens (Walker 1858). Antilles.

Eulepidotis modestula (Herrich-Schäffer 1869). Antilles.

Toxonprucha diffundens (Walker 1858). Antilles south to Brazil.

Kakopoda progenies (Guenée 1852). Florida, Antilles south to Brazil.

Parachabora abydas (Herrich-Schäffer [1869]). Mexico through the Antilles to Brazil.

Cecharismena abarusalis (Walker 1859). Florida, Antilles south to Brazil.

Cecharismena cara Möschler 1890. Antilles.

Glympis eubolialis (Walker [1866]). Antilles.

HERMINIINAE

Drepanopalpia lunifera (Butler 1878). Antilles.

Lascoria orneodalis (Guenée 1854). Florida, Antilles.

Bleptina hydrillalis Guenée 1854. Southern United States, Central America, Antilles.

Bleptina caradrinalis Guenée 1854. Southern United States, Antilles south to Brazil.

Bleptina menalcasalis Walker [1859]. Antilles south to Venezuela.

Bleptina araealis (Hampson 1901). Antilles, Florida.

HYPENINAE

Hypena lividalis (Hübner 1790). Southern Palearctic, Pantropical.

Hypena minualis (Guenée 1854). Antilles south to Brazil.

ARCTIIDAE

ARCTIINAE

Hypercompe simplex (Walker 1855). Puerto Rican Bank, Lesser Antilles.

Calidota strigosa (Walker 1855). Antilles, Southern United States.

Eupseudosoma involutum (Sepp [1855]). Southern United States, Antilles south to Argentina.

Utetheisa ornatrix (Linnaeus 1758). Southern United States, Antilles south to Argentina.

Utetheisa pulchella (Linnaeus 1758). Africa, Asia, the Neotropics.

PERICOPINAE

Composia credula (Fabricius 1775). Antilles, Puerto Rican Bank.

CTENUCHINAE

Empyreuma pugione (Linnaeus 1767). Restricted to the Puerto Rican Bank.

Horama pretus (Cramer 1777). Antilles.

Horama panthalon (Fabricius 1793). Southern United States, Antilles to southern Brazil.

Cosmosoma achemon (Fabricius 1781). Antilles south to Brazil.

Eunomia colombina (Fabricius 1793). Antilles.

Nyridela chalciope (Hübner [1831]). Antilles, Central America.

LITHOSIINAE

Afrida charientisma Dyar 1913. Antilles.

Progona pallida (Möschler 1890). Puerto Rico.

Lomuna nigripuncta (Hampson 1900). Puerto Rico.

GEOMETRIDAE

OENOCHROMINAE

Almodes terraria Guenée [1858]. Southern United States, Antilles to Colombia.

ENNOMINAE

Pero rectisectaria (Herrich-Schäffer [1855]). Puerto Rican Bank, Lesser Antilles.

Oxydia vesulia (Cramer [1779]). Southern United States, Antilles south to Argentina.

Erastria decrepitaria (Hübner [1823]). Southern United States, Antilles south to Brazil.

Sphacelodes fusilineatus (Walker 1860). Antilles.

Macaria paleolata (Guenée [1858]). Antilles.

Patalene ephyrata (Guenée [1858]). Antilles.

Cyclomia mopsaria Guenée [1858]. Antilles south to Brazil.

GEOMETRINAE

Eueana simplaria Herbulot 1986. Guana, Guadaloupe.

Phrudocentra centrifugarium (Herrich-Schäffer 1870). Florida through the Greater Antilles to Puerto Rico.

Chloropteryx paularia (Möschler 1886). Florida, Antilles.

Synchlora frondaria (Guenée [1858]). United States, Antilles south to Argentina.

Synchlora cupedinaria (Grote 1880). Florida, Greater Antilles to Virgin Islands, Nevis.

STERRHINAE

Semaeopus malefidarius (Möschler 1890). Puerto Rican Bank.

Leptostales noctuata (Guenée [1858]). Antilles.

Acratodes suavata (Hulst 1900). Southern United States, Antilles.

Lobocleta nymphidiata (Guenée [1858]). Antilles.

Scopula laresaria Schaus 1940. Puerto Rican Bank.

Idaea sp., probably *Idaea fernaria* (Schaus 1940). Antilles.

"Idaea" monata (Forbes in Ramos [1947]). Antilles.

Idaea minuta (Schaus 1901). Antilles, United States.

Idaea eupitheciata (Guenée [1858]). Antilles.

Leptostales phorcaria (Guenée [1858]). Antilles.

Leptostales oblinataria Möschler 1890. Southern United States, Antilles to South America.

LARENTIINAE

Obila praecurraria (Möschler 1890). Antilles.

Pterocypha defensata Walker 1862. Southern United States, Antilles.

HYBLAEIDAE

Hyblaea puera species complex.

COSSIDAE

Psychonoctua personalis Grote 1865. Antilles, Mexico.

CRAMBIDAE

CRAMBINAE

Argyria diplomochalis Dyar 1913. Puerto Rico.

Donacoscaptes micralis (Hampson 1919). Antilles.

Fissicrambus fissiradiellus (Walker 1863). Antilles, Panama, Colombia.

Fissicrambus minuellus (Walker 1863). Florida, Antilles, Central America.

Microcrambus atristrigellus (Hampson 1919). Jamaica.

ODONTIINAE

Cliniodes nomadalis Dyar 1912. Central America, Antilles.

Microtheoris ophionalis (Walker 1859). Southern United States, Central America, Antilles.

Mimoschinia rufofascialis (Stephens 1834). Antilles.

GLAPHYRIINAE

Alatuncusia bergii (Möschler 1891). Southern Florida, Antilles, perhaps South America.

Chalcoela iphitalis (Walker 1859). North America to Guatemala.

Chalcoela pegasalis (Walker 1859). North America, Antilles.

Contortipalpia santiagalis (Schaus 1920). Antilles.

Dichogama amabilis Möschler 1891. Southern Florida to Antilles.

Dichogama innocua (Fabricius 1793). Antilles, Central America.

Dichogama redtenbacheri Lederer 1863. Southern Florida, the Neotropics.

Galphyria badierana (Fabricius 1794). Antilles, Central America.

Schacontia new species Neotropical.

EVERGESTINAE

Evergestella evincalis (Möschler 1891). Antilles, Florida Keys.

PYRAUSTINAE

Achyra bifidalis (Fabricus 1794). Neotropical.

Achyra rantalis (Guenée 1854). New World.

Arthromastix lauralis (Walker 1859). Neotropical.

Asciodes gordialis Guenée 1854. Neotropical.

Atomopteryx serpentifera (Hampson 1913). Antilles.

Bicilia iarchasalis (Walker 1859). Neotropical.

Cryptobotys zoilusalis (Walker 1859). Neotropical.

Desmia ufeus (Cramer 1777). Neotropical.

Diaphania costata (Fabricius 1794). Antilles.

Diaphania elegans (Möschler 1890). Antilles, Central America.

Diaphania hyalinata (Linnaeus 1767). Neotropical.

Diaphantania species near *D. ceresalis* (Walker 1859). Antilles.

Ennomosia basalis (Hampson 1897). Antilles.

Epicorsia cerata (Fabricius 1795). Lesser Antilles.

Epipagis algarrobolis (Schaus 1940). Lesser Antilles.

Ercta vittata (Fabricius 1794). Neotropical.

Eulepte gastralis (Guenée 1854). Antilles.

Glyphodes sibillalis Walker 1859. Neotropical.

Herpetogramma bipunctalis (Fabricius 1794). Neotropical.

Herpetogramma phaeopteralis (Guenée 1854). Neotropical.

Loxomorpha cambogialis (Guenée 1854). Neotropical.

Marasmia cochrusalis (Walker 1859). Neotropical.

Marasmia sp. Virgin Islands.

Microthyris anormalis (Guenée 1854). Neotropical.

Oenobotys vinotinctalis (Hampson 1895). Neotropical.

Omiodes albicinctalis (Hampson 1904). Antilles.

Omiodes indicata (Fabricius 1775). Pantropical.

Palpita flegia (Cramer 1777). Neotropical.

Palpita persmilis Munroe 1959. Neotropical.

Palpita viettei Munroe 1959. Neotropical.

Palpusia species near *P. eurypalpalis* (Hampson 1912).

Phaedropsis principaloides (Möschler 1890). Antilles.

Portentomorpha xanthialis (Guenée 1854). Neotropical.

Pseudopyrausta minima (Hedemann 1894). Lesser Antilles.

Pyrausta insignitalis (Guenée 1854). Lesser Antilles.

Pyrausta laresalis Schaus 1940. Lesser Antilles.

Samea ecclesialis Guenée 1854. Antilles.

Sathria onophasalis (Walker 1859). Antilles.

Spoladea recurvalis (Fabricius 1775). Pantropical.

Steniodes mendica (Hedemann 1894). Antilles.

Synclera jarbusalis (Walker 1859). Antilles.

Syngamia florella (Stoll 1781). Neotropical.

Uresiphita reversalis (Guenée 1854). Neotropical.

PYRALIDAE
PYRALINAE

Ocrasa nostralis (Walker 1854). Pantropical.

CHRYSAUGINAE

Bonchis munitalis (Lederer 1863). Neotropical.

Murgisca subductella (Möschler 1890). Lesser Antilles.

Streptopalpia minusculalis (Möschler 1890). Neotropical.

GALLERIINAE

Pogrima palmasalis Schaus 1940. Antilles.

Thyridopyralis gallaerandialis Dyar 1901. Florida.

EPIPASCHIINAE

Phidotricha erigens Ragonot 1888. Antilles, Central America.

Phidotricha insularella (Ragonot 1888). Antilles, Central America.

Tineopaschia minuta Hampson 1916. Antilles.

PHYCITINAE

Anabasis ochrodesma (Zeller 1881). Southeast United States to northern South America.

Ancylostomia stercorea (Zeller 1848). Neotropical.

Australephestiodes stictella (Hampson 1901). Antilles.

Bema neuricella (Zeller 1848). Neotropical.

Cactoblastis cactorum (Berg 1885). Native to southern South America, now widespread because of biological control introductions.

Cassiana malacella (Dyar 1914). Mexico.

Crocidomera fissuralis (Walker 1863). Hispaniola, Puerto Rico.

Ectomyelois decolor (Zeller 1881). Antilles, Central America, South America.

Erelieva quantulella (Hulst 1887). United States, Antilles.

Etiella zinckenella (Treitschke 1832). Cosmopolitan.

Eurythmia hospitella (Zeller 1875). United States.

Fundella argentina Dyar 1919. Neotropical.

Hypargyria definitella (Zeller 1881). Neotropical.

Oryctometopia fossulatella Ragonot 1888. Texas, Antilles, Central America to Brazil.

Ozamia lucidalis (Walker 1863). Neotropical.

Sarasota furculella (Dyar 1919). Antilles.

Unadilla erronella (Zeller 1881). Neotropical.

Zamagiria laidion (Zeller 1881). Neotropical.

PTEROPHORIDAE

Adania ambrosiae (Murtfeldt 1880). North America, Antilles, Galápagos.

Adania thomae (Zeller 1877). Antilles, South America.

Exelastis cervinicolor (Barnes & McDunnough 1913). Florida, Antilles, Galápagos.

Exelastis pumilio (Zeller 1873). Neotropical.

Lantanophaga pusillidactyla (Walker 1864). Pantropical.

Megalorrhipida defectalis (Walker 1864). Pantropical.

There are literally hundreds of additional species of moths we have collected on Guana Island. They are generally small species informally lumped together as "microlepidoptera." Sorting them out is taking decades. Meanwhile Davis (1986) clarified the taxonomic status of just one of them in family Tineidae: *Acrolophus triatomellus*.

Lists of Other Insects

F OR A FEW GROUPS, I am able to expand somewhat on the information presented by Barry D. Valentine in his "An Introduction to Invertebrates," part 5, above (table 26); see that for common names of the orders and for orders not listed here.

ORDER ORTHOPTERA

The list was provided by Daniel E. Perez-Gelabert, Entomology, Smithsonian. He visited Guana from 6 to 13 October 2001. Valentine (in litt. 1 July 2003) has added another, indicated by BV in parentheses.

ACRIDIDAE: GRASSHOPPERS
Schistocerca nitens
Rhammatocerus cyanipes

GRYLLIDAE: CRICKETS
Gryllus sp.
Orocharis sp.
Amphiacusta sp.
Ectatoderus ?sp.

GRYLLACRIDIDAE: LEAF ROLLERS
A large gryllacridine not yet identified to genus (BV).

TETTIGONIIDAE: KATYDIDS
Conocephalus insularis
Microcentrum sp.
Neoconocephalus sp.

ORDER DICTYOPTERA

Lou Roth, Harvard, wrote up Guana's roaches (Roth 1994) describing *Symploce pararuficollis* as a new species. Two more species were added to Roth's (1994) list by Perez-Gelabert (see "Order Orthoptera," above) in 2001, indicated by the initials DP in parentheses.

Periplaneta americana
Panchlora sagax
Cariblatta antiguensis
Symploce ruficollis
Symploce pararuficollis
Euthlastoblatta facies
Plectoptera rhabdota
Pycnoscelus surinamensis
Hemiblabera brunneri (DP)
Eurycotis dicipiens (DP)

ORDER DIPTERA

Valentine combined efforts with Justin Runyon, Montana State University, Bozeman, to produce

TABLE 30
Bee Fly Species on Islands of the BVI

	Anthrax oedipus	Chrysanthrax gorgon	Exoprosopa cubana	Heterostylum ferrugineum	Neodiplocampta roederi	Villa lateralis	Total number of species
Tortola (5,494)			X		X	X	3
Anegada (3,872)		X			X		2
Virgin Gorda (2,120)		X				X	2
Great Camanoe (337)		X				X	2
Guana (300)	X	X		X	X	X	5
Norman (257)						X	1
Scrub (97)	X	X				X	3
Necker (30)						X	1
Eustacia (11)						X	1

SOURCE: Evenhuis and Miller (1994). Specimens in BPBM and/or USNM. The islands are arranged by size, with island areas given in parentheses in hectares.

this list of 35 families. There appear to be representatives of five more families, unidentified; this brings the total number of dipteran families on Guana Island to 40. They are here divided into suborders.

NEMATOCERA

ANISOPODIDAE, CERATOPOGONIDAE, CHIRONOMIDAE, CULICIDAE, SCATOPSIDAE

BRACHYCERA

ASILIDAE

Beameromyia virginensis Scarborough (1997) is a robber fly described from Guana. The holotype is in USNM, Smithsonian. Pinto (1994) records *Efferia stylata*, a widespread species.

BOMBYLIIDAE

The bee flies of the BVI were studied in some detail by Neal Evenhuis, B.P. Bishop Museum (BPBM), Honolulu, Hawaii, and Scott Miller (Evenhuis and Miller 1994). They collected six species total—five on Guana. The only species not collected on Guana was collected on Tortola, where only two others were found (see table 30).

DOLICOPODIDAE, EMPEDIDAE, LEPTOGASTERIDAE, STRATIOMYIDAE, TABANIDAE, THEREVIDAE

CYCLORRHAPHA

ANTHOMYIIDAE, CALLIPHORIDAE, CLUSIIDAE, DROSOPHILIDAE

The fruit fly *Drosophila simulans* may be the first Guana insect recorded in the scientific literature. Levins (1969a) tested acclimation to high temperatures in several species from the greater Puerto Rico Bank. Guana's *Drosophila simulans* were undistinguished in their abilities. Subsequently, Peter Chabora and colleagues, from the American Museum of Natural History, came to Guana and collected *Drosophila willistoni*. This species is widespread in the American tropics from Peru and Brazil to Mexico and Guana Island. It has been studied at the molecular level by Griffin and Powell (1997). Using Guana and other populations, Griffin and Powell found that *Drosophila willistoni* showed very different molecular evolutionary patterns, both at different loci within the species and—especially—with respect to other closely related *Drosophila* species. They speculate that evidence for the virtual lack of polymorphism at one specific locus—that for alcohol dehydrogenase, which is critically important in the feeding ecology of these insects because they consume fermenting fruit—indicates that the species dispersed out from one geographic area, a glacial refugium, very recently in Holocene time. They contrast *Drosophila willistoni* with other species and state: "Thus generalizing about molecular evolutionary patterns and processes . . . from a single species . . . is risky at best." This species was also used in a molecular phylogenetic study by J.M. Gleason et al. (1998): The Guana fruit flies fit right in where they were supposed to be. I am indebted to Scott Miller, Smithsonian, for calling my attention to these studies and to many other arthropod publications.

EPHYDRIDAE, HIPPOBOSCIDAE, MICROPEZIDAE, MUSCIDAE, OTITIDAE, PHORIDAE, PIPUNCULIDAE, SARCOPHAGIDAE

Blaesoxipha virgo Pape (1994) is a sarcophagid flesh fly collected from Necker and Guana islands. The holotype is at BPBM in Honolulu, Hawaii. Pape (1989) also recorded the sarcophagid *Opsidia anticlarum* from Guana, a paratypic locality.

SYRPHIDAE, STREBLIDAE, (see Parasitism, part 3, above, and Maa 1971), TACHINIDAE, TEPHRITIDAE

Anastrepha maculata Norrbom (1998) is a picture-wing tephritid fruit fly described from Guana. The holotype is in the U.S. National Museum, Smithsonian.

PART 6

Nature and Man

Human History

. . . the most unimportant colony in the Empire.

ROBB WHITE (1985)

THE EARLIEST KNOWN HUMAN immigrants to the Antilles came about 6,000 ybp, probably arriving—based on their stone tools—from Yucatan (S.M. Wilson 2001). It is unclear whether or not these people, spreading east from Cuba at least as far as Hispaniola, reached Puerto Rico prior to a wave of people moving north from South America, up through the Lesser Antilles, who arrived at the Puerto Rico Bank about 4,000 ybp (S.M. Wilson 2001). Land areas were larger then because sea level was as much as 5 m lower 4,000 ybp (Morris et al. 1977, 8). The climate was at least as hot as today's, and wetter (Curtis et al. 2001). Sea level rose rapidly, and coastal habitation sites and artifacts from this period would probably now mostly be underwater.

Beginning about 2,500 ybp, remnants of the Saladoid culture appear. Saladoid takes its name from distinctive ceramics named for a Venezuelan site, Saladera. S.M. Wilson (2001) describes the design on these ceramics as "unmistakable white-on-red painted and zone-incised-cross-hatched pottery." I have two typescript reports on Guana's early artifacts: one from Michael Gibbons, Department of Anthropology, University of Massachusetts, Boston, dated 20 December 1986; and the second from Elizabeth Righter, then of the Office of Archeology and Historic Preservation, USVI, dated November 1987. They are complementary because Gibbons excavated a cave floor whereas Righter dug pits on the flat.

The earliest artifactual remains from Guana are charcoal from the cave site and a polished stone axehead, both 2,000–1,500 ybp. Gibbons regards these as later archaic, from hunter-gatherers, not Saladoid. At 1,600–1,500 ybp, Righter got the first Saladoid ceramic remains, indicating a small settlement on the flat adjacent to the beach. Quantities of fish, turtle, and monk seal bones from this site were radiocarbon dated at 1,500 ybp—the "Great Barbecue"— as Vernon Pickering (1986) called it in his coverage published in *The Island Sun*, 23 July 1986.

A chert arrowhead, excavated from the cave, was stratigraphically dated at 1,500–1,300 ybp. In his typescript report Gibbons writes: ". . . it was my early conclusion that the material for the arrowhead, if not the arrowhead itself, was imported to . . . the BVI from elsewhere. It is commonly thought . . . from a source in central Florida." He goes on to note, however, that there is a rumor of a chert vein in the limestone of West End, Tortola, running on westward to

Great Thatch Island. As far as I know, that possibility has never been investigated.

Righter found continuing evidence of site occupation on the flat up to about 1,100–800 ybp (900–1,200 AD), after which native American cultural evidence stops: There is no evidence of people again until the colonial period. What happened? Curtis et al. (2001) provide a strong clue: Apparently rainfall was very low regionally in the period from 2,500 to 1,100 ybp, but there was a pluvial (rainy, wet) spell of several hundred years centered on 1,000 ybp. This would possibly have provided the flowing freshwater needed by people for a settlement, in a major ravine such as Quail Dove Ghut that flows out onto the flat. With the return of arid conditions, however, survival on Guana became more tenuous and, eventually, the people died or left. Perhaps visitors came over during the next 500 years, but we have no signs of them until the eighteenth century, just 300 ybp.

It is interesting to compare the evidence from Guana with the best-known site in the Virgin Islands, Tutu, in east-central St. Thomas, USVI, 2 km inland. This site was excavated by Righter (2002) and the zoarcheology was reported by Elizabeth Wing (2001). Based on burial dates, Tutu had two peak occupation periods: early, centered on 1,380 ybp, and late, centered on 560 ybp (about 620 AD and 1440 AD, respectively). These separate occupations suggest to me that during the pluvial period, centered on 1,000 ybp, the people may have moved to the coast or small islands because freshwater was more widely available. Perhaps some folks from Tutu came over to Guana Island. As on Guana, the bulk of the diet at Tutu was fish and shellfish, including sea turtles; in the early occupation, they also ate a lot of land crabs.

The diet shifted considerably from early to late occupancy of Tutu. Although marine species dominate both periods, terrestrial vertebrates, especially the large, extinct rodent *Isolobodon portoricensis*, a hutia, and stout iguana *Cyclura pinguis*, were important early, with the hutia dominating by a factor of three in terms of both calculated biomass and minimum number of individuals eaten. At late Tutu, 560 ybp, the reliance on terrestrial vertebrates increased dramatically as the larger predatory reef fishes declined. Hutias and iguanas were about equally represented as table fare, with the iguanas having the edge in biomass.

There is no evidence of people living on the islands when Columbus came by in 1493, just 50 years after the peak of late Tutu occupancy (Acevedo-Rodriguez 1996). There has been little archeological investigation of Amerindian presence in the BVI. I am told Peter Drewet, University College, London, has excavated a major site at Belmont on Tortola, but I have found no published reference to that. There were certainly Amerindians on St. Croix and Puerto Rico. The people Columbus met were called Tainos; they lived in well-organized polities and had rather high standards of culture in terms of social organization, arts, and sports. Columbus and his followers immediately set about destroying all of that. What they did not do directly, the plagues of European diseases they vectored to the native peoples did with ghastly speed (S. M. Wilson 2001). Lewisohn (1986) says of the Amerindians of the Virgin Islands: "their history begins in shadow and ends in shadow."

For almost 200 years, there is scant or no evidence of human activity on the arid islands of the BVI. Then, in 1648 the Dutch claimed Tortola (Jenkins 1923). At this date, R. White (1985) says there was an outpost of buccaneers at Sopers Hole, Tortola—the first post-Taino settlement. From then until 1672 our islands changed hands—Dutch, English, French—with the English finally winning out (Rogozinski 1992). Queen Anne's War, or the War of the Spanish Succession, 1702–13, generated a large population of unemployed seamen. Many of these became pirates; some of the best known—Edward Teach (a.k.a Blackbeard) and Edward Low, not to mention George Norman—took to hanging about the BVI (Alleyne 1986). But we have no evidence that even one of them set foot on Guana Island.

In 1671 the Leeward Islands under British control were separated from Barbados and the

Windward Island colonies. Rogozinski (1992) says that at this time the BVI mainly "served as bases for pirates and smugglers." About 1700 some British colonizers from Anguilla moved to Tortola and Virgin Gorda; in 1727 Joshua Fielding, a Quaker or Friend, visited the seat of government at Spanish Town, Virgin Gorda, and Tortola (Jenkins 1923). Thus began the best-documented period in the human history of the BVI prior to the twentieth century: the Quaker settling and occupation detailed by Charles Jenkins of Philadelphia and published in 1923. The Quakers lasted until about 1780 and the "great hurricane" of that year (Jenkins 1923; Dookhan 1971; Durham 1972; Lewisohn 1986). It is in this period that Guana Island was settled—by Quakers—partially cleared, and used for sugarcane cultivation.

There were two Quaker families, the Parkes and the Lakes. We know most about the Parkes, whose homesite later served as the foundation of Dominica House, central to today's hotel. James Parke was an important member of the Quaker Meeting held at Long Look, Tortola. He was first married to Bytha or Tabitha who died. He then married Mary Vanterpool (Jenkins 1923, 75). The Parkes lived on Guana Island at least from 1743 to 1759. The other house, over on the shoulder of Guana, looking down on the rocky shoreline southeast of White Bay toward Monkey Point, was owned by the Lakes. Jonas Lake was active in the Quaker Meeting around 1748; James Lake, about this time, made a formal complaint to the Quaker Meeting that his mother-in-law kept two of his slaves from him (in response, they appointed a committee to investigate); and John Lake, in 1747, was about to be disowned by the Quaker Meeting but "He is since dead, so ends" (Jenkins 1923, 77). Around this time, Jenkins (1923, 31) records ". . . breaches of discipline became more frequent, and they reflected the wildness and disorder of the times and place. A member from Jost Van Dykes . . . passing Guana Island . . . fired ashore in the night among his friends and wound up by 'beating in a wicked and unmanly manner, William Clandaniel, a Friend.'"

Dookhan (1971, 133) states, "Quakers fought a losing battle in the British Virgin Islands because of the unique circumstances, . . ." well listed by Jenkins (1923, 5) as: ". . . the undermining effects of slave holding, a certain laxity of morals, the apparent necessity of military preparation, the temptations of illicit trading, the ease with which wealth was accumulated, and the unhealthfulness of the climate."

This period was a tumultuous one in the history of the western world. The War of the Austrian Succession, 1740–48, overlapping King George's War with France, 1744–48, and followed by the French and Indian War, in North America, 1754–60, made it very difficult to live as an unarmed pacifist in a colony surrounded by sailors, traders, and military men of every nation. But, says Jenkins (1923, 39): "Tortola, while often alarmed, seems not to have actually suffered." Indeed, these were the golden years of wealth and prosperity in the BVI (Lewisohn 1986), leading Jenkins (1923, 45) to quote the poet Goldsmith:

> Ill fares the land, to hastening ills a prey,
> Where wealth accumulates, and men decay.

There is scant evidence that anyone lived on—or even visited—Guana Island after James Parke departed or died in 1759. After the Quaker experiment, the BVI settled back into the quiet backwaters of history for another century and more. D'Arcy (1971) records a great surge of deforestation for charcoal production "a hundred and fifty years ago"—about 1820. Righter found some nineteenth-century artifacts near the Lake House ruins; someone must have been present at least occasionally.

The great herpetologist Major Chapman Grant visited Guana Island in 1930 (C. Grant 1932). He did not describe the island or its habitats, but he was impressed by the diversity of reptiles he found in view of the size of the island. He reported iguanas present, but gave the species as *Iguana iguana*—the green iguana—which has never been seen on Guana Island and may not even be native to the BVI. Of the little boa that subsequently was named for

Major Grant, he said: "Inhabits rocky cliffs on Tortola and Guana Island." However, Grant's only actual specimen came from Tortola. At this time—1930—the late Oscar Chalwell was coming over to Guana Island occasionally to cut wood for charcoal. He was there when Louis and Beth Bigelow first arrived in 1935. He was there, tending the garden and orchard, when I arrived in 1980. He had a vast store of knowledge about the island and helped me extensively in chronicling the fauna. Oscar never saw an iguana on Guana Island until we brought them in 1984. Oscar did not recognize the boa among our snake species. Major Grant's claims remain mysterious.

After the Bigelows purchased Guana Island in 1935, they began the restoration and expansion of buildings at the site of the old Parke House. They convinced seven families over the next 17 years to buy into developing the ridgeline site into the structures that became the club—which prospered after World War II—and subsequently became the hotel. There is scant information available to me on the first decade of the Bigelows' ownership of Guana. Oscar Chalwell and the late Walter Penn worked with Louis and Beth Bigelow on construction, transportation, planting fruits and vegetables, and fishing. They would both continue working on Guana into the 1990s, as long as they were able.

I have a typescript "Guana Island History." It was received from Gloria Jarecki, the current owner. It was written by the ornithologist Erma J. ("Jonnie") Fisk and edited by Beth Bigelow, undated, but probably about 1990 before they both passed away. Jonnie Fisk first came to Guana in 1952. There must have been an active group of regular visitors prior to that because Fisk reports "The first gasoline generator was given to the Club by Sherman Pratt in 1943 on condition that it not replace the candle light used in the dining room." R. White (1985) tells some harrowing tales of adventures in the BVI during World War II. After the war, things picked up; Fisk reports: "By 1952 there were four comfortable cottages." The Bigelows maintained an orchard and garden at the southeast end of White Bay flat, where those remain today. They kept most of White Bay

FIRST ENCOUNTER

In March 1980, I came to the BVI to work for what was then the Department of Natural Resources and the Environment under the late Robert Creque (Lazell 2001). This was a temporary post funded by The Nature Conservancy through the good offices of its then–vice president Robert Jenkins. I set up at the Seaview Hotel in Road Town, Tortola, and began island surveys. Ishma Christopher, the hotel owner, kept me out of trouble with sage advice and made sure I met all the right people. From the beginning, though, I was told of a wonderful, unspoiled, largely forested island called Guana. "It's too bad you can't go there," people said. "It is a really great place and natural habitat." After several such statements, I finally asked why I could not go there. "Oh, it's a privately owned island. They won't let anybody on it." Well, there were at least 47 other islands to inventory, so I was a busy man; I did not worry much about Guana Island. One day, however, I was looking something up in the telephone book and noticed a listing for Guana Island. I dialed the number, and Mary Randall answered the phone. "I'm down here doing biological surveys of the islands," I said. "Come right over! I'll send the boat to pick you up at the dock on Beef Island," Mary said. "We would love to have a biological survey."

Mary—along with the late Oscar Chalwell, Walter Penn, and Albert Penn—was a wellspring of natural-history information. She had taken a break in managing the island, 1977–79, and the new managers had once again chopped down all the mangroves. They were already regrowing. I added four more species of reptiles to C. Grant's (1932) list of six and was sufficiently impressed with the bird life that I got Robert Chipley, then with The Nature Conservancy, to come down and initiate ornithological studies, concentrating on the quail-doves (see account in "Birds," part 5, above). Mary urged me to contact the Jareckis and tell them that what everybody said was true: Guana is a most remarkable island and natural habitat. I proposed a longer, more in-depth study for 1982, and that became the
continued

continued

foundation for what has gone on ever since. I immediately recommended extirpation—or at least control—of the sheep, control of the (still two) burros, elimination of cats, a no-dog policy, control of exotic plants such as Australian pine, and a program of species restorations (see "Restoration," part 6, below). In the main, these recommendations have been acted upon with great success. I do not believe we can save the world, but maybe—just maybe—we can save one really superb small piece of it. To quote Jonnie Fisk, writing over a decade ago now: "When, thirty-five years after I first came, I once more visited Guana I found it still a fairy tale island and I wish to thank its current owners for making it a protected natural sanctuary."

flat cleared for sheep and burro pasture and had some of it fenced as sheep paddocks. If there were ever domestic (or feral) ungulates other than burros (usually just two) and sheep (once hundreds), I have no evidence of them.

Fisk notes that Louis Bigelow "was persuaded by urban guests to oil the Guana pond in the useless and uninformed hope of controlling sandflies." From her bird notes, we glean that the mangrove swamp was cut down about 1956, presumably also because mangroves are often believed to generate biting insects. The mangroves recovered somewhat over the next couple of decades; this may account for Mirecki's (1977) sighting of the yellow warbler in 1976: see my account of this species in "Birds," part 5, above.

Henry and Gloria Jarecki bought Guana Island in 1974, and Mary Randall managed it—still largely for the vacationing families of the original club members, but increasingly for a larger clientele. Mary, Gloria, Oscar, and Walter worked on the orchard and garden and even tried to establish chickens to provide eggs. The chickens—leghorns imported from the North—succumbed to unknown ailments on Guana. In an attempt to bolster them genetically, a local rooster was brought over from Tortola. He was the last, lonely survivor when I arrived in 1980. He is long gone now.

Restoration

THIS CHAPTER IS LARGELY a reprint of Lazell (2002c), which appeared in *Ecological Restoration*, with some necessary updates.

About 4,000 years ago successive waves of human beings began invading the West Indies, causing many extirpations and extinctions (Olson 1989; Martin and Steadman 1999). The BVI did not escape this onslaught, which has continued with the most recent faunal losses perpetrated by trans-Atlantic peoples up to the present (e.g., Mirecki 1977, for the avifauna). Losses of larger, possibly keystone species are well documented: predatory birds, major seed dispersers, mammals, and the largest native herbivores—tortoises and iguanas.

We have set out to repair extirpations by introducing conspecific individuals from populations on other islands: greater flamingo *(Phoenicopterus ruber)*, white-crowned pigeon *(Columba leucocephala)*, whistling frog *(Eleutherodactylus schwartzi)*, red-legged tortoise *(Geochelone carbonaria)*, and stout iguana *(Cyclura pinguis)*. Should we now begin repairing extinctions by introducing ecologically equivalent exotic species when no other alternative exists? Possibilities might include the Hawaiian monk seal *(Monachus schauinslandi)* as a replacement for the Caribbean monk seal *(Monachus tropicalis)*, Guam rail *(Rallus owstoni)* for DeBooy's rail

(Nesotrochis debooyi), and Puerto Rican iguaca *(Amazona vittata vittata)* for the original Virgin Islands parrot *(Amazona vittata* subsp. undetermined). Could the introduction of carefully selected exotics adequately fill empty niches?

Local extirpations and species extinctions obviously leave empty ecological niches. Furthermore, if a keystone species is lost, there will be ripple effects throughout the entire community on a relatively small island (J. H. Brown 1995; Whittaker 1998; Lundberg et al. 2000; Morgan Ernest and Brown 2001; W. Bond 2001). For example, in the Antilles, local extirpation of a major seed disperser such as the white-crowned pigeon may impair recruitment in some plants resulting in their decline and eventual loss. This change may result in habitat loss for migratory birds (and declines or losses of the birds themselves). Changes in the plant community may also lead to increased soil erosion and concomitant siltation, degradation, and even death of marine communities such as coral reefs and turtle grass flats. The vast salt ponds of Anegada, BVI, supported a huge colony of flamingos in the nineteenth century (Schomburgk 1832). These ecosystems cannot be expected to function normally or naturally without flamingos today.

I seek to restore vertebrate faunas of BVI to the closest semblance possible of their prehuman functional condition (Lazell 1987a, 2000b). The following accounts provide details of efforts to restore the whistling frog, red-legged tortoise, stout iguana, greater flamingo, and white-crowned pigeon to BVI. I provide arguments for introducing exotic species to fill vacant ecological niches even when these cannot be filled by closely related species from nearby Caribbean islands.

WHISTLING FROG

Whistling frogs (*Eleutherodactylus* spp.) are poor dispersers across seawater. That a few occasionally do succeed at waif dispersal is demonstrable: They do occur on many oceanic island banks never connected to other lands (Schwartz and Henderson 1991). However, each bank characteristically supports an endemic species, presumably autochthonous. Large banks (e.g., the Puerto Rico Bank) show internal speciation. For example, *Eleutherodactylus schwartzi* is endemic to the Virgin Islands (perhaps now surviving only in BVI) and is the sister species of the Puerto Rican *Eleutherodactylus coqui* (MacLean 1982). These islands were continuous land about 12,000 years ago (H. Heatwole and MacKenzie 1967).

Two other species of whistling frogs *(Eleutherodactylus antillensis* and *Eleutherodactylus cochranae)* also occur in BVI (Ovaska et al. 2000), but these are widespread on the Puerto Rico Bank and do not show obvious inter-island differentiation. However, *Eleutherodactylus schwartzi* does: The Great Dog Island population, occurring on the smallest island in BVI that supports any frogs, is made up of notably large individuals of this species (Ovaska et al. 2000).

All whistling frogs in BVI are locally called "bo-peeps." Thus, when people report that there used to be "bo-peeps" on an island, we cannot know which species. In 1997, we chose *Eleutherodactylus schwartzi* as the one to restore to Little Thatch Island because it has been demonstrably lost from some islands (e.g., St. John, USVI; MacLean 1982), and has the smallest range of the three species. We chose stock from Tortola, the closest island to Little Thatch with a large population (the only closer island

is Frenchmans Cay where *Eleutherodactylus schwartzi* is rare). On 27 October 1997, we removed seven males, six females, and four egg clutches (8–11 eggs, or average 10 per clutch) along with the bromeliads (*Tillandsia* spp.) in which they resided on Tortola. We set the bromeliads and frogs among other bromeliads growing on Little Thatch Island. The population expanded rapidly and within a year frogs were calling "all over" this island (Jon Morely, manager, pers. comm., 24 October 1998).

Tortola may have been a poor choice for this propagule. The Great Dog population may be extirpated if development plans go forward on that island, and a unique gene pool could thus be lost. Perhaps now we should seek another protected island for relocation of some of the Great Dog Island population.

RED-LEGGED TORTOISE

The red-legged tortoise probably does not cross water well on its own, but it has been transported by humans to the extent that its status as a native on many islands is questionable (MacLean 1982; Schwartz and Henderson 1991; but see Lazell 1993). Many older local people in BVI remember well eating these tortoises and taking them from island to island. They remember populations on islands (e.g., Peter) where none have been seen in recent years; and they often remember them being pretty common. Today, tortoises are seen rarely on Virgin Gorda and Tortola. They are still common on Water Island and Little St. James in USVI. If any survive at all on St. Thomas today, then they are extremely few; one of the last known individuals, a large adult female, was brought over to Guana by the late William MacLean in 1986 to join five from Water Island, supplied by Walter Phillips in the same year. In July 1987, Nicholas Clarke, then director of the BVI National Parks Trust (NPT), brought a large male over from Tortola to Guana Island. These seven individuals were released and comprise the founders of the present population. Tortoises are common on Guana Island today (see account in "Other Reptiles," part 5, above).

TORTOISE TALES

Seven tortoises were released on Guana in 1986–87. Although I made the arrangements to acquire them, I was not in the BVI when they arrived; none was marked prior to release. From time to time a tortoise would appear and get photographed. Occasionally a baby or two would be seen. Sighting frequencies have become more and more regular, and we now know places where we can fairly predictably locate one or more individuals. In October 1995 a fortuitous combination of events enabled us to get a glimpse into the lives of Guana tortoises: Numi Mitchell was on-island preparing to radio-track iguanas; I picked up a baby tortoise on the Iguana Trail at 9 a.m., 2 October; and my wife, Wenhua Lu, brought in an adult male on 12 October. Thus we were able to outfit two tortoises with radio transmitters at least until Numi needed them for iguanas.

The baby was shell notched for subsequent identification and released at point of capture at 9:20 a.m., 4 October. Meantime it was fed flowers, fruit, beetle larvae, and raw beef. At capture, it weighed 48 g (<2 oz.) and had a carapace length of 60 mm. On release, it immediately waddled over to a *Tabebuia* blossom and began eating. We checked it periodically during the day of 4 October, up until 9:30 p.m. It moved 7 m along the trail going east-north-east, slightly down slope. For the next two days it remained within a 4-m² area, apparently rich in edibles such as *Tabebuia* flowers and ground cover vegetation. On 10 October, however, it fairly bolted 10 m farther, plunging more than a meter beyond the elbow (or hairpin) in the Iguana Trail into the wilderness. Numi reclaimed her transmitter and you probably think we never saw that tortoise again. Oh yes, we did. It had become totally anthropotropic. All one had to do was walk down to the elbow in Iguana Trail, stand there a minute or two, and out of the jungle "little thunderbolt" would come at a dead run. You see, in those nine days of captivity and radiotracking, the little guy never encountered a human that failed to bring it a hibiscus flower, or a grape, or a beetle grub. Alas, all good things must

pass, as did October of 1995. We had to leave on the 30th. On 29 October that tortoise measured 63 mm in carapace length and weighed 58 g: in four weeks it increased 5%—1.25% per week—in linear dimension and almost 21%—over 5% per week—in mass. Isn't nature wonderful? Well, human nature, sometimes. To me, the remarkable thing is that for several days, from 7 to 9 October, we did not check up on that tortoise at all, but it did not forget us.

On 12 October Wenhua brought in her adult male, spectacularly colorful, and 26.8 cm in carapace length. With a woodworking tool, we incised an 8/95 (I do not know why) in his plastral concavity; so it would not wear off, maybe. This fellow became something of a film star, and it was not until 15 October that we could wheedle Numi out of another transmitter. We released him right at the capture site and gave him a banana, which he immediately devoured, to lower his blood pressure and alleviate any hard feelings about his incarceration. For the rest of the month, tracking this tortoise became the hobby of Fred Sibley (taking breaks from his mist nets) and the job of Kim Woody, Amy MacKay, and James Rebholz, volunteers who came over from St. Croix. He moved around in a roughly equilateral triangle along the ridge above North Bay woods. He led us to several more adult tortoises that we never would have found without him. We outfitted one of these, a female, 27.5-cm carapace length, with a radio transmitter and followed her. No one was allowed to feed these two individuals (except for the one-banana consolation prize), so they did not become anthropotropic. The female was pretty much a stay-at-home, moving around in a strip 2 by 20 m—so 40 m². The male's wanderings encompassed a much larger area, over 4,743 m².

From photographs, we were fairly certain our male was the same tortoise we detained and photographed on 2 October 1994. It was then 24.5-cm carapace length and had nine clear growth laminae. On October 1995 the male had 10 growth laminae and was 26.8 cm—almost 9.4% growth—in a year in linear dimension. Not bad for an adult we did not feed (but once).

On 22 October 1997, Clive Petrovic, Stoutt Community College, Tortola, donated an adult female red-legged tortoise, 27-cm carapace (dorsal shell) length. She was set up in an enclosure with sufficient depth of loose soil to nest on Necker Island, BVI. Petrovic knew she had mated with a male housed at the Botanic Garden, Road Town, Tortola, and predicted she was gravid. She did nest, and an estimated (from egg shells) dozen hatchlings emerged on 12 July 1998. Most of the hatchlings escaped through the wire mesh of the enclosure, but two were retained in a semicaptive state, where they measured 66- and 80-mm carapace length on 19 October 1998. The adult female was released but tends to remain in close proximity to the building; she is usually easily found and regularly seen. On 13 October 2000, Stephen Durand, visiting Guana Island from the Division of Forestry and Wildlife, Dominica, found a young adult male tortoise, 22.5-cm carapace length, on Guana that I translocated to Necker Island. This individual took to his new home and its incumbent female red-legged tortoise with passionate exuberance. In July 2001, Joanne Netherwood (manager, pers. comm., 25 July 2001) saw hatchling tortoises.

STOUT IGUANA

The stout iguana is known from bones from several islands of the Puerto Rico Bank (Pregill 1981). They are seemingly poor water crossers, and we know little of their interisland geographic variation. C. Grant (1932) claimed iguanas were present on Guana Island about 1930, but he gave the species as the common or green iguana (*Iguana iguana*). This species has never been reported by anyone else on Guana Island. If there were iguanas present in the twentieth century, then they were probably the last survivors of a stout iguana population.

In 1980, I first conceived and promulgated the plan to restore stout iguanas to islands within their former range—the Greater Puerto Rico Bank—and flamingos to BVI, when I was

officially employed by what was then the Department of Natural Resources and the Environment, Government of the British Virgin Islands, under the direction of Robert Creque. Creque agreed with my plan, and I proposed it to several prominent leaders on Anegada. The offer was pretty straightforward: If I could find a suitable home for some Anegada iguanas, and capture and move them, I promised to obtain greater flamingos for reestablishment on Anegada. In getting stout iguanas from Anegada, I simply had no choice: that was the only place left that had any at all.

Over the next several years, I worked with the owners of Guana Island to establish it as a wildlife sanctuary, to remove or control exotics (such as sheep, burros, and cats), to restore vegetation, and to build a program of scientific research. During this period, NPT developed into a major quasi-governmental entity under the direction of Nicholas Clarke. The iguana transfer and flamingo importation—part and parcel of the same restoration program—were constant topics of conversation among Clarke, Louis Potter of Town and Country Planning of the BVI government (who was drawing up plans for a National Park on Anegada), and numerous other government officials. It was apparent during this period that iguanas were declining on Anegada. In 1980, when I first observed them there, they had already disappeared completely from the Citron Bush area— the same area where W. M. Carey (1975) had found the highest densities in the 1960s. The best remaining concentration was at Bones Bight, where the late Clement Faulkner maintained a feeding station for them. My colleagues on Anegada feared the worst for the population there because they saw no proximate hope of controlling or eliminating the ungulates (especially goats) that were outcompeting iguanas for food plants. There was a general consensus that a second population needed to be reestablished. In 1984, I at last brought the first iguana from Anegada to Guana Island. Over the next two years, seven more were brought.

Stout iguanas fairly exploded on Guana Island (Kirby 1986; N. Goodyear and Lazell 1994). In October 1995, I brought four hatchling stout iguanas from Guana Island to Necker Island (Lazell 1995). The animals were cage-reared until October 1996, when one escaped and the remaining three were released. All four survived and established territories where they can be regularly found and observed. On 1 May 1999, I caught and released the larger of the two females, the previous escapee, and noted that she appeared heavily gravid. On 1 October 1999, the first hatchling iguana was seen on Necker Island (Lazell 2000a). On 14–15 October 2002, John Binns, Numi Mitchell, and I visited Necker. Using water-based white latex paint, we did a mark–resight estimate indicating that there were then about 30 individuals on the island.

In detailed, long-term studies, N. C. Mitchell (1999, 2000b) reported that plants forming the mainstay of stout iguana diet on Anegada also occur on Guana. However, stout iguanas on Guana actually prefer to eat different native species. We believe this is because the forage on karst (alkaline) Anegada is relatively low quality compared to igneous (acidic) Guana. After a decade of collaborative research with a succession of five NPT directors, Mitchell concluded that vulnerability to predation, owing to fewer shelter sites, was the reason iguanas had been extirpated from islands such as Guana and Necker. In the absence of artificial predation—humans and exotic carnivores such as dogs, cats, and mongooses—the igneous rock, acid-soil islands are, many believe, the optimal iguana habitat (N. Goodyear and Lazell 1994). Large, dense populations of essentially terrestrial *Iguana delicatissima* "swarmed" on some of the small igneous cays of the Anguilla Bank 40–50 years ago (Lazell 1973). Iguanas often survive on karst, but survival is often in suboptimal habitat, and I believe that is the case with stout iguanas on Anegada. The highest densities reported for karst Anegada are well below the density on igneous Guana and Necker islands today.

GREATER FLAMINGO

Flamingos are strong long-distance fliers. Decades after the BVI breeding populations were extirpated, occasional dispersers visited these islands (Mirecki 1977). They did not recolonize. We believe they did not because they are highly gregarious nesters. There seems to be a critical minimum needed for the group to initiate reproductive activities. Finding no group to join, and being too few to initiate reproduction on their own, the dispersers apparently perished or moved on.

Schomburgk (1832) chronicled the vast numbers of greater flamingos *(Phoenicopterus ruber)* on Anegada, but he noted that they were even then declining and no longer nesting. By mid-twentieth century no resident birds remained, although small groups of flamingos occasionally visited the island (Mirecki 1977). In our restorations, we failed to achieve reproduction with seven flamingos initially imported from Bermuda in 1987. We succeeded with 18 flamingos imported in 1992 that were joined by 4 dispersers, thus totaling 22 in their first breeding season in 1995 (Lazell 2001).

In 1987, we got the first flamingos from the Bermuda Aquarium, Museum, and Zoo, through the good offices of then-director Richard Winchell. These birds came with the stipulation that they had to survive on Guana without being poached prior to placing any on Anegada. I published my plans in a local newspaper (Lazell 1987a).

Guana Island and my organization, The Conservation Agency (TCA), continued to work closely with Louis Potter, Deputy Governor Elton Georges, NPT then-Director Rob Norton, and government officials in general. On 7 March 1992, we were able to bring 18 flamingos from Bermuda to Anegada. There was a great ceremony on that occasion, involving the BVI's then-governor Peter Penfold, then–Deputy Chief Minister Ralph O'Neal, then–Education Minister Louis Walters, Guana's owners Henry and Gloria Jarecki, the prominent citizens of Anegada, NPT then-Director Rosmond DeRavariere, TCA's Vice

President Numi Goodyear (Mitchell), and many government officials. The proceedings were accurately described by N.C. Goodyear (1992) for NPT and in the local newspapers by K. Johnson (1992) and Pickering (1992), the latter explicitly detailing the long-standing—and at last fulfilled—flamingos-for-iguanas trade I had envisioned years before.

I describe these details and cite the contemporaneous media coverage because of the erroneous assertion that iguanas "were moved without the permission and involvement of the BVI government" (Garcia 2001). Left uncorrected, and without the context of the flamingos-for-iguanas restoration project, this false statement could have serious deleterious ramifications adversely affecting nongovernmental-organization (NGO) projects in BVI and even farther afield.

Internationally, Barnes (1992) provided a good account of the initial restoration, and Conyers (1996) and Colli (1996) documented the growth of the flamingo population. Unfortunately, Raffaele et al. (1998) made no mention of the Anegada (or Guana) population in their regional bird guide. Over the years, the original Guana flock dwindled as older birds died. By 1992, the remaining four individuals left Guana frequently and visited other BVI salt ponds. Far from being poached, they were extremely popular and welcome wherever they appeared. Conyers (1996) reported on the 4 birds joining the original 18 on Anegada, making 22 prior to successful nesting in 1995. These may have been the four Guana survivors, but Conyers saw no bands on them and believed all the birds from Bermuda carried bands.

There are still seven nonbreeding birds on Guana Island, all replacements for the original seven of 1987. Attempts led by Caitlin O'Connell-Rodwell of Stanford University to induce breeding with artificial stimuli in this flock are being scheduled. The Anegada population has grown. It is monitored by Rondel Smith of BVI NPT, long an active collaborator with TCA on the flamingo and iguana project. Smith reported (pers. comm., October 2004) over a hundred at time of writing.

WHITE-CROWNED PIGEON

Similar to flamingos, white-crowned pigeons (*Columba leucocephala*) are strong long-distance fliers. Mirecki (1977) reported them extirpated from BVI as a breeding species, but occasional individuals appeared. We arranged with the Puerto Rico Department of Natural Resources to bring young squabs from that island to Guana Island. We failed to get nesting with 4 in 1997, but added 10 more in 1998, achieving success with 14 total in an aviary. Unfortunately, the 1998 squabs were killed in their nests, apparently by exotic rats (*Rattus rattus*). All 14 birds were released in October 1998. We strongly suspect that these have now been joined by dispersers. In this case there is certainly no reason to worry about genetic differentiation of stocks: similar to flamingos, white-crowned pigeons in the Antilles are metapopulations with probable panmixis and slight interisland genetic divergence (Norton and Seaman 1985). I have not observed nesting on Guana Island, probably simply because I have not been present except in October, past the appropriate season. However, fledgling birds, with dull gray crowns, have been observed by others and me annually since 1999. Tortola's renowned birder, Rowan Roy (pers. comm., 12 April 2002), observed nesting on Beef Island, adjacent to the east end of Tortola, in 1999. He reports that white-crowned pigeons are now regularly seen on Beef Island and eastern Tortola. It would now be appropriate to undertake a census on all three islands.

THE NEXT SPECIES

An immediate candidate for restoration to BVI is the West Indian whistling-duck (*Dendrocygna arborea*), formerly a regular nester (Phillips 1922). Mirecki (1977) reports this species as "fairly common ... in the 1930s, but ... now extirpated." A broad view of the status of this species and current conservation effort is provided by Sorenson and Bradley (1998, 2000). These magnificent big ducks are being bred by several aviculturists in the United States and are potentially available for relocation. However, the species is listed in CITES Appendix II, so I will require a CITES export permit from the United States. There may also be import complexities beyond the usual veterinary certification for BVI, depending on whether or not BVI falls under the UK–EU CITES regulations. I can find no information on geographic variation in this species or on the origins of the captive breeders. This is a project in the preliminary research phase.

The next series of potential BVI vertebrate species restorations brings us into the more controversial realm. The practice of reestablishing an exterminated species or subspecies with a distinctly different species or subspecies is not without precedent. Of particular note is the highly successful restoration of the peregrine falcon *(Falco peregrinus)* in midcontinent North America using birds of five different subspecies (Tordoff and Redig 2001), some from as far away as Spain. Three bird species in BVI are candidates for such replacement.

The Puerto Rican woodpecker *(Melanerpes portoricensis)* is known from museum specimens in the Academy of Natural Species of Philadelphia, Pennsylvania, from St. Thomas, USVI (Cory 1889). St. Thomas and the other major Virgin Islands of the Puerto Rico Bank—St. John, Tortola, and northeast to Guana, the Camanoes, and Scrub—were a continuous landmass as recently as 6,000 years ago (Heatwole and MacKenzie 1967). I believe the woodpecker must have been a major pre-Columbian ecological influence in the forests of these islands and would be an important member of a restored fauna on a still-forested island, such as Guana.

Early European accounts of the Virgin Islands often include references to the presence of parrots there (Wiley 1991, Juniper and Parr 1998). *Amazona vittata gracilipes,* a subspecies of the iguaca or Puerto Rican parrot, survived on Culebra, east of Puerto Rico and just west of the Virgin Islands, into the twentieth century (Wetmore 1927). M. I. Williams and Steadman (2001) indicate this parrot occurred as far east

into the Lesser Antilles as Barbuda and Antigua. Today, the iguaca survives on Puerto Rico as a wild bird only at Luquillo in rain forest, in the mountains, in what was probably always a poor, peripheral habitat compared to the original lowland forest (Raffaele et al. 1998). In addition to the wild flock of about 50, about 100 are maintained in captivity at aviaries on Puerto Rico. I propose putting some captive breeders on Guana Island in subsidized, semicaptive conditions; young birds would be allowed to disperse, eventually—one hopes—adapting to wild conditions in the forest. These parrots might eventually disperse to nearby Tortola, where remnant forest would probably support a small wild population. Reforestation in BVI could ensure eventual expansion of populations to hundreds of birds.

DeBooy's rail is known from subfossil remains from many sites on Puerto Rico and the Virgin Islands (Wiley 1985). It was apparently a flightless, terrestrial species and plausibly reported as seen and surviving on Virgin Gorda, BVI, in the 1940s (Ripley 1977). The Guam rail seems ecologically similar. It has been extirpated in the wild on Guam by the brown tree snake *(Boiga irregularis)* but survives and breeds prolifically in captivity (B. Taylor 1998). There is an introduction program for the Guam rail on the island of Rota, which I disapprove of because there is no evidence that a rail was ever part of the natural ecosystem of that island. I have repeatedly suggested that an appropriate thing to do with the excess captive Guam rails is to introduce them to islands that formerly did support flightless terrestrial rails. DeBooy's rail occurred throughout what are today's Virgin Islands of the Puerto Rico Bank, and even St. Croix, just a few thousand years ago, and the species was exterminated by man (Wiley 1985). Guam rails demonstrably live well in artificial, man-dominated ecosystems, as they did on Guam prior to the snake population buildup.

I propose an experimental approach to rail restoration in BVI. I would select an island such as Norman that has been subjected to severe degradation through human (and especially feral goat) abuse. I would remove the ungulates (already under way) and control human activities, eliminating some such as woodcutting for charcoal. After a biological survey, I would introduce Guam rails and document the results. For example, the tiny terrestrial gecko *Sphaerodactylus macrolepis* reaches the highest densities of any known nonaggregated terrestrial vertebrate in the world on Guana in BVI (Rodda et al. 2001b). If this species is a major prey item for a flightless, terrestrial rail, then these densities—about 67,600 lizards per hectare—could be interpreted as an artificial product of rail extermination.

I also strongly advocate placement of Guam rails on several Pacific islands, notably Wake, formerly inhabited by *Rallus wakensis;* Laysan, formerly inhabited by *Porzana palmeri;* and especially Kahoolawe, on the Greater Maui Bank of the Hawaiian Islands, which might easily have supported four or more species of flightless terrestrial rails (Olson and James 1982). Morin et al. (1999) discuss restoration prospects—excellent—for Kahoolawe but did not consider close relatives of extinct species. I received no response to my written suggestion (Lazell in litt. 22 February 1999) that the Guam rail be considered.

The Caribbean was once densely populated by a monk seal so closely related to the surviving Hawaiian species that it has been suggested they were conspecific. An excellent account of the Hawaiian form is provided by Tomich (1986). The Caribbean form, for which the small islands in BVI called The Dogs and Seal Dogs are named (Lazell 1995), is extinct (Kenyon 1977). It is likely that this strictly carnivorous, littoral marine mammal, ranging about 175–260 kg (380–575 lb) was a keystone species in the Antillean ecosystem. I believe every effort should be made to establish a monk seal population from Hawaii in the Virgin Islands.

CONCLUSIONS

My purposes in providing this account have been to historically document our restorations in BVI, to correct errors that have been unfortunately published, and to suggest programs of introduction leading to restoration that are as yet unconsidered—or considered distinctly unpalatable—by many of my colleagues. In this latter aspect, I follow my colleagues in New Zealand, where TCA officers and members visited in 2000. Mick Clout (2001) summarizes much of this wonderfully productive, increasingly successful program, quoting Ian Atkinson "that a useful goal for restoration is 'reinstating earlier selection regimes,'" and that a strong case can be made "for replacement of some extinct species with 'ecologically appropriate' and related extant species."

Finally, the computer-generated picture of gloom-and-doom forecast by Lundberg et al. (2000), that ecosystems close down after extirpations, establishing new, depauperate levels resistant to reintroduction of the lost species, seems inapplicable to BVI, where all of seven attempts to reestablish breeding populations of four vertebrate species have been received most successfully within their ancestral ecosystems: two populations of stout iguana (Guana, Necker), two of red-legged tortoise (Guana, Necker), greater flamingo (Anegada), white-crowned pigeon (Guana), and whistling frog (Little Thatch Island).

Conservation and Biodiversity

... population growth is the greatest issue the world faces over the decades immediately ahead.... We know that eventually the world's population will have to stop
growing.

ROBERT S. McNAMARA (1977)

One constantly wonders why books and papers about the global environment and
our effects on it are so widely ignored when the arguments are ostensibly so convincing.... The impact of 6.1 billion humans on our planet is staggering and affects every aspect of our lives, whether we realize it or not.... Humans currently
consume more than 40 percent of the world's annual biological productivity.... We
are living off the principal, not the interest....

PETER RAVEN (2001)

The requisite level of public understanding ... does not yet exist.

JEFFREY D. SACHS (2004)

THERE ARE TWO POSSIBLE short-term outcomes for our species, *Homo sapiens.* The population may increase to the planet's carrying capacity and then fluctuate as a result of changes in carrying capacity induced by events such as droughts, heat waves, cold spells, volcanism, earthquakes, and war. In this scenario, essentially universal poverty will prevail and political unrest, amounting to near-permanent revolution and aggression, will dominate human activity. In the long term, the population will predictably crash, possibly—sooner or later, probably—to zero. The alternative scenario is to stop or reverse popula-tion growth at or to a level below subsistence; thus, there would be some slack in the system. If we decided, for example, that all people should have the opportunity to live at the average stan-dard of living enjoyed by citizens of the United States, then we can quickly calculate how many of us can live on Earth: We are about 300 million and we consume about 45% of Earth's total avail-able resources—renewable and nonrenewable—each year. Therefore, 300 is to 45 as *x* is to 100: About 670 million people could live on Earth at the U.S. level of average individual consumption (Lazell and Lu 1994). That is about one-tenth the number currently on Earth—a full order of mag-nitude fewer people.

However, the population of the United States is growing rapidly; we are experiencing a "baby boom" similar to that which occurred during and following World War II. When McNamara (1977), quoted above, wrote and was president of the World Bank, it was fashionable to believe in the "demographic transition": the notion that with increasing prosperity, birth rates would drop to replacement levels. Indeed, this seems to be what happened in some European countries (and possibly Japan); it seemed to be happening in the America of the 1970s. However, the slowdown did not last; historically, there has never been a demographic transition in the United States. The Great Depression of 1929, and extending into the 1930s, saw a drop in U.S. birth rates, but the prosperity induced by World War II and its following economic and technological expansion generated the great baby boom that extended through the 1950s (Easterlin 1961). What appeared to be a transition in the 1970s preceded increasing fertility in the 1980s that—with contemporaneous massive immigration—has brought us to 300 million. The terrorist strikes of 11 September 2001 precipitated a fertility explosion that, according to the broadcast news media, continues unabated at the time of writing, over three years later. The American population, with its per capita resource consumption far in excess of any other population, shows no sign of stabilizing.

The conservation movement, with our leaders' acceptance of the population problem and our emphasis on nature, began to drift before 1980. To quote Frith (1979) ". . . interest expanded and swung towards the conservation of one species, man. The wildlife conservation movement disappeared before the appeal of terms like 'The Ecology' and 'The Environment.' These, in turn, came to be almost synonymous with industrial pollution. . . . Reducing the petrol fumes and smoke in the dark canyons of the cities and recycling beer cans became more popular, and apparently more urgent, than the conservation of our dwindling wildlife resources." For more than 20 years, mention of the population problem was unfashionable even in discussions of what has come to be called "earth system science" (Lawton 2001)

or "sustainability science" (Swart et al. 2002). The magnitude of the problem and our leaders' neglect of it are concisely framed by Murray (2003) and Hannon (2004).

The whole Earth ecosystem—the biosphere—is in big trouble, and the fault is ours. *Homo sapiens* has always played a far more destructive and adversarial role with respect to the rest of the biosphere than any other species we have any knowledge of, going back tens of thousands of years (Martin and Wright 1967; Tudge 1997; Redman 1999; Flannery and Schouten 2001). The role of Americans has been especially insidious (Tucker 2000). Back in the 1980s, it became fashionable to proclaim that "man is part of nature," but this is a meaningless assertion. With the distinction between natural and artificial removed, everything is "natural": napalm, hydrogen bombs, and child rape. To have any hope of redressing Earth's situation requires a clear understanding of our role in creating the very artificial problems. As stated by Jacquette (1984): "To suppose that we can bring an end to millennia of estrangement merely by telling ourselves that we are part of the natural world is to offer an implausible solution to a very longstanding problem."

Unprecedented extinctions of animal species occurred contemporaneously with waves of humans moving out of Africa, invading other continents and islands (e.g., Tudge 1997; Roberts et al. 2001). Humans fairly exploded on the North American continent about 13,000 years ago, producing mass extinctions of large, edible animals—the megafauna—from elephants to tortoises (Alroy 2001a). The circumstantial evidence that humans literally ate up the megafauna seems incontrovertible (Dayton 2001; Kerr 2003a), but some scholars still question it (Grayson 2001; Slaughter and Skulan 2001; but see the rebuttals of Alroy 2001b). I have cited the evidence in "Restoration," part 6, above, for artificial, anthropogenic (human-caused) extinctions in the West Indies, in particular in the Virgin Islands. There can be no doubt about human responsibility here. Even among some island species that escaped total annihilation, genetic diversity was dramatically reduced (Paxinos et al. 2002), impairing adaptability and recovery.

The basic theory of anthropogenic animal extinctions is that those species that had the longest (oldest) historical exposure to humans had the best chance to keep pace with human predators evolutionarily: They adapted to us as we developed our hunting skills. Those species that were naive, who only met humans recently when those humans were more technologically adept at predation, were the most likely to succumb (Murray 2003). Even species that survived prehistoric human invasion and predation were candidates for extinction when faced with the post-1492 waves of invading Europeans with firearms (e.g., the passenger pigeon, the Carolina parakeet, and—almost—the bison). And the anthropogenic extinctions continue apace, greatly augmented by indirect destructive forces such as climate change and the introduction into natural habitats of invasive exotics (Pimm et al. 2001).

One of the most insidious engines of ecological destruction is deforestation; it typically has profound effects at a considerable distance from land actually cleared of trees (Harris 1984; Lawton et al. 2001). These effects include the death of plants and the animals that need them for food and shelter; increased erosion that produces smothering siltation of waterways, wetlands, sea bottoms, and coral reefs; local climate changes such as increased heat, light, and ultraviolet radiation; and widespread climate change that results from a decrease in rainfall that had previously been initiated by forest evapotranspiration (Lawton et al. 2001). Zwiers and Weaver (2000) underscore the point made by Stott et al. (2000) that rapid global warming seems to have occurred in the first four decades of the twentieth century, abated for the next two, and then resumed rapidly. They note: "Global warming critics have been quick to point out that the warming in the early 20th century occurred before the buildup of atmospheric greenhouse gasses...." They did not note that this warming spell followed and coincided with the most

wide-ranging and pervasive era of deforestation in human history (Williams 2003), coincident with fuel consumption producing the highest levels of soot ever attained (Chameides and Bergin 2002).

Although most climatologists recognized early on humanity's contribution to global warming (e.g., Idso 1984), the evidence was largely circumstantial, the data were still inconclusive, and other causes could not be ruled out (Stott et al. 2000). Things changed quickly early in this century. Kennedy (2000) called human production of CO_2 and other greenhouse gases "a global experiment without protocol or hypothesis," and Shaw (2002) expanded on this. Kerr (2001) headlines "... Humans Are Behind Most of Global Warming," and Philander (2001) reviewed why global warming causes have remained controversial for as long as they have. Kennedy (2001), Kerr (2002), and Forest et al. (2002) describe the continuing problems in attempting to forecast the outcomes of anthropogenic global warming.

Both in temperate North America and in Eurasia, the effects of global warming have become increasingly obvious (Zhou et al. 2001). Ponds on which we once ice-skated no longer freeze solid; southern birds and animals, from titmice to armadillos, have moved north; it rains when and where it used to snow. Scientific evaluations of change have been made. Helmuth et al. (2002) have found an unexpected relationship between tide patterns and increasing warmth: instead of intertidal marine organisms' ranges shifting north, as one might anticipate, northern populations are showing greater signs of stress and may perish first—and soon. Harvell et al. (2002) provide evidence that warming increases disease risks for both terrestrial and marine organisms. They cite examples as diverse as *Vibrio* bacteria bleaching and killing coral; mosquitoes expanding their altitudinal range upward, thus vectoring malaria and pox to endangered montane birds; and more higher latitude outbreaks of human cholera. Cifuentes et al. (2001) describe and graphically depict causes of human morbidity and mortality that result from greenhouse gases and their copollutants, including bronchitis, asthma, and cardiovascular disease. They argue that "...the substantial public health impacts we have charted here [should] become more widely recognized, and their full economic and social impacts ... integrated into discussions of climate policy...."

Another devastating side effect of the human population explosion and human globalization is the spread of exotic organisms that invade natural ecosystems and prey on or compete with the incumbents. The impact of invasive species is second only to that of human population growth and associated activities as a cause of the loss of biodiversity throughout the world (Pimentel 2002). Kennedy (2002) tells a great "cautionary tale" that involves purposeful human introduction and accidental invasions with ecology and economics: rubber, tiger mosquitoes, and West Nile virus. Briefly, rubber originated in South America and, early on, produced economic growth and wealth. However, planting rubber tree monocultures allowed a leaf-blight fungus to flourish where it had previously been rare. Unfortunately for South America, but fortunately for the rest of us, rubber trees had been purposefully introduced to southeast Asia, where they also flourished and produced prosperity. In the course of World War II, however, Japanese military expansion cut off Asian rubber exports to America and the Allies. We developed synthetic rubber, creating a financial boon for us but plunging all the real rubber harvesters into depression. All went well for us—and badly for everybody else—until radial tires became popular: They require real, not synthetic, rubber, so Asian rubber production picked up again. So far, this story is just the usual one of people getting rich and going broke in all-too-familiar cycles, but now accidental introduction and exotic species invasions give it a new twist.

A market developed in the United States for used radial tires, imported from Asia. And with the tires, in the little pools of rainwater found in virtually every one of them, came Asian tiger mosquito larvae. As Kennedy (2002) says: "...just try to empty the last pint of water" from a tire. The tiger mosquito, *Aedes albopictus,* spread rapidly from its port of accidental entry, Houston, throughout the warmer parts of the United States. It can overwinter and survive some short freezing spells. Unlike many pestiferous mosquitoes, the tiger is diurnal—active in daylight. And it is a fine vector for West Nile virus. Kennedy (2002) notes the next twist in the tale may be when someone, either accidentally or as an act of bioterrorism, introduces leaf-spot fungus into the Asian rubber plantations. I will take the American side of the story one step further and tie it back into that other horror, described above: global warming. The spread of the mosquito and its viral baggage has predictably only just begun.

The changes *Homo sapiens* has wrought on the biosphere, from outright usurpation of space, destruction of habitats, introduction of exotics, and attempts to obliterate disease organisms and pests, have brought about entirely novel selection regimes. Humanity has become "the world's greatest evolutionary force (Palumbi 2001). Since late in the eighteenth century, humans have so dominated the planet and its biosphere that geologist P. J. Crutzen (2002) has suggested that this most recent period be recognized as entirely new and different from previous geologic time, calling it the "anthropogene." Even evolutionary change proceeding as rapidly as selection can operate will apparently not be fast enough to prevent an extinction crisis on a scale "comparable to those in the fossil record..." and "...widespread depletion of biodiversity would probably be permanent on multimillion-year timescales" (Kirchner 2002) unless current trends are reversed.

A major thesis of this book is that natural biodiversity, as undiminished as possible, is critical to the stability of ecosystems with respect to high-level consumers, such as *Homo sapiens.* Literally, our comfort and welfare depend on the incredibly complex fabric of nature. Every artificial extinction diminishes our chances of long-term survival and continuing adaptive evolution.

Naturally diverse ecosystems function better than artificially degraded ones, judged by a diverse array of parameters. Lawlor and colleagues (in Kinzig et al. 2001) point out four problems in attempting to justify conservation because of an ostensibly positive relationship between diversity and stability. First, the evidence for the positive relationship is insufficient. This argument is rejected because so much of the evidence does confirm the positive relationship and exceptions are readily explicable as special cases. For example, ecosystems undergoing ecological succession from early to later seral stages—a site cleared of natural vegetation and then observed as it recovers and regenerates—will have high diversity but be "unstable" because it is ipso facto undergoing change. Second, one might argue that diversity matters at lower numbers of species but the positive relationship to stability is asymptotic: A point comes at which richer ecosystems simply cannot get any more stable or function any better. This recalls the equilibrium theory of island biogeography that claims every island has a "right" number of species determined by its area, so that every successful colonization necessitates a complementary extinction (or every extinction allows a complementary successful colonization). To the extent that there is any truth to the species : area relationship, this would indicate that spatially large ecosystems require greater diversity to function well. Third, actual species identities may not matter: Any set of the same number of species might do as well. Fourth, perhaps only dominant species matter: rare species are unimportant. I believe each of these notions is fallacious (as do Lawlor et al. in Kinzig et al. 2001).

Does species identity matter? Well, to the extent that a native species has served long coevolutionary time with its native neighbors, of course it matters. Consider the native Virgin Island iguana, *Cyclura pinguis*. It is massive, ponderous, and rather slow. Its exotic relative, *Iguana iguana,* now invasive on many Puerto Rico Bank islands (including Guana's neighbors Peter Island and Tortola), is by comparison gracile, agile, and fast. Both climb, but *Iguana*

iguana climbs much better and is thus a much more efficient predator on bird eggs and nestlings. It certainly matters to the nesting birds which of these two are *the* iguana on a given Virgin Island. Great care must be taken in rebuilding a damaged ecosystem using an exotic species when the original is extinct and unavailable (see "Restoration," part 6, above). If we bring a flightless rail back to the Virgin Islands, to replace the one exterminated by people, it certainly will be different. We will need to initiate that analogue restoration on an experimental, trial basis on one island and monitor the situation. We can only hope the substitution will prove to be better than nothing—which, as flightless rails go, is what we have now.

Do only dominant species enable smooth ecosystem functioning? Discussing ecosystem services on which humans depend, Balvanera et al. (2001) say: "Many ecosystem services may be unaffected by small losses of biodiversity, but they may deteriorate rapidly when, for instance, most of the elements of a functional group are gone." This sounds quite reasonable, but recent close work has revealed that rare species may be far more important than one might think at first glance. In controlled experiments, Lyons and Schwartz (2001) found that an introduced, exotic species invaded successfully when rare species were removed: Natural diversity was reduced. However, when rare species were left and a total biomass equal to theirs was removed from common, dominant species, resistance to invasion remained intact. This suggests the same point made above about ecosystems composed of coevolved member species: Their historical interrelationships provide cohesion and smooth ecosystem functioning not available from an artificial mix of the same number of species. Diaz et al. (2003) review this and numerous other studies and suggest that "removal experiments [are] a promising avenue for progress in ecological theory and . . . in making land-use and conservation decisions." Lawlor and colleagues (in Kinzig et al. 2001) conclude: "Studies of biodiversity and the interplay of community structure and ecosystem function are the cornerstone of applied conservation, even when the motivation for preserving nature is . . . ethical, aesthetic, or economic."

Solid, pragmatic reasons for preserving species diversity—and especially rare species—have long been advanced. Chiras (1985) notes: "Plants, microorganisms, and our fellow animals provide us invaluable services free of charge, day after day, such as controlling pests, recycling nutrients, replenishing atmospheric oxygen, maintaining local climate, . . . and reducing flooding. . . . Each time we take a prescription drug, for example, there is a 50-50 chance it came from a wild plant." He points out: "Wild plants are a boon to agriculture in its battle against insect pests, blight, and drought." Indeed, the wild relatives and ancestors of commercially cultivated agricultural crop species are now so valuable for their diversity of genes that real turf wars are breaking out over the potential genetic bounty. Fowler (2002) states: "Countries still deny access, even to plant-collecting missions organized to rescue unique populations from the threat of extinction. . . . Controversies over intellectual property rights and charges of 'biopiracy' have fueled passions and convinced many countries that they are sitting on genetic gold mines." Sometimes rare species can be too valuable for their own good.

How are conservation biologists and associated planners and managers doing in the battle to maintain biodiversity? The answers are mixed. Wetland mitigation in the United States, where developers were allowed to destroy wetlands in one place if they constructed a like-sized "wetland" elsewhere, has been a dismal failure (Kaiser 2001): The artificial wetlands do not support more than a small fraction of the biodiversity lost. When it comes to national parks and similar reserves set aside in 22 tropical countries, Bruner et al. (2001a) present evidence that most of 93 protected areas considered are as healthy today as when set aside, or have actually regained vegetative cover. Stern (2001) complained that their study did not consider a variety of social factors that will predictably impact even the most successful parks

and reserves with the passage of time. Bhagwat et al. (2001) suggest that destructive activities that are effectively stopped or controlled in parks are simply transferred elsewhere, so there is no net gain for conservation. Bruner et al. (2001b) are partially able to rebut both of these detractions, but nowhere do the total of 10 authors ever mention human population growth as the direct cause of all the problems. The original thesis may indeed be overly optimistic, but the complainants fail to offer the one solution that would mitigate the problems. Conservation without control and reduction of the human population is doomed to eventual failure (Hannon 2004).

Where is Guana Island in all of this? Can a small, 300-ha, insular reserve make a difference? Can such a small island harbor sufficient diversity to maintain full ecosystem function—reasonable stability? Guana is exemplary, I have argued, and should be in the forefront of model conservation planning (Kraus 1991). Guana's small size may make it easier to test hypotheses and perceive critical interrelationships; small refuges often have great advantages (Simberloff 1982). And, of course, Guana's biodiversity, in group after group of those organisms censused, rivals that of other Caribbean islands hundreds of times bigger. If Dominica has sufficient biodiversity to maintain ecosystem function, then Guana does too. Indeed, Guana seems to exemplify the sort of refuge Rosenzweig (2003) calls for so eloquently: significant preservation—even restoration—of unique biodiversity "in the midst of human enterprise." Guana Island is a profit-making tourist destination resort. There could easily be many more places such as Guana in the Caribbean and among the other islands of the world. Problematic—which Rosenzweig (2003) is fully aware of—are human perceptions and attitudes toward nature; this is where heavy repair work needs to be done (Sachs 2004). The notion of biophilia, human love for other living things, is today fully offset by abiding biophobia, fear and loathing of other living things beyond puppies and house-plants (Lazell 1998b). It is alarming to realize that many (most?) children in the overdeveloped Western world have a greater awareness of fictitious, animated TV "organisms" than they do of real living things in nature (Balmford et al. 2002). Apparently, most children's expression of biophilia is accommodated and absorbed by artificial caricatures; real life is both less intriguing and less satisfying. As Balmford et al. (2002) conclude, "conservationists need to reestablish children's links with nature if they are to win over the hearts and minds of the next generation." Guana Island seems to be an ideal place to do this.

What happens if we consider an even smaller insular ecosystem? Let us take another of my exemplar islands: Carrot Rock, BVI, less than 1.3 ha. It has three squamate reptile species (two endemic), only one species of termite (other terrestrial invertebrate groups have not been systematically sought), probably only 20–40 species of vascular plants, and supports at least three species of nesting birds: Zenaida Dove, Bridled Tern, and Red-billed Tropicbird. We have observed Carrot Rock before and after horrendous natural disasters including several hurricanes and one prolonged, severe drought. It has come through intact. Certainly, however, Carrot Rock lacks the resilience of Guana Island. For example, feral cats have existed and persisted on Guana despite our most valiant efforts to extirpate them. So far, they have failed to precipitate a single demonstrable extinction among the other vertebrate species present. A single, nonreproductive cat released on Carrot Rock might well and easily exterminate the Rock's endemic skink, *Maybuya macleani* (Mayer and Lazell 2000), recalling the classic, if pathetic, story of the Stephens Island Wren retold by Flannery and Schouten (2001, 83). What the cat might do to the other reptiles and the nesting birds would also predictably be catastrophic and severely destabilizing to this tiny ecosystem.

So I suggest that there is at least some, vague, imprecisely quantifiable relationship between area and ecosystem function or stability. It is no tighter than the putative species : area

relationship and essentially cooks down to the circular notion that each land area has a natural level of biodiversity that sustains ecosystem function based on the individual attributes of that particular land area—only one of which is area itself. Some islands as yet insufficiently studied appear clearly dysfunctional. For example, both Norman Island and Prickly Pear Cay, BVI, have long suffered from the ravages of feral ungulates (especially goats) and tree cutting for firewood. Neither seems to have a reasonably full complement of native terrestrial vertebrates. Neither has a frog or (as far as known) an amphisbaena (worm lizard). So far, we have not found the blind snake *Typhlops* or the little snake *Liophus exigua* on either. Prickly Pear once harbored the tiny ground gecko *Sphaerodactylus parthenopion,* endemic to the BVI, but I have been unable to rediscover it there despite considerable effort. The floras of these islands are also unstudied but appear impoverished. Certainly, in their current state of soil erosion they are contributing to coral reef destruction (although the goats and other livestock have now been removed from Norman Island and its vegetation is recovering nicely).

Guana Island also suffered tree cutting for fuel (especially charcoal), as documented by the Bigelows in the 1930s (see "Human History," part 6, above). Most of the island, except large portions of the White Bay flat, seem to have recovered; we suspect the windward side of the island, notably the largest ravine—Shangri La or Grand Ghut—was never clear-cut. It is difficult to judge the effects of global warming on this terrestrial ecosystem. To date, the most apparent warming effects have been severe bleaching and considerable death of our coral reefs (especially the patch reefs in White Bay). In addition to the sheep, cats, and rats on which war is waged constantly, Guana has some very unpleasant invasive exotics: the smothering, pink-flowered vine called *corallita,* the fire ant *Solenopsis invicta,* trombeculid mites from Florida (chiggers), and at least one species of South American household cockroach. Despite vigilance there is a constant threat of more exotic species to come.

Guana's contributions to broader area conservation are disproportionate. Our local efforts, exemplary in the BVI, highlight the point made about this entire island group—and indeed all the West Indies—by Kenneth Bain (2002) in his perceptive book *Treasured Islands:* The natural environment is our only real resource. Bain asks about the future here: "Will it be sustainable development or malignant growth?" He goes on: "It is the most complex question of all. The Government and the people of the BVI are already aware of the real dangers to the environment. . . . But it will need to be said again and again and again. To forget is to fail."

Guana is now the major breeding ground for the stout iguana, *Cyclura pinguis,* often regarded as the world's most endangered reptile (Alberts 2000; Binns 2003). We have supplied stout iguanas for successful restocking of Necker Island and plan to stock several other islands within the former range of the species in the near future. Similarly, we are breeding and restocking red-legged tortoises and (under their own steam) the once-extirpated White-crowned Pigeon. We harbor the rare, BVI endemic *Croton fishlockii* shrub and the only uncultivated population known of the bromeliad *Pitcairnia jareckii.* We provide an important way station and wintering site for North American migrant birds (McNair et al. 2002). We provide nesting habitat for many bird species including the threatened Brown Pelican, White-tailed Tropic Bird, Bridled Quail-Dove, Wilson's Plover, Black-necked Stilt, Bahama Duck, and the endangered Roseate Terns (Mirecki 1977). Our insect fauna is vast for the island's size, and it is the type locality (the original place of scientific recognition) for many species, some as yet unrecorded elsewhere.

How will Guana Island fare as the human population explosion proceeds? I have the gravest worries. At present, there is not a problem: Few people come to Guana Island, none reproduce on it, and there are no plans to increase human usage in the near future. The current owners work in partnership. When their three sons, each married, inherit the island,

however, there will be six viewpoints to consider. The number in the next generation will predictably be much larger. If land inheritance follows in the usual human pattern, I predict disaster: In a few decades, there will be so many descendants that partitioning and further development will be inevitable, with concomitant importation of all sorts of exotic species and destruction of native vegetation. Although I doubt the myriad owners would purposefully extirpate native species, I believe inadvertent annihilation would inevitably occur.

I hope I am wrong. Guana Island today is a paradise. Its continuing and potential values as a conserved, functioning ecosystem, and a site of scientific research (not to mention enthusiastic ecotourism) are infinite. The world needs thousands more places like Guana Island; we need to know and follow the Guana Island example of this generation. All Earth could be a paradise if there were just few enough of us living on it.

ENDNOTE

The importance of islands stems from three factors: (1) Islands are circumscribed, fairly closed habitats with ecosystems that may be comprehensible; (2) they frequently harbor unique, endemic organisms of major evolutionary interest; and (3) proximate islands separated by postglacial sea-level rise are ideal theaters for documenting rapid evolution. Islands are typically far more naturally diverse than abiological formulations predict, and each is unique. Meaningful conservation and avoiding tragic loss of biodiversity necessitates knowing far more than land areas, elevations, distances from sources, and latitudes. Biological interactions, physiographic and geologic structure, meteorology, and other factors combine to produce speciation, evolutionary radiation, specific adaptations, and biodiversity. Compelling evidence indicates that high levels of natural (as opposed to artificial) biodiversity underpin ecosystem resilience and stability, especially with respect to high-level consumers such as humans.

BIBLIOGRAPHY

Acevedo-Rodriguez, P. 1996. Flora of St. John, U.S. Virgin Islands. *Memoirs of the New York Botanical Garden* 78: 1–581.

Adler, G., and R.H. Tamarin. 1984. Demography and reproduction in island and mainland white-footed mice *(Peromyscus leucopus)* in southeastern Massachusetts. *Canadian Journal of Zoology* 62: 58–64.

Airy Shaw, H.K. 1988. *A Dictionary of the Flowering Plants and Ferns.* Cambridge University Press, UK.

Alberts, A. (ed.). 2000. West Indian Iguanas: Status Survey and Conservation Action Plan. International Union for the Conservation of Nature, Gland, Switzerland.

Alberts, B., J. Wilson, and T. Hunt. 1984. *Molecular Biology of the Cell.* Garland Publishing, New York.

Alexander, J., and J. Lazell. 2000. *Ribbon of Sand.* University of North Carolina Press, Chapel Hill, NC.

Allen, A., J. Brown, and J. Gillooly. 2002. Global biodiversity, biochemical kinetics, and the energetic-equivalence rule. *Science* 297 (5586): 1545–48.

Alleyne, W. 1986. *Caribbean Pirates.* MacMillan Caribbean, London, UK.

Alroy, J. 2001a. A multispecies overkill simulation of the end-Pleistocene megafaunal mass extinction. *Science* 292 (5523): 1893–96.

Alroy, J. 2001b. Responses. *Science* 294 (5546): 1459–62.

Alverson, K., R. Bradley, K. Briffa, J. Cole, M. Hughes, I. Larocque, T. Pedersen, L. Thompson, and S. Tudhope. 2001. A global paleoclimate observing system. *Science* 293 (5527): 47–48.

Amadon, D. 1949. The seventy-five percent rule for subspecies. *Condor* 51 (6): 250–58.

Amadon, D. and L.L. Short. 1976. Treatment of subspecies approaching species status. *Systematic Zoology* 25: 161–67.

Anderson, S. and K.F. Koopman. 1981. Does interspecific competition limit the sizes of ranges of species? *American Museum Novitates* 2716: 1–10.

Andrewartha, H.G. and L.C. Birch. 1982. *Selections from the Distribution and Abundance of Animals with a New Preface.* University of Chicago Press, Chicago, IL.

Andrewartha, H.G. and L.C. Birch. 1984. *The Ecological Web: More on the Distribution and Abundance of Animals.* University of Chicago Press, Chicago, IL.

Andrews, R.M. 1979. Evolution of life histories: A comparison of *Anolis* lizards from matched island and mainland habitats. *Breviora* 454: 1–51.

Andrews, R.M. 1982. Spatial variation in egg mortality of the lizard *Anolis limifrons. Herpetologica* 38 (1): 165–71.

Anonymous. 1833. A notice of the ravages of the cane fly, a small winged insect, on the sugar canes of Grenada, including some facts on its habits. *Magazine of Natural History* 6: 407–9.

Armbrecht, I., I. Perfecto, and J. Vandermeer. 2004. Enigmatic biodiversity correlations: ant diversity responds to diverse resources. *Science* 304 (5668): 284–86.

Arnett, R.H., Jr. 1985. *American Insects.* Van Nostrand Reinhold Co., New York.

Arnold, S.J. 1981. Behavioral variation in natural populations. I. Phenotypic, genetic and environmental correlations between chemoreceptive responses to prey in the garter snake, *Thamnophis elegans. Evolution* 35 (3): 489–509.

Ash, C. 2004. Horrible achievement. *Science* 303 (5657): 474.

Ashton, E. H. and S. Zuckerman. 1950. The influence of geographic isolation on the skull of the green monkey (Cercopithecus aethiops sabaeus). Part I. Proceedings of the Royal Society, Series B, 137: 212–38.

Ashton, E. H. and S. Zuckerman. 1951. The influence of geographic isolation on the skull of the green monkey (Cercopithecus aethiops sabaeus). Proceedings of the Royal Society, Series B, Part II. 138: 204–13; Part III. 138: 213–18; Part IV. 138: 354–74.

Au, S. 1974. Vegetation and Ecological Processes on Shackleford Bank, North Carolina. National Park Service, Washington, DC.

Auerbach, M. J. 1984. Stability, probability, and the topology of food webs. In Ecological Communities: Conceptual Issues and the Evidence, eds. D. R. Strong, Jr., D. Simberloff, L. G. Abele, and A. B. Thistle, 413–36. Princeton University Press, Princeton, NJ.

Auffenberg, W. 1974. Checklist of fossil tortoises (Testudinidae). Bulletin of the Florida State Museum 18 (3): 121–251.

August, P. V. 1983. The role of habitat complexity and heterogeneity in structuring tropical mammal communities. Ecology 64: 1495–1507.

Avise, J. C. 1986. Mitochondrial DNA and the evolutionary genetics of higher animals. Philosophical Transactions of the Royal Society of London, Series B 312: 325–42.

Avise, J. C. 1991. Ten unorthodox perspectives on evolution prompted by comparative population genetics findings on mitochondrial DNA. Annual Review of Genetics 25: 45–69.

Avise, J. C. 1994. Molecular Markers, Natural History and Evolution. Chapman & Hall, New York.

Bailey, N. T. 1952. Improvements in the interpretation of recapture data. Journal of Animal Ecology 21: 120–27.

Bain, K. 2002. Treasured Islands. PanOrama Press, Road Town, Tortola, BVI.

Baker, H. B. 1924. Land and freshwater molluscs of the Dutch Leeward Islands. Occasional Papers of the University of Michigan Museum of Zoology 152: 1–122.

Baker, R. J. 1979. Karyology. Biology of bats of the New World family Phyllostomatidae. Part III. The Museum, Texas Tech University Special Publications 16: 109–55.

Baker, R. J., R. L. Honeycutt, M. L. Arnold, V. M. Sarich, and H. H. Genoways. 1981. Electrophoretic and immunological studies on the relationship of the Brachyphyllinae and Glossophaginae. Journal of Mammalogy 62 (4): 665–72.

Balling, S. S., and V. H. Resh. 1984. Life history variability in the Water Boatman Trichocorixa reticulata (Hemiptera: Corixidae) in San Francisco Bay salt marsh ponds. Annals of the Entomological Society of America 77: 14–19.

Ballinger, R. E. 1973. Experimental evidence of the tail as balancing organ in the lizard, Anolis carolinensis. Herpetologica 29 (1): 65–66.

Ballinger, R. E. 1976. Evolution of life history strategies: Implications of recruitment in a lizard population following density manipulations. Southwestern naturalist 21 (2): 145–49.

Ballinger, R. E. 1977. Reproductive strategies: food availability as a source of proximal variation in a lizard. Ecology 58 (3): 628–35.

Ballinger, R. E. 1981. Can predator defense be tributive or toxins nontoxic? American Naturalist 117: 794–95.

Ballinger, R. E. 1983. Life-history variations. In Lizard Ecology: Studies of a Model Organism, eds. R. B. Huey, E. R. Pianka, and T. W. Schoener, 241–60. Harvard University Press, Cambridge, MA.

Balmer, O. 2002. Species lists in ecology and conservation: abundances matter. Conservation Biology 16 (4): 1160–61.

Balmford, A., L. Clegg, T. Coulson, and J. Taylor. 2002. Why conservationists should heed Pokémon. Science 295 (5564): 2367.

Balvanera, P., G Daily, P. Erlich, T. Ricketts, S. Bailey, S. Kark, C. Kremen, and H. Pereira. 2001. Conserving biodiversity and ecosystem services. Science 291 (5511): 2047.

Barnes, J. A. 1992. Flamingos return to the B.V.I. Forum News, NGO Forum for the U.K. Dependent Territories 7: 2.

Barnett, R., C. Burke, and M. Ziegler. 1986. Applied Mathematics. Second edition. Dellen Publishing, San Francisco, CA.

Bartlett, C. R. 2000. An annotated list of planthoppers (Hemiptera: Fulgoroidea) of Guana Island (British West Indies). Entomological News 111 (2): 120–32.

Barun, A., and G. Perry. 2003. Amphisbaena fenestrata (Virgin Islands Amphisbaena). Predation. Herpetological Review 34 (3): 244.

Bauer, A. M., and R. A. Sadlier. 2000a. The Herpetofauna of New Caledonia. Society for the Study of Amphibians and Reptiles, Ithaca, NY.

Beard, J. S. 1945. Forestry in the Leeward Islands. The British Virgin Islands report. *Development and Welfare in the West Indies Bulletin* 7A: 1–16.

Beati, L., and J. R. Keirans. 2001. Analysis of the systematic relationships among ticks of the genera *Rhipicephalus* and *Boophilus* (Acari: Ixodidae) based on mitochondrial 12S ribosomal DNA gene sequences and morphological characters. *Journal of Parasitology* 87 (1): 32–48.

Beatty, H. A. 1944a. The insects of St. Croix, V.I. *Journal of Agriculture of the University of Puerto Rico* 28 (3–4): 114–72.

Beatty, H. A. 1944b. The mammals of St. Croix, V.I. *Journal of Agriculture of the University of Puerto Rico* 28 (3–4): 181–185 (probably actually published ca. 1948).

Becker, V. O., and S. E. Miller. 1992. The butterflies of Guana Island, British Virgin Islands. *Bulletin of the Allyn Museum* 136: 1–9.

Becker, V. O., and S. E. Miller. 2002. Large moths of Guana Island, British Virgin Islands: A survey of efficient colonizers (Lepidoptera: Sphingidae, Notodontidae, Noctuidae, Arctiidae, Geometridae, Hyblaeidae, Cossidae). *Journal of the Lepidopterists' Society* 56: 9–44, 191–92.

Begon, M. 1979. *Investigating Animal Abundance: Capture-Recapture for Biologists.* University Park Press, Baltimore, MD.

Bell, G. 2001. Neutral macroecology. *Science* 293 (5539): 2413–18.

Bell, G. 2002. Response. *Science* 295 (5561): 1836–37.

Bennett, P. M., and I. P. F. Owens. 2002. *Evolutionary Ecology of Birds: Life Histories, Mating Systems, and Extinction.* Oxford University Press, Oxford, UK.

Berenbaum, M. 2003. Frass-eating grins. *American Entomologist* 49 (3): 132–33.

Bergerud, A. T. 1983. Prey switching in a simple ecosystem. *Scientific American* 249 (6): 130–36, 140–41.

Berkner, L. V., and L. C. Marshall. 1972. Oxygen: Evolution in the earth's atmosphere. *Encyclopedia of Earth Sciences* 4A: 849–61.

Beuttell, K., and J. B. Losos. 1999. Ecological morphology of Caribbean anoles. *Herpetological Monographs* 13: 1–28.

Bhagwat, S., N. Brown, T. Evans, S. Jennings, and P. Savill. 2001. Parks and factors in their success. *Science* 293 (5532): 1045–46.

Biaggi, V. 1983. *Las Aves de Puerto Rico.* Editorial de la Universidad de Puerto Rico, Rio Piedras, PR.

Binns, J. 2001. Anegada iguana *(Cyclura pinguis). Iguana Specialist Group Newsletter (IUCN)* 4 (2): 12–13.

Binns, J. 2003. Stout iguana: historical perspectives and status report. *Iguana, Journal of the International Iguana Society* 10 (2): 39–48.

Bjornstad, O. N. 2001. Chitty cycles—At last! *Trends in Ecology and Evolution* 16 (2): 72.

Blaustein, A. R. 1994. Chicken little or Nero's fiddle? A perspective on declining amphibian populations. *Herpetologica* 50: 85–97.

Bleiweiss, R. 1998a. Relative-rate tests and biological causes of molecular evolution in hummingbirds. *Molecular Biology and Evolution* 15 (5): 481–89.

Bleiweiss, R. 1998b. Phylogeny, body mass, and genetic consequences of lek-mating behavior in hummingbirds. *Molecular Biology and Evolution* 15 (5): 492–98.

Bleiweiss, R. 1998c. Slow rate of molecular evolution in high-elevation hummingbirds. *Proceedings of the National Academy of Sciences U.S.A.* 95: 612–16.

Bleiweiss, R., J. Kirsch, and J. Matheus. 1997. DNA hybridization evidence for the principal lineages of hummingbirds (Aves: Trochilidae). *Molecular Biology and Evolution* 14 (3): 325–43.

Boag, P. T. and P. R. Grant. 1981. Intense natural selection in a population of Darwin's finches (Geospizinae) in the Galapagos. *Science* 214 (4516): 82–85.

Bond, J. 1947. *Field Guide to Birds of the West Indies.* Macmillan Co., New York, NY.

Bond, J. 1961. *Birds of the West Indies.* Houghton Mifflin, Boston, MA.

Bond, J. 1971. *Birds of the West Indies,* 3rd American ed. Houghton-Mifflin Co., Boston.

Bond, M. W. 1966. *How 007 Got His Name.* Collins, London, UK.

Bond, M. W. 1971. *Far Afield in the Caribbean.* Livingston Publishing Co., Wynnewood, PA.

Bond, M. W. 1980. *To James Bond with Love.* Sutter House, Lititz, PA.

Bond, W. 2001. Keystone species—Hunting the snark. *Science* 292 (5514): 63–64.

Bonderteiner, P., M. Binder, J.-M. Moncalvo, R. Agerer, and D. S. Hibbett. 2004. Phylogenetic relationships of cyphelloid homobasidiomycetes. *Molecular Phylogenetics and Evolution* 33: 501–15.

Bowers, B. 2003. Going to bat for bats. *Audubon* 105 (4): 86.

Brescovit, A.D. 1993. Aranhas do genero *Hibana* Brescovit: Especie nova, combinacoes, sinonimas e novas ocorrences para a regiao neotropical (Araneae, Anyphaenidae). *Revista Brasiliana Entomologia* 37 (1): 131–39.

Breuil, M. 2002. Histoire naturalle des amphibiens et reptiles terrestres de L'Archipel Guadeloupeen Guadeloupe, Saint-Martin, Saint-Barthelemy. *Patrimoines Naturels* 54: 1–339.

Breure, A.S.H. 1974. Caribbean land molluscs: Bulimulidae. I. *Bulimulus. Studies on the Fauna of Curacao and Other Caribbean Islands* 45: 1–80.

Bridle, J.R., and C.D. Jiggins. 2000. Adaptive dynamics: is speciation too easy. *Trends in Ecology and Evolution* 15 (6): 225–26.

Briggs, J.C. 1984. Freshwater fishes and the biogeography of Central America and the Antilles. *Systematic Zoology* 33 (4): 428–35.

Brose, U., R. Williams, and N. Martinez. 2003. A comment on "Foraging adaptation and the relationship between food-web complexity and stability." *Science* 301 (5635): 918.

Brower, A.V.Z. 2000. Evolution is not a necessary assumption of cladistics. *Cladistics* 16 (1): 143–63.

Brown, J.H. 1981. Two decades of homage to Santa Rosalia: toward a general theory of diversity. *American Zoologist* 21 (4): 877–88.

Brown, J.H. 1984. On the relationship between abundance and distribution of species. *American Naturalist* 124: 255–79.

Brown, J.H. 1995. *Macroecology.* University of Chicago Press, IL.

Brown, J.H. and M.A. Bowers. 1984. Patterns and processes in three guilds of terrestrial vertebrates. In *Ecological Communities: Conceptual Issues and the Evidence,* eds. D.R. Strong, Jr., D. Simberloff, L.G. Abele, and A.B. Thistle, 282–96. Princeton University Press, Princeton, NJ.

Brown, W.L., and E.O. Wilson. 1956. Character displacement. *Systematic Zoology* 5 (2): 46–49.

Brownie, C., D.R. Anderson, K.P. Burnham, and D.S. Robson. 1985. *Statistical Inference from Band Recovery Data—a Handbook.* Second edition. U.S. Fish and Wildlife Service, Resource Publication 156. Washington, DC.

Bruce, R.C. 1980. A model of the larval period of the spring salamander, *Gyrinophilus porphyriticus,* based on size-frequency distributions. *Herpetologica* 36 (1): 78–86.

Bruner, A., R. Gullison, R. Rice, and G. da Fonseca. 2001a. Effectiveness of parks in protecting tropical biodiversity. *Science* 291 (5501): 125–28.

Bruner, A., R. Gullison, R. Rice, and G. da Fonseca. 2001b. Response. *Science* 293 (5532): 1046–47.

Buden, D.W. 1985. Additional records of bats from the Bahamas. *Caribbean Journal of Science* 21 (1–2): 19–25.

Burnham, K., D.R. Anderson, and J.L. Laake. 1980. Estimation of density from line transect sampling of biological populations. *Wildlife Monographs* 72: 1–202.

Butler, M.A., T.W. Schoener, and J.B. Losos. 2000. The relationship between sexual size dimorphism and habitat use in Greater Antillean *Anolis* lizards. *Evolution* 54: 259–72.

Cabanes, C., A. Cazenave, and C. Le Provost. 2001. Sea level rise during the past 40 years determined from satellite and in situ observations. *Science* 294 (5543): 840–42.

Cardinale, B., M. Palmer, and S. Collins. 2001. Species diversity enhances ecosystem functioning through interspecific facilitation. *Nature* 415 (6870): 426–29.

Carey, M. 1972. The herpetology of Anegada, British Virgin Islands. *Caribbean Journal of Science* 12 (1–2): 79–89.

Carey, W.M. 1975. The rock-iguana, *Cyclura pinguis,* on Anegada, British Virgin Islands, with notes on *Cyclura ricordi* and *Cyclura cornuta* on Hispaniola. *Bulletin of the Florida State Museum* 19 (4): 189–234.

Carlquist, S.J. 1974. *Island Biology.* Columbia University Press, New York, NY.

Carlquist, S.J. 1981. Chance dispersal. *American Scientist* 69 (5): 509–16.

Case, T.J. 2000. *An Illustrated Guide to Theoretical Ecology.* Oxford University Press, New York.

Castillo, J.A., and H.M. Mayorga. 1984. Distribucion, habitos ecologicos, reproduccion y embriologia externa de *Eleutherodactylus antillensis* (Anura). Thesis, Universidad de Panama (Facultad de Ciencias Naturales y Farmacia), Panama City, Panama.

Caughley, G., and A. Gunn. 1996. *Conservation Biology in Theory and Practice.* Blackwell Science, Cambridge, MA.

Censky, E.J. 1988. *Geochelone carbonaria* (Reptilia: Testudines) in the West Indies. *Florida Scientist* 51 (2): 108–14.

Chace, F.A., and H.H. Hobbs. 1969. The freshwater and terrestrial decapod crustaceans of the West Indies with special reference to Dominica. *Bulletin of the United States National Museum* 292: 1–258.

Chalcraft, D.R., and R.M. Andrews. 1999. Predation on lizard eggs by ants: Species interactions in a variable physical environment. *Oecologia* 119: 285–92.

Chameides, W.L., and M. Bergin. 2002. Soot takes center stage. *Science* 297 (5590): 2214–15.

Chandler, C.R., and P.J. Tolson. 1990. Habitat use by a boid snake, *Epicrates monensis,* and its anoline prey, *Anolis cristatellus. Journal of Herpetology* 24: 151–57.

Chao, A. 1984. Nonparametric estimation of the number of classes in a population. *Scandinavian Journal of Statistics* 11: 265–70.

Chao, A. 1987. Estimating the population size for capture–recapture data with unequal catchability. *Biometrics* 43: 783–91.

Chave, J., H. Muller-Landau, and S. Levin. 2002. Comparing classical community models: theoretical consequences for patterns of diversity. *American Naturalist* 159 (1): 1–23.

Chipley, R.M. 1991. Notes on the biology of the bridled quail-dove *(Geotrygon mystacea). Caribbean Journal of Science* 27 (3–4): 180–84.

Chiras, D. 1985. Why save endangered plants and animals? *Mississippi Out-of-Doors* 13 (8): 10.

Choate, J.R., and E.C. Birney. 1968. Sub-recent Insectivora and Chiroptera from Puerto Rico, with the description of a new bat of the genus *Stenoderma. Journal of Mammalogy* 49 (3): 400–412.

Chondropoulos, B.P. and J.J. Lykakis. 1983. Ecology of the Balkan wall lizard, *Podarcis taurica ionica* (Sauria: Lacertidae) from Greece. *Copeia* 1983 (4): 991–1001.

Christian, K.A. 1982. Changes in the food niche during postmetamorphic ontogeny in the frog *Pseudacris triseriata. Copeia* 1982 (1): 73–80.

Christman, S.P. 1984. Plot mapping: Estimating densities of breeding bird territories by combining spot mapping and transect technologies. *The Condor* 86: 237–41.

Church, J.A. 2001. How fast are sea levels rising? *Science* 294 (5543): 802–3.

Cifuentes, L., V. Borja-Aburto, N. Gouviea, G. Thurston, and D. Davis. 2001. Hidden health benefits of greenhouse gas mitigation. *Science* 293 (5533): 1257–59.

Clark, D.L., and J.C. Dillingham. 1984. A method for nocturnally locating lizards. *Herpetological Review* 15 (1): 24–25.

Clark, L.R., P.W. Geier, R.D. Hughes, and R.F. Morris. 1967. *The Ecology of Insect Populations in Theory and Practice.* Methuen and Co, London, UK.

Clark, P., S. Marshall, G. Clarke, S. Hostetler, J. Licciardi, and J. Teller. 2001. Freshwater forcing of abrupt climate change during the last glaciation. *Science* 293 (5528): 283–87.

Claverie, J.M. 2001. What if there are only 30,000 human genes? *Science* 291 (5507): 1255–57.

Clench, W.J. 1939. Land shells of Guana Island, Virgin Islands, West Indies. *Memorias de la Sociedad Cubana de Historia Natural* 13: 287–88.

Clout, M. 2001. Where protection is not enough: active conservation in New Zealand. *Trends in Ecology and Evolution* 16 (8): 415–16.

Cochran, W., H. Mouritsen, and M. Wikelski. 2004. Migrating songbirds recalibrate their magnetic compass daily from twilight cues. *Science* 304 (5669): 405–8.

Cohen, J.E. 1978. *Food Webs and Niche Space. Monographs in Population Biology 11.* Princeton University Press, Princeton, NJ.

Cohen, J.E. 1993. Improving food webs. *Ecology* 74 (1): 252–58.

Cohen, J.E., and C.M. Newman. 1988. Dynamic basis for food web organization. *Ecology* 69 (6): 1655–64.

Cole, B.J. 1980. Colonizing abilities, island size, and the number of species on archipelagos. *American Naturalist* 117 (5): 629–38.

Cole, G.A. 1968. Desert Limnology, B. Invertebrate Fauna. In *Desert Limnology,* Vol. 1, ed. G.W.J. Brown, 459–70. Academic Press, New York.

Colinvaux, P.A. 1978. *Why Big Fierce Animals Are Rare. An Ecologist's Perspective.* Princeton University Press, Princeton, NJ.

Colli, C. 1996. Return of the flamingos. *Welcome, BVI Tourist Guide* 25 (2): 1–4.

Collins, M., M. Haverty, and B. Thorne. 1997. The termites (Isoptera: Kalotermitidae, Rhinotermitidae, Termitidae) of the British Virgin Islands: Distribution, moisture relations, and cuticular hydrocarbons. *Sociobiology* 30 (1): 63–76.

Collins, M.S. 1991. Physical factors affecting termite distribution. *Sociobiology* 19: 283–86.

Colwell, R.K., and D.C. Lees. 2000. The mid-domain effect: Geometric constraints on the geography of species richness. *Trends in Ecology and Evolution* 15 (2): 70–76.

Colwell, R.K. and D.W. Winkler. 1984. A null model for null models in biogeography. In *Ecological Communities: Conceptual Issues and the Evidence,* eds. D.R. Strong, Jr., D. Simberloff, L.G. Abele, and A.B. Thistle, 344–59. Princeton University Press, Princeton, NJ.

Combes, C. 2001. *Parasitism. The Ecology and Evolution of Intimate Interactions.* University of Chicago Press, IL.

Condit, R., N. Pitman, E. Leigh, J. Chave, J. Terborgh, R. Foster, P. Nunez, V. Solomon Aquilar, R. Valencia, G. Villa, H. Muller-Landau, E. Losos, and S. Hubbell. 2002. Beta diversity in tropical forest trees. *Science* 295 (5555): 666–69.

Connell, J.H. 1978. Diversity in tropical rain forests and coral reefs. *Science* 199 (4335): 1302–10.

Connell, J.H. 1980. Diversity and the coevolution of competitors, or the ghost of competition past. *Oikos* 35 (2): 131–38.

Connell, J.H., and R.O. Slatyer. 1977. Mechanisms of succession in natural communities and their role in community stability and organization. *American Naturalist* 111: 1119–44.

Connor, E.F., and E.D. McCoy. 1979. The statistics and biology of the species-area relationship. *American Naturalist* 113(6): 791–833.

Conyers, J. 1996. The BVI flamingo restoration project. *Critter Talk, Newsletter of the Bermuda Zoological Society* 19 (2): 1–2.

Cory, C.B. 1889. *The Birds of the West Indies.* Estes & Lauriat, Boston.

Cousins, S.H. 1996. Food webs: From the Lindeman paradigm to a taxonomic general theory of ecology. In *Food Webs,* eds. G.A. Polis and K.O. Winemiller, 243–51. Chapman & Hall, New York.

Crawley, M.J. 1997. *Plant Ecology,* 2nd ed. Blackwell Science, Oxford, UK.

Crawley, M.J., and J.E. Harral. 2001. Scale dependence in plant biodiversity. *Science* 291 (5505): 864–68.

Cronin, M.A. 1991. Mitochondrial-DNA phylogeny of deer (Cervidae). *Journal of Mammalogy* 72 (3): 553–66.

Cronin, M.A. 1992. Intraspecific variation in mitochondrial DNA of North American cervids. *Journal of Mammalogy* 73 (1): 70–78.

Crother, B.I. 1999. Evolutionary relationships. In *Caribbean Amphibians and Reptiles,* ed. B.I. Crother, 269–334. Academic Press, San Diego, CA.

Crutzen, P.J. 2002. Geology of mankind. *Nature* 415 (6867): 23.

Cryan, J.R. and L.L. Deitz. 2002. Enigmatic treehopper genera (Hemiptera: Membracidae): *Deiroderes* Ramos, *Holdgatiella* Evans, and *Togotolania,* new genus. *Proceedings of the Entomologi-cal Society of Washington* 104: 868–83. *Deiroderes inermis* Ramos, 1957, from Guana Island.

Cryan, J., B. Wiegmann, L. Deitz, and C. Dietrich. 2000. Phylogeny of the treehoppers (Insecta: Hemiptera: Membracidae), evidence from two nuclear genes. *Molecular Phylogenetics and Evolution* 17: 317–34.

Cuba, T.R. 1981. Diversity: A two-level approach. *Ecology* 62 (1): 278–279.

Cullen, S.A. 1994. Black-necked stilt foraging site selection and behavior in Puerto Rico. *Wilson Bulletin* 106 (3): 508–13.

Curtis, J., M. Brenner, and D. Hodell. 2001. Climate change in the circum-Caribbean (late Pleistocene to Present) and implications for regional biogeography. In *Biogeography of the West Indies,* 2nd ed., eds. C.A. Woods and F.E. Sergile, 35–54. CRC Press, Boca Raton, FL.

D'Arcy, W.G. 1967. Annotated checklist of the dicotyledons of Tortola, Virgin Islands. *Rhodora* 69 (780): 385–450.

D'Arcy, W.G. 1971. The mystery *Sabal* of Anegada. *Principes* 15 (4): 131–33.

Darlington, P.J. 1938. The origin of the fauna of the Greater Antilles, with discussion of dispersal of animals over water and through the air. *Quarterly Review of Biology* 123: 274–300.

Darlington, P.J. 1957. *Zoogeography: The Geographical Distribution of Animals.* John Wiley & Sons, New York.

Darwin, C. 1859. *On the Origin of Species by Means of Natural Selection.* J. Murray, London, UK.

Darwin, C. 1873. *The Origin of Species by Means of Natural Selection, or the Preservation of Favoured Races in the Struggle for Life.* Sixth English edition with additions and corrections. John Murray, London, UK.

Daszak, P., L. Berger, A.A. Cunningham, A.D. Hyatt, D.E. Green, and R. Speare. 1999. Emerging infectious diseases and amphibian population declines. *Emerging Infectious Diseases* 5: 735–48.

David, P. 2002. *Spider-Man.* Ballantine Books, Random House, New York.

Davies, N. and D.S. Smith. 1998. Munroe revisited: A survey of West Indian butterfly faunas and their species-area relationship. *Global Ecology and Biogeography Letters* 7 (4): 285–94.

Davies, N.B. 1978. Ecological questions about territorial behavior. In *Behavioral Ecology an Evolutionary Approach,* eds. J.R. Krebs and N.B. Davies, 317–50. Blackwell Scientific Publications, Oxford, UK.

Davis, A. K. 2001. Blackpoll warbler *(Dendroica striata)* fat deposition in southern Nova Scotia during autumn migration. *Northeastern Naturalist* 8 (2): 149–62.

Davis, D. D. 1964. The giant panda: A morphological study of evolutionary mechanisms. *Fieldiana Zoology Memoirs* 3: 1–339.

Davis, D. E. and R. L. Winstead. 1980. Estimating the numbers of wildlife populations. In *Wildlife Management Techniques Manual,* ed. S. D. Schemnitz, 221–45. Fourth edition. The Wildlife Society, Washington, DC.

Davis, D. R. 1986. Neotropical Tineidae. I: The types of H. B. Moschler (Lepidoptera: Tineoidea). *Proceedings of the Entomological Society of Washington* 88 (1): 83–92.

Davis, L. R., R. K. Vandermeer, and S. D. Porter. 2001. Red imported fire ants expand their range across the West Indies. *Florida Entomologist* 84: 735–36.

Davis, W. B. 1973. Geographic variation in the fishing bat, *Noctilio leporinus. Journal of Mammalogy* 54 (4): 862–74.

Dawkins, R. 1986. *The Blind Watchmaker.* W. W. Norton, New York.

Dawson, A. G. 1992. *Ice Age Earth.* Routledge, London, UK.

Day, M., M. Breuil, and S. Reichling. 2000. Lesser Antillean iguana *Iguana delicatissima.* In *West Indian Iguanas: Status Survey and Conservation Action Plan,* ed. A. Alberts, 62–67. IUCN, Gland, Switzerland.

Dayton, L. 2001. Mass extinctions pinned on Ice Age hunters. *Science* 292 (5523): 1819.

Deevey, Jr. E. S. 1947. Life tables for natural populations of animals. *Quarterly Review of Biology* 1947: 283–314.

de la Cruz, J. 2001. Patterns in the biogeography of West Indian ticks. In *Biogeography of the West Indies,* 2nd ed., eds. C. A. Woods and F. E. Sergile, 85–106. CRC Press, Boca Raton, FL.

DeLury, D. B. 1951. On the planning of experiments for the estimation of fish populations. *Journal of the Fisheries Research Board of Canada* 8 (4): 281–307.

DeLury, D. B. 1951. The estimation of biological populations. *Biometrics* 3: 145–47.

DeLury, D. B. 1958. The estimation of population size by a marking and recapture procedure. *Journal of the Fisheries Research Board of Canada* 15 (1): 19–25.

de Mazancourt, C. 2001. Consequences of community drift. *Science* 293 (5536): 1772.

de Quieroz, K. 1987. Phylogenetic systematics of iguanine lizards. A comparative osteological study. *University of California Publications in Zoology* 118: 1–203.

Denno, R. F., D. J. Hawthorne, B. L. Thorne, and C. Gratton. 2001. Reduced flight capability in British Virgin Island populations of a wing-dimorphic insect: The role of habitat isolation, persistence, and structure. *Ecological Entomology* 26: 25–36.

DeVries, P., T. Walla, and H. Greeney. 1999. Species diversity in spatial and temporal dimensions of fruit-feeding butterflies from two Ecuadorian rainforests. *Biological Journal of the Linnaean Society* 68: 333–53.

DeVries, P. J., and T. R. Walla. 2001. Species diversity and community structure in neotropical fruit-feeding butterflies. *Biological Journal of the Linnaean Society* 74: 1–15.

Deyrup, M., L. Davis, and S. Buckner. 1998. Composition of the ant fauna of three Bahamian islands. In *Proceedings of the Seventh Symposium on the Natural History of the Bahamas,* 23–32. Bahamian Field Station, San Salvador, Bahamas.

Diamond, J. M. 1975. Assembly of species communities. In *Ecology and Evolution of Communities,* eds. M. L. Cody and J. M. Diamond, 342–444. Belknap Press, Harvard University, Cambridge, MA.

Diamond, J. M. 1984. Long-term rainforest studies. *Nature* 312: 699.

Diaz, S., A. Symstad, F. S. Chapin, D. Wardle, and L. Huenneke. 2003. Functional diversity revealed by removal experiments. *Trends in Ecology and Evolution* 18 (3): 140–46.

Dingle, H. and J. P. Hegmann, eds. 1982. *Evolution and Genetics of Life Histories.* Springer-Verlag, New York, NY.

Dixon, K. R., and J. A. Chapman. 1980. Harmonic mean measure of animal activity areas. *Ecology* 61 (5): 1040–44.

Dmi'el, R., G. Perry, and J. Lazell. 1997. Evaporative water loss in nine insular populations of the lizard *Anolis cristatellus* group in the British Virgin Islands. *Biotropica* 29 (1): 111–16.

Dookhan, I. 1971. A Pre-emancipation History of the West Indies. Collins, London, U.K.

Downhower, J. F. and L. Brown. 1977. A sampling technique for benthic fish populations. *Copeia* 1977 (2): 403–06.

Drewry, G. E., and A. S. Rand. 1983. Characteristics of an acoustic community: Puerto Rican frogs of the genus *Eleutherodactylus. Copeia* 1983: 941–53.

Drury, W. H. 1998. *Chance and Change. Ecology for Conservationists*. University of California Press, Berkeley, CA.

Duivenvoorden, J., J. Svenning, and S. Wright. 2002. Beta diversity in tropical forests. *Science* 295 (5555): 636–37.

Dunham, A. E. 1983. Realized niche overlap, resource abundance, and intensity of interspecific competition. In *Lizard Ecology: Studies of a Model Organism*, eds. R. B. Huey, E. R. Pianka, and T. W. Schoener, 261–80. Harvard University Press, Cambridge, MA.

Dunkle, S. W. 1989. *Dragonflies of the Florida Peninsula, Bermuda, and the Bahamas*. Scientific Publishers, Gainesville, FL.

Dunkle, S. W. 1990. *Damselflies of Florida, Bermuda, and the Bahamas*. Scientific Publishers, Gainesville, FL.

Dunkle, S. W. 2000. *Dragonflies through Binoculars*. Oxford University Press, Oxford, UK.

Durham, H. F. 1972. *Caribbean Quakers*. Dukane, Hollywood, FL.

Durrell, G. 1981. *The Mockery Bird*. Collins, London.

Easterlin, R. A. 1961. The American baby boom in historical perspective. *American Economic Review* 51: 869–911.

Eckert, K., J. Overing, and B. Lettsome. 1992. Sea Turtle Recovery Action Plan for the British Virgin Islands. CEP Technical Report 15. UNEP Caribbean Environment Program, Kingston, Jamaica.

Eckert, S. A. 1992. Bound for deep waters. *Natural History* 101 (3): 28–35.

Ehrlich, P. R., and A. H. Ehrlich. 1970. *Population, Resources, Environment*. W. H. Freeman Co., San Francisco, CA.

Eisner, T., H. Eisner, J. Meinwald, C. Sagan, C. Walcott, E. Mayr, E. O. Wilson, P. Raven, A. Ehrlich, P. Ehrlich, A. Carr, E. P. Odum, and C. Gans. 1981. Conservation of tropical forests. *Science* 213: 1314.

Eldredge, M. D. B., and T. L. Browning. 2002. Molecular genetic analysis of the naturalized Hawaiian population of the brush-tailed rock-wallaby, *Petrogale penicillata* (Marsupialia: Macropodidae). *Journal of Mammalogy* 83 (2): 437–44.

Elton, C. 1927. *Animal Ecology*. Sidgwick and Jackson, London, UK.

Elton, C. 1958. *The Ecology of Invasions by Plants and Animals*. Chapman & Hall, London, UK.

Endler, J. A. 1982a. Alternative hypotheses in biogeography: Introduction and synopsis of the symposium. *American Zoologist* 22 (2): 349–54.

Endler, J. A. 1982b. Problems in distinguishing historical from ecological factors in biogeography. *American Zoologist* 22 (2): 441–52.

Engel, M. S. 2001. Three new *Habralictellus* bee species from the Caribbean (Hymenoptera: Halictidae). *Solenodon* 1: 33–37.

Engelstoft, C., and K. Ovaska. 1999. The harmonic direction finder: A new method for tracking movements of small snakes. *Herpetological Review* 30: 84–87.

Engstrom, R. T. and F. C. James. 1981. Plot size as a factor in winter bird-population studies. *Condor* 83: 34–41.

Enquist, B., J. Sanderson, and M. Weiser. 2002. Modeling macroscopic patterns in ecology. *Science* 295 (5561): 1835–36.

Ernst, C. H., and T. E. J. Leuteritz. 1999. *Geochelone carbonaria*. Catalogue of American Amphibians and Reptiles 690: 1–7.

Evenhuis, N. L., and S. E. Miller. 1994. Beeflies of the British Virgin Islands (Diptera: Bombyliidae). *Florida Entomologist* 77 (3): 382–84.

Fauth, J., J. Bernardo, M. Camara, W. Resetarits, J. Van Buskirk, and S. McCollum. 1996. Simplifying the jargon of community ecology: A conceptual approach. *American Naturalist* 147: 282–86.

Feduccia, A. 2003. "Big bang" for Tertiary birds? *Trends in Ecology and Evolution* 18 (4): 172–76.

Ferner, J. W. 1979. A review of marking techniques for amphibians and reptiles. *Society for Study of Amphibians and Reptiles Herpetological Circular* 9: 1–42.

Fitch, H. S. 1970. Reproductive cycles in lizards and snakes. *University of Kansas Museum of Natural History Miscellaneous Publication* 52: 1–247.

Fitch, H. S. 1975. A demographic study of the ringneck snake *(Diadophis punctatus)* in Kansas. *University of Kansas Museum of Natural History Miscellaneous Publication* 62: 1–53. A close relative of our common snake and also a lizard eater; but a temperate ecosystem.

Flannery, T. and P. Schouten. 2001. *A Gap in Nature*. Atlantic Monthly Press, NY.

Flint, R. F. 1971. *Glacial and Quaternary Geology*. John Wiley & Sons, New York.

Floyd, H. B. and T. A. Jenssen. 1984. Prey diversity comparisons between stomach and hindgut of the lizard, *Anolis opalinus*. *Journal of Herpetology* 18(2): 204–05.

Fogarty, J. H., and F. J. Vilella. 2001. Evaluating methodologies to survey *Eleutherodactylus* frogs in montane forests of Puerto Rico. *Wildlife Society Bulletin* 29: 948–55.

Forel, A. 1911. Ameisen des Herren Prof. v. Ihering aus Brasilien (Sao Paulo usw) nebst einigen anderen aus Südamerika und Afrika. *Deutsche Entomologische Zeitschrift* 1911: 285–312.

Forest, C., P. Stone, A. Sokolov, M. Allen, and M. Webster. 2002. Quantifying uncertainties in climate system properties with the use of recent climate observations. *Science* 295 (5552): 113–17.

Fowler, C. 2002. Sharing agriculture's genetic bounty. *Science* 297 (5579): 157.

Frakes, L.A. 1980. *Climates throughout Geologic Time.* Elsevier/North Holland, New York.

Franck, O.F., and W.D. Sisson. 1980. Scorpions from the Virgin Islands (Arachnida, Scorpiones). *Museum of Texas Technological University Occasional Papers* 65: 1–19.

Frank, P.A. 1997. First record of *Molossus molossus tropidorhynchus* Gray (1839) from the United States. *Journal of Mammalogy* 78 (1): 103–05.

Frank, P.W. 1968. Life histories and community stability. *Ecology* 49 (2): 355–57.

Frankenberg, E., and Y.L. Werner. 1981. Adaptability of the daily activity pattern to changes in longitude, in a colonizing lizard, *Hemidactylus frenatus. Journal of Herpetology* 15 (3): 373–76.

Frith, H.J. 1979. *Wildlife Conservation.* Angus & Robertson, London.

Fritts, S.H., W.J. Paul, and L.D. Mech. 1984. Movements of translocated wolves in Minnesota. *Journal of Wildlife Management* 48(3): 709–21.

Futuyma, D.J. and G.C. Mayer. 1980. Non-allopatric speciation in animals. *Systematic Zoology* 29: 254–71.

Gans, C., and A.A. Alexander. 1962. Studies on amphisbaenids of the Antilles. *Bulletin of the Museum of Comparative Zoology* 128 (3): 67–158.

Garcia, M. 2001. Puerto Rico proposal. *IUCN Iguana Specialist Group Newsletter Supplement* 4 (1): 4.

Gardner, M.R., and W.R. Ashby. 1970. Connectance of large dynamic (cybernetic) systems: Critical values for stability. *Nature* 228: 784.

Garland, T. Jr., and J.B. Losos. 1994. Ecological morphology of locomotor performance in squamate reptiles. In *Ecological Morphology: Integrative Organismal Biology,* eds. P.C. Wainwright and S.M. Reilly, 240–302. University of Chicago Press, IL.

Garrison, R.W. and M.R. Willig. 1996. Arboreal invertebrates. In *The Food Web of a Tropical Rain Forest,* eds. D.P. Reagan and R.B. Waide, 183–245. University of Chicago Press, IL.

Gates, C.E. 1979. Line intersect and related issues. In *Sampling Biological Populations,* eds. R. Cormack, G. Patil, and D. Robson, 71–154. International Co-operative Publishing House, Fairfield, MD.

Gates, C.E. 1986. *LINETRAN User's Guide.* Texas A & M University, College Station.

Geist, V. 1971. *Mountain Sheep: A Study in Behavior and Evolution.* University of Chicago Press, IL.

Gentry, A.H. 1982. Bignoniaceae—Part II (Tribe Tecomieae). *Flora Neotropica Monograph* 25 (2): 1–370.

Gibbs, W.W. 2003. Virtual ecosystems. *Conservation in Practice* 4 (4): 12–19.

Gielis, C. 1992. Neotropical Pterophoridae 8: The genus *Adaina* Tutt, 1905 (Lepidoptera: Pterophoridae). *SHILAP Revista Lepidopterologia* 20 (80): 373–404.

Giger, R.D. 1973. Movements and homing in Townsend's mole near Tillamook, Oregon. *Journal of Mammalogy* 54 (3): 648–59.

Gilbert, F.S. 1980. The equilibrium theory of island biogeography: fact or fiction? *Journal of Biogeography* 7 (3): 209–35.

Gilpin, M.E. and J.M. Diamond. 1984. Are species co-occurrences on islands non-random, and are null hypothesis useful in community ecology? In *Ecological Communities: Conceptual Issues and the Evidence,* eds. D.R. Strong, Jr., D. Simberloff, L.G. Abele, and A.B. Thistle, 297–315. Princeton University Press, Princeton, NJ.

Ginsburg, R.N., and N.P. James. 1974. Holocene carbonate sediments of continental shelves. In *Geology of Continental Margins,* eds. C.A. Burk and C.L. Drake, 137–55. Springer-Verlag, New York.

Gleason, J.M., E.C. Griffith, and J.R. Powell. 1998. A molecular phylogeny of the *Drosophila willistoni* group: Conflicts between species concepts? *Evolution* 52: 1093–1103.

Godfrey, P.J. 1976. Barrier beaches of the East Coast. *Oceanus* 19: 27–40.

Goldberg, S., C. Bursey, and H. Cheam. 1998. Helminths of the lizard *Anolis cristatellus* (Polychrotidae) from the British Virgin Islands, West Indies. *Journal of the Helminthological Society of Washington* 65 (2): 259–62.

Goodyear, N.C. 1992. Flamingos return to Anegada: Status update. *B.V.I. National Parks Trust News* 1992 (August): 1.

Goodyear, N., and J. Lazell. 1994. Status of a relocated population of endangered *Iguana pinguis* on Guana Island, British Virgin Islands. *Restoration Ecology* 2 (1): 43–50.

Gorman, G.C., and R. Harwood. 1977. Notes on population density, vagility, and activity patterns of the Puerto Rican grass lizard, *Anolis pulchellus* (Reptilia, Lacertilia, Iguanidae). *Journal of Herpetology* 11 (3): 363–68.

Gotelli, N.J. 2001. *A Primer of Ecology*, 3rd ed. Sinauer Associates, Sunderland, MA.

Goto, M.M., and M.A. Osborne. 1989. Nocturnal microhabitats of two Puerto Rican grass lizards, *Anolis pulchellus* and *Anolis krugi*. *Journal of Herpetology* 23: 79–81.

Goudie, A. 1982. *The Human Impact. Man's Role in Environmental Change*. MIT Press, Cambridge, MA.

Gould, S.J. and N. Eldredge. 1983. Darwin's gradualism. *Systematic Zoology* 32 (4): 444–45.

Grace, J.B. 1999. The factors controlling species density in herbaceous plant communities: An assessment. *Perspectives in Plant Ecology, Evolution and Systematics* 2/1: 1–28.

Grace, J.B. 2001a. The roles of community biomass and species pools in the regulation of plant diversity. *Oikos* 92: 193–207.

Grace, J.B. 2001b. Difficulties with estimating and interpreting species pools and implications for understanding patterns of diversity. *Folia Geobotanica* 36: 71–83.

Grace, J.B., and G.R. Guntenspergen. 1999. The effects of landscape position on plant species density: Evidence of past environmental effects in a coastal wetland. *Ecoscience* 6 (3): 381–91.

Grace, J.B., and B.H. Pugesek. 1997. A structural model of plant species richness and its application to a coastal wetland. *American Naturalist* 149 (3): 436–60.

Graham, A. 2003. Geohistory models and Cenozoic paleoenvironments of the Caribbean region. *Systematic Botany* 28 (2): 378–86.

Grant, C. 1932. Herpetology of Tortola; notes on Anegada and Virgin Gorda, British Virgin Islands. *Journal of the Department of Agriculture of Puerto Rico* 16: 339–46.

Grant, P. and D. Schluter. 1984. Interspecific competition inferred from patterns of guild structure. In *Ecological Communities: Conceptual Issues and the Evidence,* eds. D.R. Strong, Jr., D. Simberloff, L.G. Abele, and A.B. Thistle, 201–33. Princeton University Press, Princeton, NJ.

Grant, P.R., and T.D. Price. 1981. Population variation in continuously varying traits as an ecological genetics problem. *American Zoologist* 21 (4): 795–811.

Grant, P.R. 1983. The role of interspecific competition in the adaptive radiation of Darwin's finches. In *Patterns of Evolution in Galapagos Organisms,* eds. R.I. Brown, M. Berson, and A.E. Leviton, 187–99. American Association for the Advancement of Science, Pacific Division, San Francisco, CA.

Grayson, D.K. 2001. Did human hunting cause mass extinction? *Science* 294 (5546): 1459.

Green, R.F. 1980. A note on *K*-selection. *American Naturalist* 116 (2): 291–96.

Grime, J.P. 1979. *Plant Strategies and Vegetation Processes*. John Wiley & Sons, London, UK.

Grimm, E.C., K.A. Maasch, G.L. Jacobson, Jr., W.A. Watts, and B.C.S. Hansen. 1993. A 50,000-year record of climate oscillations from Florida and its temporal correlation with the Heinrich events. *Science* 261 (5118): 198–200.

Grimm, V., and C. Wissel. 1997. Babel, or the ecological stability discussion: An inventory and analysis of terminology and a guide for avoiding confusion. *Oecologia* 109: 323–34.

Grismer, L.L. 2002. Spiny-tailed iguanas, insular evolution, and Seri Indians: How long does it take to make a new species and does it matter who makes it? *Journal of the International Iguana Society* 9 (1–2): 9–16.

Guravich, D., and J.E. Brown. 1983. *The Return of the Brown Pelican*. Louisiana State University Press, Baton Rouge, LA.

Hackney, P.A. and T.E. Linkous. 1978. Striking behavior of the largemouth bass and use of the binomial distribution for its analysis. *Transactions of the American Fisheries Society* 107 (5): 682–88.

Hagelberg, E. 2003. Recombination or mutation rate heterogeneity? Implications for mitochondrial Eve. *Trends in Genetics* 19 (2): 84–90.

Hairston, N.G. 1981. An experimental test of a guild: Salamander competition. *Ecology* 62 (1): 65–72.

Hairston, N.G., Sr. 1984. Inferences and experimental results in guild structure. In *Ecological Communities: Conceptual Issues and the Evidence,* eds. D.R. Strong, Jr., D. Simberloff, L.G. Abele, and A.B. Thistle, 19–27. Princeton University Press, Princeton, NJ.

Hall, E.R. 1981. *The Mammals of North America.* John Wiley & Sons, New York.

Hanley, W. 1977. *Natural History in America.* Quadrangle, New York Times Book Co., New York.

Hannon, B. 2004. The grand challenge of birth control. *Science* 303 (5655): 177.

Hanski, I. 1982. Dynamics of regional distribution: The core and satellite species hypothesis. *Oikos* 38: 210–21.

Harcombe, P.A., and P.L. Marks. 1976. Species preservation. *Science* 194: 383.

Hardy, I.C.W., and D.J. Kemp. 2001. Skink skirmishes: Why do owners win? *Trends in Ecology and Evolution* 16 (4): 174.

Harris, L.D. 1984. *The Fragmented Forest*. University of Chicago Press, Chicago, IL.

Hart, M., R. Reader, and J. Klironomos. 2003. Plant coexistence mediated by arbuscular mycorrhizal fungi. *Trends in Ecology and Evolution* 18 (8): 418–23.

Harvell, C.D., C.E. Mitchell, J.R. Ward, S. Altizer, A.P. Dobson, R.S. Ostfeld, and M.D. Samuel. 2002. Ecology—climate warming and disease risks for terrestrial and marine biota *Science* 296 (5576): 2158–62.

Hass, C.A., L.R. Maxson, and S.B. Hedges. 2001. Relationship and divergence times of West Indian amphibians and reptiles: Insights from albumin immunology. In *Biogeography of the West Indies,* 2nd ed., eds. C.A. Woods and F.E. Sergile, 157–74. CRC Press, Boca Raton, FL.

Hastings, A. 1988. Food web theory and stability. *Ecology* 69 (6): 1665–68.

Hastings, A. 2003. Metapopulation persistence with age-dependent disturbance or succession. *Science* 301 (5639): 1525–26.

Haug, G., K. Hughen, D. Sigman, L. Peterson, and U. Rohl. 2001. Southward migration of the intertropical convergence zone through the Holocene. *Science* 293 (5533): 1304–8.

Haverty, M., M. Collins, L. Nelson, and B. Thorne. 1997. Cuticular hydrocarbons of the termites of the British Virgin Islands. *Journal of Chemical Ecology* 23 (4): 927–64.

Haverty, M., L. Nelson, B. Thorne, M. Collins, J. Darlington, and M. Page. 1991. Cuticular hydrocarbons for species determination of tropical termites. In *USDA General Technical Report PSW-GTR-129*, pp. 58–66.

Haverty, M.I., R.J. Woodrow, L.J. Nelson, and J.K. Grace. 2005. Identification of termite species by the hydrocarbons in their feces. *Journal of Chemical Ecology,* in press.

Haverty, M., L. Nelson, B. Thorne, M. Collins, J. Darlington, and M. Page. 1992. Cuticular hydrocarbons for species determination of tropical termites. In *Proceedings of the Session on Tropical Forestry for People of the Pacific, XVII Pacific Science Congress,* eds. C.E. Conrad and

L.A. Newell, 58–65. Pacific Southwest Research Station, USDA Forest Service, Albany, CA.

Haverty, M., B. Thorne, and L. Nelson. 1996. Hydrocarbons of *Nasutitermes acajutlae* and comparison of methodologies for sampling cuticular hydrocarbons of Caribbean termites for taxonomic and ecological studies. *Journal of Chemical Ecology* 22 (11): 2081–2109.

Hawkins, B.A. 2001. Ecology's oldest pattern? *Trends in Ecology and Evolution* 16 (8): 470.

Hayes, F.E. 2002. Geographic variation, hybridization, and taxonomy of New World *Butorides* herons. *North American Birds* 56 (1): 4–10.

Hayne, D.W. 1949. Two methods for estimating population from trapping records. *Journal of Mammology* 30 (4): 399–411.

Heatwole, H. 1976. Herpetogeography of Puerto Rico. VII. Geographic variation in the *Anolis cristatellus* complex in Puerto Rico and the Virgin Islands. *Occasional Papers of the Museum of Natural History, University of Kansas,* 46: 1–17.

Heatwole, H., and I.B. Banuchi. 1966. Invenomation by the colubrid snake, *Alsophis portoricensis. Herpetologica* 22: 132–34.

Heatwole, H. and F. MacKenzie. 1967. Herpetogeography of Puerto Rico. IV. Paleogeography, faunal similarity and endemism. *Evolution* 21: 429–39.

Heatwole, H.R., R. Levins, and M.D. Byer. 1981. Biogeography of the Puerto Rican Bank. *Atoll Research Bulletin* 251: 1–55.

Heckel, D.G. and J. Roughgarden. 1979. A technique for estimating the size of lizard populations. *Ecology* 60 (5): 966–75.

Hedeen, S.E. 1984. The establishment of *Podarcis muralis* in Cincinnati, Ohio. *Herpetological Review* 15 (3): 70–71.

Hedges, S.B. 1982. Caribbean biogeography: implications of recent plate tectonic studies. *Systematic Zoology* 31 (4): 518–22.

Hedges, S.B. 1989. Evolution and biogeography of West Indian frogs of the genus *Eleutherodactylus*: Slow-evolving loci and the major groups. In *Biogeography of the West Indies: Past, Present and Future,* ed. C.A. Woods, 305–70. Sandhill Crane Press, Gainsville, FL.

Hedges, S.B. 1996a. Historical biogeography of West Indian vertebrates. *Annual Review of Ecology and Systematics* 27: 163–96.

Hedges, S.B. 1996b. The origin of West Indian amphibians and reptiles. In *Contributions to West Indian Herpetology: A Tribute to Albert*

Schwartz, eds. R. Powell and R. Henderson, 95–127. Society for the Study of Amphibians and Reptiles, Ithaca, NY.

Hedges, S.B. 2001. Biogeography of the West Indies: An overview. In *Biogeography of the West Indies*, 2nd ed., eds. C.A. Woods and F.E. Sergile, 15–33. CRC Press, Boca Raton, FL.

Hedges, S.B., C. Haas, and L. Maxson. 1992. Caribbean biogeography: Molecular evidence for dispersal in West Indian terrestrial vertebrates. *Proceedings of the National Academy of Sciences U.S.A.* 89: 1903–9.

Hedges, S.B., and R. Thomas. 1991. Cryptic species of snakes (Typhlopidae: *Typhlops*) from the Puerto Rico Bank detected by protein electrophoresis. *Herpetologica* 47 (4): 448–59.

Helmuth, B., C. Harley, P. Holpin, M. O'Donnell, G. Hofmann, and C. Blanchette. 2002. Climate change and latitudinal patterns of intertidal thermal stress. *Science* 298 (5595): 1015–17.

Helsley, C.E. 1971. Summary of the geology of the British Virgin Islands. *Transactions of the Fifth Caribbean Conference on Geology, Bulletin* 5: 69–76.

Henderson, R.W. 1982. Thermoregulation in a Hispaniolan tree snake, *Uromacer catesbyi*. *Journal of Herpetology* 16 (1): 89–91.

Henighan, S. 2003. One iguana, two solutions. *The B.V.I. StandPoint* 3 (43): 19–20.

Henshaw, R.E. and R.O. Stephenson. 1974. Homing in the gray wolf *(Canis lupus)*. *Journal of Mammalogy* 55 (1): 234–37.

Hertz, P.E. 1979. Comparative thermal biology of sympatric grass anoles (*Anolis semilineatus* and *A. olssoni*) in lowland Hispaniola (Reptilia, Lacertilia, Iguanidae). *Journal of Herpetology* 13: 329–33.

Hertz, P.E. 1992. Temperature regulation in Puerto Rican *Anolis* lizards: A field test using null hypotheses. *Ecology* 73: 1405–17.

Higgins, C.G. 1968. Beachrock. *Encyclopedia of Earth Sciences* 3: 70–73.

Highton, R., S.B. Hedges, C.A. Hass, and H. Dowling. 2002. Snake relationships revealed by slowly-evolving proteins: Further analysis and a reply. *Herpetologica* 58 (2): 270–75.

Hobson, K.A. 2002. Incredible journeys. *Science* 295 (5557): 981–83.

Hogstedt, G. 1983. Adaptation unto death: Function of fear screams. *American Naturalist* 121 (4): 562–70.

Holden, C., ed. 2001. Dinner in a mound. *Science* 291 (5504): 587.

Holling, C.S. 1973. Resilience and stability of ecological systems. *Annual Review of Ecology and Systematics* 4: 1–23.

Holmes, A., and D.L. Holmes. 1978. *Principles of Physical Geology*, 3rd ed. John Wiley & Sons, New York.

Holt, R.D. 1996. Food webs in space: An island biogeographic perspective. In *Food Webs*, eds. G.A. Polis and K.O. Winemiller, 313–23. Chapman & Hall, New York.

Houston, M.A. 1994. *Biological Diversity: The Co-existence of Species on Changing Landscapes*. Cambridge University Press, UK.

Howard, R.A. 1974–1989. *Flora of the Lesser Antilles*, Vols. 1–6. The Arnold Arboretum, Harvard University, Cambridge, MA.

Howard, R.A. and E.A. Kellogg. 1987. Contributions to a flora of Anguilla and adjacent islets. *Journal of the Arnold Arboretum* 68: 105–31.

Hubbard, H.G. 1877. Notes on the tree nests of termites in Jamaica. *Proceedings of the Boston Society of Natural History* 19: 267–74.

Hubbell, S.P. 2001. *The Unified Neutral Theory of Biodiversity and Biogeography*. Princeton University Press, NJ.

Hudson, P. 2001. Searching for synthesis in parasitology. *Trends in Ecology and Evolution* 16 (10): 589–90.

Hudson, R. 2001. July 2001 recovery plan workshop. *Iguana Specialist Group Newsletter (IUCN)* 4 (2): 14–15.

Huey, R.B. 1974. Behavioral thermoregulation in lizards: Importance and associated costs. *Science* 184: 1001–3.

Huey, R.B. and E.R. Pianka. 1983. Temporal separation of activity and interspecific dietary overlap. In *Lizard Ecology: Studies of a Model Organism*, eds. R.B. Huey, E.R. Pianka, and T.W. Schoener, 281–96. Harvard University Press, Cambridge, MA.

Humphrey, S.R., and F.J. Bonaccorso. 1979. Population and community ecology. Biology of bats of the New World family Phyllostomatidae. Part III. *The Museum, Texas Tech University Special Publications* 16: 409–41.

Hungerford, H.B. 1948. *Corixidae of the Western Hemisphere (Hemiptera)*. University of Kansas, Lawrence, KS.

Hunt, J., E. Bermingham, and R. Ricklefs. 2001. Molecular systematics and biogeography of Antillean thrashers, tremblers, and mocking birds (Aves: Mimidae). *The Auk* 118 (1): 35–55.

Hurlbut, Jr. C.S., ed. 1976. *The Planet We Live On: an Illustrated Encyclopedia of the Earth Sciences.* H.A. Abrams, New York.

Hutchinson, G.E. 1953. The concept of pattern in ecology. *Proceedings of the Academy of Natural Sciences of Philadelphia* 105: 1–12.

Hutchinson, G.E. 1959. Homage to Santa Rosalia, or why are there so many kinds of animals? *American Naturalist* 93: 145–59.

Hutchinson, G.E. 1972. *An Introduction to Population Ecology.* Yale University Press, New Haven, CT. Best book on the subject available, by the master.

Hutchinson, G.E. 1978. *An Introduction to Population Ecology.* Yale University Press, New Haven, CT.

Idso, S.B. 1984. Carbon dioxide and climate: is there a greenhouse in our future? *Quarterly Review of Biology* 59: 291–94.

Ingvarsson, P.K. 2001. Restoration of genetic variation lost—The genetic rescue hypothesis. *Trends in Ecology and Evolution* 16 (2): 62–63.

Irschick, D.J., L.J. Vitt, P.A. Zani, and J.B. Losos. 1997. A comparison of evolutionary radiations in mainland and Caribbean *Anolis* lizards. *Ecology* 78: 2191–2203.

Iturralde-Vinent, M.A., and R.D.E. MacPhee. 1999. Paleogeography of the Caribbean region: Implications for Cenozoic biogeography. *Bulletin of the American Museum of Natural History* 238: 1–95.

Jablonski, D., and J.J. Sepkowski. 1996. Paleobiology, community ecology, and scales of ecological pattern. *Ecology* 77 (5): 1367–78.

Jackson, G.C. 1986. Additional host plants of the carpenter bee, *Xylocopa brasilianorum* (L.) (Hymenoptera: Apoidea), in Puerto Rico. *Journal of Agriculture, University of Puerto Rico* 70: 255–65.

Jackson, G.C., and R.O. Woodbury. 1976. Host plants of the carpenter bee, *Xylocopa brasilianorum* (L.), in Puerto Rico. *Journal of Agriculture, University of Puerto Rico* 60: 639–60.

Jackson, J.A. 2000. Distribution, population changes and threats to least terns in the Caribbean and adjacent waters of the Atlantic and Gulf of Mexico. *Society of Caribbean Ornithology Special Publication* 1: 109–17.

Jacobson, M.K. 1968. The land Mollusca of St. Croix, Virgin Islands. *Sterkiana* 32: 18–28.

James, S.W. 1991. New species of earthworms from Puerto Rico, with a redefinition of the earthworm genus *Trigaster* (Oligochaeta: Megascolecidae). *Transactions of the American Microscopical Society* 110 (4): 337–53.

Jang, E.B., and R.E. Tullis. 1980. Hydromineral regulation in the saline water corixid *Trichocorixa reticulata* (Hemiptera: Corixidae). *Journal of Insect Physiology* 26: 241–44.

Jansma, P., G. Mattioli, A. Lopez, C. DeMets, T. Dixon, P. Mann, and E. Calais. 2000. Neotectonics of Puerto Rico and the Virgin Islands, northeastern Caribbean, from GPS geodesy. *Tectonics* 19 (6): 1021–37.

Jacquette, D. 1984. Seeking ways of entrance. *Sierra Club Bulletin* 69 (1): 139–43.

Jansen, F.J. 2005. Voice of the turtle. *Science* 307 (5707): 211.

Jarecki, L., and J. Lazell. 1987. Zur Grosse und Dichte einer Population von *Lepidodactylus lugubris* (Dumeril & Bibron, 1836) in Aiea, Hawaii (Sauria: Gekkonidae). *Salamandra* 23 (2–3): 176–78.

Jarecki, L. 1991. Hypersaline pond ecology in the British Virgin Islands. In *Proceedings of the Regional Symposium on Public and Private Cooperation in National Park Development*, 60–75. B.V.I. National Parks Trust, Road Town, Tortola.

Jarecki, L. 2003. Salt ponds of the British Virgin Islands: Investigations in an unexplored ecosystem. PhD dissertation, Durrell Institute of Conservation and Ecology, University of Kent, Canterbury, UK.

Jarvinen, O., and Y. Haila. 1984. Assembly of land bird communities on northern islands: a quantitative analysis of insular impoverishment. In *Ecological Communities: Conceptual Issues and the Evidence*, eds. D.R. Strong, Jr., D. Simberloff, L.G. Abele, and A.B. Thistle, 138–50. Princeton University Press, Princeton, NJ.

Jehl, Jr. J.R. 1984. Conservation problems of seabirds in Baja California and the Pacific Northwest. In *Status and Conservation of the World's Seabirds*, eds. J.P. Croxall, P.G.H. Evans, and R.W. Schreiber, 41–48. International Council for Bird Preservation Technical Publications 2. Paston Press, Norwich, UK.

Jenkins, C.F. 1923. *Tortola. A Quaker Experiment of Long Ago in the Tropics.* Friends' Bookshop, London.

Jiggins, C.D., and J. Mallet. 2000. Adaptive dynamics: In speciation. *Trends in Ecology and Evolution* 15 (6): 250–55.

Joglar, R.L. 1998. Los Coquís de Puerto Rico. Su Historia Natural y Conservación. Editorial De

La Universidad De Puerto Rico, San Juan, Puerto Rico.

Joglar, R.L., and P.A. Burrowes. 1996. Declining amphibian populations in Puerto Rico. In *Contributions to West Indian Herpetology: A Tribute to Albert Schwartz*, eds. R. Powell and R.W. Henderson, 371–80. Society for the Study of Amphibians and Reptiles, Ithaca, NY.

Johnson, A.C. 1978. The strange fish-eating bat of the Virgin Islands. *Virgin Islander* 4 (5): 3–6.

Johnson, K. 1992. Anegada birds in the pink. *The BVI Beacon* 8 (38): 1–14.

Jolly, G.M. 1965. Explicit estimates from capture–recapture data with both death and immigration—Stochastic model. *Biometrica* 52: 225–47.

Jones, D.L., D.G. Howell, P.J. Coney, and J.W.H. Monger. 1983. Recognition, character, and analysis of tectonostratigraphic terranes in western North America, In *Accretion Tectonics in the Circum-Pacific Region*, eds. M. Hashimoto and S. Uyeda, 21–35. Proceedings of the Oji International Seminar on Accretion Tectonics, Japan, 1981. Advances in Earth and Planetary Sciences. Terra Scientific Publishing Co, Tokyo, Japan.

Jones, T.H., M.S. Blum, R.W. Howard, C.A. McDaniel, H.M. Fales, M.B. DuBois, and J. Torres. 1982. Venom chemistry of ants in the genus *Monomorium*. *Journal of Chemical Ecology* 8: 285–300.

Juniper, T., and M. Parr. 1998. *Parrots*. Yale University Press, New Haven, CT.

Kaiser, J. 2001. Recreated wetlands no match for original. *Science* 293 (5527): 25.

Kaiser, J. 2003. Lean winters hinder birds' summertime breeding efforts. *Science* 301 (5636): 1033.

Kamil, A.C., and T.D. Sargent. 1981. *Foraging Behavior Ecological, Ethological, and Psychological Approaches*. Garland Publishing Co., New York, NY.

Karr, J.R., and K.E. Freemark. 1983. Habitat selection and environmental gradients: dynamics in the "stable" tropics. *Ecology* 64 (6): 1481–94.

Katz, M.J. 1985. On the wings of an angel. *Harvard Magazine* 88 (1): 25–32.

Kelt, D.A., and J.H. Brown. 1999. Community structure and assembly rules: Confronting conceptual and statistical issues with data on desert rodents. In *Ecological Assembly Rules*, eds. E. Weiher and P. Keddy, 71–78. Cambridge University Press, Cambridge, UK.

Kendeigh, S.C. 1961. *Animal Ecology*. Prentice-Hall, Englewood Cliffs, NJ.

Kennedy, D. 2000. New climate news. *Science* 290 (5494): 1091.

Kennedy, D. 2001. Going it alone. *Science* 293 (5533): 1221.

Kennedy, D. 2002. A tiger tale. *Science* 297 (5586): 1445.

Kenyon, K.W. 1977. Caribbean monk seal extinct. *Journal of Mammalogy* 58 (1): 97–98.

Kerr, R.A. 2001. It's official: humans are behind most of global warming. *Science* 294 (5549): 2105.

Kerr, R.A. 2001. A little sharper view of global warming. *Science* 294 (5543): 765.

Kerr, R.A. 2002. Climate change—Reducing uncertainties of global warming. *Science* 295 (5552): 29.

Kerr, R.A. 2003a. Megafauna died from big kill, not big chill. *Science* 300 (5621): 885.

Kerr, R.A. 2003b. Warming Indian Ocean wringing moisture from the Sahel. *Science* 302 (5643): 210–72.

Kierans, J.E. 1985. *Ambyloma antillorum* Kohls, 1969 (Acari: Ixodidae): Description of the immature stages from the rock iguana, *Iguana pinguis* (Sauria: Iguanidae) in the British Virgin Islands. *Proceedings of the Entomological Society of Washington* 87 (4): 821–25.

Kincaid, W.B., and E.H. Bryant. 1987. A geometric method for evaluating the null hypothesis of random habitat utilization. *Ecology* 64 (6): 1463–70.

Kinzig, A., S. Pacala, and D. Tilman. 2001 (2002). *The Functional Consequences of Biodiversity*. Princeton University Press, Princeton, NJ.

Kirby, T. 1986. Return of the monster of the Virgins. *BBC Wildlife Magazine* 4 (12): 622–23.

Kirchner, J.W. 2002. Evolutionary speed limits inferred from the fossil record. *Nature* 415 (6867): 65–68.

Klein, N., K. Burns, S. Hackett, and C. Griffiths. 2004. Molecular phylogenetic relationships among the wood warblers (Parulidae) and historical biogeography in the Caribbean basin. *Journal of Caribbean Ornithology* 17: 3–17.

Kohls, G.M. 1969. A new species of *Amblyomma* from iguanas in the Caribbean (Acarina: Ixodidae). *Journal of Medical Entomology* 6 (4): 439–42.

Kondoh, M. 2003. Foraging adaptation and the relationship between food-web complexity and stability. *Science* 299 (5611): 1388–91.

Koopman, K. F. 1975. Bats of the Virgin Islands in relation to those of the Greater and Lesser Antilles. *American Museum Novitates* 2581: 1–7.

Koopman, K. F. 1983. Two general problems involved in the systematics and biogeography of bats. In *Advances in Herpetology and Evolutionary Biology*, eds. A. Rhodin and K. Miyata, 412–15. Museum of Comparative Zoology, Cambridge, MA.

Kraus, F. 1991. Biodiversity conservation on Guana Island, British Virgin Islands. In *Proceedings of the Regional Symposium on Public and Private Cooperation in National Park Development*. 76–87. B.V.I. National Parks Trust, Road Town, Tortola.

Kraus, F. 2002. Ecology and conservation of *Sida eggersii* (Malvaceae), a rare tree of the Virgin Islands. *Caribbean Journal of Science* 38 (3–4): 184–94.

Krebs, C. J. 1978. *Ecology: The Experimental Analysis of Distribution and Abundance*. Harper & Row, New York.

Krebs, C. J. 1989. *Ecological Methodology*. Harper & Row, New York.

Krecek, J., N. Su, and R. Scheffrahn. 2000. Redescription of *Neotermes mona*, a dampwood termite (Isoptera, Kalotermitidae) from the central West Indies. *Florida Entomologist* 83 (3): 268–75.

Krutzsch, P. H., and E. G. Crichton. 1985. Observations on the reproductive cycle of the female *Molossus fortis* (Chiroptera: Molossidae) in Puerto Rico. *Journal of Zoology* 207: 137–50.

La Bastille, A., and M. Richmond. 1973. Birds and mammals of Anegada Island, B.V.I. *Caribbean Journal of Science* 13 (1–2) 91–109.

LaBerge, W. E. 1956. A revision of the bees of the genus *Melissodes* in North and Central America. Part I (Hymenoptera, Apidae). *University of Kansas Science Bulletin* 37: 911–1194.

LaBerge, W. E. 1962. Type specimens of American eucerine bees deposited in the British Museum. *Boletime da Universidade do Parana, Zoologia* 11: 1–12.

Lack, D. 1976. *Island Biology Illustrated by the Land Birds of Jamaica*. University of California Press, Berkeley, CA.

Lambeck, K., and J. Chappell. 2001. Sea level change through the last glacial cycle. *Science* 292 (5517): 679–86.

Lande, R. 1982. Rapid origin of sexual isolation and character divergence in a cline. *Evolution* 36 (2): 213–23. Excellent theoretical analysis of very rapid evolution in West Indian anole lizards.

Lande, R. 1996. Statistics and partitioning of species diversity, and similarity among multiple communities. *Oikos* 76: 5–13.

Landry, B., and C. Gielis. 1992. A synopsis of the Pterophoridae (Lepidoptera) of the Galapagos Islands, Ecuador. *Zoologische Verhandelingen* 276: 1–42.

Latta, S. C. 2002. The winter ecology of the Cape May warbler. *El Pitirre* 14 (3): 135–36.

Lawlor, L. R. 1980. Structure and stability in natural and randomly constructed competitive communities. *American Naturalist* 116: 394–408.

Lawlor, L. R., and J. M. Smith. 1976. The coevolution and stability of competing species. *American Naturalist* 110: 79–99.

Lawrence, J. F., and A. F. Newton, Jr. 1995. Families and subfamilies of Coleoptera (with selected genera, notes, references and data on family-group names). In *Biology, Phylogeny, and Classification of Coleoptera*, eds. J. Pakaluk and S. A. Slipinski, 779–1006. Muzeum I Instytut Zoologii PAN, Warszawa, Poland.

Lawton, J. H. 1981. Moose, wolves, *Daphnia*, and *Hydra*: On the ecological efficiency of endotherms and ectotherms. *American Naturalist* 117 (5): 782–83.

Lawton, J. H. 2000. *Community Ecology in a Changing World*. Ecology Institute, 21385 Oldendorf Luhe, Germany.

Lawton, J. 2001. Earth system science. *Science* 292 (5524): 1965.

Lawton, R., U. Nair, R. Pielke, and R. Welch. 2001. Climatic impact of tropical lowland deforestation on nearby montane cloud forests. *Science* 294 (5542): 584–87.

Lazarus, A. B., and F. P. Rowe. 1975. Freeze-marking rodents with a pressurized refrigerant. *Mammal Review* 5 (1): 31–34.

Lazell, J. 1961. A new species of *Sphaerodactylus* from Northern Haiti. *Breviora, Museum of Comparative Zoology*, 139: 1–5.

Lazell, J. 1962. Geographic differentiation in *Anolis oculatus* on Dominica. *Bulletin of the Museum of Comparative Zoology* 127: 466–75.

Lazell, J. 1964a. The reptiles of Sombrero, West Indies. *Copeia* 1964: 716–18.

Lazell, J. 1964b. The Lesser Antillean representatives of *Bothrops* and *Constrictor*. *Bulletin of the Museum of Comparative Zoology* 132: 245–73.

Lazell, J. 1964c. The anoles (Sauria, Iguanidae) of the Guadeloupeen Archipelago. *Bulletin of the Museum of Comparative Zoology* 131: 359–401.

Lazell, J. 1969. Nantucket herpetology. *Massachusetts Audubon* 54 (2): 32–34.

Lazell, J. 1972. The anoles (Sauria: Iguanidae) of the Lesser Antilles. *Bulletin of the Museum of Comparative Zoology* 143 (1): 1–115.

Lazell, J. 1973. The lizard genus *Iguana* in the Lesser Antilles. *Bulletin of the Museum of Comparative Zoology* 145 (1): 1–28.

Lazell, J. 1976. *This Broken Archipelago: Cape Cod and the Islands, Amphibians and Reptiles.* Quadrangle, New York Times Book Company, New York.

Lazell, J. 1979a. Deployment, dispersal, and adaptive strategies of land vertebrates on Atlantic and Gulf barrier islands. *Proceedings of the First Conference on Scientific Research in the National Parks* 1: 415–19.

Lazell, J. 1979b. Rules of thumb for field biologists. *Massachusetts Audubon Society Wildlife Management Report* 2: 1–4.

Lazell, J. 1980. *Report: British Virgin Islands, 1980.* The Conservation Agency, Jamestown, RI.

Lazell, J. 1981. Tropicbirds. *Virgin Islander* 6 (1): 23.

Lazell, J. 1983a. Biogeography of the herpetofauna of the British Virgin Islands, with description of a new anole (Sauria: Iguanidae). In *Advances in Herpetology and Evolutionary Biology*, eds. A. Rhodin and K. Myata, 99–117. Museum of Comparative Zoology, Harvard.

Lazell, J. 1983b. Rediscovery of the palm snail, *Hemitrochus nemoralinus intensus* Pilsbry (Cepolinae). *The Nautilus* 97 (3): 91–92.

Lazell, J. 1984. A population of devils *Sarcophilus harrisii* in northwestern Tasmania. *Records of the Queen Victoria Museum, Launceston,* 84: 53–56.

Lazell, J. 1987a. Flamingos, iguanas, and the restoration of rare species. *The Island Sun,* Tortola, 1307: 14, 22.

Lazell, J. 1987b. Beyond the Wallace Line. *Explorers Journal* 65 (2): 82–87.

Lazell, J. 1988. Soko Islands, South China Sea. *Explorers Journal* 66 (2): 80–85.

Lazell, J. 1989a. Guana Island: A natural history guide. *The Conservation Agency Occasional Paper* 1: 1–20.

Lazell, J. 1989b. *Wildlife of the Florida Keys.* Island Press, Washington, DC.

Lazell, J. 1991. The herpetofauna of Guana Island: Diversity, abundance, rarity, and conservation. *Publicacion Cientifica Miscellanea, Dept. Recursos Naturales de Puerto Rico,* 1: 28–33.

Lazell, J. 1993. Tortoise, *Geochelone* cf. *carbonaria,* from the Pleistocene of Anguilla, West Indies. *Journal of Herpetology* 27 (4): 485–86.

Lazell, J. 1994a. A new *Sphaerodactylus* (Sauria: Gekkonidae) from the Grenada Bank, Lesser Antilles. *Breviora* 496: 1–20.

Lazell, J. 1995. Natural Necker. *The Conservation Agency Occasional Paper* 2: 1–28.

Lazell, J. 1996a. Careening Island and Goat Islands: Evidence for the arid-insular invasion wave theory of dichopatric speciation in Jamaica. In *Contributions to West Indian Herpetology: A Tribute to Albert Schwartz,* eds. R. Powell and R. Henderson, 195–205. Society for the Study of Amphibians and Reptiles, Ithaca, NY.

Lazell, J. 1996b. Guana Island: A natural history guide. Third edition. *The Conservation Agency Occasional Paper* 1: 1–20.

Lazell, J. 1997a. Lizard letters. *Iguana Times* 6 (4): 89.

Lazell, J. 1997b. The stout iguana of the British Virgin Islands. *Iguana Times* 6 (4): 75–80.

Lazell, J. 1998a. Chiropterans, diversity and conservation. In *The Encyclopedia of Ecology and Environmental Management,* ed. P. Calow, 128–29. Blackwell Science, Oxford, UK.

Lazell, J. 1998b. Biophilia. Biophobia. In *The Encyclopedia of Ecology and Environmental Management,* ed. P. Calow, 91–92. Blackwell Science, Oxford, UK.

Lazell, J. 1999a. Giants, dwarfs, and rock-knockoffs: Evolution of diversity in Antillean anoles. *Anolis Newsletter* 5: 55–56.

Lazell, J. 1999b. The origins and evolution of the herpetofaunas of the islands on the continental shelf of South China. In *Tropical Island Herpetofauna,* ed. H. Ota, 79–96. Elsevier Science B.V., Amsterdam, Netherlands.

Lazell, J. 2000a. Necker Island update. *IUCN West Indian Iguana Specialist Group Newsletter* 3 (1): 2.

Lazell, J. 2000b. Translocation to unoccupied habitats. In *West Indian Iguanas: Status Survey and Conservation Action Plan,* ed. A. Alberts, 75–77. IUCN, Gland, Switzerland.

Lazell, J. 2000c. *Mastigoproctus transoceanicus* sp. nov. (Arachnida: Uropygida: Thelyphonidae), a genus new to the Old World, with discussion of the biogeography of the order. *Acta Zootaxonomica Sinica* 25 (3): 304–11.

Lazell, J. 2001. Restoration of the greater flamingo (*Phoenicopterus ruber*) to Anegada, British Virgin Islands. *El Pitirre* 14 (3): 113–14.

Lazell, J. 2002a. The herpetofauna of Shek Kwu Chau, South China Sea, with descriptions of two new colubrid snakes. *Memoirs of the Hong Kong Natural History Society* 25: 1–81.

Lazell, J. 2002b. Review: Biogeography of the West Indies. Patterns and Perspectives. *Copeia* 2002 (2): 552–54.

Lazell, J. 2002c. Restoring vertebrate animals in the British Virgin Islands. *Ecological Restoration* 20 (3): 179–85.

Lazell, J. 2003. Looking for lizards in all the right places. In *Islands and the Sea: Essays on Herpetological Exploration in the West Indies,* eds. R. W. Henderson and R. Powell, 213–31. Contributions to Herpetology 20, Society for the Study of Amphibians and Reptiles, Ithaca, NY.

Lazell, J. 2005. Roosevelt giant anole: Lost lizard of the Antilles. *Cryptozoology,* in press.

Lazell, J., and R. Conant. 1973. The Carolina salt marsh snake: a distinct form of *Natrix sipedon. Breviora, Museum of Comparative Zoology,* 400: 1–13.

Lazell, J., and L. Jarecki. 1985. Bats of Guana, British Virgin Islands. *American Museum Novitates* 2819: 1–7.

Lazell, J. and W. Lu. 1994. Calculations of catastrophe. *Sanctuary, Massachusetts Audubon Society,* 34 (3): 4.

Lazell, J., and W. Lu. 2000. Grayian distribution and the herpetofaunas of East Asia and eastern North America. In *Fourth Asian Herpetological Conference Abstracts,* 104. Chengdu Institute of Biology, Sichuan, China.

Lazell, J., and W. Lu. 2003. Grayian distributions: The China–America biogeographic connection. In *Studies on Biodiversity of the Guangdong Nanling National Nature Reserve,* ed. X. Pang, 65–88. Guangdong Science and Technology Press, Guangzhou, China.

Lazell, J., and N. C. Mitchell. 1998. *Anolis cristatellus wileyae* (Virgin Islands crested anole). Herbivory. *Herpetological Review* 29 (4): 237.

Lazell, J., and J. A. Musick. 1973. The kingsnake, *Lampropeltis getulus sticticeps,* and the ecology of the Outer Banks of North Carolina. *Copeia* 1973 (3): 497–503.

Lazell, J., and G. Perry. 1997. *Anolis cristatellus wileyae* (Virgin Islands crested anole). Frugivory. *Herpetological Review* 28 (3): 150.

Lazell, J., and N. C. Spitzer. 1977. Apparent play behavior in an American alligator. *Copeia* 1977 (1): 188.

Lazell, J., T. Sutterfield, and W. Giezentanner. 1984. The population of rock wallabies (genus *Petrogale*) on Oahu, Hawaii. *Biological Conservation* 30: 99–108.

Leal, M., and L. J. Fleishman. 2002. Evidence for habitat partitioning based on adaptation to environmental light in a pair of sympatric lizard species. *Proceedings of the Royal Society of London, Series B* 269: 351–59.

Leal, M., and J. A. Rodríguez-Robles. 1995. Antipredator responses of *Anolis cristatellus* (Sauria: Polychrotidae). *Copeia* 1995: 155–61.

Leal, M., and J. A. Rodríguez-Robles. 1997. Signaling displays during predator–prey interactions in a Puerto Rican anole, *Anolis cristatellus. Animal Behaviour* 54: 1147–54.

Leatherman, S. P. 1979. *Barrier Islands: from the Gulf of St. Lawrence to the Gulf of Mexico.* Academic Press, New York, NY.

LeVering, K., and G. Perry. 2003. *Cyclura pinguis* (Stout Iguana, Anegada Rock Iguana). Juvenile predation. *Herpetological Review* 34 (4): 367–68.

Levine, J. M. 2002. Species diversity and relative abundance in metapopulations. *Trends in Ecology and Evolution* 17 (2): 99–100.

Levins, R. 1969a. Thermal acclimation and heat resistance in *Drosophila* species. *American Naturalist* 103 (933): 483–99.

Levins, R. 1969b. Some demographic and genetic consequences of environmental heterogeneity for biological control. *Bulletin of the Entomological Society of America* 15: 237–40.

Lewisohn, F. 1986. *Tales of Tortola and the British Virgin Islands.* International Graphics, Hollywood, FL.

Lewis-Oritt, N., R. Van Den Bussche, and R. Baker. 2001. Molecular evidence for evolution of piscivory in *Noctilio* (Chiroptera: Noctilionidae). *Journal of Mammalogy* 82 (3): 748–59.

Lin, J. Y., and K. H. Lu. 1982. Population ecology of the lizard *Japalura swinhonis formosensis* (Sauria: Agamidae) in Taiwan. *Copeia* 1982 (2): 425–34. These lizards are very like *Anolis* in many ways; males and females perched at different heights not related to feeding. They imply the most aggressive males took the biggest tree trunks. Leaves one to wonder about the significance of perch sites in our *Anolis.*

Lindsey, K., B. Horwith, and E. Schreiber. 2000. Status of the magnificent frigatebird in the West Indies. *Society of Caribbean Ornithology Special Publication* 1: 59–64.

Lipton, P. 2005. Testing hypotheses: prediction and prejudice. *Science* 307 (5707): 219–21.

Lister, B. C. 1981. Seasonal niche relationships of rain forest anoles. *Ecology* 62: 1548–60.

Little, E., R. Woodbury, and F. Wadsworth. 1974. Trees of Puerto Rico and the Virgin Islands. Vol. 2. *Agriculture Handbook 449,* U.S. Forest Service, Washington, DC.

Little, E., R. Woodbury, and F. Wadsworth. 1976. Flora of Virgin Gorda (British Virgin Islands). *U.S. Department of Agriculture Forest Service Research Paper ITF-21:* 1–36. Institute of Tropical Forestry, Rio Piedras, Puerto Rico.

Little, E.L., Jr., and R.O. Woodbury. 1980. Rare and endemic trees of Puerto Rico and the Virgin Islands. *U.S. Department of Agriculture Conservation Research Report* 27: 1–26.

Lockwood, J.L., and S.L. Pimm. 1999. When does restoration succeed? In *Ecological Assembly Rules,* eds. E. Weiher and P. Keddy, 363–92. Cambridge University Press, UK.

Lohoefener, R., and J. Wolfe. 1984. A "new" live trap and a comparison with a pit-fall trap. *Herpetological Review* 15 (1): 25–26.

Lomolino, M.V., and R. Channell. 1995. Splendid isolation: Patterns of geographic range collapse in endangered mammals. *Journal of Mammalogy* 76 (2): 335–47.

Losos, J.B. 1990a. Ecomorphology, performance capability, and scaling of West Indian *Anolis* lizards: An evolutionary analysis. *Ecological Monographs* 60 (3): 369–88.

Losos, J.B. 1990b. A phylogenetic analysis of character divergence in Caribbean *Anolis* lizards. *Evolution* 44 (3): 558–69.

Losos, J.B. 1996a. Ecological and evolutionary determinants of the species-area relation in Caribbean anoline lizards. *Philosophical Transactions of the Royal Society of London, Series B* 351: 847–54.

Losos, J.B. 1996b. Phylogenetic perspectives on community ecology. *Ecology* 77 (5): 1344–54.

Losos, J.B. 2000. Ecological character displacement and the study of adaptation. *Proceedings of the National Academy of Sciences U.S.A.* 97 (11): 5693–95.

Losos, J.B., T.R. Jackman, A. Larson, K. de Queiroz, and L. Rodriguez-Schettino. 1998. Contingency and determinism in replicated adaptive radiations of island lizards. *Science* 279: 2115–18.

Losos, J., S. Naeem, and R. Colwell. 1989. Hutchinsonian ratios and statistical power. *Evolution* 43 (8): 1820–26.

Losos, J.B., and D. Schluter. 2000. Analysis of an evolutionary species-area relationship. *Nature* 408: 847–50.

Losos, J.B., and D.A. Spiller. 1999. Differential colonization success and asymmetrical interactions between two lizard species. *Ecology* 90 (1): 252–58.

Loucks, O.L. 1970. Evolution of diversity, efficiency, and community stability. *American Zoologist* 10: 17–25.

Lu, W., and M.A. Ivie. 1999. Tumbling flower beetles (Coleoptera: Mordellidae) of the Virgin Islands with descriptions of new species. *Annals of Entomological Society of America* 92 (5): 686–701.

Lu, W., J.A. Jackman, and P.A. Johnson. 1997. Male genitalia and phylogenetic relationships in North American Mordellidae (Coleoptera). *Annals of the Entomological Society of America* 90 (6): 742–67.

Lu, W., and J. Lazell. 1996. The voyage of the beetle. *Natural History, American Museum of Natural History,* 105 (1): 35–39.

Lundberg, P., E. Ranta, and V. Kaitala. 2000. Species loss leads to community closure. *Ecology Letters* 3: 465–68.

Lutterschmidt, W.I., and V.H. Hutchison. 1997. The critical thermal maximum: History and critique. *Canadian Journal of Zoology* 75: 1561–74.

Lydeard, C., R. Cowie, W. Ponder, A. Bogan, P. Bouchet, S. Clark, K. Cummings, T. Frest, O. Gargominy, D. Herbert, R. Hershler, K. Perez, B. Roth, M. Seddon, E. Strong, and F. Thompson. 2004. The global decline of nonmarine mollusks. *Bioscience* 54 (4): 321–30.

Lyons, K.G., and M.W. Schwartz. 2001. Rare species loss alters ecosystem function. *Ecology Letters* 4: 358–65.

Maa, T.C. 1971. An annotated bibliography of the batflies (Diptera: Streblidae, Nycteribiidae). *Pacific Insects Monograph* 28: 119–211.

MacArthur, R.H. 1955. Fluctuations of animal populations and a measure of community stability. *Ecology* 35: 533–36.

MacArthur, R.H. 1969. Patterns of communities in the tropics. *Biological Journal of the Linnaean Society* 1: 19–30.

MacArthur, R.H. 1972. *Geographical Ecology.* Harper & Row, New York.

MacArthur, R.H., and E.O. Wilson. 1967. *The Theory of Island Biogeography.* Princeton University Press, NJ.

MacFadden, B.J. 1980. Rafting mammals or drifting islands? Biogeography of the Greater Antillean insectivores *Nesophontes* and *Solenodon. Journal of Biogeography* 7 (1): 11–22.

MacFadden, B.J. 1981. Comments on Pregill's appraisal of historical biogeography of Caribbean vertebrates-vicariance, dispersal, or both. *Systematic Zoology* 30 (3): 370–72.

MacLean, W.P. 1982. *Reptiles and Amphibians of the Virgin Islands.* Macmillan, London, UK.

MacLean, W.P. 1985. Water-loss rates of *Sphaerodactylus parthenopion* (Reptila: Gekkonidae), the smallest amniote vertebrate. *Comparative Biochemistry and Physiology* 82A (4): 759–61.

MacLean, W.P. 1986. Seasonal variation in *Sphaerodactylus* water loss rates. In *Studies in Herpetology,* ed. Z. Rocek, 627–30. Societas Europaea Herpetologica, Prague.

MacLean, W.P., and R.D. Holt. 1979. Distributional patterns in St. Croix *Sphaerodactylus* lizards: The taxon cycle in action. *Biotropica* 11 (3): 189–95.

MacMahon, J.A., and K.D. Holl. 2002. Designer Communities. *Conservation Biology in Practice* 3 (1): 3–4.

Maes, V.O. 1983. Observations on the systematics and biology of a turrid gastropod assemblage in the British Virgin Islands. *Bulletin of Marine Science* 33 (2): 305–35.

Magurran, A.E. 1988. *Ecological Diversity and Its Measurement.* Princeton University Press, NJ.

Mahon, R. 1976. Effect of the cestode *Ligula intestinalis* on spottail shiners, *Notropis hudsonius. Canadian Journal of Zoology* 54 (12): 2227–29.

Malakoff, D. 2003. Researchers scramble to track virus's impact on wildlife. *Science* 299 (5610): 1176.

Maldonado Capriles, J. 1996. The status of insect alpha taxonomy in Puerto Rico after the scientific survey. *Annals of the New York Academy of Sciences* 776: 201–16.

Malfait, B.T., and M.G. Dinkelman. 1972. Circum-Caribbean tectonic and igneous activity and the evolution of the Caribbean plate. *Geological Society of America Bulletin* 83 (2): 251–71.

Malhotra, A., and R.S. Thorpe. 1991. Microgeographic variation in *Anolis oculatus* on the island of Dominica, West Indies. *Journal of Evolutionary Biology* 4: 321–35.

Malhotra, A., and R.S. Thorpe. 1997. Size and shape variation in a Lesser Antillean anole, *Anolis oculatus* (Sauria: Iguanidae) in relation to habitat. *Biological Journal of the Linnaean Society* 60: 53–72.

Malhotra, A., and R.S. Thorpe. 2000. The dynamics of natural selection and vicariance in the Dominican anole: Patterns of within-island molecular and morphological divergence. *Evolution* 54 (1): 245–58.

Malone, C.L., and R. Powell. 2002. Comments on a phylogeny of iguanid lizards. *Iguana Times* 9 (1–2): 9–11.

Malone, C., T. Wheeler, J. Taylor, and S. Davis. 2000. Phylogeography of the Caribbean rock iguana *(Cyclura):* Implications for conservation and insights on the biogeographic history of the West Indies. *Molecular Phylogenetics and Evolution* 17 (2): 269–79.

Mangel, M. 2002. The important role of theory in conservation biology. *Conservation Biology* 16 (3): 843–44.

Marcellini, D.L., and T.A. Jenssen. 1983. A character for differentiating the sympatric lizards, *Anolis cooki* and *Anolis cristatellus. Herpetological Review* 14 (4): 113–14.

Mares, M.A., M.R. Willig, and N.A. Bitar. 1980. Home range size in eastern chipmunks, *Tamias striatus,* as a function of number of captures: Statistical biases of inadequate sampling. *Journal of Mammology* 61 (4): 661–69. They found it took 25 recaptures per animal to begin to accurately predict a home range.

Margalef, D.R. 1958. Information theory in ecology. *General Systems* 3: 36–71.

Marko, P.B. 2002. Fossil calibration of molecular clocks and the divergence times of geminate species pairs separated by the Isthmus of Panama. *Molecular Biology and Evolution* 19 (11): 2005–21.

Marler, G., and L. Marler. 1984. *Sail Dive the British Virgin Islands.* Caribbean Printing Ltd., Road Town, Tortola, BVI.

Marsh, P.M. 1988. Revision of the tribe Odontobranconini in the western hemisphere (Hymenoptera: Braconidae: Doryctinae). *Systematic Entomology* 13: 443–64.

Marsh, P.M. 1993. Descriptions of new western hemisphere genera of the subfamily Doryctinae (Hymenoptera: Braconidae). *Contributions of the American Entomological Institute* 28 (1): 1–58.

Marshall, A.G. 1981. *The Ecology of Ectoparasitic Insects.* Academic Press, London, UK.

Martin, P.S. 1984. Prehistoric overkill: the global model. In *Quaternary Extinctions: a Prehistoric Revolution,* eds. P.S. Martin and R.G. Klein,

354–403. University of Arizona Press, Tucson, AZ.

Martin, P.S., and D.W. Steadman. 1999. Prehistoric extinctions on islands and continents. In *Extinctions in Near Time: Causes, Contexts, and Consequences,* ed. R.D.E. MacPhee, 17–55. Kluwer Academic/Plenum, New York.

Martin, P.S., and H.E. Wright, Jr., eds. 1967. *Prehistoric Extinctions: The Search for a Cause.* Yale University Press, New Haven, CT.

Matsuura, K. 2001. Nest mate recognition mediated by intestinal bacteria in a termite, *Reticulitermes speratus. Oikos* 92 (1): 20–26.

Matthew, W.D. 1939. Climate and evolution, with critical additions by the author and others and a bibliography of his scientific work. *Serial Publication of the New York Academy of Sciences* 1: 1–223.

Maurer, B.A. 1999. *Untangling Ecological Complexity.* University of Chicago Press, IL.

Maxson, L.R., and A.C. Wilson. 1975. Albumin evolution and organismal evolution in tree frogs (Hylidae). *Systematic Zoology* 24: 1–15.

Maxson, L.R., V.M. Sarich, and A.C. Wilson. 1975. Continental drift and the use of albumin as an evolutionary clock. *Nature* 255 (5507): 397–400.

May, R.M. 1973a. *Stability and Complexity in Model Ecosystems.* Princeton University Press, NJ.

May, R.M. 1973b. Stability in randomly fluctuating versus deterministic environments. *American Naturalist* 107: 621–50.

May, R.M. 1974. Stability and Complexity in Model Ecosystems. Second edition. *Monographs in Population Biology* 6. Princeton University Press, Princeton, NJ.

May, R.M. 1976. Simple mathematical models with very complicated dynamics. *Nature* 261: 459–67.

May, R.M. 1982. The role of theory in ecology. *American Zoologist* 22 (2): 903–10.

May, R.M. 1999. Crash tests for real. *Nature* 398: 371–72.

May, R.M. 2000. Relation between diversity and stability, in the real world. *Science* 290 (5492): 714–15.

Mayer, G.C., and R.M. Chipley. 1992. Turnover in the avifauna of Guana Island, British Virgin Islands. *Journal of Animal Ecology* 61: 561–66.

Mayer, G.C., and J. Lazell. 1988. Distributional records for reptiles and amphibians from the Puerto Rico Bank. *Herpetological Review* 19 (1): 23–24.

Mayer, G., and J. Lazell. 2000. A new species of *Mabuya* (Sauria: Scincidae) from the British Virgin Islands. *Proceedings of the Biological Society of Washington* 113 (4): 871–86.

Mayer, M. 1967. *A Boy, a Dog, and a Frog.* Dial Press, New York.

Mayhew, W.W. 1968. Biology of desert amphibians and reptiles. In *Desert Biology,* Vol. 1, ed. G.W. Brown, 196–356. Academic Press, New York.

Mayr, E. 1970. *Populations, Species and Evolution.* Belknap Press, Cambridge, MA.

McCune, B., and J.B. Grace. 2002. *Analysis of Ecological Communities.* MjM Software Design, Corvallis, OR.

McDowall, R.M. 2004. What biogeography is: a place for process. *Journal of Biogeography* 31 (3): 345–51.

McFarlane, D. 1989. Patterns of species co-occurrence in the Antillean bat fauna. *Mammalia* 53 (1): 59–66.

McFarlane, D. 1991. The species–genus relationship in Antillean bat communities. *Mammalia* 55 (3): 363–70.

McFarlane, D.A. 1999. Late Quaternary fossil mammals and last occurrence dates from caves at Barahona, Puerto Rico. *Caribbean Journal of Science* 35 (3–4): 238–48.

McFarlane, D., R. MacPhee, and D. Ford. 1998. Body size variability and a Sangamon extinction model for *Amblyrhiza,* a West Indian megafaunal rodent. *Quaternary Research* 50: 80–89.

McGuinness, K.A. 1984. The species-area relations of the communities on intertidal boulders: testing the null hypothesis. *Journal of Biogeography* 11: 439–56.

McKane, R., L. Johnson, G. Shaver, K. Nadelhoffer, E. Rastetter, B. Fry, A. Giblin, K. Kielland, B. Kwiatkowski, J. Laundre, and G. Murray. 2002. Resource-based niches provide a basis for plant species diversity and dominance in arctic tundra. *Nature* 415 (6867): 68–71.

McKinnon, J.S., and H.D. Rundle. 2002. Speciation in nature: The threespine stickleback models. *Trends in Ecology and Evolution* 17 (10): 480–88.

McLoughlin, J.C. 1979. *Archosauria.* Viking Press, New York.

McMahan, E.A. 1996. Termites. In *The Food Web of a Tropical Rain Forest,* eds. D.P. Reagan and R.B. Waide, 109–35. University of Chicago Press, IL.

McMichael, A.J. 1993. *Planetary Overload.* Cambridge University Press, UK.

McNab, B. K. 1980. Food habits, energetics, and the population biology of mammals. *American Naturalist* 116: 106–24.

McNab, B. K. 2001. Functional adaptations to island life in the West Indies. In *Biogeography of the West Indies,* 2nd ed., eds. C. A. Woods and F. E. Sergile, 55–62. CRC Press, Boca Raton, FL.

McNair, D. B., and C. D. Lombard. 2004. Population Estimates, Habitat Associations, and Management of *Ameiva polops* (Cope) at Green Cay, United States Virgin Islands. *Caribbean Journal of Science* 40 (3): 353–61.

McNair, D., F. Sibley, E. Massiah, and M. Frost. 2002. Ground-based nearctic-neotropical landbird migration during autumn in the eastern Caribbean. *Department of Life Sciences, University of the West Indies, Occasional Paper* 11: 86–103.

McNamara, R. S. 1977. How to defuse the population bomb. *Time* 110 (17): 93–94.

Medler, J. T. 1990. Types of Flatidae (Homoptera). XIV. Walker and Distant types in the British Museum. *Oriental Insects* 24: 127–95.

Mehard, C. 2001. Turtle snoop. *Nature Conservancy* 51 (5): 7.

Menke, A. S. 1986. A new *Pachodynerus* from Mayaguana Island, Bahamas, and a key to the West Indian species of the genus (Hymenoptera: Vespidae: Eumeninae). *Proceedings of the Entomological Society of Washington* 88 (4): 650–65.

Merbach, M., D. Merbach, V. Maschwitz, W. Booth, B. Fiala, and G. Zizka. 2002. Mass march of termites into the deadly trap. *Nature* 415 (6867): 36–37.

Meshaka, W. E. 2001. *The Cuban Treefrog in Florida.* University Press of Florida, Gainesville, FL.

Metzgar, L. H., and R. Hill 1971. The measurement of dispersion in small mammal populations. *Journal of Mammalogy* 52 (1): 12–20.

Meyerhoff, A., A. Boucot, D. Hull, and J. Dickins. 1996. Phanerozoic faunal and floral realms of the earth: The intercalary relations of the Malvinokaffric and Gondwana faunal realms with the Tethyan faunal realm. *Geological Society of America Memoir* 189: 1–69.

Meyerhoff, A. A., and C. W. Hatten. 1974. Bahamas salient of North America. In *Geology of Continental Margins,* eds. C. A. Burk and C. L. Drake, 429–46. Springer-Verlag, New York.

Meyerhoff, H. A. 1933. The geology of Puerto Rico. *University of Puerto Rico Monograph Series B* 1: 1–360.

Meylan, P. A. 1984. The northwestern limit of distribution of *Rhineura floridana* with comments on the dispersal of amphisbaenians. *Herpetological Review* 15 (1): 23–24.

Michael, S. F. 1997. Captive breeding of *Eleutherodactylus antillensis* from Puerto Rico (Anura Leptodactylidae) with notes on behavior in captivity. *Herpetological Review* 28: 141–43.

Miller, J. Y., and L. D. Miller. 2001. The Biogeography of the West Indian butterflies (Lepidoptera): An application of a vicariance/dispersalist model. In *Biogeography of the West Indies,* 2nd ed., eds. C. A. Woods and F. E. Sergile, 127–55. CRC Press, Boca Raton, FL.

Miller, O. K., D. J. Lodge, and T. J. Baroni. 2000. New and interesting ectomycorrhizal fungi from Puerto Rico, Mona, and Guana Islands. *Mycologia* 92: 558–70.

Miller, R. S. 1991. A revision of the Leptolycini (Coleoptera: Lycidae) with a discussion of paedomorphosis. PhD dissertation, Ohio State University, Columbus.

Miller, S. E. 1994. Dispersal of plant pests into the Virgin Islands. *Florida Entomologist* 77 (4): 520–21.

Milne, A. A. 1926. *Winnie the Pooh.* E. P. Dutton Co., New York.

Mirecki, D. N. 1977. *Report of the Cambridge Ornithological Expedition to the British Virgin Islands.* Bluebell, Cambridge, England, UK.

Miskimen, G. W., and R. M. Bond. 1970. The insect fauna of St. Croix, United States Virgin Islands. *New York Academy of Sciences Scientific Survey of Porto Rico and the Virgin Islands* 8 (1): 1–114.

Mitchell, C. 1966. *Isles of the Caribbees.* National Geographic Society, Washington, DC.

Mitchell, N. C. 1999. Effect of introduced ungulates on density, dietary preferences, home range, and physical condition of the iguana *(Cyclura pinguis)* on Anegada. *Herpetologica* 55 (1): 7–17.

Mitchell, N. C. 2000a. Anegada iguana. In *Endangered Animals: A Reference Guide to Conflicting Issues,* eds. R. P. Reading and B. Miller, 22–27. Greenwood Press, Westport, CT.

Mitchell, N. C. 2000b. Species accounts: *Cyclura pinguis.* In *West Indian Iguanas: Status Survey and Conservation Action Plan,* ed. A. Alberts, 60–62. IUCN, Gland, Switzerland.

Miyata, K. I. 1980. Patterns of diversity in tropical herpetofaunas. PhD dissertation, Harvard University, Cambridge, MA.

Monroe, B. L., Jr., and M. R. Browning. 1992. A reanalysis of *Butorides. Bulletin of the British Ornithologists' Club* 112 (2): 81–85.

Monti, L., M. Hunter, and J. Witham. 2000. An evaluation of the artificial cover object (ACO) method for monitoring populations of the redback salamander *Plethodon cinereus*. *Journal of Herpetology* 34 (4): 624–29.

Morgan Ernest, S.K., and J.H. Brown. 2001. Delayed compensation for missing keystone species by colonization. *Science* 292 (5514): 101–4.

Morgan, G.S. 2001. Patterns of extinction in West Indian bats. In *Biogeography of the West Indies*, 2nd ed., eds. C.A. Woods and F.E. Sergile, 369–407. CRC Press, Boca Raton, FL.

Morin, M., C. Atkinson, P. Banko, R. David, and M. Reynolds. 1999. Sightings of Ka-ho'olawe birds. *Elepaio* 58 (9): 55, 62–65.

Morris, B., J. Barnes, F. Brown, and J. Markham. 1977. The Bermuda Marine Environment. Special Publication 15, Bermuda Biological Station for Research.

Morrison, M., B. Marcot, and R. Mannan. 1992. *Wildlife–Habitat Relationships: Concepts and Applications*. University of Wisconsin Press, Madison.

Moul, E.T. 1969. The flora of Monomoy Island, Massachusetts. *Rhodora* 71 (785): 18–28.

Moure, J.S. 1960. Notes on the types of the neotropical bees described by Fabricius (Hymenoptera: Apoidea). *Studia Entomologica* 3: 97–160.

Muchmore, W.B. 1993. List of terrestrial invertebrates of St. John, U.S. Virgin Islands (exclusive of Acarina and Insecta), with some records of freshwater species. *Caribbean Journal of Science* 29 (1–2): 30–38.

Munroe, E.G. 1953. The size of island faunas. *Zoology, Proceedings of the Seventh Pacific Science Congress of the Pacific Science Association* 4: 52–53.

Murray, M. 2003. Overkill and sustainable use. *Science* 299 (5614): 1851–53.

Myers, N. 1976. An expanded approach to the problem of disappearing species. *Science* 193: 198.

Naeem, S. 1998. Species redundancy and ecosystem reliability. *Conservation Biology* 12: 39–45.

Naeem, S. 2002. Biodiversity equals instability? *Nature* 416: 23–24.

Naeem, S., and S. Li. 1997. Biodiversity enhances ecosystem reliability. *Nature* 390: 734–37.

Narins, P.M., and R.R. Capranica. 1976. Sexual differences in the auditory system of the tree frog *Eleutherodactylus coqui*. *Science* 192: 378–80.

Narins, P.M., and R.R. Capranica. 1978. Communicative significance of the two-note call of the treefrog *Eleutherodactylus coqui*. *Journal of Comparative Physiology* 127: 1–9.

Narins, P.M., and R.R. Capranica. 1980. Neural adaptations for processing the two-note call of the Puerto Rican treefrog *Eleutherodactylus coqui*. *Brain Behavior and Evolution* 17: 48–66.

Nellis, D., R. Norton, and W. MacLean. 1983. On the biogeography of the Virgin Islands boa, *Epicrates monensis granti*. *Journal of Herpetology* 17 (4): 413–17.

Neutel, A.-M., J. Heesterbeek, and P. Ruiter. 2002. Stability in real food webs: Weak links in long loops. *Science* 296 (5570): 1120–23.

Nichols, O.G., and D. Watkins. 1984. Bird utilization of rehabilitated bauxite minesites in Western Australia. *Biological Conservation* 30 (2): 109–131.

Norell, M.A., and K. de Quieroz. 1991. The earliest iguanine lizard (Reptilia; Squamata) and its bearing on iguanine phylogeny. *American Museum Novitates* 2997: 1–16.

Norrbom, A.L. 1998. A revision of the *Anastrepha daciformis* species group (Diptera: Tephritidae). *Proceedings of the Entomological Society of Washington* 100 (1): 160–92.

Norton, R., R. Chipley, and J. Lazell. 1989. A contribution to the ornithology of the British Virgin Islands. *Caribbean Journal of Science* 25 (3–4): 115–18.

Norton, R.L. 1993. *Alsophis portoricensis richardi* (ground snake). Feeding. *Herpetological Review* 24 (1): 34.

Norton, R.L. 1998. Field notes. West Indies region. *American Birds* 52 (1): 132–33.

Norton, R.L., and G.A. Seaman. 1985. Postfledging distribution of white-crowned pigeons banded in St. Croix, Virgin Islands. *Journal of Field Ornithology* 56 (4): 417–18.

Novotny, V., and Y. Basset. 2000. Rare species in communities of tropical insect herbivores: Pondering the mystery of singletons. *Oikos* 89 (3): 564–72.

Novotny, V., Y. Basset, S. Miller, G. Weiblen, B. Bremer, L. Cizek, and P. Drozd. 2002. Low host specificity of herbivorous insects in a tropical forest. *Nature* 416: 841–44.

Nowak, R.M. 1994. *Walker's Bats of the World*. Johns Hopkins University Press, Baltimore, MD.

Odum, E.P. 1971. *Fundamentals of Ecology*. W.B. Saunders, Philadelphia, PA.

Olmi, M. 1993. A new generic classification for Thaumatodryinae, Dryininae, and Gonatopodinae with descriptions of new species (Hymenoptera Dryinidae). *Bollettino de Zoologia Agraria e di Bachicoltura Series II* 25 (1): 57–89.

Olson, S. L. 1989. Extinction on islands: Man as a catastrophe. In *Conservation for the Twenty-First Century,* eds. D. Western and M. C. Pearl, 50–53. Oxford University Press, New York.

Olson, S. L., and H. F. James. 1982. Prodromus of fossil avifauna of the Hawaiian Islands. *Smithsonian Contributions to Zoology* 365: 1–59.

Osborne, J. 2002. Century plant devastation. *The BVI Beacon* 19 (22): 1, 24.

Otis, D. L., K. P. Burnham, G. C. White, and D. R. Anderson. 1978. Statistical inference from capture data on closed animal populations. *Wildlife Monographs* 62: 1–135.

Ottenwalder, J. 2000. Ricord's iguana *Cyclura ricordi.* In *West Indian Iguanas,* ed. A. Alberts, 51–55. IUCN, Gland, Switzerland and Cambridge, UK.

Ottenwalder, J. A. 2001. Systematics and biogeography of the West Indian genus *Solenodon.* In *Biogeography of the West Indies,* 2nd ed., eds. C. A. Woods and F. E. Sergile, 253–329. CRC Press, Boca Raton, FL.

Outlaw, D., G. Voelker, B. Milo, and D. Girman. 2003. Evolution of long-distance migration in and historical biogeography of *Catharus* thrushes: A molecular phylogenetic approach. *The Auk* 120 (2): 299–310.

Ovaska, K., and J. Caldbeck. 1997a. Courtship behavior and vocalizations of the frogs *Eleutherodactylus antillensis* and *E. cochranae* on the British Virgin Islands. *Journal of Herpetology* 31 (1): 149–55.

Ovaska, K., and J. Caldbeck. 1997b. Vocal behaviour of the frog *Eleutherodactylus antillensis* on the British Virgin Islands. *Animal Behaviour* 54: 181–88.

Ovaska, K., J. Caldbeck, and J. Lazell. 1998. *Eleutherodactylus schwartzi* (NCM). Reproduction. *Herpetological Review* 29 (2): 97.

Ovaska, K., J. Caldbeck, and J. Lazell. 2000. New records and distributional and ecological notes on leptodactylid frogs, *Leptodactylus* and *Eleutherodactylus,* from the British Virgin Islands. *Breviora* 508: 1–25.

Ovaska, K., and A. R. Estrada. 2003. *Eleutherodactylus antillensis* (coqui churi). Reproduction. *Herpetological Review* 34 (3): 229.

Overal, W. L. 1980. Host-relations of the batfly *Megistopoda aranea* (Diptera: Streblidae) in Panama. *University of Kansas Science Bulletin* 52 (1): 1–20.

Overton, W. S. 1971. Estimating the numbers of animals in wildlife populations. In *Wildlife Management Techniques,* 3rd ed., ed. R. H. Giles, 403–55. The Wildlife Society, Washington, DC.

Pacala, S., and J. Roughgarden. 1982. Resource partitioning and interspecific competition in two two-species insular *Anolis* lizard communities. *Science* 217 (4558): 444–46. Must be compared to Guana's three-species system.

Paine, R. T. 1966. Food web complexity and species diversity. *American Naturalist* 100 (910): 65–75.

Paine, R. T. 1988. Food webs: Road maps of interactions or grist for theoretical development? *Ecology* 69 (6): 1648–54.

Paine, R. T. 1996. Preface. In *Food Webs,* eds. G. A. Polis and K. O. Winemiller, ix–x. Chapman & Hall, New York.

Paine, R. T. 2002. Trophic control of production in a rocky intertidal community. *Science* 296 (5568): 736–39.

Palumbi, S. R. 2001. Humans as the world's greatest evolutionary force. *Science* 293 (5536): 1786–90.

Pape, T. 1989. Revision of *Opsidia* Coquillett (Diptera: Sarcophagidae). *Entomologica Scandinavica* 20: 229–41.

Pape, T. 1994. The world *Blaesoxipha* Lowe, 1861 (Diptera: Sarcophagidae). *Entomologica Scandinavica Supplement* 45: 1–247.

Patton, J. L., S. Y. Yang, and P. Meyers. 1975. Genetic and morphological divergence among introduced rat populations (*Rattus rattus*) of the Galápagos Archipelago, Ecuador. *Systematic Zoology* 24:296–310.

Patton, James L., and M. S. Hafner. 1983. Biosystems of the native rodents of the Galapagos archipelago, Ecuador. In *Patterns of Evolution in Galapagos Organisms,* eds. R. I. Brown, M. Berson, and A. E. Leviton, 539–68. American Association for the Advancement of Science, Pacific Division, San Francisco, CA.

Paulik, G. J., and D. S. Robson. 1969. Statistical calculations for change-in-ratio estimations of population parameters. *Journal of Wildlife Management* 33: 1–27.

Paulissen, M. A., and H. A. Meyer. 2000. The effect of toe-clipping on the gecko *Hemidactylus turcicus. Journal of Herpetology* 34 (2): 282–85.

Paull, D., E. E. Williams, and W. P. Hall. 1976. Lizard karyotypes from the Galapagos Islands: chromosomes in phylogeny and evolution. *Breviora* 441: 1–31.

Paulson, D. R. 2001. Old insects, new treatment. *Science* 293 (5537): 2005, 2007.

Paxinos, E., H. James, S. Olson, J. Ballou, J. Leonard, and R. Fleischer. 2002. Prehistoric decline of genetic diversity in the nene. *Science* 296 (5574): 1827.

Pennak, R. W. 1978. *Fresh-water Invertebrates of the United States,* 2nd ed. John Wiley & Sons, New York.

Perret, N., and P. Joly. 2002. Impacts of tattooing and PIT-tagging on survival and fecundity in the alpine newt *(Triturus alpestris). Herpetologica* 58 (1): 131–38.

Perry, G. 1996. The evolution of sexual dimorphism in the lizard *Anolis polylepis* (Iguania): evidence from intraspecific variation in foraging behavior and diet. *Canadian Journal of Zoology* 74: 1238–45.

Perry, G. 1999. The evolution of search modes: Ecological versus phylogenetic perspectives. *American Naturalist* 153: 98–109.

Perry, G., R. Dmïel, and J. Lazell. 1999. Evaporative water loss in insular populations of the *Anolis cristatellus* group (Reptilia: Sauria) in the British Virgin Islands. II. The effects of drought. *Biotropica* 31 (2): 337–42.

Perry, G., R. Dmïel, and J. Lazell. 2000. Evaporative water loss in insular populations of *Anolis cristatellus* (Reptilia: Sauria) in the British Virgin Islands. III. Response to the end of drought and a common garden experiment. *Biotropica* 32 (4a): 722–28.

Perry, G., and T. Garland, Jr. 2002. Lizard home ranges revisited: Effects of sex, body size, diet, habitat, and phylogeny. *Ecology* 87: 1870–85.

Perry, G., and J. Lazell. 1997a. *Anolis stratulus* (saddled anole). Nectivory. *Herpetological Review* 28 (3): 150–51.

Perry, G., and J. Lazell. 1997b. Review. Anolis lizards of the Caribbean: Ecology, evolution, and plate tectonics. *Copeia* 1997 (4): 906–11.

Perry, G., and J. Lazell. 2001. *Liophis portoricensis anegadae* (NCN). Night-light hunting. *Herpetological Review* 31 (4): 247.

Perry, G., K. LeVering, I. Girard, and T. Garland, Jr. 2004. Locomotor performance and social dominance in male Anolis cristatellus. *Animal Behaviour* 67: 37–47.

Perry, G., K. LeVering, and N. Mitchell. 2003. Cyclura pinguis (stout iguana, Anegada rock iguana). Juvenile behavior. *Herpetological Review,* 34 (4): 367.

Perry, G., and N. Mitchell. 2003. Guana and Necker island population assessments 2002. *Iguana, Journal of the International Iguana Society,* 10 (2): 49.

Perry, G., and E. R. Pianka. 1997. Foraging behavior: past, present, and future. *Trends in Ecology and Evolution* 12: 360–64.

Peters, R. H. 1988. Some general problems for ecology illustrated by food web theory. *Ecology* 69 (6): 1673–76.

Peters, R. H. 1991. *A Critique for Ecology.* Cambridge University Press, UK.

Petranka, J. W., and S. S. Murray. 2000. Effectiveness of removal sampling for determining salamander density and biomass: A case study in an Appalachian streamside community. *Journal of Herpetology* 35 (1): 36–44.

Philander, G. 2001. Why global warming is controversial. *Science* 294 (5549): 2105–06.

Philibosian, R., and J. A. Yntema. 1977. *Annotated List of the Birds, Mammals, Reptiles, and Amphibians of the Virgin Islands and Puerto Rico.* Information Services, Frederiksted, St. Croix, USVI.

Philips, T. K. 1998. A new genus and species of spider beetle from the Virgin Islands: *Lachnoniptus lindae* (Coleoptera: Anobiidae: Ptininae). *Florida Entomologist* 81: 112–17.

Philips, T. K., and M. Ivie. 1998. The Methiini of the West Indies (Coleoptera: Cerambycidae) with notes on the circum-Caribbean species. *Entomologica Scandinavica* 29: 57–87.

Phillips, J. C. 1922. *A Natural History of the Ducks,* Vol. 1. Houghton Mifflin, Boston.

Pianka, E. R. 1970a. Guild structure in desert lizards. *Oikos* 35: 194–201.

Pianka, E. R. 1970b. On r and K selection. *American Naturalist* 104: 592–97.

Pianka, E. R. 1973. The structure of lizard communities. *Annual Review of Ecology and Systematics* 4: 53–74.

Pianka, E. R. 1977. Reptilian species diversity. *Biology of the Reptilia* 7 (1): 1–34.

Pianka, E. R. 1981. Competition and niche theory. In *Theoretical Ecology,* ed. R. May, 167–96. Blackwell, Oxford, UK.

Pianka, E., R. Huey, and L. Lawlor. 1979. Niche segregation in desert lizards. In *Analysis of Ecological Systems,* eds. D. Horn, G. Stairs, and R. Mitchell, 67–115. Ohio State University Press, Columbus.

Pianka, E. R., and L. J. Vitt. 2003. *Lizards: Windows to the Evolution of Diversity.* University of California Press, Berkeley, CA.

Pickering, V. 1986. Arawaks' diet included local fresh-water fish. Island Sun, Road Town, Tortola, BVI, 2111: 1, 8.

Pickering, V. 1992. Flamingos restored to Anegada. *The Island Sun, Road Town, Tortola, BVI.* 1656: 1, 7.

Pielou, E. C. 1969. *An Introduction to Mathematical Ecology.* Wiley-Interscience, New York.

Pielou, E. C. 1975. *Ecological Diversity.* John Wiley & Sons, New York.

Pielou, E. C. 1977. *Mathematical Ecology,* 2nd ed. Harper & Row, New York.

Pielou, E. C. 1979. *Biogeography.* John Wiley & Sons, New York.

Pilsbry, H. A. 1928. Studies on West Indian mollusks: The genus *Zachrysia. Proceedings of the Academy of Natural Sciences of Philadelphia* 80: 581–606.

Pimentel, D., ed. 2002. *Biological Invasions.* CRC Press, Boca Raton, FL.

Pimm, S., M. Ayres, A. Balmford, G. Branch, K. Brandon, T. Brooks, R. Bustamante, R. Costanza, R. Cowling, L. Curran, A. Dobson, S. Farber, G. da Fonseca, C. Gascon, R. Kitching, J. McNeely, T. Lovejoy, R. Mittermeir, N. Myers, J. Patz, B. Raffle, D. Rapport, P. Raven, C. Roberts, J. Rodriguez, A. Rylands, C. Tucker, C. Safina, C. Samper, M. Stiassny, J. Supriatna, D. Wall, and D. Wilcove. 2001. Can we defy nature's end? *Science* 293 (5538): 2207–8.

Pimm, S. L. 1982. *Food Webs.* Chapman & Hall, London, UK.

Pimm, S. L. 1984. Food chains and return times. In *Ecological Communities: Conceptual Issues and the Evidence,* eds. D. R. Strong, Jr., D. Simberloff, L. G. Abele, and A. Thistle, 397–412. Princeton University Press, Princeton, NJ.

Pimm, S. L. 2001. *The World According to Pimm.* McGraw Hill, New York, NY.

Pimm, S. L., and R. L. Kitching. 1988. Food web patterns: Trivial flaws or the basis of an active research program? *Ecology* 69 (6): 1669–72.

Pinto, J. D. 1994. A taxonomic study of *Brachista* (Hymenoptera: Trichogrammatidae) with a description of two new species phoretic on robberflies of the genus *Efferia* (Diptera: Asilidae). *Proceedings of the Entomological Society of Washington* 96 (1): 120–32.

Plank, H. K., and M. R. Smith. 1940. A survey of the pineapple mealybug in Puerto Rico and preliminary studies of its control. *Journal of Agriculture, University of Puerto Rico* 24: 49–76.

Polis, G. A., and K. O. Winemiller. 1996. *Food Webs.* Chapman & Hall, London, UK.

Pollock, D. A. 1995. The Antillean *Physcius fasciatus* Pic (Coleoptera: Mycteridae: Lacconotinae). Redescription of the adult and description of the larva. *The Coleopterists Bulletin* 49 (4): 387–92.

Poulin, R., S. Morand, and A. Skorping (eds.). 2000. *Evolutionary Biology of Host-Parasite Relationships.* Elsevier Science, Amsterdam, Netherlands.

Power, D. M. 1972. Numbers of bird species on the California islands. *Evolution* 26 (3): 451–63.

Power, D. M. 1976. Avifauna richness on the California Channel Islands. *Condor* 78 (3): 394–98.

Pregill, G. K. 1981. An appraisal of the vicariance hypothesis of Caribbean biogeography and its application to West Indian vertebrates. *Systematic Zoology* 30 (2): 147–55.

Pregill, G. 1981. Late Pleistocene herpetofaunas from Puerto Rico. *University of Kansas Museum of Natural History Miscellaneous Publication* 71: 1–72.

Pregill, G., and S. L. Olson. 1981. Zoogeography of West Indian vertebrates in relation to Pleistocene climatic cycles. *Annual Review of Ecology and Systematics* 12: 75–98.

Price, P. W. 1980. Evolutionary Biology of Parasites. *Monographs in Population Biology* 15. Princeton University Press, Princeton, NJ.

Proctor, G. R. 1989. Ferns of Puerto Rico and the Virgin Islands. *New York Botanical Garden Memoir* 53: 1–389.

Prospero, J. M., and P. J. Lamb. 2003. African droughts and dust transport to the Caribbean: Climate change implications. *Science* 302 (5647): 1024–27.

Puente-Rolon, A. R. 2001. *Arrhyton exiguum* (Puerto Rican garden snake). Diet. *Herpetological Review* 32 (4): 261.

Pugesek, B., A. Tomer, and A. von Eye (eds.). 2003. *Structural Equations Modeling: Applications in Ecological and Evolutionary Biology Research.* Cambridge University Press, UK.

Quammen, D. 1988. *The Flight of the Iguana.* Delacorte Press, New York, NY.

Raffaele, H., J. Wiley, O. Garrido, A. Keith, and J. Raffaele. 1998. *A Guide to the Birds of the West Indies.* Princeton University Press, NJ.

Raffaelli, D. 2002. From Elton to mathematics and back again. *Science* 296 (5570): 1035–37.

Raleigh, R. F., J. B. McLaren, and D. R. Graff. 1973. Effects of topical location, branding techniques and changes in hue on recognition of cold brands in centrarachid and salmonid fish. *Transactions of the American Fisheries Society* 102 (3): 637–41.

Rand, A.S. 1964. Ecological distribution in anoline lizards of Puerto Rico. *Ecology* 45: 745–52.

Rankin, D.W. 2002. Geology of St. John, U.S. Virgin Islands. *U.S. Geological Survey Professional Paper* 1631: 1–36.

Raven, P. 2001. Why we must worry. *Science* 293 (5535): 1598.

Raxworthy, C., M. Forstner, and R. Nussbaum. 2002. Chameleon radiation by oceanic dispersal. *Nature* 415 (6873): 784–87.

Reagan, D.P. 1992. Congeneric species distribution and abundance in a three-dimensional habitat: The rain forest anoles of Puerto Rico. *Copeia* 1992: 392–403.

Reagan, D.P. 1996. Anoline lizards. In *The Food Web of a Tropical Rain Forest,* eds. D.P. Reagan and R.B. Waide, 321–45. University of Chicago Press, IL.

Reagan, D.P., G.R. Camilo, and R.B. Waide. 1996. The community food web: Major properties and patterns of organization. In *The Food Web of a Tropical Rain Forest,* eds. D.P. Reagan and R.B. Waide, 183–245. University of Chicago Press, IL.

Reagan, D.P., and R.B. Waide, eds. 1996. *The Food Web of a Tropical Rain Forest.* University of Chicago Press, IL.

Redman, C.L. 1999. *Human Impact on Ancient Environments.* University of Arizona Press, Tuscon.

Reid, H., and S. Tabor. 1920. The Virgin Islands earthquakes of 1867–1868. *Bulletin of the Seismological Society of America* 10: 9–30.

Renken, R., W. Ward, I. Gill, F. Gomez-Gomez, and J. Rodriquez-Martinez. 2002. Geology and hydrogeology of the Caribbean Islands aquifer system of the Commonwealth of Puerto Rico and the U.S. Virgin Islands. *U.S. Geological Survey Professional Paper* 1419: 1–139.

Riehl, H. 1954. *Tropical Meteorology.* McGraw-Hill, New York.

Rieppel, O. 2002. A case of dispersing chameleons. *Nature* 415 (6873): 744–45.

Richter, K.O. 1973. Freeze-branding for individually marking the banana slug *Ariolimax columbianus* G. *Northwest Science* 47 (2): 109–13.

Righter, E. (ed.). 2002. *The Tutu Archaeological Village Site: A Multidisciplinary Case Study in Human Adaptation.* Routledge Taylor & Francis Group, London, UK.

Riley, N.D. 1975. *A Field Guide to the Butterflies of the West Indies.* Collins, London, UK.

Ripley, S.D. 1977. *Rails of the World.* David R. Godine, Boston.

Roberts, R., T. Flannery, L. Ayliffe, H. Yoshida, J. Olley, G. Prideaux, and B. Smith. 2001. New ages for the last Australian megafauna: continent-wide extinction about 46,000 years ago. *Science* 292 (5523): 1888–92.

Robertson, J.G.M. 1984. A technique for individually marking frogs in behavioral studies. *Herpetological Review* 15 (2): 56–57.

Robertson, W.B. 1962. Observations on the birds of St. John, Virgin Islands. *Auk* 79: 44–76.

Robson, N. 1965. Taxonomic and nomenclature notes on Celastraceae. *Boletim da Sociedade Broteriana* 39 (2): 36–40.

Rodda, G., E. Campbell, and T. Fritts. 2001a. A high validity census technique for herpetofaunal assemblages. *Herpetological Review* 32 (1): 24–30.

Rodda, G., G. Perry, R. Rondeau, and J. Lazell. 2001b. The densest terrestrial vertebrate. *Journal of Tropical Ecology* 17: 331–38.

Rodriguez-Duran, A., and T.H. Kunz. 2001. Biogeography of West Indian bats: An ecological perspective. In *Biogeography of the West Indies,* 2nd ed., eds. C.A. Woods and F.E. Sergile, 355–68. CRC Press, Boca Raton, FL.

Rogozinski, J. 1992. *A Brief History of the Caribbean.* Facts on File, New York.

Roisin, Y., R. Scheffrhan, and J. Krecek. 1996. Generic revision of the smaller nasute termites of the Greater Antilles (Isoptera, Termitidae, Nasutitermitinae). *Annals of the Entomological Society of America* 37 (6) 776–87.

Romer, A.S. 1966. *Vertebrate Paleontology,* 3rd ed. University of Chicago Press, IL.

Root, R.B. 1967. The niche exploitation pattern of the blue-gray gnatcatcher. *Ecological Monographs* 37: 317–50.

Rosen, D.E. 1975. A vicariance model of Caribbean biogeography. *Systematic Zoology* 24: 431–64.

Rosen, D.E. 1985. Geological hierarchies and biogeographical congruence in the Caribbean. *Annals of the Missouri Botanical Garden* 72: 636–59.

Rosenzweig, M.L. 1995. *Species Diversity in Space and Time.* Cambridge University Press, UK.

Rosenzweig, M.L. 2003. *Win–Win Ecology: How the Earth's Species Can Survive in the Midst of Human Enterprise.* Oxford University Press, New York.

Roth, L.M. 1994. Cockroaches from Guana Island, British West Indies (Blattaria: Blatellidae: Blaberidae). *Psyche* 101 (1–2): 45–52.

Roughgarden, J. 1995. *Anolis Lizards of the Caribbean: Ecology, Evolution and Plate Tectonics.* Oxford University Press, Oxford, UK.

Roughgarden, J., D. Heckel, and E. Fuentes. 1983. Coevolutionary theory and the biogeography and community structure of Anolis. In *Lizard Ecology: Studies of a Model Organism,* eds. R. Huey, E. Pianka, and T. Schoener, 371–417. Harvard University Press, Cambridge, MA.

Roy, J.R.S. 1996. *British Virgin Islands National Parks Trust Bird List.* B.V.I. National Parks Trust, Road Town, Tortola.

Rubenstein, D., C. Chamberlain, R. Holmes, M. Ayres, J. Waldbauer, G. Graves, and N. Tuross. 2002. Linking breeding and wintering ranges of a migratory songbird using stable isotopes. *Science* 295 (5557): 1062–65.

Rubenstein, D.I. 1982. Review: Feeding as optimization. *Science* 217 (4562): 820–21.

Ruddiman, W.F., and A. McIntyre. 1981. The mode and mechanism of the last deglaciation: Oceanic evidence. *Quaternary Research* 16: 125–134.

Rummel, J.D., and J. Roughgarden. 1983. Some differences between invasion-structured and coevolution-structured competitive communities: A preliminary theoretical analysis. *Oikos* 41: 477–86.

Russo, C., N. Takezaki, and M. Nei. 1996. Efficiencies of different genes and tree-building methods in recovering a known vertebrate phylogeny. *Molecular Biology and Evolution* 13 (3): 525–36.

Ryvarden, L. 2000. Studies in neotropical polypores. 2: A preliminary key to neotropical species of *Ganoderma* with a laccate pileus. *Mycologia* 92: 180–91.

Sachs, J.D. 2004. Sustainable development. *Science* 304 (5671): 649.

Saliva, J.E. 2000. Conservation priorities for roseate terns in the West Indians. *Society of Caribbean Ornithology Special Publication* 1: 87–95.

Salvidio, S. 2001. Estimating terrestrial salamander abundance in different habitats: Efficiency of temporary removal methods. *Herpetological Review* 32 (1): 21–24.

Salt, G.W. 1983. Roles: their limits and responsibilities in ecological and evolutionary research. *American Naturalist* 122: 697–705.

Salt, G.W., ed. 1984. *Ecology and Evolutionary Biology: a Round Table on Research.* University of Chicago Press, Chicago, IL.

Sanders, H.L. 1968. Marine benthic diversity: a comparative study. *American Naturalist* 102: 243–82.

Sauer, J.D. 1969. Oceanic islands and biogeographic theory: a review. *Geographic Review* 59: 582–93.

Savage, J.M., and J. Villa. 1989. Introduction to the Herpetofauna of Costa Rica. Contributions to Herpetology 5, Society for the Study of Amphibians and Reptiles, Ithaca, NY.

Scarbrough, A.G. 1997. West Indian species of *Beameromyia* Martin (Diptera: Asilidae). *Insecta Mundi* 11: 237–46.

Schaus, W. 1940. Insects of Porto Rico and the Virgin Islands—Moths of the Families Geometridae and Pyralididae [sic]. *Scientific Survey of Puerto Rico and the Virgin Islands* 12 (3): 291–417.

Scheffrahn, R., J. Darlington, M. Collins, J. Krecek, and N. Su. 1994. Termites (Isoptera: Kalotermitidae, Rhinotermitidae, Termitidae) of the West Indies. *Sociobiology* 24 (2): 213–38.

Scheffrahn, R.H., and J. Krecek. 2001. New World revision of the termite genus *Procryptotermes* (Isoptera: Kalotermitidae). *Annals of the Entomological Society of America* 44 (4): 530–39.

Scheffrahn, R.H., and Y. Roisin. 1995. Antillean Nasutitermitinae (Isoptera: Termitidae): *Parvitermes collinsae,* a new subterranean termite from Hispaniola and redescription of *P. pallidiceps* and *P. wolcotti. Florida Entomologist* 78 (4): 585–600.

Scheller, U., and W.B. Muchmore. 1989. Pauropoda and Symphyla (Myriapoda) collected on St. John, U.S. Virgin Islands. *Caribbean Journal of Science* 25 (3–4): 164–95.

Scheller, W.G. 1995. Looking for lizards. *Islands* 15 (1): 154–58.

Schemnitz, D. 1975. Marine island-mainland movements of white-tailed deer. *Journal of Mammalogy* 56 (2): 535–37.

Schilthuizen, M. 2000. Ecotone: Speciation prone. *Trends in Ecology and Evolution* 15 (4): 130–31.

Schnabel, Z.E. 1938. The estimation of the total fish population of a lake. *American Mathematical Monthly* 45: 348–52.

Schoener, T. 1969. Size patterns in West Indian Anolis lizards: 1. Size and species diversity. *Systematic Zoology* 18: 386–401.

Schoener, T. 1970. Size patterns in West Indian Anolis lizards. II. Correlations with the sizes of particular sympatric species—displacement and convergence. *American Naturalist* 104 (936): 155–74.

Schoener, T. 1974. Resource partitioning in ecological communities. *Science* 185: 27–39.

Schoener, T., and A. Schoener. 1971a. Structural habitats of West Indian *Anolis* lizards I. Lowland Jamaica. *Breviora* 368: 1–53.

Schoener, T., and A. Schoener. 1971b. Structural habitats of West Indian *Anolis* lizards II. Puerto Rican uplands. *Breviora* 375: 1–39.

Schoener, T. W. 1983. Field experiments on interspecific competition. *American Naturalist* 122 (2): 240–85.

Schoener, T.W. 1984. Size differences among sympatric, bird-eating hawks: A worldwide survey. In *Ecological Communities: Conceptual Issues and the Evidence*, eds. D.R. Strong, Jr., D. Simberloff, L.G. Abele, and A. Thistle, 254–81. Princeton University Press, Princeton, NJ.

Schoener, T.W., and D.A. Spiller. 1996. Devastation of prey diversity by experimentally introduced predators in the field. *Nature* 381: 691–94.

Schomburgk, R.H. 1832. Remarks on Anegada. *Journal of the Royal Geological Society* 2: 152–70.

Schreiber, E.A. 2000. Status of red-footed, brown, and masked boobies in the West Indies. *Society of Caribbean Ornithology Special Publication* 1: 46–57.

Schreiber, E.A., and J. Burger (eds.). 2002. *Biology of Marine Birds*. CRC Press, Boca Raton, FL.

Schreiber, E.A., and D.S. Lee (eds.). 2000. *Status and Conservation of West Indian Seabirds. Special Publication Number 1*, Society of Caribbean Ornithology, Ruston, LA.

Schreiber, R., E. Schreiber, D. Anderson, and D. Bradley. 1989. Plumages and molts of the brown pelican. *Natural History Museum of Los Angeles County Contributions in Science* 402: 1–43.

Schuchert, C. 1935. *Historical Geology of the Antillean–Caribbean Region*. Braunworth, New York.

Schultz, T.R. 2003. Hyperdiversity up close. *Science* 300 (5616): 57–58.

Schumacher, F.X., and R.W. Eschmeyer. 1943. The estimation of fish populations in lakes and ponds. *Journal of the Tennessee Academy of Sciences* 18: 228–49.

Schwartz, A. 1966. Snakes of the genus *Alsophis* in Puerto Rico and the Virgin Islands. *Studies on the Fauna of Curacao and Other Caribbean Islands* 23: 177–227.

Schwartz, A. 1967. A review of the genus *Dromicus* in Puerto Rico and the Virgin Islands. *Stahlia* 9: 1–18.

Schwartz, A., and R.W. Henderson. 1991. Amphibians and Land Reptiles of the West Indies. University of Florida Press, Gainesville, FL.

Schwartz, A., and R. Thomas. 1975. A Checklist of West Indian Amphibians and Reptiles. Special Publication of the Carnegie Museum of Natural History, Pittsburg, PA.

Schwarz, E. and H.K. Schwarz. 1943. The wild and commensal stocks of the house mouse, *Mus musculus* Linnaeus. *Journal of Mammalogy* 24: 59.

Schwenk, K., and G.C. Mayer. 1991. Tongue display in anoles and its evolutionary basis. In *Anolis* Newsletter IV, eds. J.B. Losos and G.C. Mayer, 131–40. National Museum of Natural History, Washington, DC.

Scotese, C.R. 1997. Paleogeographic Atlas. PALEOMAP Progress Report 90-0497. University of Texas, Arlington.

Seaman, G.A. 1966. Foods of the bridled quail dove *(Geotrygon mystacea)* in the American Virgin Islands. *Caribbean Journal of Science* 6 (3–4): 177–79.

Seber, G.A.F. 1973. *The Estimation of Animal Abundance and Related Parameters*. Griffin, London, UK.

Seymour, J. and H. Girardet. 1986. Far from Paradise. *The Story of Man's Impact on the Environment*. British Broadcasting Corporation, London, UK.

Shackleton, N. 2001. Climate change across the hemispheres. *Science* 291 (5501): 58–59.

Shafir, S., and J. Roughgarden. 1998. Testing predictions of foraging theory for a sit-and-wait forager, *Anolis gingivinus. Behavioral Ecology* 9: 74–84.

Shannon, C.E., and W. Weaver. 1949. *The Mathematical Theory of Communication*. University of Illinois Press, Urbana.

Shaw, J. 2002. The great global experiment. *Harvard Magazine* 105 (2): 34–43, 87–90.

Short, R.V. 1976. The introduction of new species of animals for the purpose of domestication. *Symposia of the Zoological Society of London* 40: 321–33.

Sibley, F.C. 1999. Unusual invasion of dragonflies on Guana Island, British Virgin Islands. *Argia* 11 (1): 16–19.

Sibley, F.C. 2000. Additional comments on the dragonflies of the British Virgin Islands. *Argia* 12 (1): 18–19.

Sibley, F.C. 2002. Miscellaneous notes on British Virgin Island dragonflies. *Argia* 14 (1): 5–7.

Simberloff, D. 1978. Use of rarefaction and re-
lated methods in ecology. In *Biological Data in
Water Pollution Assessment: Quantitative Statis-
tical Analyses,* eds. J. Cairns, K. L. Dickson, and
R. J. Livingston, 150–65. American Society for
Testing and Materials, West Conshohocken,
PA.

Simberloff, D. 1982a. The status of competition
theory in ecology. *Annales Zoologici Fennici* 19:
241–53.

Simberloff, D. 1982b. Big advantages of small
refuges. *Natural History* 91 (4): 6–14.

Simberloff, D., and W. Boecklen. 1981. Santa Rosalia reconsidered: Size ratios and competition.
Evolution 35 (6): 1206–28.

Simberloff, D., and E. F. Connor. 1981. Missing
species combinations. *American Naturalist* 118
(2): 215–39.

Simberloff, D., and T. Dayan. 1991. The guild concept and the structure of ecological communities. *Annual Review of Ecology and Systematics*
22: 115–43.

Simberloff, D., T. Dayan, C. Jones, and G. Ogura.
2000. Character displacement and release in
the small Indian mongoose, *Herpestes javanicus. Ecology* 81 (8): 2086–99.

Simberloff, D., K. Heck, E. McCoy, and E. Conner.
1981. There have been no statistical tests of
cladistic biogeographic hypotheses. In *Vicariance Biogeography: A Critique,* eds. G. Nelson
and D. Rosen, 40–63. Columbia University
Press, New York.

Simpson, G. G. 1940. Mammals and land bridges.
Journal of the Washington Academy of Science 30:
137–63.

Simpson, G. G. 1943. Mammals and the nature of
continents. *American Journal of Science* 241: 1–31.

Simpson, G. G. 1944. *Tempo and Mode in Evolution.* Columbia University Press, New York,
NY.

Simpson, G. G. 1961. *Principles of Animal Taxonomy.* Columbia University Press, New York.

Simpson, G. G. 1964. Species density of North
American recent mammals. *Evolution* 15:
413–46.

Simpson, G. G. 1965 *The Geography of Evolution.*
Chilton Books, New York.

Sites, J., S. Davis, T. Guerra, J. Iverson, and H.
Snell. 1996. Character congruence and phylo-
genetic signal in molecular and morphological
data sets: A case study in the living iguanas
(Squamata: Iguanidae). *Molecular Biology and
Evolution* 13: 1087–1105.

Slaughter, R., and J. Skulan. 2001. Did human
hunting cause mass extinction? *Science* 294
(5546): 1460–61.

Slipinski, S. A. 1989. A review of the Passandridae
(Coleoptera, Cucujoidea) of the world. II.
genus *Catogenus* Westwood. *Polskie Pismo Ento-
mologiczne* 59: 85–129.

Smith, A., D. Smith, and B. Funnell. 1994a. *An Atlas of Mesozoic and Cenozoic Coastlines.* Cambridge University Press, Cambridge, UK.

Smith, D., L. Miller, and F. Mackenzie. 1991a. The
butterflies of Anegada, British Virgin Islands,
with descriptions of a new *Calisto* (Satyridae)
and a new *Copaeodes.* (Hesperiidae) endemic to
the island. *Bulletin of the Allyn Museum* 133:
1–25.

Smith, D., L. Miller, and J. Miller. 1994b. *The Butterflies of the West Indies and South Florida.* Oxford University Press, Oxford, UK.

Smith, D. K., and J. W. Petranka. 2000. Monitoring
terrestrial salamanders: Repeatability and validity of area-constrained cover object searches.
Journal of Herpetology 34 (4): 547–57.

Smith, O. L. 1980. The influence of environmental
gradients on ecosystem stability. *American Naturalist* 116: 1–24.

Snead R. E. 1982. *Coastal Landforms and Surface Features: a Photographic Atlas and Glossary.* Hutchinson Ross Publishing Co., Stroudsburg, PA.

Snelling, R. R. 1973. The ant genus *Conomyrma* in
the U.S. (Hymenoptera: Formicidae). *Contributions in Science, Natural History Museum of Los
Angeles County* 238: 1–6.

Snelling, R. R. 1992a. Collecting on Guana Island,
British Virgin Islands, and Puerto Rico. *Melissa*
6: 5–6.

Snelling, R. R. 1992b. Guana and Mona Islands.
Sphecos 23: 12–14.

Snelling, R. R. 1993a. Ants of Guana Island, British
Virgin Islands. *Notes from Underground* 8: 2.

Snelling, R. R. 1993b. Back to the Virgin Islands.
Sphecos 26: 12–13.

Snelling, R. R. 1995a. Systematics of Nearctic ants
of the genus *Dorymyrmex. Contributions in Science, Natural History Museum of Los Angeles
County* 454: 1–14.

Snelling, R. R. 1995b. A new spider wasp of the
genus *Psorthaspis* from the Greater Antilles
(Hymenoptera: Pompilidae, Pompilinae). *Acta
Scientifica* 6 (1–3): 103–8.

Snelling, R. R. 1996. Systematic notes on some
Bethylidae (Hymenoptera: Chrysidoidea) from
the Virgin Islands and Puerto Rico. *Memoirs of*

the Entomological Society of Washington 17: 194–208.

Snelling, R. R. 2001. Two new species of thief ants *(Solenopsis)* from Puerto Rico (Hymenoptera: Formicidae). *Sociobiology* 37: 511–25.

Snelling, R. R., and J. T. Longino. 1992. Chapter 30. Revisionary notes on the *rimosus* group of the ant genus *Cyphomyrmex* (Hymenoptera: Formicidae). In *Insects of Panama and Mesoamerica: Selected Studies,* eds. D. Quintero Arias and A. Aiello, 479–94. Oxford University Press, UK.

Snelling, R. R., and J. A. Torres. 1998. *Camponotus ustus* Forel and two similar new species from Puerto Rico (Hymenoptera: Formicidae). *Contributions in Science, Natural History Museum of Los Angeles County,* 469: 1–10.

Snyder, G. K. 1975. Respiratory metabolism and evaporative water loss in a small tropical lizard. *Journal of Comparative Physiology* 104: 13–18.

Sorenson, L. G., and P. Bradley. 1998. Update on the West Indian whistling-duck (WIWD) and wetlands conservation project—Report from the WIWD working group. *El Pitirre* 11 (3): 126–31.

Sorenson, L. G., and P. Bradley. 2000. Working group report—update on the "West Indian whistling-duck (WIWD) and wetlands conservation project." *El Pitirre* 13 (2): 57–63.

Sperling, F., J. Spence, and N. Andersen. 1997. Mitochondrial DNA, allozymes, morphology, and hybrid compatibility in *Limnoporus* water striders (Heteroptera: Gerridae): Do they all track species phylogenies? *Annals of the Entomological Society of America* 90 (4): 401–15.

Spiller, D. A., and T. W. Schoener. 1998. Lizards reduce spider species richness by excluding rare species. *Ecology* 79: 503–16.

Spitzer, N. C. 1973. Rice rat's world. *Man and Nature* 1973: 24–26.

Spitzer, N. C. 1977. The Cape Cod mammal survey: summer, 1976. *Cape Naturalist* 6 (1): 10–16.

Spitzer, N. C. 1983. Aspects of the biology of the silver rice rat, *Oryzomys argentatus*. M.S. Thesis, University of Rhode Island, Kingston, RI.

Spotila, J. R. 2004. *Sea Turtles: A Complete Guide to Their Biology, Behavior, and Conservation*. Johns Hopkins University Press, Baltimore, MD.

Staflew, F. A. 1976–88. *Taxonomic Literature: a Selective Guide to Botanical Publications and Collections with Dates, Commentaries and Types.* Second edition. Bohn, Scheltema, & Holkema, Utrecht, Netherlands.

Stallard, R. F. 2000. Possible environmental factors underlying amphibian decline in eastern Puerto Rico: Analysis of U.S. government data archives. *Conservation Biology* 15: 943–53.

Stamps, J. A., and V. V. Krishnan. 1998. Territory acquisition in lizards. IV. Obtaining high status and exclusive home ranges. *Animal Behaviour* 55: 461–72.

Stamps, J. A., V. V. Krishnan, and R. M. Andrews. 1994. Analyses of sexual size dimorphism using null growth-based models. *Copeia* 1994: 598–613.

Stamps, J. A., J. B. Losos, and R. M. Andrews. 1997. A comparative study of population density and sexual size dimorphism in lizards. *American Naturalist* 149: 64–90.

Steadman, D., G. Pregill, and S. Olson. 1984. Fossil vertebrates from Antigua, Lesser Antilles: Evidence for late Holocene human-caused extinctions in the West Indies. *Proceedings of the National Academy of Sciences* 81: 4448–51.

Stearns, H. T. 1972. Hydrology, volcanic terrain. *Encyclopedia of Earth Sciences* Reinhold, New York, NY, 4A: 555–57.

Stearns, S. C. 1982. Components of fitness. *Science* 218: 463–64. A review of a symposium on the evolution and genetics of life histories. Fine overview; always read the reviews before you tackle the symposia.

Stebbins, R. C., and N. W. Cohen. 1995. *A Natural History of Amphibians*. Princeton University Press, NJ.

Stern, M. 2001. Parks and factors in their success. *Science* 293 (5532): 1045.

Stewart, M. M. 1995. Climate driven population fluctuations in rain forest frogs. *Journal of Herpetology* 29: 369–78.

Stewart, M. M., and F. H. Pough. 1983. Population density of tropical forest frogs: Relation to retreat sites. *Science* 221: 570–72.

Stewart, M. M., and L. L. Woolbright. 1996. Amphibians. In *The Food Web of a Tropical Rain Forest,* eds. D. P. Reagan and R. B. Waide, 273–320. University of Chicago Press, IL.

Stockwell, C., A. Hendry, and M. Kinnison. 2003. Contemporary evolution meets conservation biology. *Trends in Ecology and Evolution* 18 (2): 94–101.

Stokstad, E. 2004. Songbirds check compass against sunset to stay on course. *Science* 304 (5669): 373.

Stott P., S. Tett, G. Jones, M. Allen, J. Mitchell, and G. Jenkins. 2000. External control of 20th century temperature by natural and anthropogenic forcings. *Science* 290 (5499): 2133–37.

Stoudemire, S.A. 1959. *Natural History of the West Indies.* University of North Carolina Press, Chapel Hill.

Strong, D., J. Lawton, and T. Southwood. 1984. *Insects on Plants. Community Patterns and Mechanisms.* Harvard University Press, Cambridge, MA.

Strong, D.R. 1988. Food web theory: A ladder for picking strawberries? *Ecology* 69 (6): 1647.

Su, N., Z. Hillis-Starr, P. Ban, and R. Scheffrahn. 2003. Protecting historic properties from subterranean termites: A case study with Fort Christiansvaern, Christiansted National Historic Site, United States Virgin Islands. *American Entomologist* 49 (1): 20–32.

Sugihara, G. 1980. $S = CA^z$, $z \cong 1/4$: A reply to Connor and McCoy. *American Naturalist* 117 (5): 790–93.

Sunquist, M.E., and G.G. Montgomery. 1973. Activity pattern of a translocated silky anteater *(Cyclopes didactylus). Journal of Mammalogy* 54 (3): 782.

Swanepoel, P., and H.H. Genoways. 1978. Revision of the Antillean bats of the genus *Brachyphylla* (Mammalia: Phyllostomatidae). *Bulletin of the Carnegie Museum of Natural History* 12: 1–53.

Swanepoel, P., and H.H. Genoways. 1983. *Brachyphylla cavernarum. Mammalian Species* 205: 1–6.

Swart, R., P. Raskin, and J. Robinson. 2002. Critical challenges for sustainability science. *Science* 297 (5524): 1994.

Sykes, L.R., W.R. McCann, and A.L. Kafka. 1982. Motion of Caribbean plate during last 7 million years and implications for earlier Cenozoic movements. *Journal of Geophysical Research* 87 (B13): 10656–76.

Szathmary, E., F. Jordan, and C. Pal. 2001. Can genes explain biological complexity? *Science* 292 (5520): 1315–16.

Tao, R.L., and F.A. Lewis. 2001. The broad reach of helminthology. *Science* 294 (5550): 2292.

Taylor, B. 1998. *A Guide to Rails, Crakes, Gallinules, and Coots of the World.* Yale University Press, New Haven, CT.

Terborgh, J.W., and J. Faaborg. 1980. Saturation of bird communities in the West Indies. *American Naturalist* 116 (2): 178–95.

Thomas, K.R., and R. Thomas. 1978. Locomotor activity responses to photoperiod in four West Indian fossorial squamates of the genera *Am-*

phisbaena and *Typhlops* (Reptilia). *Journal of Herpetology* 12 (1): 35–41.

Thomas, R. 1966. Additional notes on the amphisbaenids of Greater Puerto Rico. *Breviora* 249: 1–23.

Thomas, R. 1999. The Puerto Rico area. In *Caribbean Amphibians and Reptiles,* ed. B. I. Crother, 169–79. Academic Press, San Diego, CA.

Thomas, R., and J.A. Prieto Hernandez. 1984. The use of venom by the Puerto Rican snake, *Alsophis portoricensis.* Decimo Simposio de Recursos Naturales, 1983, 13–22a. Departamento de Recursos Naturales, San Juan, Puerto Rico.

Thompson, J. N. 1994. *The Coevolutionary Process.* University of Chicago Press, Chicago, IL.

Thorne, B., M. Collins, and K. Bjorndal. 1996a. Architecture and nutrient analysis of arboreal carton nests of two neotropical *Nasutitermes* species (Isoptera: Termitidae), with notes on embedded nodules. *Florida Entomologist* 79 (1): 27–37.

Thorne, B., M. Haverty, and D. Benzing. 1996b. Associations between termites and bromeliads in two dry tropical habitats. *Biotropica* 28 (4b): 781–85.

Thorne, B., M. Haverty, and M. Collins. 1994. Taxonomy and biogeography of *Nasutitermes acajutlae* and *N. nigriceps* (Isoptera: Termitidae) in the Caribbean and Central America. *Annals of the Entomological Society of America* 87 (6): 762–70.

Thorne, B., M. Haverty, and M. Collins. 1996c. Antillean termite named for a locality in Central America: Taxonomic memorial to a perpetuated error. *Annals of the Entomological Society of America* 89 (3): 346–47.

Thorne, B.L., and M.I. Haverty. 2000. Nest growth and survivorship in three species of neotropical *Nasutitermes* (Isoptera: Termitidae). *Environmental Entomology* 29 (2): 256–64.

Thorne, B.L., and J.F.A. Traniello. 2003. Comparative social biology of basal taxa of ants and termites. *Annual Review of Entomology* 48: 283–306.

Thorpe, R.S. 1987. Geographic variation: A synthesis of cause, data, pattern and congruence in relation to subspecies, multivariate analysis and phylogenesis. *Bollettino Zoologia* 54: 3–11.

Thorpe, R.S., and A. Malhotra. 1996. Molecular and morphological evolution within islands. *Philosophical Transactions of the Royal Society of London, Series B. Biological Sciences* 351: 815–22.

Tilman, D. 1980. Resources: A graphical-mechanistic approach to competition and predation. *American Naturalist* 116: 362–93.

Tinkle, D.W. 1976. Comparative data on the population ecology of the desert spiny lizard, *Sceloporus magister*. *Herpetologica* 32 (1): 1–6.

Tinkle, D., H. Wilbur, and S. Tilley. 1970. Evolutionary strategies in lizard reproduction. *Evolution* 24 (1): 55–74. A fundamental paper.

Tolson, P.J., and R.W. Henderson. 1993. *The Natural History of West Indian Boas*. R & A Publishing Limited, Taunton, Somerset, England, UK.

Tomich, P.Q. 1986. *Mammals in Hawaii*. Bishop Museum Press, Honolulu, HI.

Tordoff, H.B., and P.T. Redig. 2001. Role of genetic background in the success of reintroduced peregrine falcons. *Conservation Biology* 15 (2): 528–32.

Torres, J.A. 1981. The organization of ant assemblages in Puerto Rico: factors influencing the diversity and coexistence of species. PhD thesis, University of California, Berkeley, CA.

Torres, J.A., R.R. Snelling, and T.H. Jones. 2000. Distribution, ecology and behavior of *Anochetus kempfi* (Hymenoptera: Formicidae) and description of the sexual forms. *Sociobiology* 36: 505–15.

Townsend, D.S. 1996. Patterns of parental care in frogs of the genus *Eleutherodactylus*. In *Contributions to West Indian Herpetology: A Tribute to Albert Schwartz*, eds. R. Powell and R.W. Henderson, 229–39. Society for the Study of Amphibians and Reptiles, Ithaca, NY.

Townsend, D.S., M.M. Stewart, and F.H. Pough. 1984. Male parental care and its adaptive significance in a neotropical frog. *Animal Behaviour* 32: 421–31.

Tschinkel, W.R., and G. van Belle. 1976. Dispersal of larvae of the tenebrionid beetle, *Zophobas rugipes*, in relation to weight and crowding. *Ecology* 57 (1): 161–68.

Tucker, R. 2000. *Insatiable Appetite: The United States and the Ecological Degradation of the Tropical World*. University of California Press, Berkeley, CA.

Tudge, C. 1997. *The Time before History*. Touchstone, Simon & Schuster, New York, NY.

Tuomisto, H., K. Ruokolainen, and M. Yli-Halla. 2003. Dispersal, environment, and floristic variation of western Amazonian forests. *Science* 299 (5604): 241–44.

Turner, F.B., and C.S. Gist. 1965. Influences of a thermonuclear cratering test on close-in populations of lizards. *Ecology* 46: 845–52.

Tynan, C., D. DeMaster, and W. Peterson. 2001. Endangered right whales on the southeastern Bering Sea shelf. *Science* 294 (5548): 1894.

Uetz, P. 2000. How many reptile species? *Herpetological Review* 31 (1): 13–15.

Underwood, G., and E.E. Williams. 1959. The anoline lizards of Jamaica. *Bulletin of the Institute of Jamaica Science Series* 9: 1–48.

Valentine, B.D. 2003. A catalogue of West Indies Anthribidae (Coleoptera). *Insecta Mundi* 17: 49–67.

van Halewyn, R., and R.L. Norton. 1984. The status and conservation of seabirds in the Caribbean. *International Council of Bird Preservation Technical Publication* 2: 169–222.

Vandermeer, J., I. Granzow de la Cerda, D. Boucher, I. Perfecto, and J. Ruiz. 2000. Hurricane disturbance and tropical tree species diversity. *Science* 290 (5492): 788–91.

Veit, R.R., and W.R. Petersen. 1993. *Birds of Massachusetts*. Massachusetts Audubon Society, Lincoln, MA.

Vinnikov, K.Y., and N.C. Grady. 2003. Global warming trend of mean tropospheric temperature observed by satellites. *Science* 302 (5643): 249–72.

Vitt, L.J. 1983. Ecology of an anuran-eating guild of terrestrial tropical snakes. *Herpetologica* 39: 52–66.

Vogel, P. 1984. Seasonal hatchling recruitment and juvenile growth of the lizard *Anolis lineatopus*. *Copeia* 1984 (3): 747–57.

Vuilleumier, F., and D. Simberloff. 1980. Ecology versus history as determinants of patchy and insular distributions in high Andean birds. *Evolutionary Biology* 12: 235–379.

Walker, F. 1851. List of the specimens of Homopterous insects in the collection of the British Museum. 2: 261–636.

Walsh-McGehee, M. 2000. Status and conservation priorities for white-tailed and red-billed tropicbirds in the West Indies. *Society of Caribbean Ornithology Special Publication* 1: 31–38.

Walter, H.S. 1990. Small viable population: The red-tailed hawk of Socorro Island. *Conservation Biology* 4: 441–43.

Webb, Jr. J.P., and R.B. Loomis. 1977. *Ectoparasites*. Special Publications 13, Museum of Texas Technology University, Lubbock, TX.

Weber, N.A. 1972. Gardening ants. The attines. *Memoirs of the American Philosophical Society* 92: 1–146.

Weiher, E., and P. Keddy (eds.). 1999. *Ecological Assembly Rules: Perspectives, Advances, Retreats*. Cambridge University Press, UK.

Weissman, D.B., and D.C. Rentz. 1976. Zoogeography of the grasshoppers and their relatives

(Orthoptera) on the California Channel Islands. *Journal of Biogeography* 3:105–14.

Wenzel, R.L., and V.J. Tipton. 1966. Some relationships between mammal hosts and their ectoparasites. In Ectoparasites of Panama, Publications of the Field Museum of Natural History, Chicago, IL.

Wenzel, R.L., V. Tipton, and A. Kiewlicz. 1966. The streblid batflies of Panama. In *Ectoparasites of Panama*, eds. R.L. Wenzel and V. Tipton, 405–675. Field Museum of Natural History, Chicago, IL.

Wester, M., P. Marra, S. Haig, S. Bensch, and R. Holmes. 2002. Links between worlds: Unraveling migratory connectivity. *Trends in Ecology and Evolution* 17 (2): 76–83.

Wetherbee, D. K., R.P. Coppinger, and R.E. Walsh. 1972. *Time Lapse Ecology: Muskeget Island, Nantucket, Massachusetts.* MSS Educational Publications, New York, NY.

Wetmore, A. 1927. The birds of Puerto Rico and the Virgin Islands. *New York Academy of Sciences Scientific Survey of Puerto Rico and the Virgin Islands* 9 (4): 409–598.

Whidden, H.P., and R.I. Asher. 2001. The origin of the Greater Antillean insectivores. In *Biogeography of the West Indies,* 2nd ed., eds. C. A. Woods and F.E. Sergile, 237–52. CRC Press, Boca Raton, FL.

White, G. 1774. *The Natural History and Antiquities of Selborne, in the County of Southampton.* T. Bensley, London, UK.

White, G., D. Anderson, K. Burnham, and D. Otis. 1982. *Capture–Recapture and Removal Methods for Sampling Closed Populations.* Los Alamos National Laboratory, NM.

White, J.L., and R.D.E. MacPhee. 2001. The sloths of the West Indies: A systematic and phylogenetic review. In *Biogeography of the West Indies,* 2nd ed., eds. C.A. Woods and F.E. Sergile, 202–35. CRC Press, Boca Raton, FL.

White, R. 1985. *Two on the Isle.* W. W. Norton & Co., New York.

Whitford, W. 2002. *Ecology of Desert Systems.* Academic Press, San Diego, CA.

Whittaker, R. J. 1998. *Island Biogeography: Ecology, Evolution, and Conservation.* Oxford University Press, Oxford, UK.

Wiechert, U., A. Halliday, D. Lee, G. Snyder, L. Taylor, and D. Rumble. 2001. Oxygen isotopes and the moon-forming giant impact. *Science* 294 (5541): 345–48.

Wiens, J.A. 1984. On understanding a non-equilibrium world: myth and reality in community patterns and processes. In *Ecological Communities: Conceptual Issues and the Evidence,* eds. D. R. Strong, Jr., D. Simberloff, L.G. Abele, and A. B. Thistle, 439–57. Princeton University Press, Princeton, NJ.

Wiens, J.J., and B.D. Hollingsworth. 2000. War of the iguanas: conflicting molecular and morphological phylogenies and long-branch attraction in iguanid lizards. *Systematic Biology* 49 (1): 143–59.

Wilbur, H.M. 1997. Experimental ecology of food webs: complex systems in temporary ponds. *Ecology* 78 (8): 2279–2302.

Wiley, J.W. 1985. Bird conservation in the United States Caribbean. *Bird Conservation* 2: 107–59.

Wiley, J.W. 1991. Status and conservation of parrots and parakeets in the Greater Antilles, Bahama Islands and Cayman Islands. *Bird Conservation International* 1: 187–214.

Williams, D., D. Oi, S. Porter, R. Pereira, and J. Briano. 2003. Biological control of imported fire ants (Hymenoptera: Formicidae). *American Entomologist* 49 (3): 150–63.

Williams, D.F., and S.E. Braun. 1983. Comparison of pitfall and conventional traps for sampling small mammal populations. *Journal of Wildlife Management* 47 (3): 841–45.

Williams, E.E. 1969. The ecology of colonization as seen in the zoogeography of anoline lizards on small islands. *Quarterly Review of Biology* 44: 345–89.

Williams, E.E. 1976. West Indian anoles: A taxonomic and evolutionary summary. 1. Introduction and a species list. *Breviora* 440: 1–21.

Williams, E.E. 1983. Ecomorphs, faunas, island size, and diverse end points in island radiations of *Anolis.* In *Lizard Ecology: Studies of a Model Organism,* eds. R.B. Huey, E.R. Pianka, and T.W. Schoener, 326–70. Harvard University Press, Cambridge, MA.

Williams, M. 2003. *Deforesting the Earth: From Prehistory to Global Crisis.* University of Chicago Press, IL.

Williams, M.I., and D.W. Steadman. 2001. The historic and prehistoric distribution of parrots (Psittacidae) in the West Indies. In *Biogeography of the West Indies,* 2nd ed., eds. C.A. Woods and F.E. Sergile, 175–89. CRC Press, Boca Raton, FL.

Williams, R.J., and N.D. Martinez. 2000. Simple rules yield complex food webs. *Nature* 404: 165–68.

Williamson, M. 1981. *Island Populations.* Oxford University Press, UK.

Willig, M. R., and M. R. Gannon. 1996. Mammals. In *The Food Web of a Tropical Rain Forest,* eds. D. P. Reagan and R. B. Waide, 399–431. University of Chicago Press, IL.

Willink, A., and A. Roig-Alsina. 1998. Revision del genero *Pachodynerus* Saussure (Hymenoptera: Vespidae, Eumeninae). *Contributions of the American Entomological Institute* 30 (5): 1–117.

Wilson, D. E. 1979. Reproductive patterns. In *Biology of Bats of the New World Family Phyllostomatidae. Part III,* eds. R. J. Baker, J. K. Jones, Jr., D. C. Carter, 317–78. Special Publications 16, Museum of Texas Technology University, Lubbock, TX.

Wilson, E. O. 2003. *Pheidole in the New World: A Dominant, Hyperdiverse Ant Genus.* Harvard University Press, Cambridge, MA.

Wilson, J. B. 1999. Assembly rules in plant communities. In *Ecological Assembly Rules,* eds. E. Weiher and P. Keddy, 130–64. Cambridge University Press, UK.

Wilson, L. D., and L. Porras. 1983. The ecological impact of man on the south Florida herpetofauna. *Museum of Natural History Special Publication no. 9.* University of Kansas, KS.

Wilson, S. M. 2001. The prehistory and early history of the Caribbean. In *Biogeography of the West Indies,* 2nd ed., eds. C. A. Woods and F. E. Sergile, 519–27. CRC Press, Boca Raton, FL.

Windley, B. F. 1977. *The Evolving Continents.* Wiley, New York, NY.

Winemiller, K. O., and G. A. Polis. 1996. Food webs: What can they tell us about the world? In *Food Webs,* eds. G. A. Polis and K. O. Winemiller, 1–22. Chapman & Hall, New York.

Wing, E. S. 2001. Native American use of animals in the Caribbean. In *Biogeography of the West Indies,* 2nd ed., eds. C. A. Woods and F. E. Sergile, 481–518. CRC Press, Boca Raton, FL.

Wittmer, W. 1992. Zur Kenntnis der Gattung Tytthonyx LeConte, 1851. Beitrag 2. (Coleoptera: Cantharidae, Subfam. Silinae, Tribus Tytthonyxini). *Entomologica Basiliensia* 15: 333–78.

Wojcik, D., C. Allen, R. Brenner, E. Forys, D. Jouvenaz, and R. S. Lutz. 2001. Red imported fire ants: Impact on biodiversity. *American Entomologist* 47 (1): 16–23.

Wolcott, G. N., and L. F. Martorell. 1937. The ant, *Monomorium carbonarium ebeninum* Forel, in a new role: As a predator on the egg-cluster of *Diatraea saccharalis* F. in Puerto Rican cane

fields. *Journal of Agriculture, University of Puerto Rico* 21: 577–79.

Woodroffe, C., D. Stoddart, R. Harmon, and T. Spencer. 1983. Coastal morphology and late Quaternary history, Cayman Islands, West Indies. *Quaternary Research* 19: 64–84.

Woodruff, R. E., and W. H. Pierce. 1973. *Scyphorus acupunctatus,* a weevil pest of yucca and agave in Florida (Coleoptera: Curculionidae). *Florida Department of Agricultural and Consumer Services, Division of Plant Industry Entomology Circular* 135: 1–2.

Woods, C. A. 1990. The fossil and recent land mammals of the West Indies: An analysis of the origin, evolution, and extinction of an insular fauna. *Atti Convegni Lincei, Accademia Nazionale dei Lincei, Rome* 85: 641–80.

Woolfenden, G. E. 1975. Florida Scrub Jay helpers at the nest. *Auk* 92: 1–15.

Yorath, C. J. 1990. *Where Terranes Collide.* Orca Books, Victoria, BC.

Zardoya, R., and A. Meyer. 1996. Phylogenetic performance of mitochondrial protein-coding genes in resolving relationships among vertebrates. *Molecular Biology and Evolution* 13 (3): 933–42.

Zhou L. M., C. J. Tucker, R. K. Kaufmann, D. Slayback, N. V. Shabanov, and R. B. Myneni. 2001. Variations in northern vegetation activity inferred from satellite data of vegetation index during 1981 to 1999. *Journal of Geophysical Research-Atmospheres* 106 (D17): 20069–83.

Zimmerman, B. L. 1994. Audio strip transects. In *Measuring and Monitoring Biological Diversity: Standard Methods for Amphibians,* eds. W. R. Heyer, M. A. Donelly, R. W. McDiarmid, L.-A. C. Hayek, and M. S. Foster, 92–97. Smithsonian Institution Press, Washington, DC.

Zona, S. 1990. A monograph of *Sabal* (Arecaceae: Coryphoideae). *Aliso* 12: 583–666.

Zug, G. R. 1971. The distribution and patterns of the major arteries of the iguanids and comments on the intergeneric relationships of iguanids (Reptilia: Lacertilia). *Smithsonian Contributions to Zoology* 83: 1–23.

Zug, G. R. 1984. Review: Advances in Herpetology and Evolutionary Biology. Essays in Honor of Ernest E. Williams. *Copeia* 1984 (2): 554–56.

Zwiers, F. W., and A. J. Weaver. 2000. The causes of 20th century warming. *Science* 290 (5499): 2081–82.

Zwickel, F. C., and A. Allison. 1983. A back marker for individual identification of small lizards. *Herpetological Review* 14 (3): 82.

INDEX

dodo, 92, 114, 124. *See also* extinction

Dominica, 22–23, 72, 78, 332; reptiles on, 89, 137, 138, 141, 148

Dookhan, I., 316

Dorantes skipper *(Urbanus dorantes cramptoni),* 302

doves. *See* pigeons and doves

dragonflies (Odonata), 19–20, 48, 270–273

Drewet, Peter, 315

drought, 145, 151, 153, 230; censusing, 33, 50, 53; dry season, 65, 70, 72. *See also* rain (precipitation)

Drury, William H., 126

Drury's hairstreak *(Strymon acis mars),* 301

Dunham, A. E., 88

Dunkle, S. W., 271

Durand, Stephen, 225, 322

earthquakes, 106, 113, 117, 136, 298

ecogenetics, 141, 148–149. *See also* genetics

Ecological Restoration (Lazell), 319

ecology, 1–5, 12–17, 59; theoretical, 126

Ecology of Invasions by Plants and Animals, The (Elton), 1

ecomorphism, 186–188

ecosystems, 1, 4, 8–10, 13, 331; as information, 6; of islands, 27, 30; resilience of, 24–25

ectoparasites, 94, 95, 98

eggs, 73, 86, 125, 176–177, 191; snake, 65–66, 129, 197–198

egrets, 214, 235, 236

Einstein, Albert, 135

Eldredge, M. D. B., 145, 146

elevation, 14, 16, 127, 138, 151; and biogeography, 113–114; and diversity, 133–134, 206; of islands, 116, 130, 133

El Niño, 19–20

Elton, Charles, 1

El Verde Field Station, 70

endangerment, 182, 184, 221, 334. *See also* extinction

endemism, 94, 125, 130, 298, 333–334; in Caribbean, 159; and evolution, 112, 131; and frogs, 175

Endler, J. A., 113–116

energetics, 78, 90, 151

Engstrom, R. T., 12

ensembles, 21, 80–92, 137

entropy, 143

enzymes, 151

Epidendron ciliare, 112

equilibrium: punctuated equilibrium, 24, 78, 87, 92, 132, 142, 146; theories, 10, 30, 137

equilibrium theory of island biogeography (ETIB), 136, 138, 331

Ernst, C. H., 192

erosion, 108, 136, 319, 329

Eschmeyer, R. W., 50

Estrada, A. R., 177

Evenhuis, Neal, 312

evolution, 6, 113–14, 140–54, 171, 225, 258; and adaptation, 145–46, 148; and character divergence, 149–50; coevolutionary theory, 88, 331; on Guana, 129, 134, 150–54; and humans, 330; mutation, 140, 143, 151; of parasites, 93–94; and radiation, 125; rate of, 64, 126, 141, 143–47; and sand-barrier islands, 130–31; and selection, 89, 92, 149, 151, 330; and speciation, 135, 142, 144–45. *See also* natural selection

Evolutionary Ecology of Birds (Bennett & Owens), 208

exoskeleton, 243, 265

extinction, 4, 14, 30, 136, 331; of amphibians, 175, 180; of bats, 203, 205; and climate, 108; and evolution, 114, 125; on Great Guania, 129–30; human-caused (anthropogenic), 89, 125, 128, 142, 319, 328–29; and sand-barriers, 132

extirpation, 24–25, 203, 214, 220, 318, 326

Faaborg, J., 208

falcons. *See* hawks and falcons

false barred sulphur *(Eurema elathea),* 299

family, 156

fangs, 148–49

Faulkner, Clement, 322

Fauth, J., 21

Feduccia, A., 208

feral cats, 78, 333

Ferner, J. W., 45

Field Guide to the Butterflies of the West Indies, A (Riley), 298

Fielding, Joshua, 316

fiery skipper *(Hylephilia phyleus),* 301

finches: on Galapagos, 114–15, 145; Lesser Antillean bullfinch *(Loxgilla noctis),* 232, 236

fish, 47, 50, 116, 145

fishing bat *(Noctilio leporinus),* 199–201

Fisk, Bradley, 208

Fisk, Erma Jonnie, 192, 222, 317, 318

Fitch, H. S., 62

flamingoes, 4, 109; greater flamingo *(Pheonicopterus ruber),* 215, 235, 319, 323–24, 326

Flannery, T., 333

flies (Diptera), 84. *See also* fruit flies

flightlessness, 137, 258–60, 325. *See also* adaptation; birds

Flint, Richard Foster, 107–8

Florida white *(Appias drusilla boydi),* 299

Floyd, H. B., 69

Fogarty, J. H., 179

food-niche, 74, 75, 84, 88, 89

food webs, 21, 59, 68–71, 73–79, 175; diversity and, 68, 69, 70, 73–77

Food Webs and Niche Space (Cohen), 72

foraging, 71–72, 76

Forel, Auguste, 285, 286

Forest, C., 329

forests, 92, 121, 135, 136, 176, 180

fossil records, 13, 92, 108, 124, 146

fossorial (burrowing) species, 35, 43, 55, 125

Fowler, C., 332

Frank, P.W., 5

Freemark, K. E., 91

free-tail bat *(Tadarida brasiliensis),* 203, 204, 205

Frenchcap Island, 152

Frith, H. J., 328

Fritts, S. H., 119

frogs, 3, 70, 123, 133; abundance of, 175–76, 179; censusing, 41, 45–46, 48; grass frogs *(Rana taiphensis),* 137; leopard frog *(Rana pipiens),* 180; piping frog *(Eleutherodactylus antillensis),*

overspill islands, 111
Overton, W. S., 50
owls, 22, 68–69, 130, 134; screech-owls, 134, 222–23, 235

Pacala, S., 60, 85, 87
Paine, Robert T., 68, 69, 76, 137
painted lady *(Vanessa cardui)*, 300
Palumbi, Stephen, 125
Panama, 123, 187
Pape, T., 312
parasites, 33, 71, 93–99, 293, 296; Diptera, 94, 310, 312; disturbance transfers, 94; ectoparasites, 94, 95, 98; evolution of, 142; helminths, 98–99; planarians, 238; streblid batfly, 93–97; ticks, 97–98, 241
Parkes family, 153, 316, 317
parrots, 94, 319, 325
path analysis (Miyata), 25–26, 28–29
Patton, James, 145
Paul, Debbie, 207, 218, 219, 226
Paulik, G. J., 56
Paulissen, M. A., 46
Paulson, Dennis R., 270
Payne, Ann, 95–96
pearly-eyed thrashers *(Margarops fuscatus)*, 221, 225–26, 233
pelicans, 210, 212, 233
Penn, Albert, 196, 317
Penn, Walter, 317, 318
peregrine falcon *(Falco peregrinus)*, 216, 236, 325
Perez-Gelabert, Daniel E., 310
Perfecto, Ivette, 1
Perry, Gad, 72, 112, 186, 190, 198, 216
pesticides, 210, 212, 216
Peter Island, 82, 144, 147, 149, 150, 152
Peters, R. H., 78
Petranka, J. W., 35, 37
Petrovic, Clive, 196, 322
Philander, G., 329
Philibosian, R., 221
Phillips, Walter, 320
phyla, 156
phylogenetics, 141, 148, 182. *See also* genetics
Pianka, E. R., 60, 62, 72, 81, 194
Pickering, Vernon, 314, 324

pied-billed grebe *(Podilymbus podiceps)*, 208–9, 234
Pielou, E. C., 5, 10, 11, 115, 118–19
pigeons and doves, 4, 94; bridled quail-dove *(Geotrygon mystacea)*, 39–41, 43, 54–55, 207, 221–22, 234; common ground-dove *(Columbina passerina)*, 221, 233; rock dove *(Columba livia)*, 220, 236; scaly-naped pigeon *(Columba squamosa)*, 221, 233; white-crowned pigeon *(Columba leucocephala)*, 221, 235, 319; zenaida dove *(Zenaida aurita)*, 221, 233, 333
Pimm, S., 70–71, 74, 78
piracy, 315
planar areas, 127, 133
planthoppers (Fulgoridae), 133, 145, 210, 255–61; *Melormenis basalis*, 260; salt grass hopper *(Toya venilia)*, 256, 258, 259–60; sugar eater *(Saccharosydne saccharivora)*, 256–57
plants. *See* vascular plants
plate tectonics, 106, 113, 117, 136, 298
pluvials, 107–8, 315. *See also* climate; rain
Poisson distribution, 32–33
Polis, G. A., 77
pollution, 179–80, 210, 328
polydamas swallowtail *(Battus polydamas)*, 299
population, 14, 140, 142–43, 175; density, 107, 120, 136; and extinction, 129
population biology, 62–67, 127; and life histories, 64–67; and reproduction, 62–63. *See also* censusing; trophic levels
population growth, 121, 327, 330, 332, 334; of lizards, 182, 188
porcelain crab *(Petrolisthes quadratus)*, 252
potatoes *(Solanum)*, 145–46
Potter, Louis, 322, 323
Powell, J. R., 312
Powell, Robert, 194
Power, D. M., 133
Pratt, Sherman, 317
precipitation. *See* rain (precipitation)
predation, 64, 82, 141, 175, 184, 259; in local guilds, 88–89; predator-prey relations, 90, 91; and trophic levels, 68–72, 76–78

Pregill, G. K., 123, 150–51
Prendini, Lorenzo, 240, 241
Preston curve, 11
prey, 71, 72, 76, 90, 91, 175
Price, Peter W., 93–94, 96
Prickly Pear Cay, 333
principal components analysis (PCA), 28
Proctor, George, 157
proportions, 3, 9–10, 12, 56–57, 78. *See also* ratios
Protista, 157
Puente-Rolon, A. R., 198
Puerto Rican screech-owl *(Otus nudipes newtoni)*, 222–23, 235
Puerto Rican woodpecker *(Melanerpes portoricensis)*, 325
Puerto Rico, 13, 27, 70, 99, 180; bats of, 14–16, 133, 203, 204; and distribution, 91, 111, 115, 118, 122–23; frogs of, 175, 178; geological history of, 101–2, 105–6; Puerto Rico Bank, 3, 63, 82, 128, 137; reptiles of, 121, 147–48, 181–83, 187, 194, 196
Puerto Rico-northern Virgin Islands (PRVI) microplate, 105–6
Pugesek, B., 26, 29
punctuated equilibrium, 24, 78, 87, 92, 132, 142, 146
purple land crab *(Gecarcinus ruricola)*, 251

quail dove ghut, 39–41, 55, 222, 224. *See also* bridled quail-dove

radiation (species production), 113, 115, 118, 125, 140, 154; of amphisbaena, 195; and dispersal, 120; on islands, 123
radiotracking, 52. *See also* population biology
Raffaele, H., 222, 223, 229, 232, 324; and aquatic birds, 210, 213, 215, 216, 217, 219; bird lists of, 207
Raffaelli, D., 76–77
rails, 53, 258, 331; clapper rail *(Rallus longirostris)*, 216, 234; common moorhen *(Gallinula chloropus)*, 217, 234; DeBooy's rail *(Nesotrochis debooyi)*, 4, 319,

DATE DUE